COMPUTER SCIENCE AND STATISTICS:
Proceedings of the Seventeenth Symposium on the Interface

COMPUTER SCIENCE AND STATISTICS

PROCEEDINGS OF THE SEVENTEENTH SYMPOSIUM ON THE INTERFACE

Lexington, Kentucky, March 1985

Edited by

D.M. ALLEN

University of Kentucky
Lexington, Kentucky, U.S.A.

86

ORTH-HOLLAND – AMSTERDAM · NEW YORK · OXFORD · TOKYO

ISBN: 0 444 70018 8

Published by:
ELSEVIER SCIENCE PUBLISHERS B.V.
P.O. BOX 1991
1000 BZ AMSTERDAM
THE NETHERLANDS

Sole distributors for the U.S.A. and Canada:
ELSEVIER SCIENCE PUBLISHING COMPANY, INC.
52 VANDERBILT AVENUE
NEW YORK, N.Y. 10017
U.S.A.

This work relates to the Department of the Navy Grant Number N00014-85-G-0157 issued by the Office of Naval Research. The United States Government has a royalty-free license throughout the world in all copyrightable material contained herein.

Library of Congress Cataloging-in-Publication Data

Symposium on the Interface (17th : 1985 : Lexington, Ky.)
Computer science and statistics.

1. Mathematical statistics--Data processing--Congresses. I. Allen, David M., 1938- . II. Title.
QA276.S95 1985 519.5'0285 86-8904
ISBN 0-444-70018-8 (U.S.)

PRINTED IN THE NETHERLANDS

Preface

The Seventeenth Symposium on the Interface of Computer Sciences and Statistics was held in the Radisson Plaza Hotel, Lexington, Kentucky on March 17-19, 1985. The conference was hosted by the University of Kentucky. The format for the Symposium was very similar to the preceding symposia in the series. Dr. John Nash presented the keynote address on Monday morning. This was followed by two sets of three parallel sessions and workshops. On Tuesday there were three sets of three parallel sessions.

The sessions encompassed a broad range of topics. A number of sessions dealt with computational methods for traditional statistical areas. These included Time Series, Nonlinear Models, Repeated Measures Data Analysis, and Categorical Data Analysis. Some sessions were in the relatively new areas of Statistics such as Artificial Intelligence, the Metadata of Computational Processes, Statistical Computing Languages, and Statistical Workstations. There were also sessions in Numerical Methods, Density Estimation, Teaching of Statistical Computing, Statistical and Mathematical Software and Graphics. During one session the entire audience participated in a round table discussion on the Performance of Statisticians with Statistical Software. Written versions of nearly all these papers are in this volume. A few papers were not included because of prior copyright elsewhere or because the manuscript was not received from the authors.

A large number of people helped make the Seventeenth Symposium a big success. The organizing committee was Gary Anderson, Kenneth Berk, Thomas J. Boardman, Daniel B. Carr, William F. Eddy, Alan B. Forsythe, Richard J. Heiberger, Sally E. Howe, Robert E. Kass, William Kennedy, J. Richard Landis, John Nash, Wesley L. Nicholson, Gordon Sande, Victor Solo and Constance L. Wood.

The office staff of the Department of Statistics, particularly Debra Arterburn and Brian Moses, oversaw the correspondence and bookkeeping, maintained a participant data base, assembled registration packets, and manned the registration desk. Wimberly C. Royster, Dean of the Graduate School, M. A. Baer, Dean of the College of Arts and Sciences, and Joseph M. Gani, Chairman of the Department of Statistics, were all very supportive and made many resources of the University available for the Symposium.

The facilities of the Radisson Plaza Hotel were extremely nice. Thanks are extended to Cindy Edwards and the rest of the Radisson staff. The Greater Lexington Convention and Vistors Bureau welcomed participants at the airport, provided literature on things to do and places to eat, and also helped with the registration.

The American Statistical Association was helpful in many ways. The efforts of Randall Spoeri and Jean Smith are particularly appreciated. Financial support for the Symposium came from the Office of Naval Research and the University of Kentucky.

David M. Allen
Lexington

Financial Supporters of the Seventeenth Interface Symposium

U. S. Office of Naval Research,

U. S. Department of the Navy

University of Kentucky

Cooperating Organization

American Statistical Association

Contents

Availability of Proceedings

15, 16th (1983,84)	North-Holland Publishing Company Distributor for U.S.A. and Canada: Elsevier Science Publishing Company, Inc. 52 Vanderbilt Avenue New York, New York 10017
13, 14th (1981,82)	Springer-Verlag New York, Inc. 175 Fifth Avenue New York, New York 10010
12th (1979)	Jane F. Gentlemen Department of Statistics University of Waterloo Waterloo, Ontario Canada N2L 3G1
11th (1978)	Institute of Statistics North Carolina State University Post Office Box 5457 Raleigh, North Carolina 27650
10th (1977)	David Hogben Statistical Engineering Laboratory Applied Mathematics Division National Bureau of Standards U. S. Department of Commerce Washington, D.C. 20034
9th (1976)	Prindle, Weber, and Schmidt, Inc. 20 Newbury Street Boston, Massachusetts 02116
8th (1975)	Health Sciences Computing Facility, AV-111 Center for Health Sciences University of California Los Angeles, California 90024
7th (1974)	Statistical Numerical Analysis and Data Processing Section 117 Snedecor Hall Iowa State University Ames, Iowa 50010
4,5,6th (1971, 1972,73)	Western Periodicals Company 13000 Raymer Street North Hollywood, California 91605

The Keynote Session

Organizer: *David M. Allen*

Keynote Address:

Taking It With You - Portable Statistical Computing

John C. Nash

COMPUTER SCIENCE AND STATISTICS:
The Interface, D.M. Allen (ed.)

Keynote Address

TAKING IT WITH YOU -- PORTABLE STATISTICAL COMPUTING

John C. Nash

Faculty of Administration
University of Ottawa
Ottawa, Ontario, K1N 6N5
Canada

The subject of the presentation is the needed or wanted basis for portable statistical computing -- the infrastructure statisticians should have in order to carry out desired statistical computations wherever they happen to be. Expanding on this theme, we will examine what this basis implies for statistical software, the data sets we examine, our own practices and "documentation" in the widest sense, the computing hardware and software environments useful to support this activity, and the standards needed to assist us in rendering our work portable.

INTRODUCTION

As an active user and promoter of small computer solutions to both scientific and general administrative problems, and as scientific computing editor for Byte magazine, I am clearly identified with that proliferating technology collectively called the "microcomputer revolution". However, the main objective of this presentation does NOT concern microcomputers, except where the gadgetry illustrates how obstacles to portability of statistical computing arise or may be overcome. In working to make our work as free from ties to geographic locations as possible, I firmly believe that clear thinking and a wide perspective are far more important than brilliance in the design of a specific piece of hardware or software.

STATISTICAL COMPUTING -- DEFINITIONS

The basis of statistical computing has, in my opinion, five facets:

1) methods for data analysis and statistical interpretation

2) data which is to be the subject of analysis or computation

3) documentation of what WE -- the statisticians -- do, that is, of statistical practice

4) tabulation and display mechanisms, which are separated from methods to reflect the necessary involvement of machinery to effect the desired outputs

5) the training, education and research (self-education of the profession) to improve the overall technology of statistical computing as practised.

Here we do not consider the analysis of the results of computations as part of the task at hand. However, this distinction is blurred by the development of expert systems for particular areas of statistics.

The basis of statistical computing listed above is in the domain of ideas. Their realization is the work upon which many of us labour. We endeavor first to render the ideas in greater detail as generalized software -- computer programs, data files and structures, books, research papers, presentations, and designs of graphics. Second, we try to put the ideas into the "hardware" forms -- disks and tapes, paper, integrated circuits, audio/visuals. The juxtaposition of these software/hardware ideas is deliberate, in that it focuses attention on the possibility that there may be several renderings of an idea in different "languages" of expression and different media of recording.

PORTABILITY

One can think of several routes to make serious statistical computing portable. Portable personal computers of considerable power are now available, some of which are battery powered and need no AC power supply. The hardware, however, needs to be complemented by suitable software, and our data must be at hand in a useful form. Neither of these latter requirements is currently satisfied, but the availability of the machinery will entice developments to appear over the next few years.

To gain access to more powerful computers and software, and to larger data sets than may be accommodated on a portable microcomputer, we may look to long-distance communication via terminals. Here the major limitations are on the flexibility and convenience of data and command input and of displays or printed output. Few voice or data communication facilities have sufficient capacity for detailed graphics, either input or output.

Despite their present limitations, various communications technologies available now do allow the sharing of software and data sets, but only if the program or data files are in some sense "standard" so that the recipient may make use of them. To date, standards for these statistical, as opposed to computational, constructs are not in place.

Finally, even when the ideas behind a particular statistical computation have been transmitted between practitioners, we may observe that the results obtained by the different workers are not the same. Ultimately, we need a commonality of approach and methods at a relatively detailed level. Simply specifying a method, for example linear regression, is far from sufficient.

We now examine some of these ideas in more detail.

DATA

Data has many attributes: format, medium, content (or lack thereof), timeliness, volume (of data), history (author, origin, methods of gathering, notes and opinions), imputation methods, sampling design, aggregation procedures, whether "raw" or "cooked", security or confidentiality status (I owe this addition to a conversation with Gordon Sande). Other workers, particularly John Tukey, have presented similar categorization lists. In transferring data from one set of workers to another, we must take account of some or all of the above attributes. The task of developing a generalized format to accommodate these needs is not a trivial one. With Fred Brown, a research assistant, I have tried to develop such a format, but do not yet feel satisfied that it is ready to publish.

TABULATION AND DISPLAY

The aspects of tabulation and display which render them useful as tools for statistical analysis are the very features which are obstacles to portability. These can be summarized as form, style and practice. Form will reflect the overall type of design followed. Cleveland ([1], [2], [3]) has made a number of observations on form which also reflect on style -- how the particular form is translated to the object seen. The impact of available machinery on form and style chosen is obvious if one considers but one example, the Chernoff face. This display translates elements of a multivariate observation into features loosely resembling a human face. I have personally found it a useful mechanism for demonstrating results but a rather poor exploratory data analysis tool. Nevertheless, if one wishes to use "faces", then some way of drawing them must be found.

Traditional approaches (Flury & Riedwyl, 1981) use plotters of various types. One can envisage bit-map displays of modern microcomputers (e.g. Macintosh) being suitable, but conventional computer terminals lack the flexibility to "draw" the necessary graphs. An alternative approach is to change the style, and to some extent the form, of the "face" and use printer-plot ideas. Turner & Tidmore (1981) developed a FORTRAN program for this which was relatively easily transferred to the Amdahl mainframe at the University of Ottawa by Mr. P. Beynon, one of my students. Later Fred Brown designed a face-drawing program in BASIC for an Osborne 1, in the process applying some ideas from portraiture to improve the "facial" proportion.

In transporting their analyses, statisticians are unlikely to be satisfied with just one of the above alternatives being available. When graphical devices are available, the printer-plot is unlikely to satisfy. Therefore, a range of software is going to be needed, all pieces of which should interface easily to the data and to the command processor, thereby allowing the statistician to control the computations.

As a footnote to this discussion, I would like to point out that statistical displays of a relatively advanced nature are being used outside the profession. On Monday, March 11, 1985, on page B6 of the Toronto Globe and Mail (Report on Business) is a quite nicely executed set of star displays with an interesting choice of axes directions and scalings. This serves to underline the need for standardization of the practice of tabulation and display so that readers moving from one set of displays to another are not fooled by a simple change in the conventions.

METHODS

Methods are the translation of statistical thought into procedures. The greatest obstacle here to portability is the many levels of choice in transferring the general idea into a specific and unambiguous procedure. For instance, in considering the general method of regression, 100 years old this year, we must first decide between the usual least squares loss function or other metrics, second (assuming least squares) whether conventional linear, ridge or nonlinear approaches should be used, and third (assuming conventional linear l.s.) which algorithm to implement. Even having chosen a particular algorithm in general, for example, solution of normal equations, QR decomposition or singular value decomposition of the independent variable matrix (Nash, 1984, p. 166ff), we may have to select an implementation approach.

So far, we have no executable program code. Software is the realization of methods, and once again it is the diversity of options which hampers the portability of the statistical computations. We may choose to organize our statistical software as individual programs which stand alone, as a collection or library of related programs and/or subroutines, or as an integrated package not requiring the user to provide controls or operating system commands. Clearly the current trend is toward packages, even though this may make it more difficult to perform particular computations in particular computing environments. The usual form in which packages are distributed is as an ensemble of code executable on a particular computer configuration, since it runs against the producers' interests to have users transport (steal?) the code to other machines. Libraries are usually available only in machine (object) code form, while the individual programs of statistical software may be found as source code.

Source code must be expressed in some programming language, and most object code reflects some of the constraints implicit in all programming languages. The languages themselves echo features of the hardware which is available -- floating-point arithmetic, graphical devices, memory management. At the hardware level, we note that there are many established international, national or institutional standards which have been agreed and adopted. (I specifically exclude the so-called "industry standards" created by advertising copy writers.) Programming language standards are gradually having an influence on the software being written, but to my knowledge there are no standards yet being considered for the design and expression of program packages. For the user to be able to begin using one package after experience with another, some reasonably simple guidelines are clearly needed for the user interface, for the meaning of commonly used words, and for accessing data, devices, or other computing resources.

As statisticians we should be more aggressive in supporting existing standards, even as we begin the search for new ones to cover our particular area of work. Our lack of awareness of programming standards is illustrated by code published by Frank (1981) in the Journal of the American Statistical Association. In a program barely one page in length, practically each line has some construct or other which is non-standard, a typographical error, or a stylistic fault. If the purpose in publishing this code is to allow its use by other statisticians, then the editors, even more than the author, have missed the target!

HANDLING CHOICE

To render our computations portable to other computing environments and practitioners, I suggest four main routes:

1) Documentation of sufficient quality is needed so that all relevant details of the implementation of a method or the characteristics of a data set or approach to an analysis are clearly discernible. Special features -- the exceptions to the rules -- need to be noted.

2) Statisticians need to agree, either formally or informally, on the procedures and ideas of standard algorithms and practices. While the effort to formalize agreement may appear to be enormous, there is a growing body of work which is carried out by specific methods attributed to workers by name, for example, Marquardt's method for nonlinear least squares parameter estimation. Such methods can be written down clearly (Nash, 1979) in step-and-description form, and modifications can be noted in suitable documentation. However, the will is needed to perform activities seemingly peripheral to statistics.

3) For most statistical analysis the computations may be considered conventional. To avoid disagreements over the results, standard computer programs and data handling procedures are needed. Again, the effort to obtain formal agreement may not be required, since many statisticians are using a relatively small set of packages such as Minitab, SAS, SPSS or BMDP. There is a considerable interest in the development of test problems (see the workshop session "Measuring the performance of statisticians with statistical software" of these proceedings) and it is likely the producers of packages will align their major programs to produce similar results in order to avoid criticism and consequent marketing headaches. Once again, variations on a theme need to be documented. Moreover, the existence of a standard method should not prevent researchers from attempting different approaches.

4) Mechanisms need to be established for resolving real or apparent inconsistencies in results. Statisticians are in the forefront in this regard, since our journals have adopted a practice of presenting papers followed by discussions. This presents one avenue for airing differences of opinion. For discussions at a more detailed level, workers may want to consider establishing electronic mail conferences, moderated by knowledgeable researchers who can focus discussion.

DOCUMENTATION

My firm opinion is that good documentation is the core of advances in portability, and should mention the following:

- the data or type of data which can be/was analyzed
- the methods, algorithms, software used
- the time/date when each entry in the documentation was made
- all edits (of data / methods / documentation)
- observations / comments / hunches
- the name(s) of persons adding to or changing documentation.

TRAINING, EDUCATION AND RESEARCH

Portability of statistical computing concerns the transfer of ideas, which at present is plagued by our academic traditions. These have led to delays in publication because of the financial pressures on journals and the slowness of refereeing and review. Worse, since academic workers' career development depends in part on journal articles, there is little credit for non-traditional forms of idea transfer -- computer conferencing, software development, computer aided instruction development. It is also clear that use is going to be made of statistical computation by those who have had no part in developing the tools -- new statisticians, professionals in other disciplines, and the general public. The last group is an increasing "user" in developing business or public policy, where it is important to argue the consequences of decisions rather than the validity of the data or methods. Consequently, impatience with results which cannot be repeated is to be expected, and the codification and standardization of statistical practice can have a large payoff.

A by-product of such codification is that it permits expert systems, either tactical (for specific types of computations) or strategic (to recommend global approaches to data analysis), to be developed.

REALIZATION OF PORTABILITY

The discussion above has a possible concrete realization which can be begun immediately. The technical requirements to allow statistical data and software to be transferred from location to location via communications technologies can be met, even if not with great ease. At a minimum, these requirements are

1) file formats for programs and data, which I would currently recommend be simple text files (code may have to be transferred as hexadecimal digits).

2) file transfer mechanisms, such as electronic mail with suitable file server(s). Byte magazine already allows users to download programs which have appeared in the magazine, but access is at the moment via long-distance voice lines, which are much more expensive than the packet-switched data networks.

3) standards for data and programs. While not yet established, one can imagine a relatively simple, limited standard for small to medium sized data sets and for the expression of programs in source code in one or more programming language for a restricted class of target machines.

The technical requirements, as delineated above, will not be translated into a reality without investments. First, entrepreneurs will need to foresee sufficient rewards to justify the expenditure for a "head-end" file store to maintain the base of data and software with attendant telecommunications hardware and software to allow easy access for (possibly) naive users. The hardware for telecommunications at the present time should probably link to one or more of the public packet switched networks rather that the usual voice-line telephone. Software must handle both the database as well as the user interface. Simple but effective charging algorithms are needed so that revenues can be recorded and collected without undue difficulty for subscribers.

The development of standards requires investments of time and money on the part of those involved in statistical computing. Except in the quality control area, statisticians have not yet participated (as statisticians) in these types of activities.

The third "investment" needed is in the development of the intellectual property to be transferred and shared among statisticians. Developers will have to receive academic credit for such work, or it will have to be remunerated in the marketplace. The latter remuneration requires royalties to be paid, suitable cooperative enforcement of ownership of the intellectual property, and attractive pricing and service by the vendors to encourage users to obtain the material from the authorized source. Indeed, software vendors such as Borland International have demonstrated that a good product at an attractive price will not be "stolen" to an appreciable extent.

PROGNOSIS

The above recipe for permitting portability of statistical computing via a central database of data, programs and documentation is feasible to try now. I believe that the time is ripe to begin some experiments in restricted areas of statistical computation to discover the details of design which will facilitate further progress. Standards for computer programs for statistical computations are overdue, particularly for those which are published in journals. In order to move from the domain of research to generally available reality, analyses of the risks and benefits of commercial investment will need to be prepared, and consortia formed to market (partial) implementations of such systems. This last point represents the end-goal of the ideas presented here, and believing that the concepts presented are feasible to carry out, I have started to seek business alliances to realize them. However, I hope that those in the audience who do not accept the total parcel presented will still find valuable points within the discussion. Finally, while I have focussed on moving ideas rather than people and machinery, it should be kept in mind that there are often reasons why it is necessary to travel and transport in order to take our statistical computations with us.

REFERENCES

[1] Cleveland, W.S., Graphs in scientific publications, American Statistician 38 (4) (November 1984) 261-269.

[2] Cleveland, W.S., Graphical methods for data presentation: full scale breaks, dot charts and multibased logging, American Statistician, 38 (4) (November 1984) 270-280.

[3] Cleveland, W.S. and McGill, R., The many faces of a scatterplot, J. Amer. Statistical Assoc., 79 (388) (December 1984) 807-822.

[4] Flury, B. and Riedwyl, H., Graphical representation of multivariate data by means of asymmetrical faces, J. Amer. Statistical Assoc., 76 (376) (December 1981) 757-776.

[5] Franck W.E., The most powerful invariant test of normal versus Cauchy with applications to stable alternatives, J. Amer. Statistical Assoc., 76 (376) (December 1981) 1002-1005.

[6] Nash, J.C., Compact numerical methods for computers (Adam Hilger, Bristol, 1979)

[7] Nash, J.C., Effective scientific problem solving with small computers (Reston Publishing Co., Reston, 1984)

[8] Nash J.C., Design and implementation of a very small linear algebra algorithms package, Communications of the ACM, 28 (1) (January 1985) 89-94.

[9] Turner, D.W. and Tidmore, F.E., A FORTRAN program for generating Chernoff-type faces on a line printer, Mathematics Department, Baylor University, Waco, Texas (October, 1981)

Computation in Time Series

Organizer: *Victor Solo*

Invited Presentations:

On Bootstrap Estimates of Forecast Mean Square Error for Autoregressive Processes

David F. Findley

The EM Algorithm in Time Series Analysis

Robert H. Shumway

Designing an Intelligent System for Spectral Analysis

Donald B. Percival, et al

COMPUTER SCIENCE AND STATISTICS:
The Interface, D.M. Allen (ed.)
Elsevier Science Publishers B.V. (North-Holland), 1986

ON BOOTSTRAP ESTIMATES OF FORECAST MEAN SQUARE ERRORS FOR AUTOREGRESSIVE PROCESSES

David F. Findley

Statistical Research Division
Bureau of the Census
Washington, D.C. 20233

This paper presents several analyses which suggest that the bootstrap procedure used by Freedman and Peters to simulate errors in forecasting future values of an econometrically modelled process is of limited usefulness for estimating mean square forecast errors.

1. INTRODUCTION

Freedman and Peters (1984) recently applied a resampling procedure (the "bootstrap") to obtain estimates of mean square error for the forecasts from an autoregression with exogeneous terms. In this paper, we start with a theoretical analysis of their suggested procedure for the case of (not necessarily stationary) autoregressive models without exogenous terms and later describe two situations in which the same conclusions hold in the presence of exogenous variables.

The theoretical mean square forecast error from an estimated model is the sum of two components, the mean square forecast error of the optimal predictor and the mean square difference between the optimal forecast and the estimated model's forecast. This latter component is of order 1/T, where T is the length of the observed series, and so is negligible with large samples. Our theoretical analysis in Section 2 shows that the bootstrap estimate of mean square forecast error is the sum of the usual (naive) large-sample estimate of the first component, easily obtainable without the bootstrap, and a small-sample estimate of the second. A gaussian Monte Carlo value of the second component is obtained in Section 3 for series of length 25 from the AR(2) models used in the study of Ansley and Newbold, along with the value of the root mean square error (rmse) of the large-sample estimator of the m-step-ahead forecast error, for m = 1, 2 and 5. In these examples, the rmse is always substantially larger than the O(1/T) component, supporting the observation of Stine (1982) that estimates of the second component are of little use in estimating mean square forecast error unless better estimators of the first component are available. In the final section, we discuss conditional forecast mean square errors associated with predictions of the future of the observed sample path, and conclude that in this context as well, the bootstrap's potential contribution seems limited.

2. BOOTSTRAP ESTIMATES OF UNCONDITIONAL MEAN SQUARE FORECAST ERROR

The simple bootstrap procedure of Freedman and Peters we described below would appear to be appropriate when observations $y_1,\ldots,y_T$ are available from a time series obeying a general p-th order autoregression (p<T) of the form

$$(2.1)\qquad y_t = \delta + \phi_1 y_{t-1} + \ldots + \phi_p y_{t-p} + e_t \quad (t \geq p+1),$$

where e_t $(t \geq p+1)$ are independent, identically distributed random variables with mean 0 and variance σ^2 which are independent of earlier y's; that is, for k>0, e_t and y_{t-k} are independent. It is assumed that the order p is known and, only for simplicity of notation, that all of the parameters $\phi_1,\ldots,\phi_p$ and δ are unknown. Define $\underline{\theta} = (\delta,\phi_1,\ldots,\phi_p)$. For any m>0 we can use back substitution in (2.1) to obtain

$$(2.2)\qquad y_{T+m} = \sum_{j=0}^{m-1} \psi_j e_{T+m-j} + f_m[\theta](y_T,\ldots,y_{T-p+1}),$$

where the coefficients $\psi_0(=1)$, ψ_1, $\psi_2,\ldots$ satisfy

$$(2.3)\qquad \sum_{k=0}^{\min(j,p)} \phi_k\psi_{j-k} = 0 \qquad (\phi_0 = -1),$$

and where $f_m[\theta](y_T,\ldots,y_{T-p+1})$ is linear in $y_T,\ldots,y_{T-p+1}$ and δ. For example, if p=1, then $\psi_j = \phi_1^j$ and $f_m[(\delta,\phi_1)](y_t) = \delta(1 + \phi_1 + \ldots + \phi_1^{m-1}) + \phi_1^m y_t$. The two expressions on the right hand side of (2.2) are stochastically independent since e's are independent of earlier y's. It follows from this that $f_m[\theta](y_T,\ldots,y_{T+p-1})$ describes the optimal forecast (the conditional mean of y_{T+m} given $y_1,\ldots,y_T$) and that $\sum_{j=0}^{m-1} \psi_j e_{T+m-j}$ is the resulting forecast error.

This optimal forecast cannot be precisely determined because $\underline{\theta}$ is unknown. If $\hat{\underline{\theta}} = (\hat{\delta}, \hat{\phi}_1,\ldots,\hat{\phi}_p)$ is any estimate of $\underline{\theta}$ obtained using $y_1,\ldots,y_T$, then $f_m[\hat{\underline{\theta}}](y_T,\ldots,y_{T-p+1})$ is a forecast of y_{T+m} with forecast error

$$(2.4)\qquad y_{T+m} - f_m[\hat{\underline{\theta}}](y_T,\ldots,y_{T-p+1})$$

$$= \sum_{j=0}^{m-1} \psi_j e_{T+m-j} + \{f_m[\underline{\theta}](y_T,\ldots,y_{T-p+1})$$

$$- f_m[\hat{\underline{\theta}}](y_T,\ldots,y_{T-p+1})\}.$$

Since the e_{T+m-j}, $j=0,\ldots,m-1$ are independent of $\hat{\underline{\theta}}$, the two terms on the right hand side of (2.4) are independent. Consequently, using E to denote expectation, the mean square m-step-ahead forecast error when the forecast is given by $f_m[\hat{\underline{\theta}}](y_T,\ldots,y_{T-p+1})$ satisfies

$$(2.5)\qquad E\{y_{T+m} - f_m[\hat{\underline{\theta}}](y_T,\ldots,y_{T-p+1})\}^2$$

$$= \sigma^2 \sum_{j=0}^{m-1} \psi_j^2 + E\{f_m[\underline{\theta}](y_T,\ldots,y_{T-p+1})$$

$$- f_m[\hat{\underline{\theta}}](y_T,\ldots,y_{T-p+1})\}^2 .$$

If T is large, and $\hat{\underline{\theta}}$ is a consistent estimator of $\underline{\theta}$ (e.g. from least squares, if $E|e_t|^\alpha < \infty$ for some $\alpha>2$, see Lai and Wei (1983)), then the second term on the right in (2.5) can be ignored and the mean square forecast error can be adequately approximated by

$$(2.6)\qquad \hat{\sigma}^2(T-p) \sum_{j=0}^{m-1} \hat{\psi}_j^2$$

where the $\hat{\psi}$'s are obtained by using $\hat{\phi}$'s in (2.3), and $\hat{\sigma}^2(T-p)$ is given by

$$(2.7)\qquad \hat{\sigma}^2(T-p) = (T-p)^{-1} \sum_{t=p+1}^{T}$$

$$\{y_t - \hat{\delta} - \hat{\phi}_1 y_{t-1} - \ldots - \hat{\phi}_p y_{t-p}\}^2.$$

If T is small, however, then the second term on the right in (2.5) need not be negligible. Also, the quantity (2.6) may be an inadequate approximation to $\sigma^2 \sum_{j=0}^{m-1} \psi_j^2$. For the situation in which T is small, Freedman and Peters (1983) propose the following bootstrap procedure. Define

$$\hat{e}_t = y_t - \hat{\delta} - \hat{\phi}_1 y_{t-1} - \ldots - \hat{\phi}_p y_{t-p} ,$$

$$t = p+1,\ldots,T$$

Since we are concerned with the situation in which only one realization of the series y_t is observed, we will now regard the $\hat{e}_t$'s and $\hat{\underline{\theta}}$ as fixed. We will assume that the sample mean $\bar{\hat{e}}$ of the $\hat{e}$'s is 0, as happens, for example, when $\hat{\underline{\theta}}$ is chosen to minimize $\hat{\sigma}^2(T-p)$ in (2.7). (Otherwise, use $\hat{e}_t - \bar{\hat{e}}$ in place of $\hat{e}_t$ below.) Then if we define e_t^*, $t>p$, by successive independent draws with replacement from $\{\hat{e}_{p+1},\ldots,\hat{e}_T\}$, we obtain a series of identically distributed random variables with mean 0 and variance $\hat{\sigma}^2(T-p)$ whose common distribution is the empirical distribution of $\{\hat{e}_{p+1},\ldots,\hat{e}_T\}$. Now we define the so-called psuedo-data series, y_t^*, by means of $y_t^* = y_t$, $1\leq t\leq p$ and

$$(2.8)\qquad y_t^* = \hat{\delta} + \hat{\phi}_1 y_{t-1}^* + \ldots + \hat{\phi}_p y_{t-p}^*$$

$$+ e_t^* \quad (t>p).$$

The e^*'s are independent of earlier y^*'s. Let $\underline{\theta}^*$ denote the value corresponding to $\hat{\underline{\theta}}$ when $y_1^*,\ldots,y_T^*$ are used in place of the original values $y_1,\ldots,y_T$: For example, if $\hat{\underline{\theta}}$ was obtained by least squares, we choose $\underline{\theta}^*$ so that

$$\sum_{t=p+1}^{T} \{y_t^* - \delta^* - \phi_1^* y_{t-1}^* - \ldots - \phi_p^* y_{t-p}^*\}^2 .$$

is minimized.

We have now created an analogue of the original situation, but one in which we can use a (psuedo-) random number generator to simulate draws with replacement from $\{\hat{e}_{p+1},\ldots,\hat{e}_T\}$ and so obtain as many (psuedo-) independent realizations of $y_1^*,\ldots y_{T+m}^*$ as we like. With these realizations, finally, we can approximate the distribution of the forecast error process $y_{T+m}^* - f_m[\underline{\theta}^*](y_T^*,\ldots,y_{T-p+1}^*)$ to any desired degree of accuracy. To the extent that this resembles the distribution of $y_{T+m} - f_m[\hat{\underline{\theta}}](y_T,\ldots,y_{T+p-1})$, we thereby gain information about the error process in which we are actually interested.

For example, following Freedman and Peters (1983), given realizations $y_1^{*(n)},\ldots,y_{T+m}^{*(n)}$, $n=1,\ldots,N$, we can approximate

$$(2.9) \qquad E^*\{y_{T+m}^* - f_m[\underline{\theta}^*](y_T^*,\ldots,y_{T-p+1}^*)\}^2$$

by means of

$$N^{-1}\sum_{n=1}^N\{y_{T+m}^{*(n)} - f_m[\underline{\theta}^{*(n)}](y_T^{*(n)},\ldots,y_{T-p+1}^{*(n)})\}^2 .$$

(In (2.9) and below, we use E^* to denote expectation with respect to the distribution of the series e_t^*.)

The question is, what is the relationship between the quantity (2.9) and $E\{y_{T+m} - f_m[\hat{\underline{\theta}}](y_T,\ldots,y_{T-p+1})\}^2$? To obtain a partial answer, we note that, by analogy with (2.5), the quantity (2.9) is equal to

$$(2.10) \qquad \hat{\sigma}^2(T-p)\sum_{j=1}^{m-1}\hat{\psi}_j^2 + E^*\{f_m[\hat{\underline{\theta}}](y_T^*,\ldots,y_{T-p+1}^*) - f_m[\underline{\theta}^*](y_T^*,\ldots,y_{T-p+1}^*)\}^2 .$$

Thus, this bootstrap procedure inflates the naive estimate of mean square prediction error, (2.6), by an amount

$$(2.11) \qquad E^*\{f_m[\hat{\underline{\theta}}](y_T^*,\ldots,y_{T-p+1}^*) - f_m[\underline{\theta}^*](y_T^*,\ldots,y_{T-p+1}^*)\}^2$$

which is clearly a proxy for the mean square deviation of $f_m[\hat{\underline{\theta}}](y_T,\ldots,y_{T-p+1})$ from $f_m[\underline{\theta}](y_T,\ldots,y_{T-p+1})$,

$$(2.12) \qquad E\{f_m[\hat{\underline{\theta}}](y_T,\ldots,y_{T-p+1}) - f_m[\underline{\theta}](y_T,\ldots,y_{T-p+1})\}^2 ,$$

appearing as the second component on the right hand side of (2.5). Since the quantity (2.6) is known independently of the bootstrap procedure, we conclude that an estimate of (2.11) is, in fact, the only contribution made by this procedure. Further, to estimate (2.11) it is clear that psuedo-future data $y_{T+1}^*,\ldots,y_{T+m}^*$ are not required, but only realizations of $y_1^*,\ldots,y_T^*$. Thus, in place of Freedman and Peters' procedure to estimate the mean square m-step-ahead forecast error, it seems appropriate to only consider quantities

$$(2.13) \qquad N^{-1}\sum_{n=1}^N\{f_m[\hat{\underline{\theta}}](y_T^{*(n)},\ldots,y_{T-p+1}^{*(n)}) - f_m[\underline{\theta}^{*(n)}](y_T^{*(n)},\ldots,y_{T-p+1}^{*(n)})\}^2,$$

using these to estimate (2.12), the component of mean square forecast error due to the use of $\hat{\underline{\theta}}$ instead of $\underline{\theta}$ in the forecast function.

Somewhat analogous observations can be made for the model selection procedure proposed in Freedman and Peters (1983): Suppose two different autoregressive models, of orders p(A) and p(B), are fit to the observed data $y_1,\ldots,y_T$, resulting in estimated parameters $\underline{\theta}_A$ and $\underline{\theta}_B$, residual populations $\{e_{p(A)+1}^A,\ldots,e_T^A\}$ and $\{e_{p(B)+1}^B,\ldots,e_T^B\}$, and psuedo-data series y_t^{A*} and y_t^{B*} as above. Freedman and Peters suggest that each model be fit to, and then used to forecast, the psuedo-data from the other model, and that bootstrap estimates of the mean square forecast error be calculated. The model having the smaller estimated mean square forecast error is to be preferred. Thus, using an obvious notational scheme, the idealized quantities to be compared are

$$E^{A*}\{y_{T+m}^{A*} - f_m^B[\underline{\theta}_B^{A*}](y_T^{A*},\ldots,y_{T-p(B)}^{A*})\}^2$$

and

$$E^{B*}\{y_{T+m}^{B*} - f_m^A[\underline{\theta}_A^{B*}](y_T^{B*},\ldots,y_{T-p(A)}^{B*})\}^2 .$$

By the argument used to derive (2.5), these idealized quantities are equal, respectively, to

(2.14) $$\sigma_A^2(T-p(A)) \sum_{j=0}^{m-1} (\psi_j^A)^2 +$$

$$E^{A*}\{f_m^A[\theta_A](y_T^{A*},\ldots,y_{T-p(A)}^{A*}) -$$

$$f_m^B[\theta_B^{A*}](y_T^{A*},\ldots,y_{T-p(B)}^{A*})\}^2$$

and

(2.15) $$\sigma_B^2(T-p(B)) \sum_{j=0}^{m-1} (\psi_j^B)^2 +$$

$$E^{B*}\{f_m^B[\theta_B](y_T^{B*},\ldots,y_{T-p(B)}^{B*}) -$$

$$f_m^A[\theta_A^{B*}](y_T^{B*},\ldots,y_{T-p(A)}^{B*})\}^2 .$$

Since the leading expressions in (2.14) and (2.15) can be calculated independently of the bootstrap, we see, as before, that the bootstrap's only contribution is to compare forecasts and that pseudo-data at times later than T are not needed for this.

All of the arguments given above also apply to the case of vector autoregressions, and thus also to the case of autoregressions with exogeneous variables, provided that endogeneous and exogenous variables are simultaneously forecasted from a combined vector autoregression. They also apply if all needed values of the exogenous variables are assumed to be nonrandom and known, as in Freedman and Peters (1984)

3. THE SIZE OF (2.12) IN SOME EXAMPLES

Again using an obvious notation, let us rewrite (2.5) as

(3.1) $$\sigma_{m,T}^2 = \sigma_m^2 + E\hat{\Delta}_{m,T}^2$$

The analogous formula for the bootstrap estimate (see (2.10)) can be written

(3.2) $$\sigma_{m,T}^{*2} = \hat{\sigma}_m^2(T-p) + E^*\Delta_{m,T}^{*2}$$

For estimating $\sigma_{m,T}^2$, the practical significance of having an estimate $E^*\Delta_{T,m}^{*2}$ of $E\hat{\Delta}_{m,T}^2$ depends upon the size of $E\hat{\Delta}_{m,T}^2$ relative to σ_m^2 and to the root mean square estimation error of the large-sample estimate $\hat{\sigma}_m^2(T-p)$ of σ_m^2,

$$\text{rmse}(\hat{\sigma}_m^2(T-p)) = \{E(\hat{\sigma}_m^2(T-p) - \sigma_m^2)^2\}^{1/2} .$$

In Table (3.1) below, we present Monte Carlo estimates of the ratios $E\hat{\Delta}_{m,T}^2/\sigma_m^2$ and

(3.3) $$E\hat{\Delta}_{m,T}^2/\text{rmse}(\hat{\sigma}_m^2(T-p))$$

for the observation length T=25 for some gaussian AR(2) processes

(3.4) $$y_t = \delta + \phi_1 y_{t-1} + \phi_2 y_{t-1} + e_t$$

utilized in the study of Ansley and Newbold (1981). We note that these quantities are relevant for the estimation of $\sigma_{m,T}$ as well, since, for example,

$$\sigma_{m,T} = \sigma_m\{1 + (E\hat{\Delta}_{m,T}^2/\sigma_m^2)\}^{1/2} ,$$

which is well approximated by

$$\sigma_m\{1 + \frac{1}{2}(E\hat{\Delta}_{m,T}^2/\sigma_m^2)\}$$

if $(E\hat{\Delta}_{m,T}^2/\sigma_m^2)^2/8$ is negligible (Taylor polynomial approximation). For each pair of coefficients ϕ_1, ϕ_2 in the Table, we estimated the quantities $E\hat{\Delta}_{m,T}^2$ and $\text{rmse}(\hat{\sigma}_m^2(T-p))$ as the mean of sample estimates obtained from 1000 stationary pseudo-Gaussian series satisfying (3.4) with $\delta = 0$, using least squares to estimate δ, ϕ_1 and ϕ_2. (The IMSL pseudo-Gaussian generator GGNML was utilized.) The tabled results suggest that estimation of $E\hat{\Delta}_{m,T}^2$ is of little consequence when $\hat{\sigma}_m^2(T-p)$ is used to estimate σ_m^2.

Table 3.1 Values of $E\hat{\Delta}^2_{m,T}/\sigma^2_m$ and (3.3) for M=1, 2 and 5, for selected Gaussian AR(2) processes, with T=25.

ϕ_1	ϕ_2	m	$E\hat{\Delta}^2_{m,T}/\sigma^2_m$	(3.1)
.40	-.15	1	.01	.02
		2	.01	.01
		5	.00	.01
.80	-.65	1	.01	.05
		2	.04	.04
		5	.02	.02
.80	-.16	1	.03	.04
		2	.02	.03
		5	.02	.04

We have not included results for those of Ansley and Newbold's AR(2) models whose characteristic polynomials have a root in the annulus $1.0<|z|<1.24$. With T=25, simulations for such models produced large numbers of explosive series (the estimated characteristic polynomials had a root in $|z|<1.0$).

4. CONDITIONAL MEAN SQUARE FORECAST ERROR

In the preceding sections, we investigated unconditional mean square forecast error. However, it is the error associated with predicting a future point on the observed sample path (realization) which usually is most of interest.

4A. Mean Square Error Formulas

Since, by (2.1), the value of y_{T+m} depends on the data $y_1,\ldots,y_T$ only through the last p observations, it is easy to check that we can simply reinterpret the expectation operator E in (2.5) as designating expectation conditional upon $y_T, y_{T-1},\ldots,y_{T-(p+1)}$ and thereby obtain the fundamental decomposition of the mean square forecast error conditional upon the observed sample path. The $y_T, y_{T-1},\ldots,y_{T-(p+1)}$ in the second term on the right in (2.5) are now held constant, with the result that this second term simplifies into a linear expression in the higher order moments of $\hat{\underline{\theta}} - \underline{\theta}$. The mean-zero first order case is illustrative: If

$$y_t = \phi y_{t-1} + e_t \quad (\phi\neq 0) \tag{4.1}$$

with e_t, $t>1$, i.i.d. having mean 0 and variance σ^2, and with e_t independent of y_{t-k} whenever $k>0$, then $f_m[\phi](y_T) = \phi^m y_T$. From the the Taylor polynomial expansion of $f_m[\hat{\phi}](y_T)$ about $\hat{\phi} = \phi$, we have

$$f_m[\hat{\phi}](y_T) - f_m[\phi](y_T) = y_T \sum_{j=1}^{m} C_{m,j}\phi^{m-j}(\hat{\phi} - \phi)^j , \tag{4.2}$$

where $C_{j,m} = m(m-1)\ldots(m-j+1)/j!$.

Taking the mean square of (4.2) conditional on y_T, we obtain

$$E\{f_m[\hat{\phi}](y_T) - f_m[\phi](y_T)\}^2 = y_T^2 \sum_{j,k=1}^{m} C_{m,j}C_{m,k}\phi^{2m-j-k}E\{\hat{\phi} - \phi\}^{j+k} \tag{4.3}$$

To estimate (4.3) via the bootstrap, we replace y_T^* in (2.11) by y_T (ideally generating the pseudo-data in such a way that $y_T^* = y_T$, but see 4B. below). By analogy with (4.3), we then have

$$E^*\{f_m[\phi^*](y_T) - f_m[\hat{\phi}](y_T)\}^2 = y_T^2 \sum_{j,k=1}^{m} C_{m,j}C_{m,k}\hat{\phi}^{2m-j-k}E^*\{\phi^* - \hat{\phi}\}^{j+k} . \tag{4.4}$$

The efficacy of the bootstrap procedure is usually related to the extent to which the distribution of $\underline{\theta}^* - \hat{\underline{\theta}}$ resembles that of $\hat{\underline{\theta}} - \underline{\theta}$ and to how insensitive this latter distribution is to the true parameter value $\underline{\theta}$. However, for our problem, the situation illustrated by (4.3) and (4.4) obviously holds generally: the expected mean square of $f_m[\hat{\underline{\theta}}](y_T,\ldots,y_{T-p+1}) - f_m[\underline{\theta}](y_T,\ldots,y_{T-p+1})$ conditional on $y_T,\ldots,y_{T-p+1}$ depends on the true value of $\underline{\theta}$ as well as on the distribution of $\hat{\underline{\theta}} - \underline{\theta}$, suggesting that the quality of the bootstrap approximation will be influenced by the accuracy of $\hat{\underline{\theta}}$ as an estimate of $\underline{\theta}$.

4B. Bootstrapping Conditional Sample Paths

It would seem like an attractive idea, when, as in this section, statistics associated with the distribution of y_t conditional on $y_T,\ldots,$ y_{T-p+1} are being approximated, to generate pseudodata y_t^* for the bootstrap in such a way that $y_t^* = y_t$ holds for $T-p+1 \leq t \leq T$.

For example, it would be appealing to estimate ϕ^* in (4.1) from sample paths passing through y_T.

To illustrate a first approach to accomplishing this, suppose we have bootstrapped residuals $e_{p+1}^*,\ldots,e_T^*$ from an estimate $\hat{\phi}$ of ϕ in (4.1). To generate y_t^* satisfying

$$y_t^* = \hat{\phi} y_{t-1}^* + e_t^* , \quad 2 \leq t \leq T$$

with $y_T^* = y_T$, we could obviously set $y_T = y_T^*$ and recursively define

$$y_t^* = \hat{\phi}^{-1} y_{t+1}^* - \hat{\phi}^{-1} e_{t+1}^* , \quad 1 \leq t \leq T-1 . \tag{4.5}$$

In this case, however, y_t^* is neither independent of nor even uncorrelated with e_{t+1}^* for $1 \leq t \leq T-1$. Thus the bootstrapped data fail to have a basic property of the original data, and the consequences of this for the estimation of $\hat{\phi}$ from $y_1^*,\ldots,y_T^*$ are an unresolved issue. Furthermore, (4.5) is numerically unstable when $|\hat{\phi}|<1$.

When the series y_t is stationary, a second approach, which avoids the difficulties just encountered, would seem to recommend itself. To illustrate with the first order case again, if y_t satisfying (4.1) is stationary, then it is easy to verify that the random variables a_t defined by

$$a_t = y_t - \phi y_{t+1} \tag{4.6}$$

are uncorrelated with one another, satisfy $Ea_t^2 = Ee_t^2$, and each a_t is uncorrelated with y_{t+j} for all $j \geq 1$. (This equation is sometimes called the time-reversed representation of the process y_t.) We can therefore use, as an estimate of ϕ, the value $\tilde{\phi}$ minimizing $\sum_{t=1}^{T-1} (y_t - \tilde{\phi} y_{t+1})^2$, then define $\tilde{a}_t = y_t - \tilde{\phi} y_{t+1}$, $t=1,\ldots,T-1$, draw randomly with replacement from this set of residuals (after centering about their sample mean) to obtain $a_1^*,\ldots,a_{T-1}^*$ and, finally, define $y_T^* = y_T$ and

$$y_t^* = \tilde{\phi} y_{t+1}^* + a_t^* \tag{4.7}$$

for $t = T-1,\ldots,1$, thus generating a pseudodata sample path containing y_T. This procedure is appropriate only if the a_t defined by (4.6) are i.i.d., since this is a property of the a_t^*.

We will now show, however, that the white noise noise series a_t can be independent only if the cumulants of y_t (or, equivalently, those of e_t) are those of a Gaussian series, i.e., are 0 for orders higher than 2. Indeed, let κ_r denote the r-th order cumulant $\text{cum}(e_t,\ldots,e_t)$ of e_t for some $r>2$ (assumed to exist). Since, from (4.6),

$$y_t = \sum_{j=0}^{\infty} \phi^j a_{t+j}$$

it is easy to see that the a_t's are independent if and only if a_t is independent of y_{t+j} for each $j \geq 1$. In this case, the r-th order cumulants $\text{cum}(a_t, y_{t+j},\ldots,y_{t+j})$ will be 0; see Brillinger (1975, p. 19) for the fundamental properties of cumulants. For $j=1$, in particular, since we can write

$$y_{t+1} = e_{t+1} + \phi \sum_{j=0}^{\infty} \phi^j e_{t-j}$$

and

$$a_t = y_t - \phi y_{t+1} = -\phi e_{t+1} + (1-\phi^2)\sum_{j=0}^{\infty}\phi^j e_{t-j},$$

we are then led to

$$0 = \text{cum}(a_t, y_{t+1},\ldots,y_{t+1}) =$$

$$-\phi\ \text{cum}(e_{t+1},\ldots,e_{t+1})$$

$$+ (1-\phi^2)\phi^{r-1}\sum_{j=0}^{\infty}\phi^{jr}\text{cum}(e_{t-j},\ldots,e_{t-j})$$

$$= \kappa_r\{(\phi^{r-1}-\phi)/(1-\phi^r)\} \quad .$$

Since $0<|\phi|<1$, it follows that $\kappa_r = 0$, as asserted. If the distribution of e_t is determined by its moments and if all moments exist, then e_t, and hence also y_t, is therefore Gaussian. For Gaussian time series, however, pseudo-Gaussian Monte Carlo simulations seem like a more natural device to use to generate sample paths than the bootstrap.

We conclude from the preceding discussion that generally satisfactory methods are lacking for obtaining bootstrap sample paths through the final observations $y_{T-p+1},\ldots,y_T$.

Remark. The calculation used above, showing that assuming one-step forward and backward prediction are i.i.d. is tantamount to assuming that the observations are Gaussian, can be extended to stationary autoregressive processes of arbitrary order. A much more general assertion is made in Result 2.2 of Donoho (1981), namely, more that a strictly stationary non-Gaussian time series with finite second moments can have (ignoring rescalings) at most one invertible representation as a moving average of an i.i.d. white noise process. Some important details are missing in the proof which is given there, however.

CONCLUSION

Our results suggest that the estimates of mean square forecast error which result from the bootstrap procedure proposed by Freedman and Peters are not significantly more rereliable than the large sample estimates, which are ill-behaved, in small samples. This does not exclude the possibility that other methods of bootstrapping these statistics could prove useful.

ACKNOWLEDGEMENT

We thank Craig Ansley for suggesting that conditional mean square errors be investigated and William Schucany for stimulating us to consider the time-reversed representation.

REFERENCES

[1] Ansley, C. F. and P. Newbold (1981). "On the Bias in Estimates of Forecast Journal of Multivariate Analysis 13, 1-23. Mean Squared Error." Journal of the American Statistical Association, 76, 569-578.

[2] Donoho, D. L. (1981). "On Minimum Entropy Deconvolution" in Applied Time Series Analysis II (ed. D. F. Findley) 565-608, New York: Academic Press.

[3] Freedman, D. A. and S. F. Peters (1983). "Using the Bootstrap to Evaluate Forecasting Equations." University of California/Berkeley, Department of Statistics Technical Report No. 20.

[4] Freedman, D. A. and S. F. Peters (1984). "Bootstrapping an Economic Model: Some Empirical Results." Journal of Business and Economic Statistics 2, 150-158.

[5] Lai, T. L. and C. Z. Wei (1983). "Asymptotic Properties of General Autoregressive Models and Strong Consistency of Least Squares Estimates of Their Parameters." Journal of Multivariate Analysis 13, 1-23.

[6] Stine, R. A. (1983). "Prediction Intervals for Autoregressions." Paper presented at the 3rd International Symposium on Forecasting.

COMPUTER SCIENCE AND STATISTICS:
The Interface, D.M. Allen (ed.)

THE EM ALGORITHM IN TIME SERIES ANALYSIS

R. H. Shumway

Division of Statistics
University of California
Davis, CA 95616

The EM algorithm is ideally suited for maximizing likelihood functions arising in time series models involving stochastic signals embedded in noise. Successive steps involve simple regression computations, and the likelihood is nondecreasing at each step. Furthermore, the algorithm provides a simple and natural approach to handling problems caused by irregularly observed time series data. The simplicity of the approach is illustrated by applying the EM algorithm to the problem of estimating parameters in the state-space model. Examples involving biomedical data, economic data and data collected from the soil sciences are presented to illustrate the general procedure. A review is given of past experience in applying the algorithm, using both minimally configured microcomputers and large-scale mainframes.

1. INTRODUCTION

One of the benefits resulting from the explosive growth of microcomputer technology is that research workers now have easy access to computer programs for applying some of the computer intensive methods of time series analysis. Two examples are the Kalman filtering and smoothing recursions for the state-space model and iterative methods for maximum likelihood estimation using Newton-Raphson or EM algorithms.

A very general model which subsumes a whole class of special cases of interest in much the same way that linear regression does is the state-space model introduced in Kalman (1960) and Kalman and Bucy (1961). Although the model was originally utilized in aerospace related research, it has recently been applied to modeling data from economics (Harrison and Stevens (1976), Harvey and Pierse (1984), Kitagawa (1981), Kitagawa and Gersch (1984), Shumway and Stoffer (1982)), medicine (Jones (1984)) and in the soil sciences (Shumway (1985)).

The general form of the multivariate state-space model involves assuming that the rx1 observation vector $\underline{y}_t = (y_{1t},\dots,y_{rt})'$ can be written in the form

$$\underline{y}_t = A_t\underline{x}_t + \underline{v}_t \;, \qquad (1.1)$$

for t=1,2,...,n, where A_t is an rxp design matrix which specifies how the unobserved state vector $\underline{x}_t = (x_{1t},x_{2t},\dots,x_{pt})'$ can be converted into the observation vector $\underline{y}_t$ at any time point t. The additive rx1 observation noises $\underline{v}_t$ are assumed to be independent with $E\underline{v}_t = \underline{0}$ and covariance

$$R = E(\underline{v}_t\underline{v}_t') \qquad (1.2)$$

The form of (1.1) is almost identical to the standard regression model with $\underline{x}_t$ corresponding to a vector of random regression coefficients.

The behavior of the state vector $\underline{x}_t$ is determined by its initial value $\underline{x}_0$, and the state equations

$$\underline{x}_t = \Phi\underline{x}_{t-1} + \underline{w}_t \;, \qquad (1.3)$$

defined for t=1,...,T, where Φ is a pxp transition matrix and $\underline{w}_t$ is another independent model noise process with $E\underline{w}_t = 0$ and rxr model noise covariance matrix

$$Q = E(\underline{w}_t\underline{w}_t') \;. \qquad (1.4)$$

This is, of course, closely related to the first order autoregressive model defined previously, although no restrictions are imposed to guarantee stationarity. The specification is completed by assuming that the initial vector $\underline{x}_0$ has mean $\underline{\mu}$ and covariance matrix

$$\Sigma = E(\underline{x}_0 - \underline{\mu})(\underline{x}_0 - \underline{\mu})' \;. \qquad (1.5)$$

An important feature of the multivariate state-space formulation is that it provides one with a great flexibility in tailoring models to special circumstances. For example, suppose that we observe

$$y_t = x_t + v_t$$

where the unobserved series x_t is the second-order autoregressive process

$$x_t = \phi_1 x_{t-1} + \phi_2 x_{t-2} + z_{1t} \;.$$

This autoregressive "signal plus noise" model can be easily put into the state-space format (1.1) and (1.3) by writing

$$y_t = (1,0)\begin{pmatrix} x_t \\ x_{t-1} \end{pmatrix} + v_t$$

where

$$\begin{pmatrix} x_t \\ x_{t-1} \end{pmatrix} = \begin{pmatrix} \phi_1 & \phi_2 \\ 1 & 0 \end{pmatrix}\begin{pmatrix} x_{t-1} \\ x_{t-2} \end{pmatrix} + \begin{pmatrix} w_{1t} \\ 0 \end{pmatrix} ,$$

with the obvious identifications for A_t, $\underline{x}_t$ and Φ in Equations (1.1) and (1.3). Many different specific models can be expressed in state-space form as we shall see in later sections.

The introduction of the state-space approach as a tool for modeling data in the social and biological sciences requires that one be able to handle the model identification and parameter estimation problems since there will rarely be a well defined differential equation describing the state transitions. Furthermore, we would like to be able to handle general versions of (1.1) and (1.3) which provide for the possibility of missing data which occurs so often in the biological sciences. The problems of interest for the state-space model relate to estimating the state-vector $\underline{x}_t$ and the unknown parameters μ, Σ, Φ, Q and R. The problem of estimating $\underline{x}_t$ recursively under the assumption that the parameters are known was originally solved by Kalman (1960) and Kalman and Bucy (1961) and is the celebrated Kalman filter.

2. FILTERING, SMOOTHING AND FORECASTING

The problem of estimating $\underline{x}_t$ in the state-space model (1.1)-(1.5) can be approached by noticing that the linear estimator with minimum mean square error is the expectation conditioned on the observed data $\underline{y}_1,\ldots,\underline{y}_n$. In order to specify this procedure, consider the general conditional mean

$$\underline{x}_t^s = E(\underline{x}_t | \underline{y}_t,\ldots,\underline{y}_s), \qquad (2.1)$$

defined as a function of t, the point at which we need the value, and the span, s, of data which is used to determine the estimator. The general mean squared covariance function of the estimator (2.1) will be denoted by

$$P_{tu}^s = E[(\underline{x}_t - \underline{x}_t^s)(\underline{x}_u - \underline{x}_u^s)' | \underline{y}_1,\ldots,\underline{y}_s]. \qquad (2.2)$$

Several cases of interest can be distinguished depending on the span of the data and the point t at which the estimator is desired. For example, the one-step predictors x_t^{t-1} are the Kalman filter estimators whereas the conditional means x_t^T, based on the complete data span $y_1,\ldots,y_T$, are the Kalman smoothed estimators. Forecasting can be defined as the computation of x_t^T for t>T.

The computation of the quantities in equations (2.1) and (2.2) is a formidable undertaking if approached by straightforward methods. The dimensions of the vectors specified by the model are at least rT x 1 or pT x 1 where T denotes the number of data points observed in time. However, the recursions developed by Kalman (1960) and Kalman and Bucy (1961) require only that matrix computations of order rxr or pxp be performed recursively to develop the conditional means and covariances. The process of finding the Kalman filter (x_t^{t-1}) and smoother (x_t^T) estimators again involves using the linearity assumption to determine the minimizers of the mean square errors P_t^{t-1}. The derivation requires using the projection theorem recursively in conjunction with the model equations (1.1) and (1.3). The reader is referred to Jazwinski (1970) or Anderson and Moore (1979) for details.

The calculation of the Kalman filter estimators proceeds by the so-called forward recursions

$$\underline{x}_t^{t-1} = \Phi \underline{x}_{t-1}^{t-1} \qquad (2.3)$$

$$\underline{x}_t^t = \underline{x}_t^{t-1} + K_t(\underline{y}_t - A_t\underline{x}_t^{t-1}) \qquad (2.4)$$

for t=1,...,T with $\underline{x}_0^0 = \mu$. The one-step forecast $\underline{x}_t^{t-1}$ is a strict update of the previous estimated value whereas the best estimator involving current data $\underline{x}_t^t$ is a weighted average of $\underline{x}_t^{t-1}$ and the error that one makes in predicting $\underline{y}_t$. The pxr weight or <u>gain matrix</u> K_t is defined as

$$K_t = P_{tt}^{t-1}A_t'(A_tP_{tt}^{t-1}A_t' + R)^{-1}, \qquad (2.5)$$

where the covariances are updated recursively

using the recursions

$$P_{tt}^{t-1} = \Phi P_{t-1,t-1}^{t-1}\Phi' + Q \qquad (2.6)$$

and

$$P_{tt}^{t} = P_{tt}^{t-1} - K_t A_t P_{tt}^{t-1} \qquad (2.7)$$

with $P_{00}^{0} = \Sigma$.

If the estimator for $\underline{x}_t$ is to be based on all of the data $\underline{y}_1,\ldots,\underline{y}_T$, we need the Kalman smoother estimators. These can be developed by solving successively the backward recursions for $t=T,T-1,\ldots,1$ using the equations

$$\underline{x}_{t-1}^{T} = \underline{x}_{t-1}^{t-1} + J_{t-1}(x_t^T - x_t^{t-1}) \qquad (2.8)$$

where

$$J_{t-1} = P_{t-1,t-1}^{t-1}\Phi'(P_{tt}^{t-1})^{-1} . \qquad (2.9)$$

The mean square error covariance for the smoothed estimator satisfies the recursions

$$P_{t-1,t-1}^{T} = P_{t-1,t-1}^{t-1} + J_{t-1}(P_{tt}^{T} - P_{tt}^{t-1})J_{t-1}' \qquad (2.10)$$

If a forecast is needed it is clear that one only needs to extend the forward recursions (2.3)-(2.7) into the future under the convention that $K_t=0$ in (2.4) and (2.7).

The Kalman filter and smoother recursions give a convenient means for calculating the conditional expectations which are of greatest interest in solving problems in smoothing and forecasting for time series. The data are not required to be regularly spaced so that the smoothed estimators x_t^T can be used in lieu of missing values (see Section 3). The main problem which remains, however, is in specifying values for the unknown parameters $\underline{\mu}$, Σ, Φ, Q and R which are needed in order to apply the recursions.

3. ESTIMATION OF PARAMETERS

The estimation of the parameters involved in specifying the state-space model (1.1)-(1.5) can be accomplished using maximum likelihood if we are willing to assume that $\underline{x}_0,\underline{w}_1,\ldots,\underline{w}_T$ and $\underline{v}_1,\ldots,\underline{v}_T$ are jointly normal and uncorrelated random vectors.

The usual likelihood is the "innovations" form of Schweppe (1968), which involves writing the joint likelihood of the innovations

$$\underline{\varepsilon}_t = \underline{y}_t - A_t\underline{x}_t^{t-1}, \qquad (3.1)$$

conditional on $\underline{x}_0$, where x_t^{t-1} is defined in (2.1). The innovations, conditional on $\underline{y}_1,\ldots,\underline{y}_{t-1}$, have zero means and covariance

$$\Sigma_t = A_t P_{tt}^{t-1} A_t' + R . \qquad (3.2)$$

The log likelihood for estimating the parameter $\theta = (\Phi,Q,R)$ is essentially

$$\log L(Y;\theta) \stackrel{\Omega}{=} -\frac{1}{2}\sum_{t=1}^{T}\log|\Sigma_t| - \frac{1}{2}\sum_{t=1}^{T}\underline{\varepsilon}_t'\Sigma_t^{-1}\underline{\varepsilon}_t, \qquad (3.3)$$

which is a highly nonlinear function of the unknown parameters. The usual procedure is to fix $\underline{x}_0$ and then develop a set of recursions for the log likelihood function and its first two derivatives. Then, a Newton-Raphson algorithm can be used to successively update the parameter values until the log likelihood (3.3) is maximized. This approach is advocated, for example, by Gupta and Mehra (1974), Ansley and Kohn (1984), or Jones (1980).

We give a simpler approach here, based on the EM or expectation-maximization algorithm of Dempster et al (1977). The EM algorithm was adapted to this time series model in Shumway and Stoffer (1982). The EM algorithm proceeds by successive maximizing the current conditional expectation of the complete (but unobserved) data log likelihood based on $X = (x_0,w_1,\ldots,w_T, v_1,\ldots,v_T)$ conditional on the incomplete (but observed) data $Y = (\underline{y}_1,\ldots,\underline{y}_T)$. This complete-data log likelihood, given in Shumway and Stoffer (1982), involves the parameters $\theta = (\mu,\Sigma,\Phi,Q,R)$ in a convenient form but cannot be maximized directly since the $\underline{x}_t$ process is not observed. However, if the current value of θ is θ_i and E_i denotes the expectation under θ_i the EM algorithm proceeds by maximizing

$$Q(\theta|\theta_i) = E_i[\log L(X,\theta)|Y] \qquad (3.4)$$

at each step. Equation (3.4) can be written in terms of the Kalman smoothed outputs. The maximization of the resulting function with respect to the parameters Φ, Q and R then is exactly analogous to maximizing the usual multivariate normal likelihood function and yields the regression estimators

$$\Phi(i+1) = S_t(1)[S_{t-1}(0)]^{-1}, \qquad (3.5)$$

$$Q(i+1) = T^{-1}\{S_t(0) - S_t(1)(S_{t-1}(0))^{-1}S_t'(1)\}, \qquad (3.6)$$

where

$$S_t(j) = \sum_{t=1}^{T} \left(P_{t,t-j}^T + \underline{x}_t^T \underline{x}_{t-j}^{T\,\prime}\right) \quad (3.7)$$

for j=0,1, and

$$R(i+1) = T^{-1} \sum_{t=1}^{N} \left(\underline{\varepsilon}_t^T \underline{\varepsilon}_t^{T\prime} + A_t P_{tt}^T A_t'\right), \quad (3.8)$$

with

$$\underline{\varepsilon}_t^T = \underline{y}_t - A_t \underline{x}_t^T \ . \quad (3.9)$$

The term involving $\underline{\mu}$ and Σ has only a single observation and we arbitrarily fix Σ and take

$$\underline{\mu}_{i+1} = \underline{x}_0^T \quad (3.10)$$

The Kalman smoother can be used to compute all the terms in (3.7) except $S_t(1)$, which involves

$$P_{t,t-1}^T = \text{cov}(\underline{x}_t, \underline{x}_{t-1} | \underline{y}_1, \ldots, \underline{y}_T) \ . \quad (3.11)$$

Shumway and Stoffer (1982) have given the following backward recursions for determining $P_{t,t-1}^T$ for t=T,T-1,...,2. The basic recursion uses

$$P_{t-1,t-2}^T = P_{t-1,t-1}^{t-1} J_{t-2}' + J_{t-1}(P_{t,t-1}^T - \Phi P_{t-1,t-1}^{t-1}) J_{t-2}', \quad (3.12)$$

where we start with

$$P_{T,T-1}^T = (I - K_T A_T)\Phi P_{T-1,T-1}^{T-1} \ . \quad (3.13)$$

The overall procedure can be regarded as simply alternating between the Kalman filtering and smoothing recursions and the multivariate normal maximum likelihood equations (3.5)-(3.10). We summarize the iterative procedure as follows:

1. Initialize μ_0, Φ_0, Q_0, R_0 and fix Σ.
2. Use the Kalman recursions (2.3)-(2.9) to calculate x_t^T, P_{tt}^T and $P_{t,t-1}^T$.
3. Evaluate the log likelihood (3.3).
4. Update parameters to μ_1, Φ_1, Q_1, R_1 using Equations (3.5)-(3.10).
5. Return to step 2.

One of the advantages of the EM algorithm results from the simplicity of standard multivariate normal calculations which depend only on output from the forward and backward Kalman recursions. Successive steps of the form (3.4) never decrease the likelihood function and one is guaranteed to converge to at least a local maximum of the log likelihood function under fairly mild regularity conditions (see Wu (1984)). While the convergence rate of the EM algorithm is somewhat slower than that possible with Newton-Raphson or scoring algorithms (in the neighborhood of the maximum), one may be able to avoid the large divergent step corrections which are characteristic of these latter two procedures in the multiparameter situation.

An attractive feature available within the state-space framework relates to the ability to treat series which have been observed irregularly over time. The EM algorithm allows one to have parts of the observation vector $\underline{y}_t$ missing at a number of observation times without invalidating the computational procedures described in the previous two sections. An especially simple procedure results for the special case where the unobserved and observed parts of the error vector $\underline{v}_t$ are uncorrelated.

Suppose that at a given step, we define the partition of the rx1 observation vector $\underline{y}_t = (\underline{y}_t^{(1)\prime}, \underline{y}_t^{(2)\prime})'$, where $\underline{y}_t^{(1)}$ is the $r_1 \times 1$ observed portion and $\underline{y}_t^{(2)}$ is the $r_2 \times 1$ unobserved portion leading to the partitioned form

$$\begin{pmatrix} \underline{y}_t^{(1)} \\ \underline{y}_t^{(2)} \end{pmatrix} = \begin{pmatrix} A_t^{(1)} \\ A_t^{(2)} \end{pmatrix} \underline{x}_t + \begin{pmatrix} \underline{v}_t^{(1)} \\ \underline{v}_t^{(2)} \end{pmatrix} \quad (3.14)$$

where $A_t^{(1)}$ and $A_t^{(2)}$ are $r_1 \times p$ and $r_2 \times p$ matrices and

$$\text{cov} \begin{pmatrix} \underline{v}_t^{(1)} \\ \underline{v}_t^{(2)} \end{pmatrix} = \begin{pmatrix} R_{11} & R_{12} \\ R_{21} & R_{22} \end{pmatrix} \ . \quad (3.15)$$

Stoffer (1982) established that Equations (2.3)-(2.10) hold for the missing data case given above if one makes the replacements $\underline{y}_t' = (\underline{y}_t^{(1)\prime}, \underline{0}')$ and $A_t' = (A_t^{(1)\prime}, 0')$, and $R_{12} = R_{21} = 0$. That is, if $\underline{y}_t$ is incomplete, the filtered and smoothed estimators can be calculated from the usual equations by entering zeroes in the observation vector $\underline{y}_t$ where data is missing and by zeroing out the corresponding row of the design matrix A_t. This leads to the smoothed estimators $\underline{x}_t^{(T)}$ and the covariance functions $P_t^{(T)}$, $P_{t,t-1}^{(T)}$ in the missing data case.

The maximum likelihood estimators, as computed in the EM procedure, require that one take the conditional expectation of (3.4) under the assumption that $\underline{y}_t$ is incompletely observed. Now, defining the incomplete data as $Y_T^{(1)} = (\underline{y}_1^{(1)}, \underline{y}_2^{(1)}, \ldots, \underline{y}_T^{(1)})$, the expectation of the third term can be computed by conditioning first on both $Y_T^{(1)}$ and $\underline{x}_t$ and then on $Y_T^{(1)}$ which leads to (cf. Shumway and Stoffer (1982), Shumway (1984))

$$R(i+1) = n^{-1} \sum_{t=1}^{N} D_t G_t D_t' \qquad (3.16)$$

where

$$G_t = \begin{pmatrix} G_t^{(1)} & G_t^{(1)} F' \\ FG_t^{(1)} & FG_t^{(1)} F' + R_{22.1} \end{pmatrix} \qquad (3.17)$$

with

$$F = R_{21} R_{11}^{-1}, \qquad (3.18)$$

$$R_{22.1} = R_{22} - R_{21} R_{11}^{-1} R_{12}, \qquad (3.19)$$

and

$$G_t^{(1)} = \underline{\varepsilon}_t^T \varepsilon_t^{T'} + A_t^{(1)} P_{tt}^{(T)} A_t^{(1)'} . \qquad (3.20)$$

where

$$\underline{\varepsilon}_t^T = \underline{y}_t^{(1)} - A_t^{(1)} \underline{x}_t^T . \qquad (3.21)$$

The matrix D_t is a permutation matrix which reorders the variables in their original form. This is necessary because the application of (3.17)-(3.20) requires that the variables be ordered so that the observed values appear in $\underline{y}_t^{(1)}$.

A simplification introduced in Shumway and Stoffer (1982) is to assume that the errors relating the unobserved and observed components are uncorrelated, i.e. $R_{12} = 0$, so that the correction (3.17) reduces to

$$G_t = \begin{pmatrix} G_t^{(1)} & 0 \\ 0 & R_{22} \end{pmatrix} . \qquad (3.22)$$

If the vector observation has all components missing, the correction reduces to adding R from the previous iterate.

4. EXAMPLES

4.1 An Irregularly Observed Biomedical Series

In order to give an illustration of an incomplete series, consider the problem of modeling the level of several biomedical parameters monitored after a cancer patient undergoes a bone marrow transplant. The data in Figure 3.1, presented by Jones (1984), are measurements made for 92 days on the three variables log(white blood count), log(platelet) and HCT(hematocrit). Approximately 40% of the values are missing, with the missing values mainly occurring after the 35th day. (The missing values are shown along the time axis on the plotted series). The main objectives in this example are to model the three variables using the state-space approach and to smooth the data. According to Jones (1984), "Platelet count at about 100 days post transplant has previously been shown to be a good indicator of subsequent long term survival."

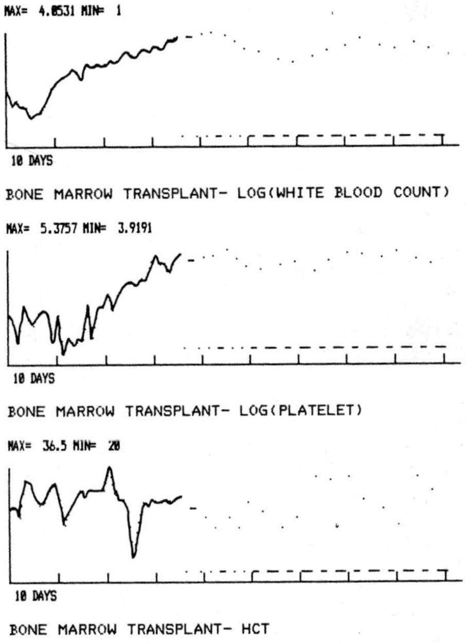

Figure 1 - Bone marrow transplant data (Jones (1984)).

The simple state-space model with three components was chosen with the observed log(WBC), log(platelet) and HCT denoted by y_{1t}, y_{2t}

and y_{3t} and the unknown true levels denoted by x_{1t}, x_{2t} and x_{3t}. The true vector process satisfies the state equation

$$\begin{pmatrix} x_{1t} \\ x_{2t} \\ x_{3t} \end{pmatrix} = \begin{pmatrix} .981 & -.035 & .008 \\ .059 & .925 & .006 \\ -1.078 & 1.811 & .823 \end{pmatrix} \begin{pmatrix} x_{1,t-1} \\ x_{2,t-1} \\ x_{3,t-1} \end{pmatrix} + \begin{pmatrix} w_{1t} \\ w_{2t} \\ w_{3t} \end{pmatrix}$$

where the transition matrix was extimated after 30 iterations of the EM algorithm. The state and observation covariance matrices were estimated as

$$Q = \begin{pmatrix} .014 & -.002 & .013 \\ -.002 & .003 & .027 \\ .013 & .027 & 3.485 \end{pmatrix}, \quad R = \begin{pmatrix} .007 & 0 & 0 \\ 0 & .017 & 0 \\ 0 & 0 & .631 \end{pmatrix}$$

Again, the coupling between the first two series and the third series is relatively weak. The regression relating x_{3t}(HCT) to the other two series seems to be fairly strong, i.e.

$$x_{3t} = -1.078x_{1,t-1} + 1.811x_{2,t-1} + .823x_{3,t-1} + w_{3t}$$

The smoothed values, as evaluated using the Kalman recursions, are shown in Figure 2 below. The approximate standard errors $\sqrt{P_{tt}^T}$ of the interpolated missing values in the latter parts of the series are in the ranges .11-.13, .08-.09 and 1.7-2.0 for the three series respectively.

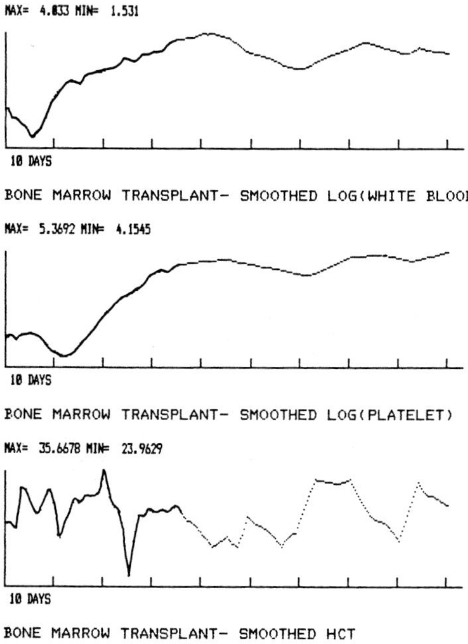

Figure 2 - Smoothed bone marrow transplant data

4.2 Signal Extraction for Soil Sciences Data

As an example of a simple signal extraction problem consider the following example from Shumway (1985) involving salt content values measured at intervals of one meter over a line transect. Figure 3 shows the average of five such transects (parallel samples) taken from Morkoc et al (1984).

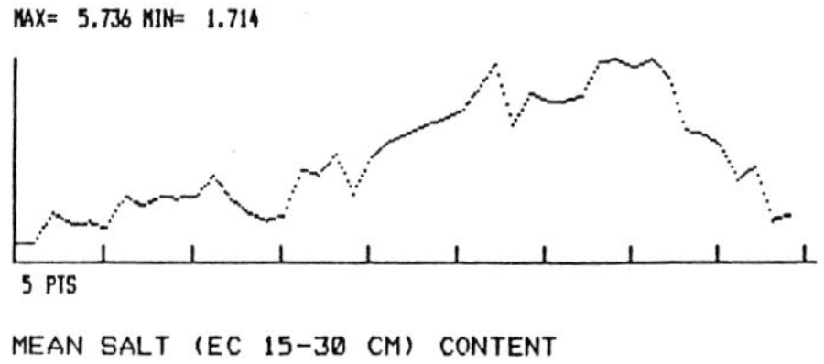

Figure 3 - Average salt content over five transects (1 pt = 1 m). (Morkoc et al (1984)).

It is plausible that the salt content can be represented as a non-stationary trend function superimposed on noise. We might assume (see Shumway (1985)) that the observed salt content at the spatial point s, say y_s, can be represented as

$$y_s = x_s + v_s \tag{4.1}$$

where x_s is the smooth trend function and v_s is the irregular white noise component with variance σ_v^2. The basic objective is to produce an estimator for the nonstationary trend function x_s. In order to specify smoothness constraints for the trend function x_s we might assume that the second difference (derivative) is small, say

$$\nabla^2 x_s = w_{1s} \tag{4.2}$$

where ∇ is the usual difference operator and w_{1s} is a noise with variance σ_w^2. There is an obvious similarity here to spline smoothing (see Wecker and Ansley (1984)). Now, since

$$\nabla^2 x_s = x_s - 2x_{s-1} + x_{s-2}, \tag{4.3}$$

it is clear that by defining the state vector $\underline{x}_s = (x_s, x_{s-1})'$, the model in Equations (4.1) and (4.2) can be written in the state-space form

$$y_s = (1,0) \begin{pmatrix} x_s \\ x_{s-1} \end{pmatrix} + v_s \tag{4.4}$$

where

$$\begin{pmatrix} x_s \\ x_{s-1} \end{pmatrix} = \begin{pmatrix} 2 & -1 \\ 1 & 0 \end{pmatrix} \begin{pmatrix} x_{s-1} \\ x_{s-2} \end{pmatrix} + \begin{pmatrix} w_{1s} \\ 0 \end{pmatrix} \tag{4.5}$$

and the obvious identifications can be made in (1.1) and (1.3). The transition matrix Φ is fixed in this case and we have only to estimate the variances σ_v^2 and σ_w^2 associated with the observation and model noises respectively. The estimator for σ_w^2 comes from q_{11} in

$$Q = T^{-1}\left(S_t(0)-S_t(1)\Phi' - \Phi S_t'(1) + \Phi S_{t-1}(0)\Phi'\right) \quad (4.6)$$

where Φ is the fixed transition matrix. The estimator for σ_v^2 follows directly from (3.8) as usual. The final estimators for the variances are $\hat{\sigma}_v^2 = .102$, $\hat{\sigma}_w^2 = .021$.

The smoothed values $\underline{x}_s^T$ under this model are plotted in Figure 4 and it is clear that the smoothed values follow the major turns in the data quite well. The resulting smoothed series has a prediction standard error of .16.

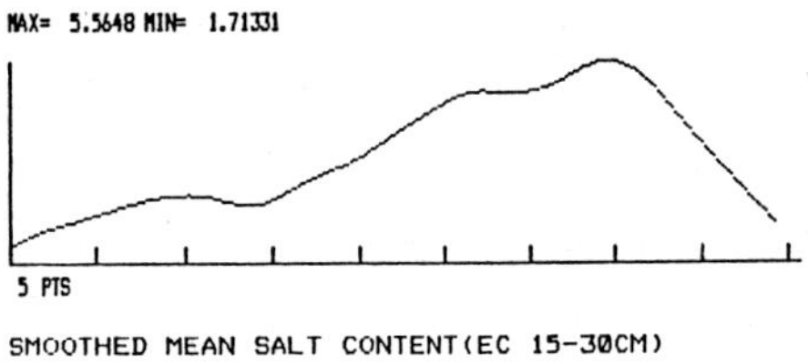

Figure 4 - Smoothed salt content using (4.1) and (4.2) with $\hat{\sigma}_v^2 = .102$, $\hat{\sigma}_w^2 = .021$.

4.3 Forecasting and Seasonal Adjustment of Economic Series

The inherent flexibility of the state-space model can be exploited for developing additive models for economic time series. The use of state-space methods for analyzing additive models of importance in economics has been proposed by Kitagawa (1981), Kitagawa and Gersch (1984) and Harvey (1983). As an example, consider the quarterly data on earnings-per-share shown in Figure 5 for the U.S. company, Johnson and Johnson. The general character of the series seems to emerge as an exponential trend with a seasonal kind of oscillation superimposed on this trend; the seasonal oscillation tends to repeat every four quarters.

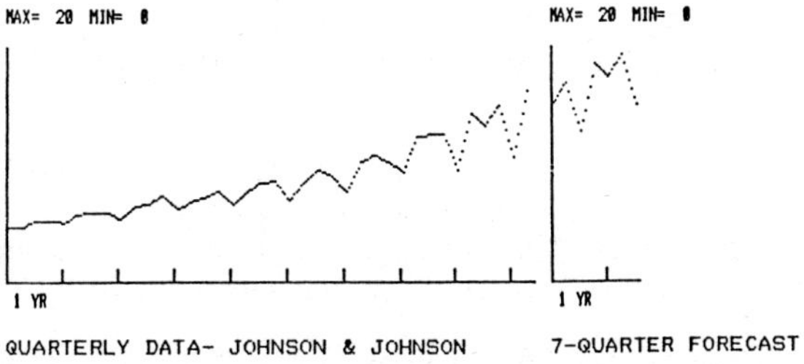

Figure 5 - Quarterly earnings per share (1970(4) to 1980(1) and 7 quarter forecast for Johnson and Johnson.

In order to develop an additive model for this particular kind of data, suppose that we regard the observed series y_t as being composed of trend, seasonal and irregular components, denoted by x_{1t}, x_{2t} and v_t respectively. The observed data can be modeled as

$$y_t = x_{1t} + x_{2t} + v_t\ , \quad (4.7)$$

where the exponential trend component might be modeled as

$$x_{1t} = \phi x_{1,t-1} + w_{1t}\ , \quad (4.8)$$

where $\phi>1$ represents the growth rate. The quarterly seasonal component might be modeled as

$$x_{2t} = -x_{2,t-1} - x_{2,t-2} - x_{2,t-3} + w_{2t}, \quad (4.9)$$

reflecting the fact that the sum of the four quarters should be approximately 0 for the seasonal factor. The problems of interest for the model can be reduced first to estimating the parameters and then the unobserved components x_{1t} and x_{2t}. One would also like to be able to forecast y_t. A problem of some interest in economic applications is in estimating the series with seasonal effects excluded, i.e. $\left(x_{1t}^T + x_{2t}^T\right)$, sometimes termed <u>seasonal adjustment</u>.

The model specified by (4.7), (4.8) and (4.9) can be put into state-space form by defining the state-vector $\underline{x}_t = (x_{1t}, x_{2t}, x_{2,t-1}, x_{2,t-2})'$, so that the observation Equation (1.1) becomes

$$y_t = (1,1,0,0)\begin{pmatrix} x_{1t} \\ x_{2t} \\ x_{2,t-1} \\ x_{2,t-2} \end{pmatrix} + v_t \quad (4.10)$$

with the state Equation (1.3) given by

$$\begin{pmatrix} x_{1t} \\ x_{2t} \\ x_{2,t-1} \\ x_{2,t-2} \end{pmatrix} = \begin{pmatrix} \phi & 0 & 0 & 0 \\ 0 & -1 & -1 & -1 \\ 0 & 1 & 0 & 0 \\ 0 & 0 & 1 & 0 \end{pmatrix} \begin{pmatrix} x_{1,t-1} \\ x_{2,t-1} \\ x_{2,t-2} \\ x_{2,t-3} \end{pmatrix} + \begin{pmatrix} w_{1t} \\ w_{2t} \\ 0 \\ 0 \end{pmatrix} \quad (4.11)$$

where

$$R = r_{11}, \quad Q = \begin{pmatrix} q_{11} & 0 & 0 & 0 \\ 0 & q_{22} & 0 & 0 \\ 0 & 0 & 0 & 0 \\ 0 & 0 & 0 & 0 \end{pmatrix} \quad (4.12)$$

gives the two covariance structures. Harvey (1981, p. 180) shows that this model with $\phi=1$ is essentially an ARIMA (0,1,1) x $(0,1,1)_4$ which has been applied to accounting data by Griffin (1977).

The computational modifications required for this state-space model are minor since $\hat{q}_{11}$ and $\hat{q}_{22}$ can now be obtained as the first two diagonal elements in Q defined by (4.6). The estimated transition parameter ϕ is just the ratio of the upper left corner elements of $S_t(1)$ and $S_{t-1}(0)$. That is

$$\Phi(i+1) = \frac{[S_t(1)]_{11}}{[S_{t-1}(0)]_{11}} \tag{4.13}$$

where $[A]_{ij}$ denotes the ij^{th} element of the matrix A.

Table 1 shows the successive estimators for the four parameters as applied to the Johnson & Johnson data.

Table 1 - Successive parameter estimates for earnings-per-share for Johnson & Johnson using additive model

Iter	ϕ	q_{11}	q_{22}	r_{11}	2ℓogL
1	1.028	.010	.010	.033	-93.96
2	1.036	.012	.029	.062	- 5.31
3	1.037	.012	.047	.068	3.55
4	1.037	.011	.061	.066	6.26
5	1.037	.011	.072	.062	7.34
6	1.037	.010	.080	.057	7.85
7	1.037	.010	.085	.054	8.13
8	1.037	.010	.088	.051	8.30
9	1.037	.010	.090	.048	8.42
10	1.037	.010	.092	.046	8.50
11	1.037	.010	.097	.038	8.74
12	1.037	.010	.096	.037	8.77
13	1.037	.010	.096	.036	8.78
14	1.037	.010	.096	.035	8.80
15	1.037	.010	.096	.035	8.80

The log likelihood converges nicely to a local maximum although, at the tenth iteration, the process was stopped and the seasonal and irregular component variances were incremented strongly in the directions that were suggested on examination of previous iterations. The final value of 1.037 for the parameter ϕ implies that the exponential growth rate is approximately 3.7 percent per quarter.

The values of the parameters given in Table 1 were then used to estimate the trend x_{1t} and seasonal components x_{2t} of the model. These are shown in Figure 6 and we note that the estimated "trend plus seasonal," say $x_{1t} + x_{2t}$, produces credible version of the original series. The estimated trend might be taken as a seasonally adjusted version of the series.

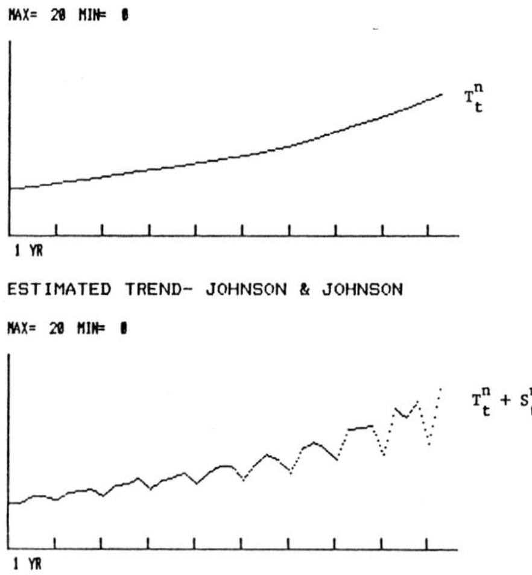

Figure 6 - Estimated trend, x_{1t}^T, and "trend plus seasonal," $x_{1t}^T + x_{2t}^T$, for the earnings data.

A fundamental question of interest here would be in producing forecasts for the series, say

$$y_t^T = x_{1t}^T + x_{2t}^T$$

for t>T. It is clear that adding the Kalman smoother outputs for the first two components of $\underline{x}_t^T$ will generate these forecasts and that the mean square error for the forecasts can be computed as

$$(\sigma_t^T)^2 = [P_{tt}^T]_{11} + 2[P_{tt}^T]_{12} + [P_{tt}^T]_{22} ,$$

where $[P_{tt}^T]_{ij}$ denotes the ij^{th} element of P_{tt}^T. Table 2 shows a three-quarter forecast for the second through fourth quarters of 1980 compared with the actual values. There seems to be quite good agreement between the observed and predicted values and all three prediction intervals include the true values.

Table 2 - Comparison of observed earnings and forecasts for Johnson & Johnson

Qtr	Obsvd	Frcst	*Error	approx 95% PI
1980(2)	14.67	14.97	.02	13.97-15.97
1980(3)	16.02	16.77	.05	15.75-17.79
1980(4)	11.61	12.24	.05	11.22-13.26

*Error = |observed-forecast|/observed

The seven quarter forecasts are appended to the original observed series in Figure 5 and can be seen to provide a very plausible forecast of the underlying earnings series.

5. DISCUSSION

The application of the state-space approach to modeling time series data in economics and in the biological and physical sciences has been hindered by the lack of accessible computing power and software. Although the model is inherently appealing, the process of developing software for the computationally intensive recursive and iterative procedures for smoothing and for parameter estimation has been slow and painful. This paper has described a proposed procedure by which the Kalman filter and EM algorithm can be combined to solve simultaneously the problems of smoothing, forecasting and parameter estimation.

The software for performing these computations is available in BASIC for microcomputers and in FORTRAN for large scale mainframes. The BASIC version for microcomputers is currently available and running on a Tandy 1200HD (or IBM PC, PC-XT) using MSDOS. For a sample of 61 observations from three time series, each iteration required 14 minutes. An earlier version running on a TRS-80, Model III required 34 minutes per iteration. The availability of FORTRAN and BASIC compilers combined with an 8086 chip should reduce these running times significantly. A version in FORTRAN written for the CDC-6600, required only one minute per iteration for a sample involving 600 observations from each of seven time series. A listing of a FORTRAN program which uses a quasi Newton-Raphson algorithm instead of the EM algorithm appears in Jones (1984).

REFERENCES

[1] Anderson, B.D.O. and Moore, J.B., Optimal Filtering (Prentice Hall, New Jersey, 1979).

[2] Ansley, C.F. and Kohn, R., On estimation of ARIMA models with missing values, in Parzen, E. (ed.), Time Series Analysis of Irregularly Observed Data (Springer-Verlag, New York, 1984).

[3] Dempster, A.P., Laird, N.M. and Rubin D.B., Maximum likelihood from incomplete data via the EM algorithm, J. of the Royal Statist. Soc., Ser B, 39, (1977) 1-38.

[4] Gupta, N.K. and Mehra, R.K., Computational aspects of maximum likelihood estmation and reduction in sensitivity function calculations, IEEE Trans. on Aut. Control AC-19, (1974) 774-783.

[5] Harrison, P.J. and Stevens, C.F., Bayesian forecasting, Jour. Royal Statist. Soc. B 38 (1976) 205-247.

[6] Harvey, A.C. and Pierse, R.G., Estimating missing observations in economic time series, J. Amer. Statist. Assoc. 79 (1984) 125-131.

[7] Harvey, A.C., Forecasting economic time series with structural and Box-Jenkins models: A case study, J. of Business and Economic Statistics 1, No. 4 (1983).

[8] Harvey, A.C., Time Series Models (Halsted Press, New York, 1981).

[9] Jazwinski, A.H., Stochastic Processes and Filtering Theory (Academic Press, New York, 1970).

[10] Jones, R.H., Fitting multivariate models to unequally spaced data, in Parzen, E. (ed.), Lecture Notes in Statistics 25 (Springer-Verlag, New York, 1984).

[11] Jones, R.H., Maximum likelihood fitting of ARMA models to time series with missing observations, Technometrics 22 (1980) 389-396.

[12] Kalman, R.E., A new approach to linear filtering and prediction theory, Trans. ASME J. of Basic Eng. 8 (1960) 35-45.

[13] Kalman, R.E. and Bucy, R.S., New results in linear filtering and prediction theory, Trans. ASME J. of Basic Eng. 83 (1961) 95-108.

[14] Kitagawa, G., A nonstationary time series model and its fitting by a recursive technique, J. Time Series Analysis 2 (1981) 103-116.

[15] Kitagawa, G. and Gersch, W., A smoothness priors modeling of time series with trend and seasonality, J. Amer. Statist. Assoc. 79 (1984) 378-389.

[16] Morkoc, F., Biggar, Nielsen, D.R. and Rolston, D.E., Analysis of surface soil water content and surface soil temperature using a state-space approach. Submitted: Soil Sci. Soc. Am. J.

[17] Schweppe, F.C., Sensor array data processing for multiple signal sources, IEEE Trans. on Information Theory IT-4 (1968) 29-305.

[18] Shumway, R.H., Som applications of the EM algorithm to analyzing incomplete time series data, in Parzen, E. (ed.), Time Series Analysis of Irregularly Observed Data (Springer-Verlag, New York, 1984).

[19] Shumway, R.H. and Stoffer, D.S., An approach to time series smoothing and forecasting using the EM algorithm, J. Time Series Anal. 3 (1982) 253-264.

[20] Shumway, R.H., Time series in the soil sciences: Is there life after Kriging? To appear in Proc. of Int. Soc. of Soil Sci. and Soil Sci. Soc. of Amer. Workshop on Soil Spatial Variability, Centre for Agri. Publ., Wageningen, The Netherlands.

[21] Stoffer, D.S., Estimation of parameters in a linear dynamic system with missing observations, Ph.D. Dissertation, Univ. of California, Davis.

[22] Wu, C.F., On the convergence properties of the EM algorithm, Ann. Statist. 11 (1981) 95-103.

COMPUTER SCIENCE AND STATISTICS:
The Interface, D.M. Allen (ed.)

Designing An Intelligent System for Spectral Analysis

D. B. Percival, A. Buja, R. D. Martin, E. O. Belcher,
R. K. Kerr, S. D. Yee, and C. B. Hurley

Applied Physics Laboratory (HN-10)
Department of Statistics (GN-22)
University of Washington
Seattle, Washington 98105

ABSTRACT

The design of a software package to help a user perform spectral analysis is described.

1. Introduction

Spectral analysis is widely used in the engineering and physical sciences, but, because of its complexity, there are many pitfalls to its successful application. There are currently a number of software packages that can do the numerical computations that are required for spectral analysis, but none of them offer extensive guidance for the user. Recent developments in computer science have made it feasible to construct intelligent software in the form of expert systems that mimic the actions of a human expert in such diverse fields as medicine, geology, and computer installation. Moreover, Gale and Pregibon[3] have made a first attempt at constructing an expert system for statistical analysis, namely, the REX system for regression analysis.

Because of these developments and the recent availability of powerful computer workstations with high resolution graphics, we are developing a software package on such a workstation to help scientists perform spectral analysis. The research questions that our project addresses are: 1) what is a good way to incorporate intelligence into a software package? 2) what help can a software package provide a user for organizing the results of a spectral analysis? 3) is it possible to develop a systematic strategy for spectral analysis such that, given a time series that may be regarded as a realization of a stationary process and given some or no *a priori* knowledge on the shape of its underlying spectrum, no important features of the data are missed? and 4) what new tools for spectral analysis are possible on a state-of-the-art workstation? In this report we concentrate on the first two of these questions.

2. Desired Features for an Ideal Software Package

What exactly do we feel is lacking in available software for doing spectral analysis? For heavy users of interactive statistical packages such as S and ISP, one deficiency is a lack of a data base management system. In the course of a spectral analysis, a user can produce a large number of new auxiliary data sets that are formed by manipulating the original time series. (In a recent analysis of some wind speed data, one user produced over 50 auxiliary data sets.) Keeping track of all these new data sets is a real problem. It is a common experience amongst analysts to be unable to recall with the passage of time where all the auxiliary data sets came from. An ideal software package would provide some way to organize these data sets automatically.

A second desirable feature is more extensive graphical capabilities than current software packages generally provide. The availability of workstations with enough power to quickly update a graphical display (so-called real-time graphics) opens up a whole new category of displays that a user would like to have available.

A third area in which software can aid a user is to provide help in the specification of parameters for sophisticated methods such as robust fitting of autoregressive models. Here the statistical methodology has become so complex that even the designers of the methods have difficulty in applying them without constantly referring to their own technical reports.

For inexperienced users, the main problem with current software is the lack of in-depth help. An ideal software package should do, guide, explain, and even teach the techniques of good spectral analysis. Loosely speaking, augmenting software to provide such help is called making the software more "intelligent".

3. An Example of Spectral Analysis

In order to incorporate intelligence into spectral analysis software, it is helpful to develop a model of how a human expert does spectral analysis. To focus our discussion below, let us quickly step through an example of a spectral analysis (the reader is referred to Priestley[6] and Bloomfield[1] for a complete discussion of the statistical theory used here). The time series for our example is monthly average values of the daily water flow of the Willamette River at Salem, Oregon. We begin by examining a plot of the data versus time (figure 1a). We note immediately the marked cyclical behavior of the data. There is, however, a problem with regarding this series as a realization of a stationary process, namely, there is much less variability in the series at the low points of each cycle than at the high points.

Since the data are all positive, we might consider looking at the logarithm of this data in an attempt to stabilize its variance over time. (For some purposes for which spectral analysis is used, such a transformation would not be desirable even if it did stabilize the variance; we assume that this is not the case here.) This transformation is shown in figure 1b. We see that the variability of this series is much more uniform.

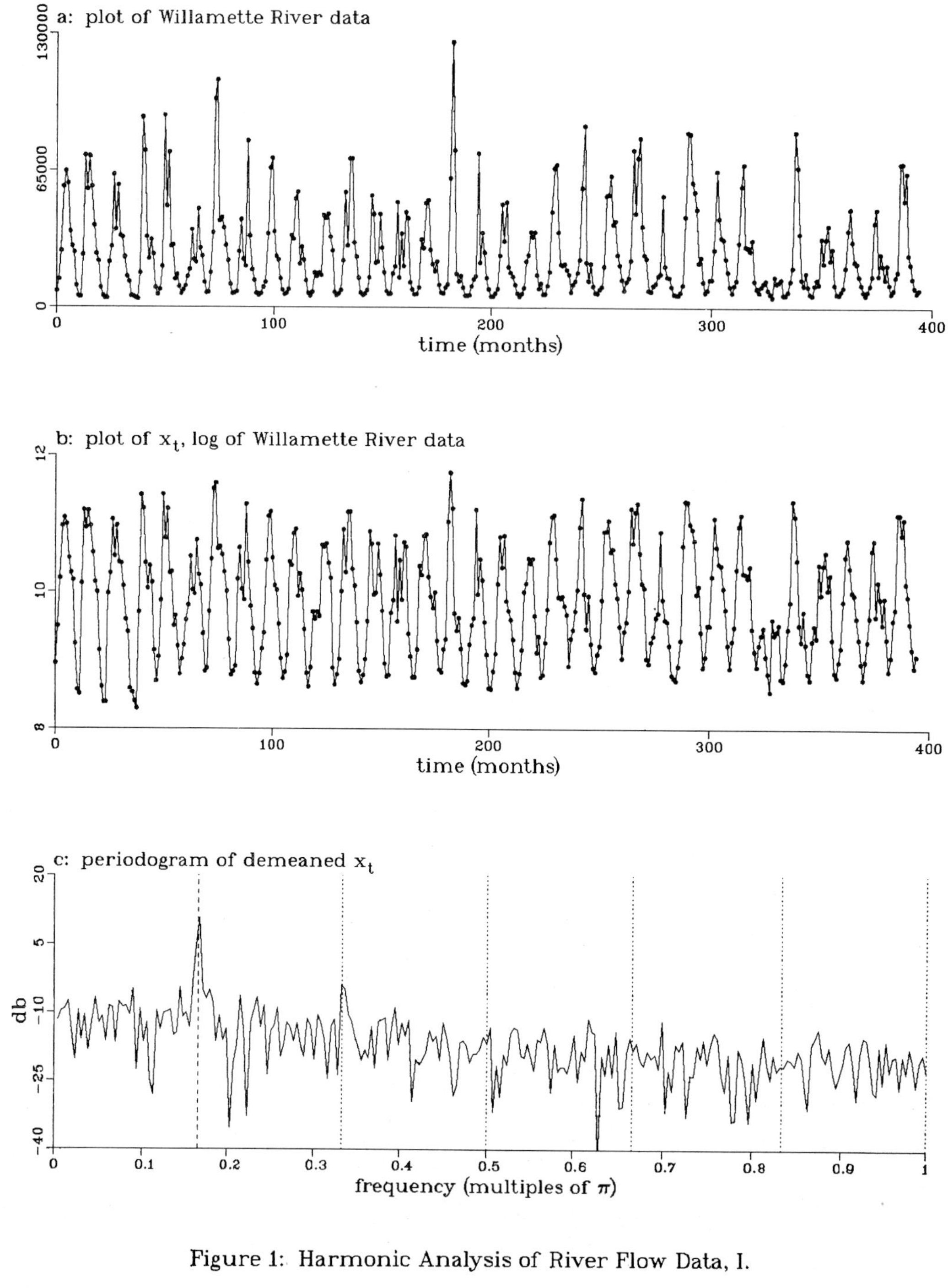

Figure 1: Harmonic Analysis of River Flow Data, I.

Since the sampling time is one month, figure 1b shows that the period of the phenomena is about one year (as one would suspect from physical considerations). This plot suggests that this time series may be modeled by a harmonic process of the form

$$X_t = \mu_X + \sum_{k=1}^{K} \{A_k \cos(\omega_k t) + B_k \sin(\omega_k t)\} + \varepsilon_t \ , \quad (1)$$

where μ_X, K, $\{A_k\}$, $\{B_k\}$ and $\{\omega_k\}$ are unknown constants and $\{\varepsilon_t\}$ is a zero mean stationary process with variance σ_ε^2 and spectral density function $h_X()$. If $\{\varepsilon_t\}$ were a white noise process, the spectrum for $\{X_t\}$ would be completely determined by $\{A_k\}$, $\{B_k\}$, $\{\omega_k\}$, and σ_ε^2.

Our first task is to estimate K, the number of sinusoids with distinct frequencies in the model, and the corresponding ω_k's. The standard way to do this is to look for peaks in the periodogram of $\{X_t - \overline{X}\}$, where $\overline{X}$ is the sample mean. Figure 1c shows that there is one prominent peak in the periodogram near the angular frequency with a period of one year ($\pi/6 \approx .16667\pi$ radians per month, indicated by the dashed vertical line). This peak is 10 db above all other peaks, so we should include a term in our model to account for it (if there were any doubt as to the significance of the peak, we could appeal to a formal statistical test such as Fisher's g or Siegel's test[7]).

Besides the peak corresponding to an annual period, there are numerous other bumps in the periodogram that may or may not be due to other sinusoidal components. If we assume that the expected variation in the river flow is periodic with a period of one year but is not necessarily sinusoidal, we would expect to see peaks at frequencies that are harmonics of $\pi/6$. These harmonics are indicated in figure 1b by vertical dotted lines. We see that the second largest peak in the periodogram does occur at the first harmonic ($\pi/3$). There are no other peaks that seem to be particularly prominent. (Again Siegel's test can help us judge the significance of questionable peaks.)

To see if we can identify some components that may be hidden due to leakage from the dominant peaks, figure 2a shows the periodogram for the data after it has been tapered with a 100% cosine taper. Again there are lots of bumps besides the dominant two we have already identified, none of which seem to be particularly prominent.

Based upon our examination of the plots in figure 1, let's assume a model given by equation (1) with $K=2$ and $\omega_k = k\pi/6$ for which $\{\varepsilon_t\}$ is a white noise process. This is a simple linear regression model which we can fit to our data using least squares. Figures 2b and 2c show the residuals from this fitted model plotted versus time and offset from the beginning of a year, respectively. To continue the analysis of this data, we would carefully study these residual plots to judge the adequacy of our model.

There are two comments we should make about this analysis. First, the actions that we have outlined are not a literal record of what an expert did. Some false starts and "snooping around" have been removed. Second, for this time series, if our assumed model were true, we would have only one estimate for the spectrum (ignoring minor variations such as fitting the model by some criterion other than least squares). For time series that must be modelled by a purely continuous stationary process (i.e., the spectrum is determined by a spectral density function), there is a subjective element introduced by the choice of such things as data tapers, prewhitening filters, window smoothing parameters, and order of autoregressive models. These choices result in a wide variety of different spectral estimates. Unless we have some external information about a time series, there is no way of telling which estimate is closest to the "truth." Moreover, since, to quote Tukey[8], "... most spectrum analysis is exploratory in character," it is often not the goal to pick one of these estimates as the best estimate, but rather we want to look at many different spectral estimates to try to understand our data and to look for interesting features in it.

4. Prototype Expert System for Spectral Analysis

Our first attempt to incorporate intelligence into spectral analysis software was to develop a prototype expert system. We built the system using computer hardware and software available to us in 1984, namely, a VAX 750 with primitive graphics terminals running under the 4.2 BSD UNIX operating system with Franz LISP and OPS5, a programming language for a production system. Such a system requires that the knowledge of an expert be summarized in production rules of the general form "if A, B, ... are true, then assert action C." Our first task was to extract the knowledge of an expert in this form.

To do so, we followed an expert through the analysis of several "typical" time series such as the river flow data. We were able to come up with a "script" that represented the decisions and actions that the expert took at each stage of the analysis. Each portion of the script was initially coded into production rules. As an example, a production rule that we could have included based upon the river flow analysis is "if the data is positive and if the variability of the series is proportional to the height of the series, then make a log transformation."

We learned several things from this exercise. First, it is difficult to capture the expertise involved in spectral analysis using just production rules. Much of our script was purely procedural in nature, and this was rather clumsy to code with production

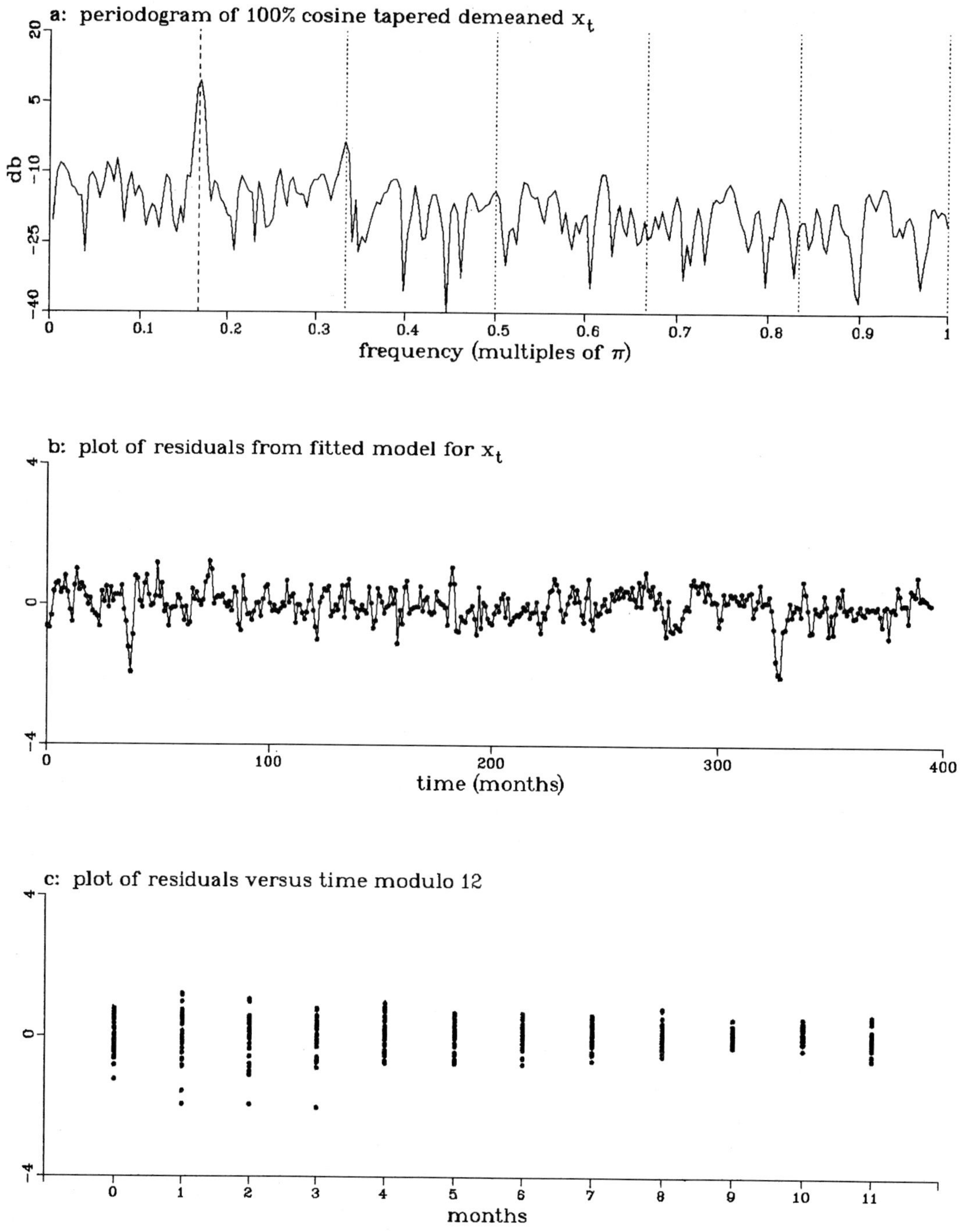

Figure 2: Harmonic Analysis of River Flow Data, II.

rules. For example, in the river flow analysis, once we had noted the strong cyclical variation in the log of the original data (figure 1a), there was a procedure that we followed: we identified the frequencies of the sinusoidal components in the model using the periodogram, fitted the model to the data, and examined the residuals. We found it easier to write some of the purely procedural parts of the system in the C programming language.

Second, graphical displays play a critical role in spectral analysis. There are many features of data that are difficult to extract by a statistical measure but that are readily apparent to the trained eye. To obtain this visual information from an untrained user, the expert system was programmed to carry out a dialog between itself and the user. It presented a series of graphs to the user and queried him or her about the presence or absence of certain features in the graphs. If the user was unable to answer the system's questions, the system would attempt either to help the user by supplying examples or to answer the questions by itself based upon some test statistics. This approach exploits the superior human visual ability to find structure in graphs.

Third, rather simple automatic mechanisms were found for keeping track of an analysis and of the auxiliary data sets created during a spectral analysis. The OPS5 code and C procedural routines did their numerical work by calling task programs. The collection of these tasks programs is by itself a primitive system for carrying out spectral analysis. For example, suppose the values of the log of the river flow series reside in a data file called "lrf". To taper this series with a 100% cosine data taper and calculate a periodogram for it (as was done in figure 2a), we would give the following commands to the UNIX operating system:

```
taper -p 1.00 lrf lrf.tpr
pgram lrf.tpr lrf.tpr.pgm
```

The tapered time series and its periodogram would now be in the auxiliary files "lrf.tpr" and "lrf.tpr.pgm", respectively. (The names of these two files can be arbitrarily chosen.) Part of the action of both commands is to place a copy of the commands themselves at the end of a special file named "hist.tsa". A list of this file at the end of an analysis gives a complete history of all commands that were executed during the course of an analysis.

In addition, the formats of "lrf.tpr" and "lrf.tpr.pgm" are special in that they contain not only data values but also a copy of the UNIX command that created them. A special task program called "genesis" could then be evoked at any later date to find out how these two auxiliary files were created. Thus the command

```
genesis lrf.tpr lrf.tpr.pgm
```

would yield the output

```
lrf.tpr: taper -p 1.00 lrf lrf.tpr
lrf.tpr.pgm: pgram lrf.tpr lrf.tpr.pgm
```

This simple automatic mechanism has proven quite useful for keeping track of auxiliary data sets and could form the basis of a more elaborate data base management system. (A report that describes this software system in detail is available upon request.)

The final lesson that we learned is that our approach was painfully inadequate. The chief complaint from those who observed the system in action was that it was too rigid and did not allow the user to "snoop around" easily when interesting features of the data were displayed by the system: the script became a straight jacket that forced the user to follow a certain course of actions. In effect, our script modelled only what the expert did on the majority of occasions and failed to capture what was done when some unexpected feature of the time series is revealed. Our system is unfortunately just another example of a "feeble prototype" (to use the words of Tukey[8] in describing efforts to date in creating expert systems for statistics).

We believe that a useful expert system can be built for spectral analysis but not with an off-the-shelf production system such as OPS5. The problems that must be overcome are the following. First, a better way must be found to extract information from graphs. This is critical since so much of the information that an analyst uses comes from graphs. For example, one possible solution to the straight-jacket problem is to enrich the expert system by including many more rules to represent *all* possible conclusions that an expert could draw from a graph. Under our current approach, this would mean that the expert system would have to guide the user through an exhaustive list of questions about the presence or absence of certain features. This is not feasible since such a scheme would quickly exhaust the patience of the user.

Second, some mechanism has to be incorporated in the system to allow it to "forget" certain "facts" that it has learned and all conclusions that it has deduced from these "facts." (This problem is called "truth maintenance" in the expert system literature.) This is probably the chief difference between statistical analysis and medical diagnosis for which production systems have been successful. In the latter discipline tests are performed on a patient, and from their results conclusions are drawn. The results of the tests themselves are never really questioned. In statistical analysis, certain hypotheses are assumed to be true until it becomes obvious that they are wrong. To site the river flow data as an example, if we hadn't noticed the relationship between variability and value of the series in figure 1a, we might have carried out a harmonic analysis on the original data. When we got to the point of plotting the residuals, we hopefully would have noticed a cyclical variability in the residuals that would have lead us back to concentrate on figure 1a. (To quote Chambers[2], "... data analysis is a more heterogeneous, quantitative and iterative process than ... medical diagnosis ... .")

Finally, creating an expert system that is primarily for non-experts vastly limits the number of potential users of the system. Experts are not interested in using it because they want to ignore all of the "help" facilities. Non-experts may find them initially useful, but, after several runs through such a system, they will rapidly acquire the expertise built into the system and will become bored with using it.

5. Display Oriented System for Spectral Analysis

In January of 1985, we received four state-of-the-art LISP machines for use in our project through a grant from the Department of Defense University Research Instrumentation Program with matching funds from the University of Washington. The availability of these machines and the experience we obtained in designing our prototype expert system caused us to design a new system from scratch. Our new approach is to produce a system for spectral analysis that is useful for experts in such a way that it can be augmented with various "help" facilities for less experienced users.

In order to produce a system that is useful to experts, we need to have a model of how experts do spectral analysis. Since following a script is obviously not what an expert does, we have attempted to come up with a more reasonable model. Our new model is a rather simple one, namely, that an expert does spectral analysis by carefully examining a sequence of graphics displays. At each stage of the analysis the features that the expert observes in a display prompt him or her to look at another display to learn something more about the time series.

With this model for spectral analysis, a rather simple design for more intelligent software is possible. Our first task is to create a set of independent graphics displays that an expert finds useful. The expert can make use of such a display as is, but the less sophisticated user can obtain help by requesting a list of features that he or she should be looking for. Alternatively, the user could go though an interactive "miniscript" that refers to only the one display at hand and that is designed to force him or her to note as much about the time series as possible from that display. Anything that the user learns about the time series from such a miniscript can be stored in a data object that represents the time series. (For our purposes we can define a data object for a time series as a computer representation of both the values of the time series and all other information that is known or has been deduced about the series.)

To clarify these ideas, let us look at a mock-up of one display in our proposed system (figure 3). Each display consists of one or more graphics windows and four "mouse" sensitive windows to control what is visible in the graphics windows and to allow the user to advance to other displays. The mock-up shows the periodogram display as it would be applied to the data object that contains the log of the river data. For this display there is only one graphics window. It shows the values of the periodogram for the time series versus frequency.

The "goodies" window allows the user to do several things: to reset parameters that control exactly how the periodogram is calculated and plotted; to augment the basic plot; to perform some statistical tests that are associated with the periodogram; to manipulate the data object under study; and to create a new data object from the values shown in the plot. In the mock-up, the first five items in this window show the user in bold letters the current values of the settable parameters. Thus the periodogram was calculated from a demeaned time series and by applying a cosine data taper to 20% of the time series. It was then evaluated on a finer grid of frequencies than the standard frequencies. The results of these computations were plotted on a decibel versus linear scale. All of the settable parameters can be changed by moving a "mouse" controlled pointer to the appropriate place and by either clicking a button on the "mouse" (to, say, select a linear "y" scale) or by clicking and entering a value from the keyboard (to change the proportion of data tapered from 20% to some other value). As soon as a parameter is reset, the plot in the graphics window is automatically updated.

Three augmentations to the plot are possible in this version of the periodogram display. One or more user-specified fundamental frequencies can be indicated on the plot by vertical dashed lines, and any number of associated harmonics can be shown by vertical dotted lines. In the mock-up a fundamental frequency corresponding to a period of one year and its first five harmonics are shown. The third augmentation allows the user to plot one or more copies of the kernel associated with the data taper. This option allows the user to identify peaks in the periodogram that are due solely to window leakage.

A list of all data objects in the current analysis is given in the data objects window. The first data objects in the list are those that are being examined in the current display and are marked "active". For the periodogram display there is only one, namely the data object that contains the log of the river flow data. The user can manipulate these data objects and create new ones by selecting (by means of the "mouse") one of the final three items in the "goodies" window. The item "make new data object" allows the user to create a new data object from the values plotted in the graphics window. The "add comments" item lets the user add any comments desired to any of the data objects in the current analysis. Finally, the "examine data object" item is used to look at all the auxiliary information that has been stored along with the actual data values.

Included with each graphics display is a directory of all other displays. In the mock-up, after the user is finished looking at the periodogram display, he or she may select one of six graphics displays to see next and may optionally choose any of the listed data objects to serve as the input to that display if he or she does not want to use the default "active" data object.

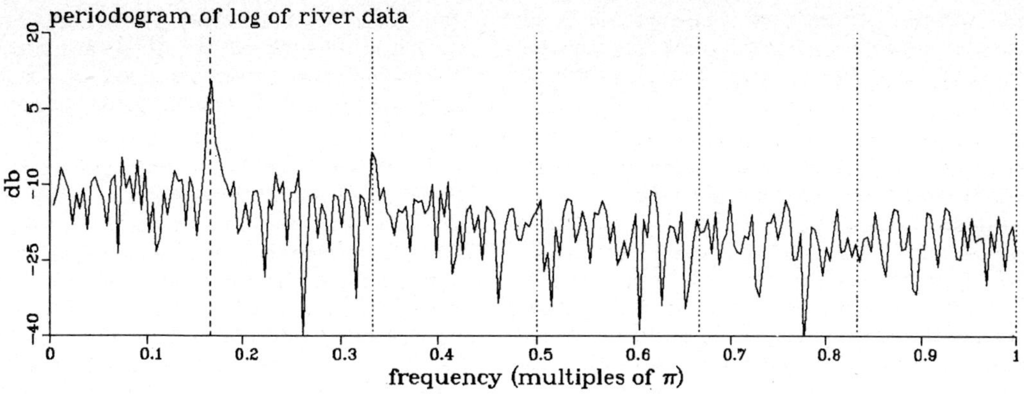

Goodies

- demean data: **yes** no
- "x" units: db **linear**
- "y" units: **db** linear
- standard frequencies only: yes **no**
- data taper: **cosine;** dpss
  proportion of data tapered: **20%**

- show fundamental frequency
- show harmonics: **5**
- show kernel
- do Fisher's test
- do Siegel's test

- make new data object
- add comments to data object
- examine data object

Directory of Displays

- harmonic regression
- wrap around plot
- white noise test
- time series plot
- make transformation (log, etc.)
- periodogram

Data Objects

- log of river flow data (active)
- river flow data
- square root of river flow data

Help

- What should I be looking for in this display?
- What do the goodies do?
- Why should I look at other displays?

Figure 3: Mockup of Periodogram Display

For the less sophisticated user, the help window offers three types of guidance. The first help item gives the user a list of features (and examples if so desired) that he or she should be looking for in the current graphics display. The system queries the user concerning the presence or absence of each feature and stores the results of this interaction in the "active" data object. The second help item explains in detail (with examples if necessary) what each of the items in the "goodies" window does. The third item in the help window tells the user why he or she might want to look at other displays. Based upon what display the user is currently looking at and what information is known about the time series (as stored in its corresponding data object), the system will order the items in the directory of displays to reflect what it thinks would be the most informative displays to look at next.

Each graphics display has a small set of production rules that allows the system to order the directory of displays and explain to the user the rationale for the order. For example, the fact that the harmonic regression display is listed first in the directory in the mock-up may be due to some knowledge supplied by the user from one of two sources: a previously examined display such as the time series plot display (where the user might have noted "strong periodic variation"); or the feature extraction question-and-answer session in the current display (where the user might have noted that the periodogram has "one or more prominent narrow peaks").

The system that we are designing around this display-oriented model of spectral analysis cannot be called an expert system since it only provides local (i.e., from one display to the next) guidance and not global guidance. Its chief advantages are: modularity of design (each display is independent of all other displays); help to the user is added in a well-defined way after each display has been designed; and the help facilities are non-intrusive and can be completely ignored. We also feel that our design helps alleviate the well-known knowledge transfer bottleneck common to expert systems since here the expert need only answer a few well-defined questions to make the system "intelligent" ("What do you hope to learning by looking at that graph?", "What other graphs would help you clarify questions raised by this graph?", etc.).

6. Future Directions

We are currently implementing the spectral analysis system described in the previous section. After the rudiments of the system are in place and a prototype of the system has been critiqued, we plan to incorporate as many graphics displays as time, resources, and interest allow. We also plan to augment the system by exploring the following research topics.

6.1 Classification of Time Series

We recognize that there are many users who require more global help than our proposed system can give them. One possibility to provide such help is suggested by Schank's cognitive model approach to AI problems, in which he defines understanding as the ability to relate the problem at hand to one's past experience. Gersch[4] has recently published some results on nearest neighbor rule classification of time series. His idea is to have a data base of time series and a measure of dissimilarity between time series (he used the Kullback-Leibler information number). Any new time series is then classified by comparing it to each of the time series in the data base. The nearest neighbor to the new time series is defined as that time series in the data base which is least dissimilar.

These ideas can be used to produce global help for a user. The first step is to have an expert do a spectral analysis on a large number of different time series. For each time series, the expert will use some combination and ordering of graphics displays and will create a certain collection of data objects. When an inexperienced user comes in with a new time series, it is classified using Gersch's scheme, and the user is told to follow the actions the expert took in analyzing the time series in the data base that is least dissimilar. (If there are several time series in the data base that are close in dissimilarity, the user could select visually that one series that he or she feels to be closest to the new series.)

What we need to investigate is 1) whether Gersch's classification scheme is adequate and, if not, whether we can come up with one that is (Gersch's scheme is a time domain one; there is a corresponding frequency domain one that we plan to explore); 2) what is the most effective way of telling the user to follow a set of actions in our system; and 3) how we can automatically update the data base of time series (this will involve some issues in machine learning).

6.2 Automated Creation of Graphics Displays

One of the nice features of the S and ISP interactive statistical packages is the ease with which a user can expand the system by adding new functions of his or her own creation. If our system is to be widely used, we need to develop some way for the user to add new graphics displays. One of us (Kerr) will be exploring this problem of a "program-writing" program in a complex system.

6.3 New Data Analysis Tools

In a future report[5] we will give some answers to the fourth question of the introduction, namely, "what new tools are available for spectral analysis on a state-of-the-art workstation?". We have several promising ideas to exploit the unique graphical capabilities of these machines.

7. Acknowledgement

This work was sponsored by the Office of Naval Research under contract number N00014-81-K-0095

References

1. P. Bloomfield, *Fourier Analysis of Time Series: An Introduction*, John Wiley & Sons, New York (1976).
2. J. M. Chambers, "Some Thoughts on Expert Software," *Computer Science and Statistics: Proceedings of the 13th Symposium on the Interface* (March 1981).
3. W. A. Gale and D. Pregibon, "An Expert System for Regression Analysis," *Computer Science and Statistics: Proceedings of the 14th Symposium on the Interface* (July 1982).
4. W. Gersch, "Nearest Neighbor Rule Classification of Stationary and Non-stationary Time Series," pp. 221-270 in *Applied Time Series Analysis II*, ed. D. F. Findley,Academic Press, New York (1981).
5. D. B. Percival, A. Buja, R. D. Martin, E. C. Belcher, R. K. Kerr, S. D. Yee, and C. B. Hurley

"A Display Oriented System for Spectral Analysis," *192nd Western Regional Joint Meeting of the Institute of Mathematical Statistics and the Biometric Society* (June 1985).

6. M. B. Priestley, *Spectral Analysis and Time Series*, Academic Press, London (1981).
7. A. F. Siegel, "Testing for Periodicity in a Time Series," *Journal of the American Statistical Association* **75**(370) pp. 345-348 (June 1980).
8. J. W. Tukey, "Data Analysis: History and Prospects," pp. 183-202 in *Statistics: An Appraisal*, ed. H. T. David,Iowa State University Press, Ames, Iowa (1984).

Statistics and Artificial Intelligence

Organizer: *William F. Eddy*

Invited Presentations:

Artificial Intelligence and Statistics: Do We Have the Cart Before the Horse? (Abstract Only)

William F. Eddy

Bayesian Image Restoration (Abstract Only)

Stuart Geman

Knowledge Representation for Expert Data Analysis Systems

Ronald A. Thisted

Production Systems and Belief Functions

Gail Gong

COMPUTER SCIENCE AND STATISTICS:
The Interface, D.M. Allen (ed.)

ARTIFICIAL INTELLIGENCE AND STATISTICS: DO WE HAVE THE CART BEFORE THE HORSE?

William F. Eddy

Department of Statistics
Carnegie-Mellon University

The last two decades have seen a growing interest in production systems, or rule-based expert systems. Originally, production rules were statements of the form "if A then B" and reasoning in these systems was simple (albeit tedious) and exact. Recently, a number of rule-based expert systems have been used on inexact reasoning (that is, on uncertain knowledge). This talk will provide a comparative review of some of the best-known methods of inference used in expert systems and will argue that most of these methods are hopeless as models of human reasoning.

BAYESIAN IMAGE RESTORATION

Stuart Geman

Division of Applied Mathematics
Brown University

We develop a class of probability image models that accommodate smoothness, edges, textures, and other, "higher level", image attributes. These are Markov Random Fields with a three dimensional graph structure. The "bottom" level of the graph is the pixel process, corresponding to the actual digitized image. Successively higher levels correspond to increasiningly complex attributes, including locations and orientations of edges, line segments, and polygonal regions. The constructed distribution is employed as a prior distribution on images. Given a degraded picture, we seek the image that maximizes the posterior distribution (the so-called MAP estimator). Maximization is performed by a highly parallel computational technique called stochastic relaxation.

We will present the results of experiments with some simple pictures. These demonstrate: (1) parameter estimation for the prior; and (2) blure and noise removal, segmentation, and boundary-finding at extremely low signal to noise ratios.

COMPUTER SCIENCE AND STATISTICS:
The Interface, D.M. Allen (ed.)

Knowledge Representation for Expert Data Analysis Systems

Ronald A. Thisted

Department of Statistics
The University of Chicago
Chicago, Illinois 60637

An expert system is a computer program which performs a task at the level of performance of a human expert with some years of experience at the task. In this paper we examine what it would mean for a computer program to be an expert system for data analysis, why there is some hope that such a system could be developed, and what makes an expert system different from other sorts of statistical software with which statisticians are familiar. Standard programs implement algorithms for computations on data, which in turn are represented using data structures. The choice of a suitable data structure often determines the form an algorithm will take, and such a choice may be crucial to the efficiency or feasibility of the computation. In expert systems the primary "data" are the fact, heuristics, and strategies used by experts to solve problems in their domain of expertise. An appropriate form for representing statistical knowledge is a prerequisite for a successful expert data analysis system. We examine some alternatives for knowledge representation in this context. Quite apart from its potential contribution to expert systems, such investigations shed light on the nature of data-analytic expertise and how such expertise can be taught.

1. What is an expert system?

This paper is an introduction to the issues involved in designing and implementing an expert system that might be useful in data analysis, with particular attention to aspects of the problem of representing statistical knowledge in a form suitable for computation. Expert systems differ in substantial respects from "ordinary" statistical software systems, and the differences are fundamental to an understanding of the role that expert knowledge plays.

1.1. General definition and examples.

Expert systems are defined partly in terms of what they do, partly in terms of how they do it, and partly in terms of the principles that led to their construction. There is some agreement (see Chapter 2 of Hayes-Roth, Waterman, and Lenat (1983), for instance) that an expert system must perform a complex task at the level of a human expert who has several years of experience at that task. Several attributes shared by expert systems have emerged. An expert system must embody expertise, in the sense that it is based upon rules which correspond to what human experts do; it must employ symbolic reasoning, rather than purely numerical computation in solving problems; it must exhibit intelligence in the sense that it can reason from basic principles – and can recognize which principles are applicable – rather than being able to deal only with situations narrowly specified in advance; it must be dealing with a problem of sufficient complexity that human experts are generally required; it should have some ability to reformulate a problem from the form originally presented into a form more suitable for analysis; and finally, it must have some ability to reason (or at least to explain) about its own reasoning process. This last attribute of having an explanation facility seems crucial and, to some extent, defining.

Some examples of successful expert systems, which are consulted by experts in practice, are DENDRAL (Buchanan, Sutherland, and Feigenbaum, 1969; Lindsay, et al, 1980) which identifies organic chemical compounds based on spectrographic data; MYCIN (Shortliffe, 1976) which diagnoses infectious blood diseases; and CADUCEUS (Pople, 1981), a system for diagnosis in internal medicine.

1.2. Expert systems for data analysis

What role could an expert system play in the practice of statistics? Several different "role models" have been suggested, and they lead to very different kinds of programs, performing very different kinds of tasks. Oldford and Peters (1984) developed a prototype expert system to recognize collinearity in regression problems. This program was designed to be the *Guardian of the Novice,* in effect, to prevent the unexperienced user of regression from stumbling blindly into hazardous terrain. The REX system of Pregibon and Gale (1984), on the other hand, might be termed a *Guide for the Perplexed.* REX was designed to guide its user through an appropriate regression analysis, in effect taking on the role of instructor as well as expert. Both of these systems assume users with little background in statistics or data analysis.

Another role that experts systems could play in statistical work is that of an *intelligent assistant,* with the knowledge required to examine all of those things which the competent data analyst knows he or she should look at, but for which there is often little time (or patience).

On this view, a program with quite limited intelligence could be widely useful; it would not even have to be able to deal with problems, it would simply have to be able to recognize the problems and bring them to the attention of the expert human statistician. In the absence of a plea for help from the program, the statistician could assume that no difficulties requiring special expertise were present, freeing him or her to devote more time and energy to problems of greater difficulty or complexity.

A final role model for expert systems in data analysis, perhaps the most ambitious of all, is that of an *apprentice consultant.* In this view the system would interact with a practiced, if not expert user, say a PhD student in statistics consulting with a scientist on a problem in data analysis. It would "look over the shoulder" of the user, making suggestions and noting possible problems. The goal here is once again to assist a user with some background in statistical analysis to make a better, more thorough analysis, and to bring to the fore situations which may require more expertise than either the program or its user possess.

The statistical consulting program at the University of Chicago is not unlike that at many universities. Under the direction of two faculty members, all PhD students must participate in statistical consulting with members of the university community, to whom consulting services are offered without fee. A major problem is that the program directors are booked with a solid three-week backlog of cases. Many of these cases turn out to be (for the statistician) routine. The possible role of expert systems here is to kill the three-week backlog by not wasting the human expert's time on routine matters, while at the same time, providing some assurance that major difficulties are not simply being overlooked.

In the remainder of the paper, the intelligent assistant and the apprentice consultant models will be of primary interest.

1.3. Expert data-analysis systems differ from standard statistical software.

A natural question that arises is whether expertise could be built in to existing statistical packages such as Minitab, SAS, SPSS, and the like. To answer this question it is important to understand how expert software differs from the standard software that statisticians are used to writing and interacting with.

Statistical computer packages increasingly offer on-line "help" facilities, but none of the models of expert systems outlined in the previous section could adequately be built upon these facilities. Today, in order to receive help, the program user must know that help is needed and must know when and how to ask. In return, the program generally can give assistance only so far as the syntax of the program's command language. Advice concerning what the next step to be taken in the analysis should be, or whether a proposed step is appropriate, would require not only monitoring the sequence of commands entered by the user, but also some ability to reconstruct the reasoning behind those commands.

The user-system interaction is also different. Statistical computer packages are designed to give lots of answers for a few economically worded questions generated by the user. The expert systems discussed here, on the other hand, are more adept at raising questions rather than answering them. In effect, their role is to note aspects of the data set that may render *all* of the answers produced by a standard package inappropriate, misleading, or meaningless.

Finally, the internal construction of expert software is likely to be quite different from that of standard statistical software in terms of control structure. While flow of control in the latter is often a matter of sequential invocation of routines explicitly or implicitly requested by the user's typed commands, the flow of control in expert systems will depend more upon the characteristics of the particular data set under consideration. The internal construction of the expert system will be suitable for more symbolic than numerical computation (although today's numerical computations will necessarily be invoked as subroutines to obtain intermediate results), which suggests that the code will include substantial chunks of LISP or Prolog. The greater the extent to which the data themselves determine the statistical computations to be applied, the more one's view of what constitutes a statistical algorithm becomes distorted. This leads us to some consideration of the roles played by data, algorithms, and knowledge in expert systems.

2. Algorithms, data structures, and knowledge bases.

The essence of standard programming as we understand it today is neatly summarized in the title of Niklaus Wirth's book, *Algorithms+Data Structures=Programs.* It is now well-understood that the choice of data structure can greatly influence the suitability of alternative algorithms for particular tasks, and can also greatly affect the performance of algorithms, and even their feasibility. (For instance, it is rather difficult to carry out a binary search in a linearly-linked list.)

In expert systems we may have a parallel formula: "Knowledge+ Inference=Expertise," reflecting the common-sense notion that experts both know a lot, and know when and how to apply their knowledge. The term "knowledge" as used here represents the collection of facts, heuristics, and strategies that experts use to solve problems. A *knowledge base* is a structured collection of symbolically-represented expert knowledge.

The power of an expert system depends on its knowledge base. It must have adequate coverage, that is, it must contain facts, heuristics, and strategies sufficient to cover a the wide range of problems in its domain. It must also have an adequate *representation* for that knowledge, suitable for an appropriate search algorithm to find those components of the knowledge base which are relevant in the current context. There are several schemes for knowledge representation that have been developed in the AI literature, of which a few seem to be particularly well-suited to knowledge about data analysis.

The most promising candidates are *production systems* (discussed by William Eddy and by Gail Gong in their presentations in this session), *augmented transition networks,* and *frame systems.* Production systems are collections of rules ("productions") of the form, "If *condition-A* then *action-B.*" Taken together, the collection of productions can be thought of as defining a tree and a way of traversing its branches. In the data-analysis context, each node in this tree corresponds to a stage in the data analysis, and moving from one node to another would generally correspond to performing a small piece of the data analysis. Augmented transition networks can most easily be thought of in this setting as adding information to the tree which records the relationship between any two connected nodes. Finally, frames are quite general ways of organizing knowledge; both production systems and ATNs can be embedded in the frame paradigm. In our setting we can think of a frame as being a set of productions which preserves the context in which the productions are employed.

The inferential machinery, or the method by which the knowledge base is searched to apply to a situation at hand, is related to the adequacy of coverage and adequacy of representation of the knowledge base in much the same way that algorithms are related to data structures in conventional programming. With these ideas as background, we now turn to consideration of some issues involve in building a suitable knowledge base for data analysis.

3. Knowledge engineering.

From the scientific standpoint, knowledge is representational, in the sense that we cannot say that we know something (or that we understand a phenomenon) until we can represent it using a model which embodies what it is that we think we know. One of the major benefits of publishing scientific papers is that in the act of writing, authors are forced to come to grips with the difficulties, inconsistencies, gaps, and inadequacies that were simply not apparent to them before. The theorem whose proof was sketched on a napkin may prove to be more delicate than first thought; the iron-clad argument may reveal a chink in the argument. What is more, the concrete representation makes it possible to transmit this knowledge to others in a way that is more feasible and more certain than through observation and apprenticeship.

A concrete representation is not a prerequisite to having knowledge, however. Human experts by definition possess abilities which others do not, and these abilities are based on facts and methods which they have assimilated and refined over time, whether they have done so consciously or not. Experts often cannot articulate the relevant knowledge they possess which they use on a daily basis, and what they do say they know may in fact conflict with what they actually use in practice. Many experts are ill-prepared by training or by inclination to articulate the knowledge they use in rendering expert judgments accurately. This makes it difficult to teach new people to become experts in the field.

At this point enters the *knowledge engineer.* The term has been coined by AI workers in expert systems to denote an individual who is trained in expertise elicitation and articulation, a psycho-analyst of expertise. Knowledge engineers typically are grounded both in computer science and cognitive psychology, and what they do is to work with a human expert in his or her domain of knowledge to elicit, and then to fashion a concrete representation of, the knowledge that the expert brings to bear to solve difficult problems that arise in the expert's domain. There is a shortage of people with the qualifications and experience to do this work.

Note that the knowledge engineers themselves are experts in a field, too – that of knowledge elicitation. To distinguish this top-level domain of expertise from the domains of experts to which it is applied, following Gale and Pregibon, we refer to the top-level area as the *it subject domain,* and the areas of application we refer to as the *ground domains.*

Statistical consulting is very similar to knowledge engineering. Statistical consultants are expert in statistical analysis (the subject domain), and they apply their knowledge by collaborating with experts in other fields of inquiry (the ground domains). Moreover, the first job of the statistical consultant is to help the client to articulate what he or she knows that is relevant to the problem (but may not have realized consciously). We help our colleagues to question assumptions they make implicitly. We help them to turn from matters of little consequence ("Do I use n or n-1 here?") and to focus on those matters that turn out to be essential ("Can you remember anything at all about the experiment that might distinguish these two halves of the data?" "Oh, yes. They were run in different years by different technicians."). We know that the questions people bring to us are usually not the appropriate questions which ultimately get addressed, and we assist in the process of getting the right questions formulated so that they can be addressed.

As a consequence of these similarities to knowledge engineering, statistics as a discipline has something to contribute to artificial intelligence work in general, and to expert systems research in particular. We have been about parts of the knowledge engineering business for at least half a century. (At the same time, however, we have devoted little attention to understanding very thoroughly how we accomplish what we do in this enterprise.) Statistics can contribute some of the basic ideas and methods of data analysis, experience in statistical consultation, and techniques for elaboration and for display. It may even be that, despite the shortage of trained statisticians, we may even end up contributing bodies to knowledge engineering front (since the pay is better).

Constructing an expert system which embodies knowledge about data analysis or about statistical consulting involves much that would be required in an expert system to construct expert systems, in that the ground domain for the system (statistical consulting) is itself a high-level domain of expertise which can in turn be applied in a number of ground domains. The current effort by Gale and Pregibon (1984) to construct *Student,* an expert system capable of learning to do data analysis in a variety of contexts by working through a sequence of problems in those contexts, is in effect, an expert system for building expert systems. It is an ambitious endeavor, which nonetheless shows signs of great promise.

How should we go about the process of studying what knowledge we bring to bear on statistical problems so that we can construct a suitable representation for it? Pregibon and Gale and others have used the device of constructing worked examples, carefully annotated, and diaries of the analysis process. These devices can be coupled with explanation to colleagues who can be expected to ask penetrating questions when the reasoning process is not entirely clear, and can be assisted by automatic means such as statistical packages which keep "journal files" of the sequence of commands used in analyzing a data set. Thisted (1984) has described the role that computing software environments can play in learning about how data analysts behave and what strategies they adopt. On this view, a considerable amount can be learned about the process of statistical analysis without actually attempting to implement any of it in an actual expert system to be run on a computer. A similar view has been expressed in the artificial intelligence literature by Doyle (1984).

4. Statistical consulting as a model.

A few words are in order concerning knowledge about statistical consulting as a basis for expert systems in data analysis. The questions of what *facts* consultants know, what *heuristics* they employ, and the *strategies* that they adopt are all understudied problems. There has been a surge of interest within the statistical community in the last five years on the topic of teaching statistical consulting, and the resulting reflections on the process of statistical consulting are valued contributions to this secondary endeavor of building consulting expertise into usable computer systems. At the same time, the emphasis has been more on apprenticeship and supervision of trainees rather than on the special skills that expert data analysts have and how those skills might be transmitted. We know of no study, thorough or otherwise, of the process by which successful consultants in data analysis approach their work and achieve their results.

This said, we can begin to outline the areas in which research is likely to be fruitful. Data analysts consulting with scientists expert in their (ground) domain are general-purpose scientific detective/psycho-analysts. They proceed by asking questions, and often these questions are suggested by what they see in the data. The answers to these questions lead both to alternative ways of looking at the data and to new questions. The important work of the consultant seems to get done through the questions he asks of the client. It is important, then to investigate how these questions are structured, what plans of inquiry are adopted, and what it is that leads to the formulation of these plans.

The natural way to learn about these issues is to observe experts at work (as knowledge engineers do), perhaps even to conduct experiments involving them. Some years ago, I received a telephone call from a colleague in pediatric neurology; he had a quick question. "I can't remember," he said, "whether I should use standard deviation or standard error. Which do you suggest?" We began to talk, and over the course of a few weeks, it became clear that the answer was, "None of the above." We ultimately used a three-factor unbalanced mixed model with covariates–and we learned more about the disease process under study by doing so. Unfortunately, I have no idea what sequence of events led from the innocuous question on his part to the ultimately more complex solution at which we arrived. This is the process which requires scrutiny and study.

5. Representing knowledge about question-asking.

What must be considered in building a concrete representation for the knowledge about question-asking that data analysts seem to possess and use to such good effect? Questions are asked both of the data and the expert in the ground domain. These questions often alternate, the data suggesting questions to ask of the client, whose response suggest new questions to ask of the data. We can distinguish four levels of questioning: asking questions of the data, asking questions of the experts, using answers to formulate questions, and asking questions about questioning. We now turn to just the first of these levels, as it is the level which we are currently closest at being able to explicate. Some of the issues raised in the remainder of this section are dealt with more thoroughly in Thisted (1985).

"Asking questions of the data" can be broken down into three rough stages which together describe a single step in the analysis of a data set: focus, selection, and transformation/reassessment. In *focusing* the analyst concentrates on a relatively small subset of the data perhaps a handful of variables (or cases) of interest at the moment. *Selection* is the process by which a collection of appropriate transformations of the data is identified; transformation here meaning nearly any computation on the data set, including computing a regression (producing estimated coefficients, fitted values, residuals), computing and displaying a scatterplot of two variables, constructing a confidence interval, etc

Finally, *transformation* and *reassessment* is the process of carrying out the computation, and then reassessing the situation. Reassessment may lead to a change of focus, to a change in the class of appropriate transformations, or to new questions of the client.

6. On carts and horses.

Bill Eddy's opening remarks were entitled "Artificial intelligence in statistics: Do we have the cart before the horse?" This provocative title prompts a few observations about the AI cart and the statistical horse.

There is no cart. It should be clear from the outset that expert systems for general use in data analysis don't exist, although a few demonstration systems have been built. Moreover, there is no general methodology for building expert systems. And at least for the kinds of systems we have been discussing, there are no general-purpose expert systems of any kind which incorporate the higher-level meta-knowledge of a domain which interacts with a variety of ground domains.

There is no horse. What makes a particular data analysis a good one is little studied–and even less understood. At the moment, we teach data analysis and consulting by example, and we hope that some of it will rub off on our students.

We need both carts and horses (in either order). The combination of the two is certainly more useful than either separately. What is more, understanding horses may help us to mass-produce carts, and vice-versa. A better understanding of useful heuristics and successful strategies (from expert systems) will lead to improvements in statistical teaching and practice. What is more, with even rudimentary expert data-analysis systems, the human experts can be reserved for the important problems, since there are so many problems and so few experts.

Neither cart nor horse may be possible. This is a fact, and we must live with it. But many useful things have been learned by striving for the impossible. Hence,

We must attempt to build both carts and horses. There is much to be learned solely from the attempt. Except perhaps for John Tukey's personal tour-de-force (Tukey, 1977) which records what Tukey senses from his own experience and reflection to be important and useful strategies and techniques for data analysis, there has been no serious attempt to represent what data analysts do, and hence, what knowledge they possess.

We cannot wait until data analysis is more fully understood to begin work on expert systems for data analysis, primarily because there is not much work going on trying to understand what it is that good data analyst's do, and how it can be taught. The major benefit from work in expert systems for data analysis may well be a better understanding of the process of data analysis. It is useful to recall a brief bit of recent history. We wrote programs before we appreciated the role of data structure, top-down construction, information hiding, loop invariants, and the rest. Indeed, much of what we know about these ideas was learned through reflection on what made some programs better than others and some programmers better than others. Even if no generally useful expert systems are built, we may still make great strides in improving the general quality of data analysis because we better understand what goes into data analysis of high quality, so that we can convey it more directly and more successfully to budding data analysts.

At the same time, much of expert systems work is closely related to what we think data analysts actually do. Both good data analysis and successful knowledge engineering involve drawing out an expert, evoking what he knows but does not say about a problem. Both the statistical consultant and the knowledge engineer must be adept at asking the right question which brings into focus the critical aspect of what is being done. Thus, work in expert systems for data analysis may well bring new paradigms for knowledge articulation to the attention of workers in AI, and at the same time may help to make the techniques of knowledge engineering needed to construct general expert systems more readily available.

Acknowledgement. This material is based upon research supported by National Science Foundation Grant No. DMS-8412233 to the University of Chicago.

References

[1] Buchanan, B. G., Sutherland, G. L., and Feigenbaum, E.. A. (1969). "Heuristic DENDRAL: A program for generating explanatory hypotheses in organic chemistry," in *Machine Intelligence,* B. Melzer and D. Michie, editors, **4**, 209--254.

[2] Doyle, Jon (1984). "Expert systems without computers or theory and trust in artificial intelligence," *The AI Magazine,* **5(2)**, 59--63.

[3] Gale, William A., and Pregibon, Daryl (1984). "Constructing an expert system for data analysis by working examples," in *COMPSTAT 1984: Proceedings in Computational Statistics,* (Prague, Czechoslovakia), T. Havránek, Z. Šidák, and M. Novák, editors. Physica-Verlag: Vienna. 227--236.

[4] Hayes-Roth, Frederick, Waterman, Donald A., and Lenat, Douglas B., editors (1983). *Building Expert Systems.* Addison-Wesley: Reading, Massachusetts.

[5] Lindsay, R. K., Buchanan, B. G., Feigenbaum, E. A., and Lederberg, J. (1980) *Applications of Artificial Intelligence for Organic Chemistry: The DENDRAL Project.* McGraw-Hill: New York.

[6] Oldford, R. Wayne, and Peters, Stephen C. (1984). "Building a statistical knowledge based system with Mini-Mycin," *Proceedings of the Statistical Computing Section,* American Statistical Association: Washington, 85--90.

[7] Pople, H. E., Jr. (1981). "Heuristic methods for imposing structure on ill-structured problems: The structuring of medical diagnostics," in *Artificial*

Intelligence in Medicine, P. Solovitz, editor. Westview Press: Boulder, Colorado, 119--185.

[8] Pregibon, Daryl, and Gale, William A. (1984). "REX: An expert system for regression analysis," in *COMPSTAT 1984: Proceedings in Computational Statistics,* (Prague, Czechoslovakia), T. Havránek, Z. Šidák, and M. Novák, editors. Physica-Verlag: Vienna. 242--248

[9] Shortliffe, E. H. (1976). *Computer-Based Medical Consultation: MYCIN.* American Elsevier: New York.

[10] Thisted, Ronald A. (1984). "Computing environments for data analysis," Technical Report Number 166, Department of Statistics, The University of Chicago.

[11] Thisted, Ronald A. (1985). "Representing statistical knowledge and search strategies for expert data analysis systems," Technical Report Number 171, Department of Statistics, The University of Chicago.

[12] Tukey, John W. (1977). *Exploratory Data Analysis,* Addison-Wesley:Reading, Massachusetts.

COMPUTER SCIENCE AND STATISTICS:
The Interface, D.M. Allen (ed.)

PRODUCTION SYSTEMS AND BELIEF FUNCTIONS

Gail Gong

Statistics Department
Carnegie-Mellon University

Expert systems are computer programs which use domain-specific knowledge to make inferences about problems arising in that domain. Some expert systems must handle knowledge which is uncertain, and a popular tool for handling such uncertain knowledge has been the adhoc uncertainty factors found in MYCIN. We explore another tool, belief functions, introduced by Art Dempster and Glenn Shafer.

1. Production Systems

Suppose a customer wants to buy a VAX computer. He has some idea of what he wants: his VAX should have so much disk space; it should support so many micorn lines; he wants it to connect to this kind of tape drive and that kind of printer; and so on. However, there are still many details that need to be decided. What kind of wires should be used to connect this to that? What kind of boards are necessary? The customer needs a VAX expert to insure that the order is consistent and complete.

Actually, DEC has a computer program that configures VAX's. The program, called R1, uses production rules. An example of a production rule might be:

DISTRIBUTE-MB-DEVICES-3

IF:

THE MOST CURRENT ACTIVE CONTEXT IS DISTRIBUTING MASSBUS DEVICES

AND THERE IS A SINGLE PORT DISK DRIVE THAT HAS NOT BEEN ASSIGNED TO A MASSBUS

AND THERE ARE NO UNASSIGNED DUAL PORT DISK DRIVES

AND THE NUMBER OF DEVICES THAT EACH MASSBUS SHOULD SUPPORT IS KNOWN

AND THERE IS A MASSBUS THAT HAS BEEN ASSIGNED AT LEAST ONE DISK DRIVE AND THAT SHOULD SUPPORT ADDITIONAL DISK DRIVES

AND THE TYPE OF CABLE NEEDED TO CONNECT THE DISK DRIVE TO THE PREVIOUS DEVICE ON THE MASSBUS IS KNOWN

THEN:

ASSIGN THE DISK DRIVE TO THE MASSBUS

A rule contains a left-hand-side (LHS) and a right-hand-side (RHS). THe LHS is a set of conditions which must be satisfied before the conclusions or actions in the RHS can be accepted. To get an idea of how this production program might run, suppose that each customer order results in a meeting. At the meeting are representatives of the rules (one representative for each rule), a secretary, and an arbiter. The secretary begins by writing the specifications of the customer order on the blackboard; each representative watches carefully to see if the LHS of his particular rule has yet been satisfied by the specifications on the blackboard; when a representative sees his rule satisfied, he signals the arbiter; more than one rule can be satisfied at any one instant, so the arbiter must decide which of the satisfied rules can "fire"; the secretary changes the specifications on the blackboard according to the RHS of the fired rule. As more rules are fired, the blackboard changes and other representatives find their rules satisfied. For each set of conditions on the blackboard, a representative can have his rule fired at most once. The meeting continues until no representative finds his rule satisfied. The blackboard at the end of the meeting describes the completed specifications of the customer order.

In R1, the rules and conditions are assumed to be deterministic. Either the customer wants a printer or he doesn't. Given that he wants a printer, he may or may not need this kind of board, but if we have enough conditions about what he wants, we can be quite sure of what kind of board he needs. In, say, a medical diagnosis problem, we are often not sure if the patient has a particular set of symptoms. Also, deterministic rules are harder or impossible to obtain. We cannot say that a person with this list of symptoms is surely to have this disease. The best we can say is that given these symptoms, the person is <u>likely</u> to have this disease. The problem then becomes that of expressing and reasoning with these uncertainties.

The computer scientists are convinced that using probabilities is too hard if not impossible, so they have turned to rather adhoc procedures, such as the certainty factors found in MYCIN. Recently, however, some computer scientists have discovered belief functions, which were first proposed by Dempster, and later developed by Shafer. (See Shafer (1976).) Belief functions are appealing to the computer scientists because they are less restrictive than probabilities, they can express ignorance, and they have some mathematical backing. Several artificial intelligence groups are trying to implement belief functions into their expert systems, and this is reason enough for statisticians to become actively involved in belief functions for expert systems.

2. A Small example

To introduce the ideas of belief functions and their relationships to production systems, we will consider the following tiny example. This is not, of course, a real expert system, but it uses if-then rules to help obtain a desired conclusion.

Suppose I go away on a trip for a week. During that time, I am forced to leave my house unoccupied and unguarded. Upon my return, I discover that the television set is missing. I also notice that there are dried-up muddy footprints leading to and from the back door. Who was the thief?

The house is surrounded by clean sidewalks, so an ordinary passerby would not have had muddy boots unless he had been walking in the garden and unless the garden was wet. An idea flashes. Maybe it was the gardener. When I left on Sunday, the garden was dry. The gardener comes on Monday. Therefore I construct the rule: <u>If it rained on Monday, then the gardener had muddy boots.</u>

I don't know my gardener very well, but I have the feeling that he is not a professional thief. He would not have entered the house had the door been locked. I construct another rule: <u>If the gardener had muddy boots, and the door was unlocked, then the footprints belong to the gardener.</u> Another rule that obviously follows is: <u>If the footprints belong to the gardener, then the gardener is guilty.</u>

I might have some other evidence that corroborates with the footprint evidence. Fingerprints are found in the house that do not match any of the fingerprints of the members in my family. I construct one more rule: <u>If the fingerprints match those in the gardener's toolshed, then the gardener is guilty.</u>

Figure 1 summarizes the four rules.

We emphasize here that we are allowing for the possibility that each of these rules needs not have 100 percent certainty of holding. Even though it was raining on Monday, we allow the possibility that the gardener did <u>not</u> have muddy boots. Perhaps it was raining so hard that he decided to wait until Thursday to work on the garden. Also there is uncertainty on the left-hand-side conditions. The weather reports in the

Figure 1.

A = "It rained on Monday."
B = "The gardener had muddy boots."
C = "The door was unlocked."
D = "The footprints belong to the gardener."
E = "The fingerprints match those in the gardener's toolshed."
F = "The gardener is guilty."

r_1 : A -> B
r_2 : B & C -> D
r_3 : D -> F
r_4 : E -> F

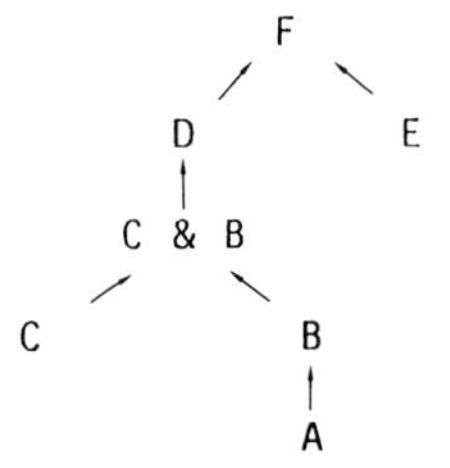

newspaper are perhaps somewhat reliable, but misprints are possible; and I'm not sure whether I checked the back door before I left because it is used so infrequently.

Our goal is to quantify our uncertainties both of the left-hand-side conditions and of the rules and then be able to calculate the resulting uncertainties of the right-hand-side conditions. That is, if we have a measure of belief on A and a measure of belief on the rule A -> B, then what is our measure of belief on B?

3. An Introduction to Belief Functions

The material in this section is from Shafer (1976) and Shafer and Tversky (1984). A frame of discernment Θ is a set of all possibilities under consideration. For example, if we were concentrating just on the question of whether or not A were true, we might consider the frame

$$\Theta_A = \{a_0, a_1\},$$

where a_0 denotes "A is not true", and a_1 denotes "A is true". The frame of discernment can be much more complex of course. For example, if we were concentrating on the rule A -> B, we would consider the frame

$$\Theta_{AB} = \{(a,b) : a = 0,1;\ b = 0,1\}, \qquad (1)$$

where a = 0 or 1 according to whether A is false or true, and similarly for b and B.

Just as it is easier to introduce probabilities through probability density functions, it is easier, here, to introduce belief functions through <u>basic probability assignments</u>. Shafer defines the function m : 2^{Θ} -> [0,1] to be a basic probability assignment if

$$m(\Phi) = 0,$$
$$\sum_{A \subset \Theta} m(A) = 1.$$

We call the subsets A of Θ which have positive m-value assignments, $m(A) > 0$, the focal elements of m. The belief function, Bel : $2^{\Theta} \to [0,1]$, associated with m is defined by

$$Bel(A) = \sum_{B \subset A} m(B).$$

Before discussing the properties or interpretations of basic probability assignments and belief functions, let us consider a simple numerical example:

$$\Theta = \{\theta_1, \theta_2, \theta_3\},$$
$$m\{\theta_1\} = .32$$
$$m\{\theta_1,\theta_2\} = .08$$
$$m\{\theta_1,\theta_3\} = .48$$
$$m(\Theta) = .12.$$

Notice that, unlike probability density functions, the domain of m is not restricted to singletons; also the focal elements are not disjoint. We think of our belief as being divisible into chunks. The m above says we put .32 of our belief on θ_1; .08 of the mass is "free to wander" among the elements in $\{\theta_1,\theta_2\}$; and so on. A mass which is free to wander on all of Θ reflects ignorance. The more specific a chunk of mass is, the more information it reflects. An m function which puts all its mass on Θ reflects total ignorance. An m function which puts all its mass on a singleton reflects total certainty for that singleton.

The m function describes the measure of belief that we commit exactly to each set; the total amount of belief committed to each set is described by the associated belief function:

$$Bel\{\theta_1\} = .32$$
$$Bel\{\theta_1,\theta_2\} = .40$$
$$Bel\{\theta_1,\theta_3\} = .80$$
$$Bel(\Theta) = 1.00$$

For example, the belief on $\{\theta_1,\theta_2\}$ is gotten my adding the m on $\{\theta_1\}$ to the m on $\{\theta_1,\theta_2\}$, these two sets being the subsets of $\{\theta_1,\theta_2\}$ with positive m values. The mass .40 is the amount of mass that is confined somewhere in $\{\theta_1,\theta_2\}$, and it represents our total belief on $\{\theta_1,\theta_2\}$.

What do the numbers .32 or .08 mean? In answer to this question, Shafer and Tversky (1984, p. 23) propose some thought experiments. The simplest involves simple support functions whose m functions have the form

$$m(A) = s,$$
$$m(\Theta) = 1\text{-}s,$$

where A is a subset of Θ and $0 \le s \le 1$. We describe such a belief function as the simple support function with mass s on the focal element A. Simple support functions result from a piece of evidence that offers support for a single subset A. For example, in the gardener example, if we are concentrating our attention on whether or not A is true (so that we are looking at the frame of discernment Θ_A), the newspaper reporting rain on Monday should give some support on $\{a_1\}$. The amount of support depends on the following thought experiment.

Imagine a sometimes-reliable truth machine. In its "truth" mode, it tells the truth, but in its "unreliable mode" it generates totally random statements which give us no added information. The probability of being in the truth mode is s, while the probability of being in the unreliable mode is 1-s. The truth machine spouts out "A is true". Shafer and Tversky propose that the resulting belief function should be the simple support function with mass s on the focal element A. In the gardener example, we think of the newspaper as the sometimes-reliable truth machine with probability s of telling the truth, and probability 1-s of printing a totally random statement. The newspaper reporting "Rain on Monday" leads to a simple support function with mass s on the focal element $\{a_1\}$.

Shafer and Tversky propose other thought experiments for belief functions which are more complicated than simple support functions. We will not describe them here.

It may turn out that we have another piece of evidence for rain on Monday. A neighbor recalls that it rained on Monday. We would like to combine our evidence supplied by the newspaper with that supplied by the neighbor. We will use Dempster's combination rule. Given the basic probability assignments m_1 with focal elements $A_1, ..., A_k$ and m_2 with focal elements $B_1, ..., B_n$, if K, given by

$$K = (1 - \sum_{A_i \cap B_j = \Phi} m_1(A_i)m_2(B_j))^{-1},$$

is positive, then the belief function resulting from the Dempster combination has m function $m = m_1 \oplus m_2$, defined by

$$m(A) = K \sum_{A_i \cap B_j = A} m_1(A_i)m_2(B_j).$$

The formula appears more complicated than the concept. To illustrate, suppose

$$\Theta = \{\theta_1, \theta_2, \theta_3\},$$
$$m_1\{\theta_1,\theta_2\} = .40$$
$$m_1(\Theta) = .60$$

$$m_2\{\theta_1,\theta_3\} = .08$$
$$m_2(\Theta) = .20$$

The Dempster combination $m = m_1 \oplus m_2$ is easily gotten by considering the following table.

m_2 \ m_1	$\{\theta_1,\theta_2\}$ .40	Θ .60
$\{\theta_1,\theta_3\}$ .80	$\{\theta_1\}$ .32	$\{\theta_1,\theta_3\}$ .48
Θ .20	$\{\theta_1,\theta_2\}$ .80	Θ .12

Since there are no null intersections of focal elements, K = 1. Also each intersection of a focal element of m_1 with a focal element of m_2 leads to a distinct set, so we can just read off m from the body of the table:

$$m\{\theta_1\} = .32$$
$$m\{\theta_1,\theta_2\} = .08$$
$$m\{\theta_1,\theta_3\} = .48$$
$$m(\Theta) = .12$$

It is instructive to compare the belief functions of m_1, m_2, and m:

$$Bel_1\{\theta_1,\theta_2\} = .40$$
$$Bel_1(\Theta) = 1.00$$

$$Bel_2\{\theta_1,\theta_3\} = .80$$
$$Bel_2(\Theta) = 1.00$$

$$Bel\{\theta_1\} = .32$$
$$Bel\{\theta_1,\theta_2\} = .40$$
$$Bel\{\theta_1,\theta_3\} = ..80$$
$$Bel(\Theta) = 1.00$$

Since $Bel_1\{\theta_1,\theta_2\} = Bel\{\theta_1,\theta_2\}$, the belief on $\{\theta_1,\theta_2\}$ has remained constant, but in Bel, some of the mass that contributes to total belief on $\{\theta_1,\theta_2\}$ is constrained to lie in $\{\theta_1\}$.

Continuing with the gardener example, suppose that we believe that the newspaper gives the simple support function

$$m_{news}\{a_1\} = .6$$
$$m_{news}(\Theta_A) = .4$$

(the newspaper is not very reliable), and the neighbor gives simple support function

$$m_{neighbor}\{a_1\} = .3$$
$$m_{neighbor}(\Theta_A) = .7$$

(our neighbor is old and often forgets the day of the week).

m_{news} \ m_{neigh}	$\{a_1\}$ .30	Θ .40
$\{a_1\}$ .60	$\{a_1\}$ .18	$\{a_1\}$ .42
Θ .40	$\{a_1\}$ .12	Θ .28

As in the example above, K = 1. However the set $\{a_1\}$ is the result of several distinct intersections of focal elements of m_{news} with focal elements of m_{neigh}, and so getting the Dempster combination $m = m_{news} \oplus m_{neighbor}$ requires a summation:

$$m\{a_1\} = .18 + .42 + .12 = .72$$
$$m(\Theta_A) = .28$$

The corroborating pieces of evidence have resulted in a fairly high support for $\{a_1\}$, even though the individual pieces of evidence each gave only meager support.

4. Belief functions for uncertain rules

Of the goals stated at the end of Section 2, we have discussed methods to attain our first goal, to quantify our uncertainties in the left-hand-side conditions. We now consider the second and third goals, to quantify our uncertainties of the rules themselves and to propagate the uncertainties to the right-hand-side conditions. Let us focus on the rule r_1 : A -> B. Our relevant frame of discernment is Θ_{AB}, defined in (1). Since r_1 is logically equivalent to the elements in the se {(0,0),(0,1),(1,1)} being true, it seems reasonable to represent our belief on the rule r_1 by a simple support function with focal element {(0,0),(0,1),(1,1)} :

$$m_{r_1}\{(0,0),(0,1),(1,1)\} = p_{r_1} \qquad (2$$

$$m_{r_1}(\Theta_{AB}) = 1 - p_{r_1}.$$

How should we interpret the mass p_{r_1} that we assign to th focal element {(0,0),(0,1),(1,1)}? To answer this, we need t see how our belief on A propagates through our belief on th rule to give a belief on B.

In Section 3, when we were considering evidence on A, w restricted our attention to the frame of discernment Θ_A. No considering the rule A -> B, we have a different frame Θ_{A} Actually the two frames are not unrelated. Θ_A is a coarsenin of Θ_{AB}. The elements in Θ_A can be put into a one-to-on correpondence with a partition of the elements in Θ_{AB}. Th correspondence, which we denote by equality, is

$$a_0 = \{(0,0),(0,1)\},$$
$$a_1 = \{(1,0),(1,1)\}.$$

Notice that both sides in the first equation, a_0 and {(0,0),(0,1) represent A being false, and both sides in the secon equation, a_1 and {(1,0),(1,1)}, represent A being true. Th belief function

$$m_A\{a_1\} = p_A$$
$$m_A(\Theta_A) = 1 - p_A$$

defined on the subsets of Θ_A can be considered equivalent the belief function

$$m_A\{(1,0),(1,1)\} = p_{A'}$$
$$m_A(\Theta_{AB}) = 1 - p_{A'}$$

defined on the subsets of Θ_{AB}.

To propagate our belief on A, which is described by m_A, through our belief on the rule, which is described by m_{r_1}, we can simply use the Dempster combination $m = m_A \oplus m_{r_1}$.

m_A \ m_{r_1}	{(0,0),(0,1),(1,1)} p_{r_1}	Θ $1-p_{r_1}$
{(1,0),(1,1)} p_A	{(1,1)} $p_A p_{r_1}$	{(1,0),(1,1)} $p_A(1-p_{r_1})$
Θ $1-p_A$	{(0,0),(0,1),(1,1)} $(1-p_A)p_{r_1}$	Θ $(1-p_A)(1-p_{r_1})$

The resulting m function can be easily read from the table. Letting Bel be the corresponding belief function, we are interested in

$$\begin{aligned} \text{Bel}\{\text{"B is true"}\} &= \text{Bel}\{(0,1),(1,1)\} \\ &= m\{(1,1)\} \\ &= p_A p_{r_1}, \end{aligned} \tag{4}$$

We return now to the interpretation of the number p_{r_1}. Suppose that we are absolutely sure that A is true. This leads to m_A defined in (3) with $p_A = 1$, and substituting this value into the (4) gives Bel{"B is true"} = p_{r_1}. Therefore, p_{r_1} is our belief on B if we are absolutely sure that A is true.

Up to this point, we have been concentrating on the rule r_1 : A->B. This is the bottom right branch of the tree in Figure 1. Given some evidence on A and some belief on the rule r_1, we have calculated a belief $p_A p_{r_1}$ on B. We can take this belief on B, combine it with evidence on C and belief on r_2 to get a belief $p_A p_C p_{r_1} p_{r_2}$ on D. In turn, this belief on D together with belief on r_3 gives belief $p_A p_C p_{r_1} p_{r_2} p_{r_3}$ on F. Also, evidence on E together with belief on the rule r_4 gives additional and independent belief $p_E p_{r_4}$ on F. Combining these two pieces of support on F gives a total belief

$$p_A p_C p_{r_1} p_{r_2} p_{r_3} + p_3 p_{r_4} - (p_A p_C p_{r_1} p_{r_2} p_{r_3})(p_3 p_{r_4})$$

on F.

5. Discussion

We have seen a very simple and rather tentative introduction into production systems and belief functions. The hope here was a germ from which grow deeper thoughts about the problems of dealing with uncertainty in expert systems. There are many questions that need to be addressed: Is the belief function m_{r_1} chosen in (2) of the appropriate form for reflecting beliefs on rules? The combination rule requires that the two belief functions entering into the combination be based on independent evidence. How do we handle dependent pieces of evidence? We have only considered a very small example. The combination rule potentially involves intersections and multiplications of all subsets of the frame? In a large problem, how do we handle the computational explosion?

6. References

Dempster, A. and A. Kong, "Belief Functions and Communications Networks", Harvard University Technical Report (1984).

Shafer, G., A Mathematical Theory of Evidence (Princeton University Press, 1976).

Shafer, G. and A. Tversky, "Weighing Evidence: The Design and Comparison of Probability Thought Experiments", University of Kansas Technical Report (1984).

Graphics

Organizer: *Wesley L. Nicholson*

Invited Presentations:

Computer Graphics: State of the Art for Data Analysis (Abstract Only)

Richard J. Littlefield

Graphics for Specification: A Graphic Syntax for Statistics

Paul F. Velleman

Grand Tour Methods: An Outline

Andreas Buja and Daniel Asimov

COMPUTER SCIENCE AND STATISTICS:
The Interface, D.M. Allen (ed.)

COMPUTER GRAPHICS: STATE OF THE ART FOR DATA ANALYSIS

R.J. Littlefield

Battelle Northwest
Richland, WA 99352

The field of computer graphics (CG) is now over 20 years old. In that time, a rich variety of techniques have been developed for graphical display and interaction. These techniques have been applied to such diverse areas as computer aided design and manufacturing, flight simulation, advertising, big-budget movies, video games, and of course, data analysis. Compared to other applications, the CG techniques used for data analysis are usually quite primitive. This presentation surveys the current capabilities and limitations of CG, discusses how these affect its application to data analysis, and suggests ways in which more sophisticated CG techniques could be applied to data analysis. Particular emphasis is given to graphical interaction and the role of workstations.

COMPUTER SCIENCE AND STATISTICS:
The Interface, D.M. Allen (ed.)

Graphics for Specification: A Graphic Syntax for Statistics

Paul F. Velleman
Cornell University

The Data Desk is a full-function statistics package for the Macintosh personal computer. It employs a new graphic-based syntax for specifying statistics operations and data manipulation. This article describes the principles behind the design of this interface and discusses some of the consequences of this design.

Computer graphics have traditionally been important parts of statistical analyses. (Of course, "traditional graphics" in statistical computing means "used for a decade or more by those who could afford the hardware.") Graphics were used primarily for presentation of results and as tools in analyzing data.

With improving technology came animation and interactive control of graphics. These were great advances in principle, but the only contact most data analysts had with them was watching video tapes and movies enviously at conferences.

Recently, interactive graphics have begun to come out of the laboratory. We are seeing more displays in which the viewer/analyst interacts in real time with the display. For example, PRIM's of various kinds and origins, Brushing scatterplots, and other ways to perceive higher dimensions are becomming more widespread.

There has also been a growing interest in the graphical control of computer operating systems. The most widespread (and one of the cheapest at today's prices) is found on the Apple Macintosh personal computer. The ideas behind the Macintosh operating environment are by no means new, but in the Mac they have been made accessable and affordable.

Jerry Lefkowitz and I have been engaged in a project to develop a statistics environment that uses graphic control as the means of communication between the data analyst and a statistics program. The program is called ***The Data Desk***, and is currently running on a Macintosh computer. This article is the initial report on that project.

Graphics:

For the purposes of this discussion, I define *graphics* in a very general way.

• Any display whose meaning or function relies to some important degree on the physical position of things on the screen (rather than, for example, on the numerical value or verbal meaning of things on the screen) I will include under the rubric of graphics. This means that if an operation is performed by pointing to a word rather than typing it, I consider it to be a graphic operation. If the word moves on the page, or is made to appear, or disappear, or change font or style, I consider that a graphic operation. One reason for this eclectic definition is that I can see no reasonable way to draw a line between graphic symbols that happen to be numerals or letters and other graphic symbols. The definition is thus an operational one; if it is used like a graph then it *is* a graph (even if it looks like text at a glance).

The Environment:

We have implemented this design on a Macintosh computer. The relevant technical specifications are:

• Graphics hardware: A high-resolution, fast, monochrome graphics screen (372x512 pixels). A mouse with a single button.

• Computing hardware: 8MHz MC68000 with 128K (or 512K) RAM and 64K ROM programmed with highly specialized support functions. Full IEEE floating point numerics via software emulation. One (or more) 400K disk.

• Language: All programming in an extended ISO Pascal. Program resides on a Macintosh XL (née Lisa 2/10) and is cross-compiled for the Mac. Currently the program is about 20,000 lines, but it makes extensive use of the support provided in the Mac ROM for menus, windows, controls, etc.

• User's environment: The environment is a "Desktop metaphor". The user sees objects on an imaginary desktop. The objects can be moved, grouped, or discarded by dragging them with the mouse. These objects open into windows to reveal their contents. The windows can overlap each other and can be repositioned freely.

Syntax

The basic syntax of a command is *object(, object, ...) verb*. This syntax obviates the need for a "Do it" button and provides the opportunity to avoid many syntax errors by inactivating commands (verbs) that would be inappropriate for the arguments (objects) selected.

Principles:

• **Object-oriented**. The screen shows graphic objects (usually as icons) that represent data analytic objects. For example, each variable has an icon, so a particular variable is not usually referred to by name, but rather by pointing to its icon.

issues: The major issues here are in identifying the appropriate set of objects. For example, one could consider making each case an icon and graphically gathering samples. One could consider different icons for integer, real, text, and mixed type variables so that their nature would be immediately obvious on the screen. However, we need to balance additional information against the chance of overwhelming the user. We have settled on a relatively sparse set of objects: Variables (of a few types), collections of variables (of a few types), output objects (plots, tables, etc), and a few special objects.

It is also important to establish consistent behavior among objects. For example, the same physical action should have similar consequences for all objects. For example, Opening an object (on ***The Data Desk***, by double-clicking on it or using the Open command) always reveals its internal contents. An opened variable exhibits its data elements, and opened plot is drawn in its window, and opened bundle of variables exhibits the icons of the variables collected together and their order. Windows must also behave consistently. A window exhibiting data is relocated and resized in the same way as one exhibiting a plot.

• **WYSIWYG.** What You See Is What You Get. At any time, the screen shows the current state of the data. That is not to say that the screen is cluttered with a spreadsheet of data values. (Rather, the data are arranged however the user wishes.) But one can immediately discover the contents of a variable or the state of an analysis by opening the approprite icon. Even data editing is semi-graphical in the sense that the user opens a variable icon, points to an errant data value, and types the correction.

issues: One of the problems with WYSIWYG operation is that WYDSIWYDG: What You Don't See Is What You Don't Get. To operate on an icon, the icon must be visible or reachable as part of a collection of icons whose icon is visible. Data cannot be edited out-of-sight. This is either a restriction (if you like UNIX-style operations that can change everything on the disk with one keystroke) or an advantage (if you want to be protected from unanticipated consequences of global operations.)

• **User-Driven operation:** The user is in charge of the interaction. Any operation is available whenever it is reasonable (but see the next item). Dialogs in which the user is asked questions are limited to specific details, and have defaults that can be accepted by pressing a single button whenever possible.

issues: We have taken a specific stand against "menu-driven" packages in which the program takes control of the dialog and the user supplies responses to a long sequence of questions. Menu trees in our design are intentionally short and are actively pruned to cut away branches that would make no sense in the current context.

• **Error Avoidance:** The menus (being graphical) are dynamic. Only those operations that make sense for the arguments selected are available. For example, if only one variable has been selected, the "scatterplot" command cannot be selected. If tests or confidence intervals are requested, the "pooled t for μ_1-μ_2" is not offered unless two variables have been specified as arguments.

issues: This is a very powerful way to avoid many errors that would otherwise require error messages. It simplifies interactions with the user, and it is a valuable pedagogical technique. Menu items that are not active are still visible, but in a gray type. To avoid restricting sophisticated users, the design of commands, defaults, and dialogs must be made with an understanding of the statistical properties of the procedures involved.

• **Customized Controls:** Controls are graphic images on the screen that serve to control the environment or the behavior of the program. They are manipulated with the mouse. Thus, you can push a button by pointing to a picture of a button on the screen with the mouse and pressing the mouse button. Because they are graphic structures, controls can be designed to suit a specific purpose. Why should you press a button labeled F5 when you could press a button labeled "Delete Data File"? Controls can also be positioned intelligently. For example, buttons can appear directly under the cursor when the cursor's position has been otherwise fixed.

issues: The design and positioning of controls is a specialized area worthy of further consideration. At present, have copied work done by others (mostly Apple) for the Mac, but an argument could be made for designing controls customized for some statistically-based operations. For example a control might slide or turn smoothly to control the turning of a three-dimensional scatterplot. This is an area of future research.

Some Consequences:

Some of the consequences of this graphic syntax have become clear to us only in the course of executing the design. Others, only in the course of teaching 100 undergraduates to use the program and learning from their experiences. Among the conclusions worth noting:

• There is no need for unique variable names or for restrictions on characters or length (within reason). Variables are identified by pointing to them. The screen is graphically dynamic, so (for example) long variable names are ordinarily shortened to avoid cluttering the screen. To see the full name, point to the variable and click the mouse button. Thus, for example,

Temperature °C
Σx^2
123
Things I never told my father

are all legal variable names.

• Commands can be verbose (and, consequently, more statistically precise) because the user is not typing them, but rather is pointing to them. Thus, for example, the alternative hypothesis in a test can be stated very explicitly as, for example: $\mu_1 < \mu_2$.

• Operation speed is greately improved. (Empirically, we have observed that even touch typists who are experienced users of interactive statistics packages can work much faster on ***The Data Desk***. Certainly students doing similar assignments are completing them faster on our program than on the widely used interactive statistics package we have taught with to date.)

• Learning speed is greately improved. Computer-naive undergraduates were given a single one-hour lecture and hands-on drill. After that they were on their own with very little additional support needed. (Teaching assistants were available, but were not asked computer questions very often.)

Note: These last two points have usually been thought to be mutually exclusive. Tutorial programs that are easy to learn usually get in the way of experienced users. Some programs offer a "Do you want verbose prompts?" question early in the session to try to alleviate the problem.

We have found that this environment is both easy to learn and easy to use with no changes whatever. It appears that this stems from the fundamental simplicity of the interactions on the desktop. One way of viewing this is to consider the (folk) "Law of Complexity Conservation" which states that there is a fixed amount of complexity in a given type of program, but it can be shifted among the designer, writer, novice, and expert. We have tried to shift as much of the complexity as possible onto our shoulders and off of the shoulders of the users.

Problems:

• It is difficult to write programs (macros) in a language that lacks a written syntax. One possiblity is to "record" actions to play back later, but that has its own problems. While we have a design completed for macros, this is still an area for further research.

• This style of user interface is computing- intensive. We find that we are driving the Mac fairly hard; anything with less power than a 68000 might not be able to keep up. One absolute requirement is sharp graphics. (We haven't felt a need for color yet at all.) The chief bottleneck (as with many Mac programs) is the disk drive.

Pleasant Surprises:

• You can really do quite alot on a $2000 microcomputer. The Mac is a very powerful machine, even in its 128K size. The 512K machine should handle substantial size datasets.

• On a fully integrated system, many things come for free. For example, it took no effort whatever to interface our program to most communications packages for the Mac to make up and down-loading of data possible. It was straightforward to provide the ability to paste output and plots into word processing documents, or to move them to graphics programs for further enhancement.

• The environment offers some unanticipated pedagogical advantages. For example, commands and output can be sufficiently verbose to be statistically precise. Greek and math symbols are available to write things in standard notation.

Whither?

The Data Desk is now a reasonably stable environment with a standard collection of statistical capabilties. We have been using the program in a second-term statistics class of 100 computer-naive sophomores with success, and will make it available for general use by Fall term 1985. The next research area is extensions to interactive graphics. Much of the design of these ideas is completed, but they have not yet been implemented, and are thus a subject for a future talk.

COMPUTER SCIENCE AND STATISTICS:
The Interface, D.M. Allen (ed.)

Grand Tour Methods: An Outline

Andreas Buja

Statistics Department GN-22,
University of Washington, Seattle, WA 98195

Daniel Asimov

Computer Science Department,
University of California at Berkeley, Berkeley, CA 94720

We would like to report on research about some advanced methods for exploratory data analysis based on dynamic computer graphics. These methods are now feasible because current hardware allows us to recompute and redisplay scatter plots of up to 1000 data points five to thirty times per second, thus creating the illusion of continuous motion in a plot. Our methods are based on the simple idea of moving projection planes in high (4-10) dimensional data spaces. That is, we design 1-parameter families of 2-planes in p-space, with the parameter being thought of as time. We then project p-dimensional quantitative data onto these planes in rapid succession while increasing the time parameter in small steps, which generates movies of data plots that convey a tremendous wealth of information.

We call these dynamic graphics "grand tour" methods. In our presentation, we will show a short (5 minutes, 16mm) film featuring two artificial data sets (five circles in 10-space and a 3-dimensional torus in 6-space), and two well known real data sets: the Boston Housing data [1] and the Particle Physics data (see [2], the well-known PRIM-9 movie). The film can be requested from the authors.

It may be true that any single aspect of structure in data can be isolated and somehow displayed in a number of static plots, but the grand tour offers a multitude of aspects simultaneously and in relation to each other. It can frequently replace hours of staring at plots by a short inspection of a movie and dramatically reduce the probability of missing structure as well. In our experience, the usefulness of this type of display depends less on the dimension of data space than on the intrinsic dimension of the data. If the data form 0-,1- or 2-dimensional manifolds (i.e. clusters, curves, or surfaces), the human eye is able to pick up the "gestalt" almost instantly due to motion. If, however, the intrinsic structure is of four or higher dimensions, grand tour methods alone will not necessarily be successful, and other tools will have to be used, perhaps in conjunction with the grand tour.

We would like to point out an important aspect of the grand tour whose impact is not apparently understood in a current discussion of projection pursuit contained in P.J.Huber and discussants [3]. Projection pursuit in its original version is the search for informative projections through optimization of information indices as functions of data projections. Thus the output consists of one or several data plots corresponding to global or local maxima of some index chosen by the data analyst. In contrast, the grand tour is NOT just another vehicle for finding interesting static plots, and it is not simply a competitor of projection pursuit. The output of the grand tour is a MOVIE with all the information encoded in the smooth motion of the scatter plots. We argue that the speed vectors of data points in a grand tour provide two additional dimensions of information in addition to the two dimensions of location, thus letting us perceive a full 4-dimensional space at any given point in time. In comparison to the grand tour, three-dimensional rotation is degenerate in that one of the two infinitesimal rotations is held fixed, resulting in the loss of one dimension of information.

Dynamic features must be carefully considered in the design of a grand tour. To mention a few desiderata:

- A basic requirement is (at least piecewise) smoothness of motion to avoid jitter in the movie and prevent fatigue of the human eye. The smoother the motion, the clearer will be the perception of the information encoded in the velocities. Ideal smoothness is achieved by so called geodesics, a notion which is applicable to our context in the precise sense of differential geometry. Our favorite implementation is actually based on piecewise geodesic motion.

- It is important to avoid distraction due to excessive within-screen-spin. By this we mean rotation which takes place within the projection plane rather than in the embedding space, and which is hence uninformative if not disturbing. As it turns out, any given grand tour can be modified such that it avoids within-screen-spin completely, although the additional computational expense may well slow it down to an unbearable extent.

- Another desideratum is the following: the 2-plane in data space which encodes velocity should be kept orthogonal to the projection 2-plane to avoid confounding of location and speed of the dynamic scatter plot points. This is satisfied by the above mentioned geodesic motion, but one can show that this requirement confines the grand tour to a fixed 4-space, and hence must be abandoned if the tour is to scan 5- and higher-dimensional space. In our implementation we use only piecewise geodesics, which allows us to scan any dimension of space.

We have developed a set of tools for designing and implementing grand tours. They can be divided roughly into two classes:

1) Parametrization of planes by Euler angles, and design of paths which scan parameter space.

2) Interpolation between randomly selected planes by "shortest paths", and analogues of splines.

At this point we need to introduce some terminology from differential manifolds. Since the actual computer implementation requires a pair of orthogonal vectors in data space for the calculation of horizontal and vertical screen coordinates, we need the Stiefel manifolds $S_{2,p}$ of orthonormal 2-frames in p-space. Similarly, since we would often like to equivalence all data projections which can be transformed into each other through screen rotations, we also introduce the Grassmann manifolds $G_{2,p}$ of 2-planes in p-space. For implementation purposes, we consider a grand tour as a curve on a Stiefel manifold, but for theoretical and conceptual considerations we prefer to look at it as a curve on a Grassmannian.

The parametrization class of techniques mentioned above parametrizes either manifold by angles, similar to the way longitude and lattitude parametrize a 2-sphere. Angles are reals mod 2π, i.e. elements of the circle $T^1 = \mathbb{R} \bmod 2\pi$, and a p-dimensional product of circles is a torus T^p . We use tori as parameter spaces because they allow natural curves of great smoothness and flexibility, namely the ones obtained by pushing straight lines from $\mathbb{R}^p$ into $\mathbb{T}^p$. If the coordinates of a vector in $\mathbb{R}^p$ are linearly independent over the rationals, then the straight line generated by this vector is dense in $\mathbb{T}^p$; hence the resulting grand tour is dense in the Stiefel or Grassmann manifold if the parametrization is onto. We have examples of parametrizations of the Stiefel variety as well as the Grassmannian. For topological reasons they cannot be 1-1. The techniques for parametrization are borrowed from numerical analysis and they are based on concatenations of planar rotations (Givens transformations) and/or reflections on hyperplanes (Householder transformations). Underlying these constructions is the fact that any orthogonal mapping can be decomposed into a sequence of Givens and/or Householder transforms.

The interpolation class of techniques for grand tour construction is based on successively sampling planes and connecting them by motion along suitable interpolation paths. In the tour version we will show in our movie, these paths are geodesics on the Grassmannian, which are described in an article by Wong [4]. They correspond to the simultaneous interpolation of the principal angles between two 2-planes. This scheme results in a tour which lacks smoothness at the endpoints of interpolation paths, but geodesics enjoy many favorable properties, some of which we mentioned above in our discussion of dynamic aspects of grand tours. Another nice feature is the low computational cost which is not greater than that of ordinary 3d-rotations, at least when the tour proceeds on a geodesic path. At the endpoint of a geodesic segment, there is a pause of a fraction of a second due to sampling a new random plane and setting up the parameters for the corresponding interpolation segment. In practice, viewers do not find these pauses unpleasant, on the contrary, they perceive ceaseless motion as overwhelming and tiring.

Currently, we are in the process of constructing analogues of spline interpolators on the Grassmannian. The geodesic tour just described can be considered as a spline tour of order zero. Splines of higher order will lead to perfectly smooth motion, but will lose some of the simplicity of the geodesic tour.

D.Asimov discusses desirable properties of grand tours in a forthcoming paper [5]. He states that asymptotically a tour should form a dense subset of $G_{2,p}$, whereas in terms of finite time it should spread out quickly on the Grassmannian. This latter requirement is formalized by the notion of "minimal amount of time needed to get within an ε- neighborhood of any 2-plane." Theoretical lower bounds can be given by comparing the volume of an ε-neighborhood with the total volume of the Grassmannian. It is clear that this ratio becomes less favorable for higher dimensional data spaces. (For volume computations on Grassmannians, see, e.g., Santalo [6]. Asimov's paper contains tables and displays which indicate what can be expected in various dimensions. It is apparently possible to come within 12 degrees of any plane by watching 1800 randomly sampled planes in 4 dimensions, whereas 28 degrees are possible with the same number of planes in 6 dimensions. In 8 dimensions one can expect only 39 degrees, and in 10 dimensions 44 degrees. Although these figures appear very discouraging at first, we should remember that this type of discussion is somewhat academic, as it neglects the dynamic nature of the grand tour which lets us perceive four rather than two dimensions at a time. Second, the dimension of the data space is less of a factor than the intrinsic dimension of the data in determining how well we can perceive structure in data (see above). Recognizing the difficulty of finding structure of low codimension by tour methods, we plan to combine an interactive and dynamic projection pursuit version with the grand tour as this will permit the grand tour to remain in neighborhoods of local and global extrema of information indices on the Grassmannian.

In the previous paragraph, we referred implicitly to metrics on the Grassmannian when we mentioned ε-neighborhoods of 2-planes. We seem to have an intuitive notion of what is meant by "distance between two 2-planes", but there are ramifications which we will explain briefly. The best formalization of our intuitive notion is probably given by the maximal angle between a vector in one plane and its projection onto the other plane. A proof is necessary to show that this leads to a metric on the Grassmannian, and a simple way to go about it is via an interpretation in terms of the Hausdorff metric on the unit circles in p-space, which are in 1-1-correspondence with the 2-planes. This metric can also be defined as the larger of the two principal angles θ_1 and θ_2 between two 2-planes. In some sense this is an L_∞-metric because it turns out that $(\theta_1^p + \theta_2^p)^{1/p}$ define metrics on the Grassmannian, too, which we call L_p-metrics for obvious reasons. Wong mentions the L_2-case as the one which creates the Riemannian structure on the Grassmannian. The other metrics for $1 < p < \infty$ generate Finsler geometries but these all lead to the same geodesics. Notice that the L_∞-case does not lead to a Finsler space due to its non-differentiable nature, but it is obtained as the limiting case of a 1-parameter family of Finsler geometries.

In what follows we present a few ideas which greatly increase the flexibility of the grand tour as a viewing method for multivariate data. The grand tour described so far would scan too many projections of modest interest in many situations. For example, in the case of predictor-response data, one would like to concentrate on plots of linear combinations of responses versus linear combinations of predictor variables, while in the case of repeated-measures data, one would like to concentrate on contrasts of treatment responses, i.e., linear combinations whose coefficients sum up to zero. In the same situation, one could also be interested in the dependence of contrasts on linear combinations of covariates. We conclude that, for practical data analysis, one needs modified grand tours which offer more flexibility in the choice of data projections to be scanned. For predictor-response data the modification consists of confining a grand tour to pairs of normalized vectors which scan the unit sphere of predictor space and response space respectively. The manifold to be toured simplifies to a product of spheres. This is a submanifold of dimension p-2 as compared to 2p-3, the dimension of the full Stiefel manifold. We will show an implementation of this type of tour in our movie. For repeated-measures data, one would confine the scanning vectors to the space of contrasts, i.e., the vectors which are orthogonal to (1,1,1,...).

Grand tour techniques can also be brought to bear in contexts which are rather different from those we have considered so far. A basic data analytic operation is the comparison of several plots of one given data set. The problem is to identify cases and groups of cases across two or more plots. To support this operation, one can use geodesic interpolation of two projection planes to transform one scatter plot into the other dynamically. This makes use of the fact that our visual system keeps track of the identity of moving objects.

Obviously, there are many more possibilities of applying motion graphics to data analysis. We hope that the grand tour will be recognized as a useful tool and a natural extension of 3d-graphics. Conceptually, higher dimensional motion graphics are at least as "intuitive" or "counter-intuitive" as 3-dimensional ones, and some important capabilities of the visual system seem to work in higher dimensions as well. Partial supportive evidence for this claim will be provided by our film.

REFERENCES:

[1] Belsley, D.A., Kuh, E., Welsch, R.E., Regression Diagnostics (1980), Wiley.

[2] Fisherkeller, M.A., Friedman, J.H., Tukey, J.W., "PRIM-9", a movie, 1974, Stanford, CA.

[3] Huber, P.J., and discussants, Projection Pursuit Methods, Ann. of Statistics (1985), to appear.

[4] Wong, Y.-C., Differential Geometry of Grassmann Manifolds, Proc. Nat. Acad. Sci., vol. 57 (1967), p.589f.

[5] Asimov, D., The Grand Tour, SIAM Jrnl. on Sci. and Stat. Computing, 6:1 (1985), p.128f.

Nonlinear Models

Organizer: *Robert E. Kass*

Invited Presentations:

Nonlinear Least Squares and First-Order Kinetics
Douglas M. Bates and Dennis A. Wolf

Computational Experience with Confidence Regions and Confidence Intervals for NonLinear Least Squares
Janet R. Donaldson and Robert B. Schnabel

Curvatures for Parameter Subsets in Nonlinear Regression
R. Dennis Cook and Miriam L. Goldberg

COMPUTER SCIENCE AND STATISTICS:
The Interface, D.M. Allen (ed.)

Nonlinear Least Squares and First-Order Kinetics

Douglas M. Bates
Dennis A. Wolf

University of Wisconsin - Madison

Donald G. Watts

Queen's University at Kingston

One of the persistent problems with the use of nonlinear least squares programs is specifying and coding model functions and partial derivatives then incorporating this code into the program. We show how these difficulties can be bypassed for the important class of models defined by linear systems of differential equations. Not only are the model functions easily specified but the partial derivatives can be automatically generated to allow the use of sophisticated optimization algorithms without an additional burden on the user. These models are widely used in pharmacokinetics and chemical kinetics.

An additional problem that occurs in pharmacokinetic analysis is incorporation of non-homoscedastic error structures. We show how the "transform both sides" approach due to Carroll and Ruppert can be used with this model specification strategy.

1. Introduction

One common difficulty with using nonlinear regression programs is specifying and coding the model function and, possibly, its derivatives. Specifying the model function, particularly in the case of implicit models defined by systems of differential equations, can provide an opportunity for the user to make syntax and transcription errors which take a long time to detect and correct. An even more fertile ground for errors is specifying and coding derivatives of the model function with respect to the model parameters. In our experience, this is the single most error-prone stage in a nonlinear regression analysis. Empirical evidence of this difficulty is the popularity of derivative-free methods whether based on finite difference approximations to the derivatives or other schemes such as DUD (Ralston and Jennrich, 1978).

For one important class of models, the first-order kinetic models defined by linear systems of differential equations, Jennrich and Bright (1976) demonstrated that these difficulties can be avoided. They gave a representation of the solution of the differential equations in terms of the matrix exponential and showed that the model derivatives can be computed simultaneously with the model function. We provide a different derivation with greater generality in section 2 and discuss some of the implementation considerations in section 3.

Linear kinetics models are widely used in pharmacokinetics where they are called "linear compartment models" or, simply, compartment models. A straightforward application of nonlinear least squares to pharmacokinetic data is often inappropriate, though, because the assumption of homoscedasticity (constant variance) is not warranted. Weighted least squares methods are sometimes used but we have found the transformation method of Carroll and Ruppert (1984) to be simple and effective. In section 4 we describe the method and its implementation, then give some examples in section 5.

2. Linear Kinetics

A first-order kinetics system, such as a compartment model, is one described by a set of linear differential equations. In the compartment models, an organism is considered as composed of homogeneous, well-mixed compartments which communicate with each other by the exchange of material. A drug administered to the bloodstream could pass from the blood to body tissues, back into the blood, and finally be eliminated from the system through the kidneys, for example. The blood would be considered as one compartment, other body tissues as a second compartment, while the exterior of the system would be an implicit, third compartment. Such a system and its communication paths would be represented as in Figure 1.

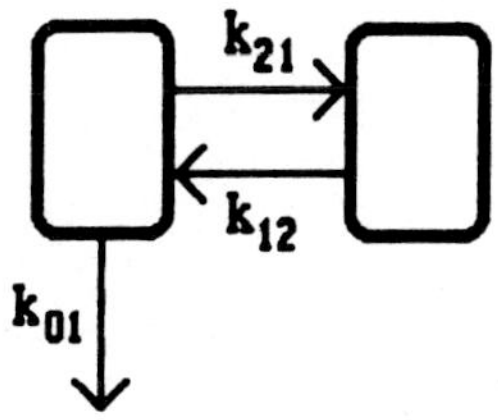

Figure 1: A 2-compartment model

The concentration of the drug in the various compartments at any time t would be written

$$\underline{\gamma}(t) = (\gamma_1(t), \gamma_2(t))^T$$

In a system with K compartments, $\underline{\gamma}$ would be K-dimensional. The kinetics of the system, which describe how the concentrations change with time, are linear if we can represent the derivatives of $\underline{\gamma}$ with respect to time as a linear function of $\underline{\gamma}$. That is, the system is governed by the system of differential equations

$$\dot{\underline{\gamma}}(t) = \frac{d\underline{\gamma}(t)}{dt} = A\,\underline{\gamma}(t) + \underline{\iota}(t) \tag{2.1}$$

where A is the $K \times K$ *system matrix* which does not depend on $\underline{\gamma}$ or t and $\underline{\iota}(t)$ is the *driving function* for the system which indicates how material is being added to the system.

In pharmacokinetics, the driving function is usually a bolus injection into a compartment, corresponding to an impulse or δ-function in that compartment, or an intravenous infusion into a compartment, corresponding to a constant input function in that compartment from time t_0 to t_f. With a bolus injection, we usually consider the injection as determining initial conditions

$$\underline{\gamma}_0 = (\gamma_{0,1}, \gamma_{0,2}, \ldots, \gamma_{0,K})^T = \underline{\gamma}(0)$$

but we will find it convenient to consider general driving functions in this section.

These systems are often described in terms of *rate constants* denoted k_{ij} which give the multiplier for the communication from compartment j to compartment i as shown in Figure 1. (By convention, a rate constant k_{0i} is the rate constant for elimination from compartment i). The system in Figure 1 would correspond to the linear differential equations

$$\frac{d\gamma_1(t)}{dt} = -(k_{01}+k_{21})\gamma_1(t) + k_{12}\gamma_2(t)$$

$$\frac{d\gamma_2(t)}{dt} = k_{21}\gamma_1(t) - k_{12}\gamma_2(t)$$

If we set $\theta_1 = k_{01}$, $\theta_2 = k_{21}$, and $\theta_3 = k_{12}$, the system matrix is then

$$A = \begin{pmatrix} -(\theta_1+\theta_2) & \theta_3 \\ \theta_2 & -\theta_3 \end{pmatrix}$$

The solution to the system (2.1) with driving function $\underline{\iota}(t)$ is

$$\underline{\gamma}(t) = e^{At} * \underline{\iota}(t) \tag{2.2}$$

where e^{At} is the matrix determined by the convergent power series

$$e^{At} = I + At + \frac{A^2t^2}{2!} + \frac{A^3t^3}{3!} + \cdots$$

and the * denotes convolution. That is,

$$e^{At} * \underline{\iota}(t) = \int_0^t e^{A(t-u)}\underline{\iota}(u)\,du$$

In the case of a bolus injection where $\underline{\iota}(t)$ is an impulse function, the solution (2.2) collapses to

$$\underline{\gamma}(t) = e^{At}\underline{\gamma}_0 \tag{2.3}$$

Using (2.2) or the special form (2.3), we can determine the state of the system at any time t and hence determine the N-dimensional expected response vector $\underline{\eta}$ for a nonlinear regression model where the response being considered is the concentration in one compartment and the experimental conditions are the times $t_i, i=1, \ldots, N$ at which this concentration is measured. However, we can also use the same technique to determine the derivatives

$$\frac{\partial \underline{\eta}}{\partial \theta_p} \quad p = 1, \ldots, P$$

where P is the total number of parameters. To avoid cumbersome expressions, we will adopt the convention that a subscript p denotes differentiation with respect to θ_p. We obtain the derivatives by differentiating the system (2.1) to obtain

$$\dot{\underline{\gamma}}_p(t) = A\underline{\gamma}_p(t) + A_p\underline{\gamma}(t) + \underline{\iota}_p(t) \tag{2.4}$$

which is simply another linear system of differential equations with system matrix A and driving function $A_p\underline{\gamma}(t) + \underline{\iota}_p(t)$. The solution is thus

$$\underline{\gamma}_p(t) = e^{At} * [A_p\underline{\gamma}(t) + \underline{\iota}_p(t)] \tag{2.5}$$

where $\underline{\gamma}(t)$ can be obtained from (2.2).

Returning to the system of Figure 1, suppose that the input function was a bolus injection of known amount into compartment 1, the blood. Since the "volume of distribution" for the blood would generally be unknown, the initial concentrations would be represented as

$$\underline{\gamma}_0 = (\theta_4, 0)^T \tag{2.6}$$

and the solution would be given by (2.3). Equation (2.5) collapses to

$$\underline{\gamma}_p(t) = e^{At} * A_p e^{At}\underline{\gamma}_0 + e^{At}\underline{\gamma}_{0,p} \tag{2.7}$$

This may still seem complicated but the pieces are rather simple. Here

$$\underline{\gamma}_{0,1} = \underline{\gamma}_{0,2} = \underline{\gamma}_{0,3} = \underline{0}$$

$$\underline{\gamma}_{0,4} = (1,0)^T$$

$$A_1 = \begin{pmatrix} -1 & 0 \\ 0 & 0 \end{pmatrix}$$

$$A_2 = \begin{pmatrix} -1 & 0 \\ 1 & 0 \end{pmatrix}$$

$$A_3 = \begin{pmatrix} 0 & -1 \\ 0 & 1 \end{pmatrix}$$

$$A_4 = \begin{pmatrix} 0 & 0 \\ 0 & 0 \end{pmatrix}$$

so

$$\underline{\gamma}_1(t) = e^{At} * A_1 e^{At}\underline{\gamma}_0$$

$$\underline{\gamma}_2(t) = e^{At} * A_2 e^{At}\underline{\gamma}_0$$

$$\underline{\gamma}_3(t) = e^{At} * A_3 e^{At}\underline{\gamma}_0$$

and

$$\underline{\gamma}_4(t) = e^{At}\underline{\gamma}_{0,4}$$

There is another way in which parameters can enter the kinetic system and that is as a "dead time" or lag time. The measured time, t, may not correspond to the effective time in the system and it may be more realistic to describe the kinetics in terms of

$$\tau = (t - t_0)_+$$

where t_0 is an unknown parameter. This modification is easily incorporated into (2.2) and (2.4) to generate the required expected responses and derivatives.

3. Implementation

The implementation of these methods involves two considerations: specifying the model, and performing the calculations in (2.2) and (2.4).

The model can be specified by indicating the roles of the parameters as rate constants, initial conditions, dead times, etc. through a parameter-use matrix. We have chosen to use a matrix with 3 columns, the first containing the parameter number. If the parameter is a rate constant the second and third columns indicate the source and sink compartments with a sink of 0 indicating elimination. Initial conditions or other forms of driving functions are specified with negative values in the third column and the number of the affected compartment in the second column. A -1 in the third column indicates the level of an impulse, -2 indicates the level of a constant infusion, -3 indicates the slope of a linear infusion, etc. This specification scheme, combined with the restartable property of linear kinetic systems, can be used to model a driving function using splines. To indicate a lag time, we use a zero in the second column.

As an example, the parameter-use matrix for the system described in Figure 1 with the initial conditions (2.6) is

$$\begin{matrix} 1 & 1 & 0 \\ 2 & 1 & 2 \\ 3 & 2 & 1 \\ 4 & 1 & -1 \end{matrix}$$

Using this information and the current parameter values, a program can generate A and $\underline{\iota}(t)$.

Notice that this scheme allows a single parameter to have multiple uses. Changing the parameter-use matrix to

$$\begin{matrix} 1 & 1 & 0 \\ 2 & 1 & 2 \\ 2 & 2 & 1 \\ 3 & 1 & -1 \end{matrix}$$

and re-fitting the model will allow testing of the hypothesis that $k_{12} = k_{21}$.

Once A and $\underline{\iota}(t)$ have been determined, the expressions in (2.2) and (2.5) must be evaluated. Moler and Van Loan (1978) give an extensive survey of methods for the matrix exponential and conclude that methods based on an eigenvalue-eigenvector decomposition of A should be used when evaluations for many different t's are required. If the eigenvalues of A are real and there is a complete set of eigenvectors so we can write

$$A = U \Lambda U^{-1} \tag{3.1}$$

with

$$\Lambda = diag(\lambda_1, \ldots, \lambda_K) \tag{3.2}$$

then

$$e^{At} = U e^{\Lambda t} U^{-1} \tag{3.3}$$

where

$$e^{\Lambda t} = diag(e^{\lambda_1 t}, \ldots, e^{\lambda_K t}) \tag{3.4}$$

which immediately gives an evaluation for impulse driving functions through (2.3).

One difficulty here is that the decomposition in (3.1) does not always exist, even for non-pathological cases, and the detection of those cases is quite difficult. Standard eigenvalue-eigenvector routines such as those in Eispack (Smith et al., 1976) will usually return a decomposition even in degenerate cases and the only clue that the decomposition doesn't exist is that U has a huge condition number. Bavely and Stewart (1979) provide a method to reduce A to a block-diagonal form which can be used to evaluate the matrix exponential in these cases. The method can be implemented in a straightforward fashion but is too lengthy to describe here.

Assuming then that software such as Eispack code can produce the decomposition (3.1) with a well-conditioned U, it is convenient to pre-multiply all the system vectors by U^{-1} to produce

$$\xi(t) = U^{-1}\gamma(t)$$

$$\xi_0(t) = U^{-1}\gamma_0(t)$$

$$\kappa(t) = U^{-1}\iota(t)$$

$$\xi_p(t) = U^{-1}\gamma_p(t)$$

$$\kappa_p(t) = U^{-1}\iota_p(t)$$

and, finally,

$$C_p = U^{-1} A_p U$$

The notation for $\xi_p(t)$ and $\kappa_p(t)$ is not consistent with earlier usage since, for example, $\xi_p(t)$ is not the derivative with respect to θ_p of $\xi(t)$. It is convenient though.

Expressions (2.2) and (2.5) now become

$$\xi(t) = e^{\Lambda t} * \kappa(t) \tag{3.6}$$

and

$$\xi_p(t) = e^{\Lambda t} * [C_p \xi(t) + \kappa_p(t)] \tag{3.7}$$

Because $e^{\Lambda t}$ is diagonal, the convolutions can be evaluated as scalar convolutions. For example, with an impulse driving function, (3.6) and (3.7) reduce to

$$\xi(t) = e^{\Lambda t} \xi_0 \tag{3.8}$$

and

$$\xi_p(t) = e^{\Lambda t} * C_p e^{\Lambda t} \xi_0 + e^{\Lambda t} \xi_{0,p} \tag{3.9}$$

Each element in the convolution matrix is

$$\{e^{\Lambda t} * C_p e^{\Lambda t}\}_{i,j} = c_{p;i,j} e^{\lambda_i t} * e^{\lambda_j t} \tag{3.10}$$

where

$$e^{\lambda_i t} * e^{\lambda_j t} = e^{\lambda_j t}[e^{\lambda_i t - \lambda_j t} * 1] \tag{3.11}$$

$$= \begin{cases} \dfrac{e^{\lambda_i t} - e^{\lambda_j t}}{\lambda_i - \lambda_j} & \lambda_i \neq \lambda_j \\ t e^{\lambda_i t} & \lambda_i = \lambda_j \end{cases}$$

In practice, the condition $\lambda_i = \lambda_j$ is determined by comparing $|(\lambda_i - \lambda_j)t|$ to the relative machine precision.

Since this implementation uses the rate constants directly and the constants must be non-negative, the actual parameters that we use are the logarithm of the rate constants and of the unknown initial concentrations. This avoids having to use constrained optimization methods for physically meaningful parameter estimation. It does produce a minor difficulty when a particular path is not needed for the model since the estimate of the log rate constant tends to negative infinity. This situation is easily detected by the user and the model re-specified.

4. Heteroscedasticity

With measurements of physical quantities, such as drug concentrations, it is not uncommon to have the level of the noise increase with the level of the signal so nonlinear regression modelling with a constant variance assumption is inappropriate. A realistic fitting of compartment models should include some method of allowing for changing variances in the noise. Some weighted least-squares methods have been used (Jennrich and Bright, 1976, Kramer et al., 1974, Wagner and coworkers, 1977) but the weights are often chosen on an ad-hoc basis and, more importantly, the weights are often based on the observed concentrations rather than the predicted concentrations.

Several related transformation methods, which model the changing variance as a function of the response level and thus account for heteroscedasticity, have been proposed (Box and Cox, 1964, Carroll and Ruppert, 1984, Pritchard, Downie, and Bacon, 1977). We find the Carroll and Ruppert approach to be reasonable and easy to implement. This uses the Box-Cox transformation family

$$y^{(\lambda)} = \begin{cases} \dfrac{y^{\lambda}-1}{\lambda} & \lambda \neq 0 \\ \log(y) & \lambda = 0 \end{cases} \qquad (4.1)$$

in what Carroll and Ruppert call "transforming both sides".

For a given value of λ, the estimates $\underline{\theta}_\lambda$ are determined by fitting the transformed data $y^{(\lambda)}$ to the transformed model function $f^{(\lambda)}(t,\underline{\theta})$ resulting in a loglikelihood, up to a constant, of

$$l(\lambda) = \lambda \sum_{i=1}^{N} \log(y_i) - \frac{n}{2}\log(S(\underline{\theta}_\lambda)) \qquad (4.2)$$

which is then optimized over λ. Since the derivatives of $f^{(\lambda)}(t,\underline{\theta})$ with respect to $\underline{\theta}$ are easily calculated from $df/d\underline{\theta}$, we can use the methods of the previous section to calculate models and derivatives for transformed compartment models.

The loglikelihood function over a range of λ can give an indication of what are "reasonable" values for λ. In some cases, as shown in the following section, there is very little sensitivity of the data to transformation and λ is essentially irrelevant. In other cases, the value of λ is sharply determined and the need for transformation clearly defined. We examine the plot of the loglikelihood versus λ to determine a reasonable and "natural" value of λ (usually 0, 1/2, or 1) and, using the rationale of Box and Cox (1982) or Hinkley and Runger (1984), condition the subsequent analysis on that value of λ.

5. Examples

We consider three examples from the literature to demonstrate the application of the transformation approach for homoscedasticity and the flexibility of model description. The Brunhilda data from Jennrich and Bright (1976) shown in Table 1 are blood concentrations of sulphate measured by a radioactive assay. The results are quoted as counts. Jennrich and Bright fit a three-compartment catenary model (Figure 2) to these data using weighted least squares with the weights proportional to y_i^{-2} and assuming an initial concentration corresponding to a count of 2×10^5. We fit the same model but with a sixth parameter of the initial count in compartment one and using the power transformations.

time min.	activity counts
2	151117
4	113601
6	97652
8	90935
10	84820
15	76891
20	73342
25	70593
30	67041
40	64313
50	61554
60	59940
70	57698
80	56440
90	53915
110	50938
130	48717
150	45996
160	44968
170	43602
180	42668

Table 1: Data from Jennrich and Bright (1976)

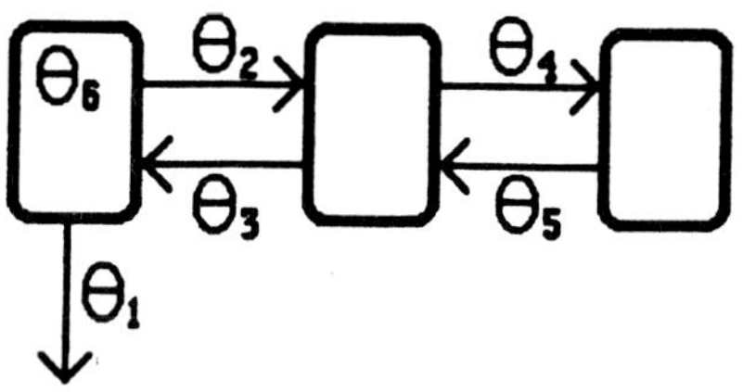

Figure 2: A 3-compartment catenary model

The loglikelihood of λ, along with the data in the origina count scale, is shown in Figure 3. For λ, the MLE was about -0. with wide 95% confidence limits of -2 to 1.75 so we selected λ = (log transformation). The fitted parameters, confidence limits an parameter use matrix are shown in Table 2. In addition, th parameter estimates for $\lambda = 1$ are included for comparison.

The parameter estimates are very insensitive to transformation primarily because the relative range of the responses is nc large. The ratio of the largest to the smallest observation is 3.54: and even the logarithm transformation is fairly linear over thi range as shown in Figure 4b.

We also show the observed and predicted responses and som of the residual analysis in Figure 4. The residuals for this mode do not show suspicious patterns but fitting these data with a tw compartment model did produce noticeable patterns in the residu als. The need for a three compartment model to adequatel represent these data was confirmed with an F-test.

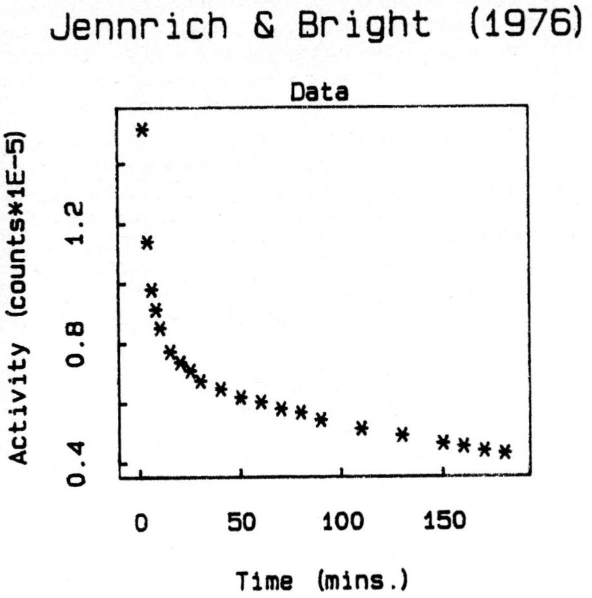

Figure 3a

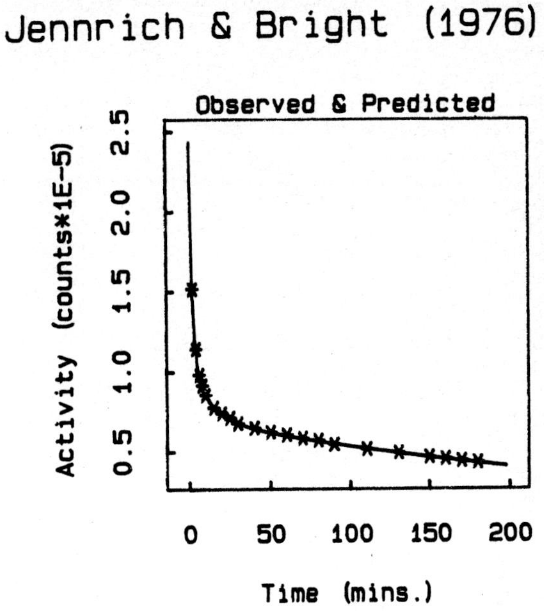

Figure 4a

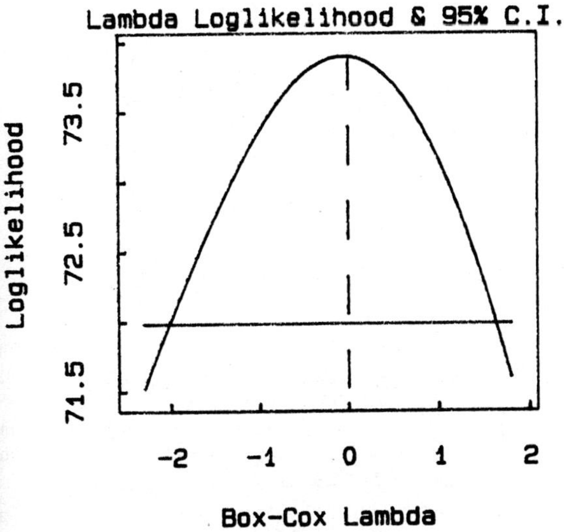

Figure 3b

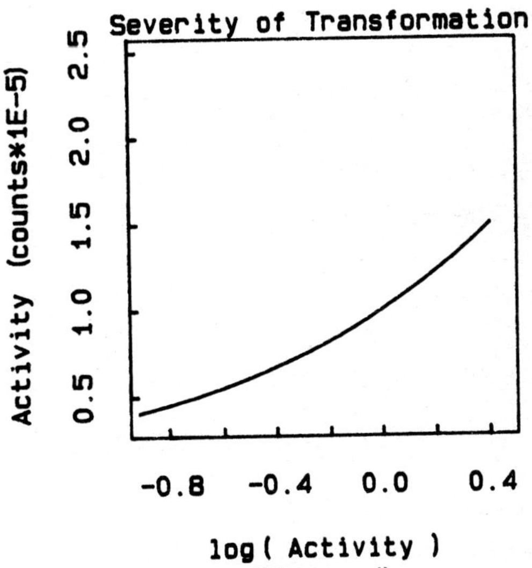

Figure 4b

Par.	Use		Est.(0)	95% conf. int.	Est.(1)
1	1	0	0.00941	0.0085,0.0104	0.00972
2	1	2	0.2848	0.2324,0.3491	0.3011
3	2	1	0.1923	0.1642,0.2253	0.2022
4	2	3	0.0342	0.0244,0.0481	0.0384
5	3	2	0.0627	0.0525,0.0749	0.0667
6	1	-1	2.434	2.228,2.659	2.489

Table 2: Parameter estimates for Brunhilda data
θ_6 is scaled by 10^{-5}

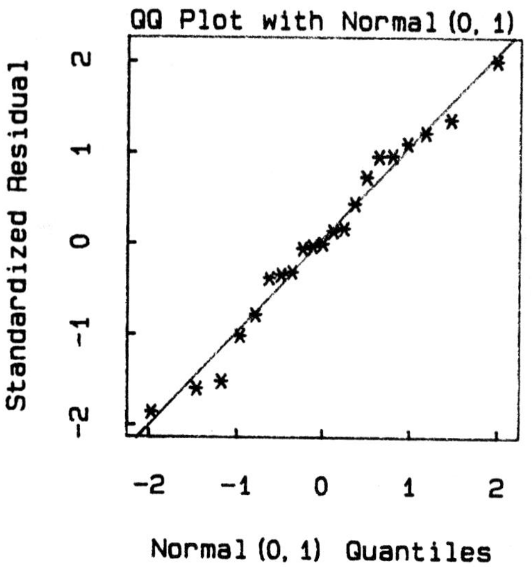

Figure 4c

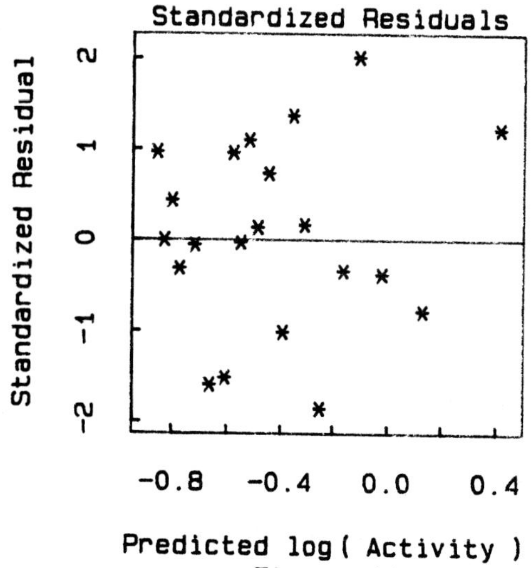

Figure 4d

One point of interest about the fitted parameters is that the initial activity assumed by Jennrich and Bright (1976), 2×10^5, is not included in the confidence limits for θ_6. If the model is fitted on the log scale with an initial activity of 2×10^{-5}, the residual sum-of-squares is 0.00287 with 16 degrees of freedom. Including θ_6 in the model produces a residual sum-of-squares of 0.000878 so the calculated F statistic for a test of $\theta_6 = 2\times10^5$ is 34.06 with 1 and 15 degrees of freedom. Besides the formal F-test demonstrating that 2×10^5 is a poor value of θ_6, we also found that the residuals for that fit exhibited poor behavior.

As a second example, we consider the digoxin data from Kramer et al. (1974). A rapid (bolus) intravenous injection of 1 mg. of this drug was administered to five healthy male volunteers and blood samples were periodically withdrawn and assayed using a ^{125}I radioimmunoassay. The data from person DL, consisting of serum digoxin concentrations, is shown in Table 3. Kramer et al. (1974) fit the data to the three-compartment mammillary model of Figure 5 using weighted least squares with weights proportional to $\exp(-0.294y_i)$. These weights were obtained from a separate experiment.

We again found that $\lambda = 0$ appeared to be a suitable choice but this time the plot of the loglikelihood versus λ (Figure 6b) indicates a fairly short range of acceptable λ values. The MLE is at about 0.1 with approximate 95% confidence limits of -0.1 to 0.35 . The fitted parameters, confidence limits and parameter use matrix are shown in Table 4 along with the parameters estimated with $\lambda = 1$. In this example the difference between the parameters estimated using an unweighted analysis and those obtained from an

time hr.	concentration ng./ml.
0.035	20.50
0.068	17.50
0.102	14.50
0.135	12.50
0.168	13.00
0.235	12.00
0.302	11.00
0.368	9.10
0.502	9.60
0.753	5.60
1.003	4.90
2.005	3.20
3.008	2.00
4.030	1.80
7.833	0.90
15.800	0.85
23.717	0.70
36.450	0.45
47.183	0.43
71.750	0.39

Table 3: Data from Kramer et al. (1974) Person DL

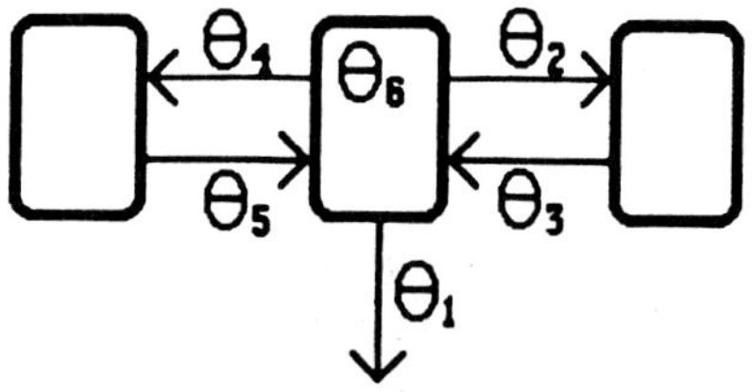

Figure 5: A 3-compartment mammillary model

analysis of the logs is striking but here the ratio of maximum observation to the minimum observation is greater than fifty so the log transformation is quite nonlinear as shown in Figure 7b. The fitted values reported in the original paper differ only slightly from those here. The residuals, displayed in Figures 7c and 7d, do not demonstrate disturbing patterns.

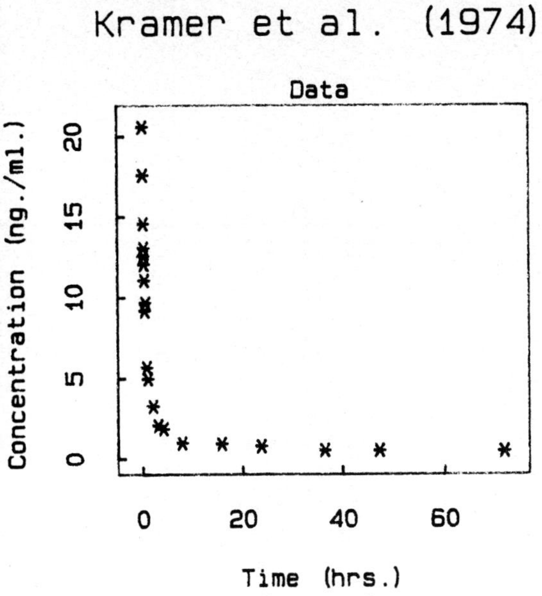

Figure 6a

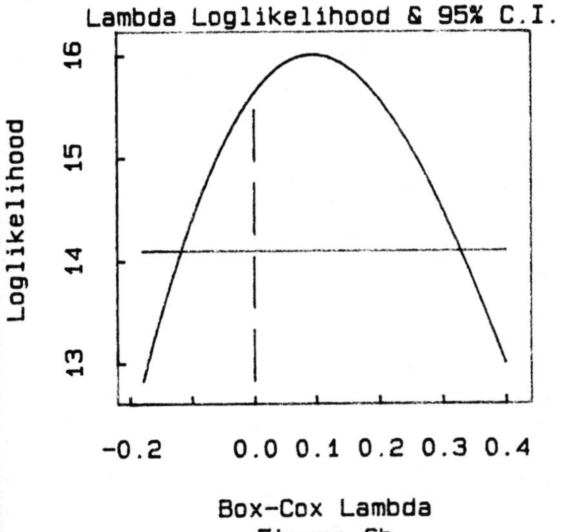

Figure 6b

Par.	Use		Est.(0)	95% conf. int.	Est.(1)
1	1	0	0.2344	0.1764,0.3115	0.5387
2	1	2	1.250	0.5546,2.819	9.517
3	2	1	1.453	0.5797,3.643	12.92
4	1	3	0.7964	0.5967,1.063	1.736
5	3	1	0.06493	0.0451,0.0935	0.2174
6	1	-1	19.42	16.08,23.44	29.05

Table 4: Parameter estimates for Person DL
Kramer et al. (1974)

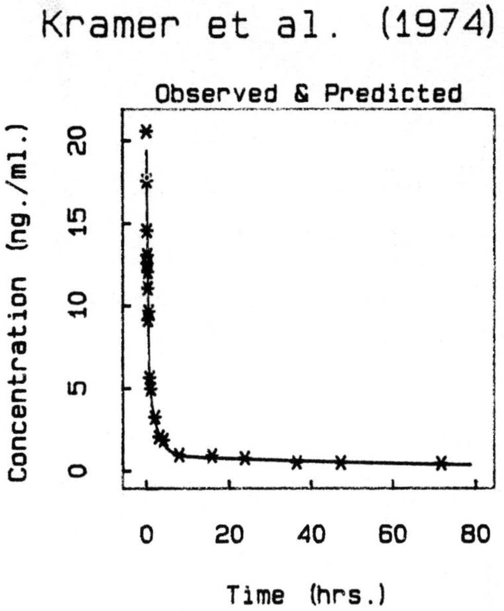

Figure 7a

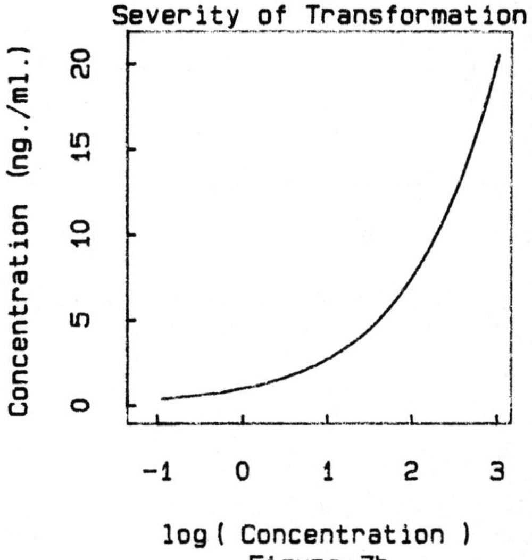

Figure 7b

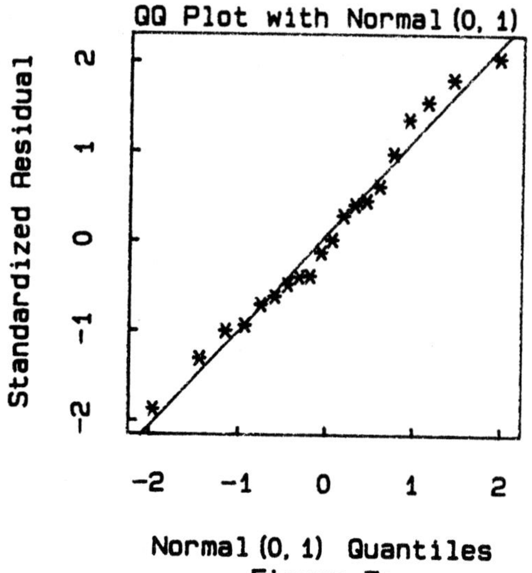

Figure 7c

time hr.	concentration mcg./ml.
0.25	215.6
0.5	189.2
0.75	176.0
1.0	162.8
1.5	138.6
2.0	121.0
3.0	101.2
4.0	88.0
6.0	61.6
12.	22.0
24.	4.4
48.	0.1

Table 5: Data from Kaplan et al. (1972) Subject 5

Par.		Use	Est.(0.5)	95% conf. int.	Est.(1)
1	1	0	0.2252	0.2092,0.2451	0.2285
2	1	2	0.2995	0.1784,0.5028	0.3145
3	2	1	0.8536	0.5787,1.259	0.9248
4	1	-1	242.7	227.0,259.3	243.8

Table 6: Parameter estimates for Subject 5 Kaplan et al. (1972)

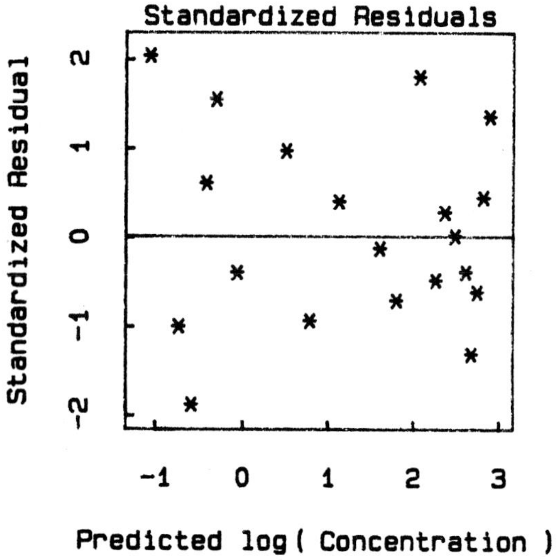

Figure 7d

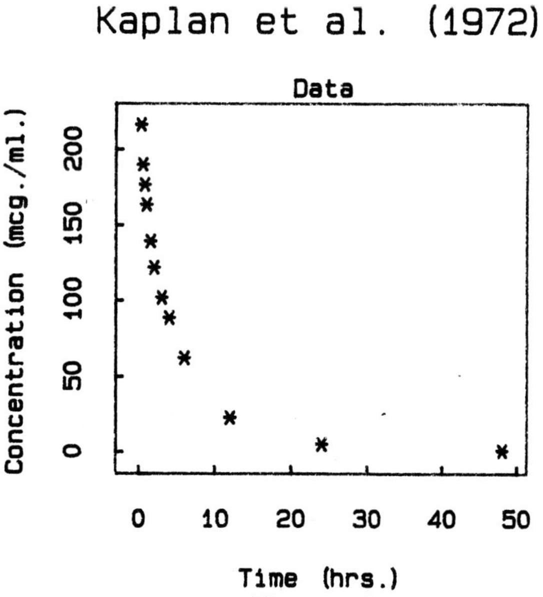

Figure 8a

Both these examples indicate the need for a three-compartment model. In practice, the use of two-compartment models is much more common such as the example from Kaplan et al. (1972) who studied the pharmacokinetic profile of sulfisoxazole in man after a bolus 2 g. intravenous injection. The data from Table 5 were fit to a two-compartment model with the results shown in Table 6.

The loglikelihood curve, plotted in Figure 8, achieves a maximum at about 0.7 with approximate 95% confidence limits of 0.35 to 0.95 so we chose a convenient λ of 0.5. The estimates obtained from the untransformed data fit fall within the confidence limits obtained using the optimal λ. The transformation is quite linear over most of the range of concentrations, but as the severity of the transformation increases, i.e. as λ decreases, the last observation becomes more important in determining the fit. Again, the residual analysis in Figure 9 does not reveal suspicious patterns.

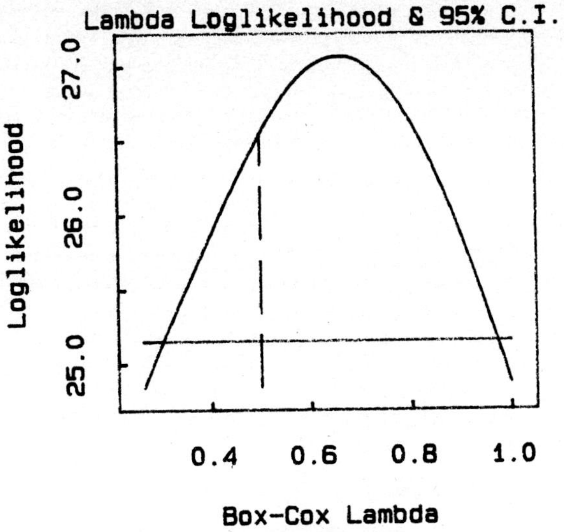

Figure 8b

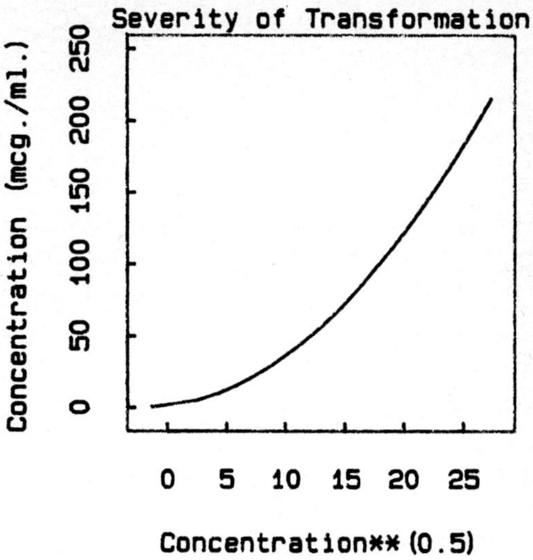

Figure 9b

Kaplan et al. (1972)

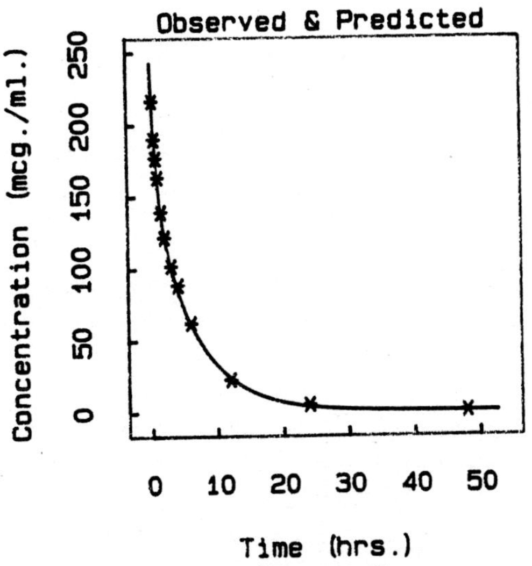

Figure 9a

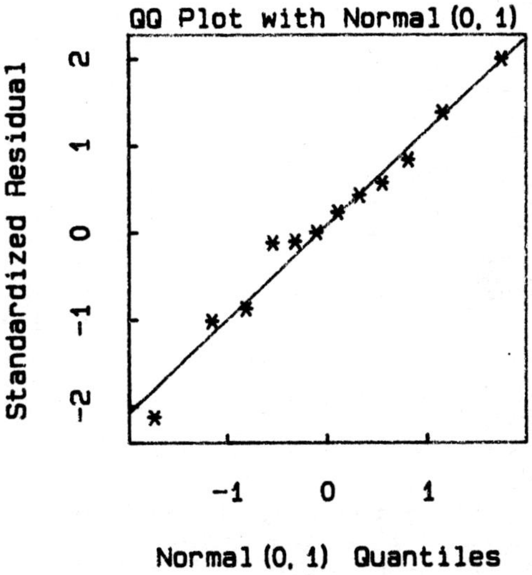

Figure 9c

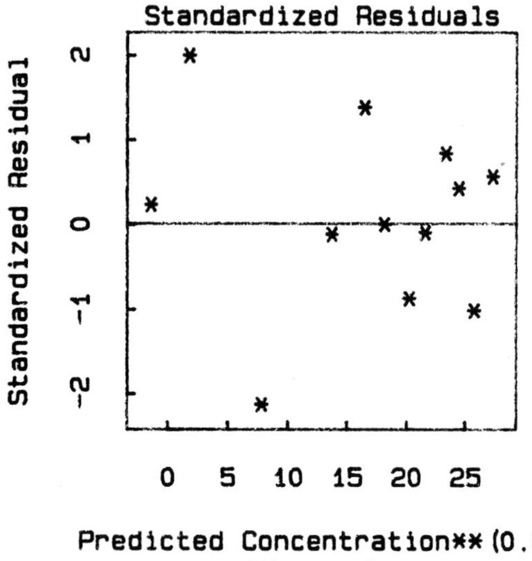

Figure 9d

6. Discussion

The calculation of γ_p in section 2 can easily be generalized. For example, second derivatives represent solutions to systems of the form

$$\dot{\gamma}_{q,p} = A\gamma_{q,p} + A_q\gamma_p + A_{q,p}\gamma + A_p\gamma_q + \iota_{q,p} \quad (6.1)$$

which, again, have a solution through convolution as

$$\gamma_{q,p} = e^{At} * [A_q\gamma_p + A_p\gamma_q] \quad (6.2)$$

since, in our representation,

$$A_{q,p} = 0$$

and

$$\iota_{q,p} = 0$$

As mentioned in section 2, the method of Bavely and Stewart (1979) allows generalization to systems where there are degenerate eigenspaces so U does not exist. These methods can also extend to the case of complex eigenvalues which, though rare, can occur in practice.

In some chemical modelling situations, the rate constants may be given as functions of other experimental settings such as temperature and pressure. The Arrhenius model is often used for this. The chain rule can be used to obtain the derivatives with respect to the Arrhenius parameters given the derivatives for the rate constants. The important area of modelling pharmacokinetic parameters, such as elimination rate constants, for entire populations is addressed by NONMEM (Beal and Sheiner, 1984). Many of the pharmacokinetic parameters of interest are functions of the rate constants and driving functions so the derivatives with respect to these parameters can be obtained through the results of sections 2 and 3.

Another situation that occurs in chemical modelling is the availability of measurements on more than one response. The derivatives of the model functions from section 2 can be used in the generalized Gauss-Newton algorithm (Bates and Watts, 1984, Bates and Watts, 1985a) to minimize the Box-Draper estimation criterion (Box and Draper, 1965) which takes in account correlations between responses. Applications of multi-response estimation for systems of linear differential equations are given in Bates and Watts (1985b).

The approach of differentiating the differential equations to obtain the "sensitivity functions" or derivatives with respect to model parameters has been used by Caracotsios and Stewart (1985) in more general reactor modelling. Their methods apply to mixed systems of differential and algebraic systems as well as to certain types of partial differential equations.

Using transformations to deal with heteroscedasticity, as described in section 4, is a powerful technique but it can result in using too many parameters. Many of the data sets for which compartment models are used consist of a dozen or fewer observations. Even adding one parameter to account for heteroscedasticity could result in "over-fitting" the data. It also opens the possibility of masking deterministic inadequacies of the model, using a 2-compartment model where a 3-compartment model is appropriate say, by changing the stochastic part, that is altering λ. The sensitivity of the deterministic model to the transformation for homoscedasticity is considered in Wolf (1985).

Acknowledgements

This research has been supported by the National Sciences Foundation under grant number DMS-8404970 and by the Natural Sciences and Engineering Research Council of Canada and aided by access to the Statistics Research Computer at the University of Wisconsin. We had helpful discussions with Professor Ray Carroll and Dr. Lewis Sheiner about this material.

References

Bates, D. M., and Watts, D. G. (1984), "A Multi-Response Gauss-Newton Algorithm," *Communications in Statistics Simulation and Computation*, 13, 705-715.

Bates, Douglas M., and Watts, Donald G. (1985a), "A General ized Gauss-Newton Procedure for Multi-Response Paramete Estimation," *SIAM Journal of Scientific and Statistical Compu ing*, 6, 000-000.

Bates, Douglas M., and Watts, Donald G. (1985b), "Mul tiresponse Estimation with Special Application to Systems c Linear Differential Equations," *Technometrics*, 27, 000-000 (with discussion)

Bavely, Connice A., and Stewart, G. W. (1979), "An Algorith for Computing Reducing Subspaces by Block Diagonaliz tion," *SIAM Journal of Numerical Analysis*, 16, 359-367.

Beal, Stuart, and Sheiner, Lewis (1984), *NONMEM Users' Guid* San Francisco: U. of California.

Box, George E. P., and Cox, David R. (1964), "An Analysis Transformations," *Journal of the Royal Statistical Society, Sc B*, 26, 211-252.

Box, G. E. P., and Draper, N. R. (1965), "The Bayesian Estim tion of Common Parameters from Several Responses *Biometrika*, 52, 355-365.

Box, G. E. P., and Cox, D. R. (1982), "An Analysis of Transf mations Revisited, Rebutted," *Journal of the American Stati cal Association*, 77, 209-210.

Caracotsios, M., and Stewart, W. E. (1985) "Software Sensitivity Analysis of Mathematical Models." Chemi Engineering Department, U. of Wisconsin - Madison.

Carroll, R. J., and Ruppert, D. (1984), "Power Transformations When Fitting Theoretical Models to Data," *Journal of the American Statistical Association*, 79, 321-328.

Hinkley, D. V., and Runger, G. (1984), "The Analysis of Transformed Data," *Journal of the American Statistical Association*, 79, 302-309. (Discussion 309-320)

Jennrich, R. I., and Bright, P. B. (1976), "Fitting Systems of Linear Differential Equations Using Computer Generated Exact Derivatives," *Technometrics*, 18, 385-392. (Discussion 393-399)

Kaplan, Stanley A., Weinfeld, Robert E., Abruzzo, Charles W., and Lewis, Margaret (1972), "Pharmacokinetic Profile of Sulfisoxazole Following Intravenous, Intramuscular, and Oral Administration to Man," *Journal of Pharmaceutical Sciences*, 61, 773-778.

Kramer, William G., Lewis, Richard P., Cobb, Tyson C., Forester, Jr., Wilbur F., Visconti, James A., Wanke, Lee A., Boxenbaum, Harold G., and Reuning, Richard H. (1974), "Pharmacokinetics of Digoxin: Comparison of a Two- and Three-Compartment Model in Man," *Journal of Pharmacokinetics and Biopharmaceutics*, 2, 299-312.

Moler, Cleve, and Van Loan, Charles (1978), "Nineteen Dubious Ways to Compute the Exponential of a Matrix," *SIAM Review*, 20, 801-836.

Pritchard, D. J., Downie, J., and Bacon, D. W. (1977), "Further Considerations of Heteroscedasticity in Fitting Kinetic Models," *Technometrics*, 19, 227-236.

Ralston, M. L., and Jennrich, R. I. (1978), "DUD, a Derivative-Free Algorithm for Nonlinear Least Squares," *Technometrics*, 20, 7-14.

Smith, B. T., Boyle, J. M., Dongarra, J. J., Garbow, B. S., Ikebe, Y., Klema, V. C., and Moler, C. B. (1976), *Matrix Eigensystem Routines - EISPACK Guide*, Springer Verlag.

Wagner, John G., and coworkers (1977), "Pharmacokinetic Parameters Estimated from Intravenous Data by Uniform Methods and Some of Their Uses.," *Journal of Pharmacokinetics and Biopharmaceutics*, 5, 161-182.

Wolf, Dennis A. (1985), *Nonlinear Least Squares for Linear Compartment Models*, Ph.D. thesis, Univ. of Wisconsin-Madison.

COMPUTER SCIENCE AND STATISTICS:
The Interface, D.M. Allen (ed.)
Elsevier Science Publishers B.V. (North-Holland), 1986

COMPUTATIONAL EXPERIENCE WITH CONFIDENCE REGIONS AND CONFIDENCE INTERVALS FOR NONLINEAR LEAST SQUARES[1]

Janet R. Donaldson and Robert B. Schnabel[2]

Center for Applied Mathematics, National Bureau of Standards, Boulder, Colorado 80303
and
Department of Computer Science, University of Colorado, Boulder, Colorado 80309

We present the results of a Monte Carlo study of several methods for constructing confidence regions and confidence intervals about parameters estimated by nonlinear least squares. We compare the estimates produced by the most commonly discussed methods, namely the lack-of-fit method, the likelihood method, and three variants of the linearization method. The linearization method is computationally inexpensive and produces easily understandable results, while the likelihood and lack-of-fit methods both are much more expensive and more difficult to report. In our tests, both the lack-of-fit and likelihood procedures perform very reliably, but all three linearization methods often produce gross underestimates of confidence regions and sometimes produce significant underestimates of confidence intervals. Among the three variants of the linearization method, the variant based solely on the Jacobian appears preferable to the two variants that utilize the full Hessian, because it is cheaper to compute, and is always as reliable as the other two variants and sometimes more reliable. Cases when the linearization method confidence regions will be poor appear to be reliably predicted by the Bates and Watts parameter effects curvature diagnostic.

1. Introduction

This paper presents the results of an empirical study comparing several methods for constructing confidence regions and confidence intervals about parameters estimated by nonlinear least squares. The methods compared are the lack-of-fit method, the likelihood method, and three variants of the linearization method.

The need for confidence regions and intervals commonly arises in data fitting applications, where a response variable y_i observed with unknown error $\dot{e}_i$ is fit to m fixed predictor variables $\mathbf{x}_i$ using a function $f(\mathbf{x}_i;\boldsymbol{\theta})$ which can be either linear or nonlinear in the p parameters $\boldsymbol{\theta}$. The function $f(\mathbf{x}_i;\boldsymbol{\theta})$ is linear in $\boldsymbol{\theta}$ if it can be written

$$f(\mathbf{x}_i;\boldsymbol{\theta}) = \mathbf{x}_i\boldsymbol{\theta} = \sum_{j=1}^{p} x_{i,j}\theta_j,\ i=1,\ldots,n.$$

Otherwise, it is nonlinear. The methods analyzed in this study are identical when $f(\mathbf{x}_i;\boldsymbol{\theta})$ is linear in $\boldsymbol{\theta}$; otherwise they are not.

When the error $\dot{e}_i$ is additive, the response variable can be modeled by

$$y_i = f(\mathbf{x}_i;\dot{\boldsymbol{\theta}}) + \dot{e}_i,\ i=1,\ldots,n,$$

where $\dot{\boldsymbol{\theta}}$ denotes the true but unknown value of the parameters. The least squares estimator of $\dot{\boldsymbol{\theta}}$ is the parameter value, denoted $\hat{\boldsymbol{\theta}}$, which minimizes the sum of the squares of the residuals, where the residuals, $r_i(\boldsymbol{\theta})$, are estimates of the random error, $\dot{e}_i$,

$$r_i(\boldsymbol{\theta}) = y_i - f(\mathbf{x}_i;\boldsymbol{\theta}).$$

Thus,

$$\hat{\boldsymbol{\theta}} = \arg\min S(\boldsymbol{\theta})$$

where $S(\boldsymbol{\theta})$ is the residual sum of squares,

$$S(\boldsymbol{\theta}) = \sum_{i=1}^{n} r_i(\boldsymbol{\theta})^2 = \mathbf{R}(\boldsymbol{\theta})^T\mathbf{R}(\boldsymbol{\theta})$$

with $\mathbf{R}(\boldsymbol{\theta})$ denoting a column vector with i^{th} component $r_i(\boldsymbol{\theta})$, and $\mathbf{R}(\boldsymbol{\theta})^T$ denoting the transpose of $\mathbf{R}(\boldsymbol{\theta})$.

In our study, we assume that the model is correct and that the errors are normal, independent, identically distributed random variables with zero mean and variance $\dot{\sigma}^2$, i.e., distributed as $N(\mathbf{0},\dot{\sigma}^2\,\mathbf{I})$. Then, the least squares estimator $\hat{\boldsymbol{\theta}}$ is the maximum likelihood estimator of the parameters $\dot{\boldsymbol{\theta}}$ of the p-variate normal density function,

$$L(\mathbf{Y}) = (2\pi\dot{\sigma}^2)^{-n/2}\ \mathbf{e}^{(-\dot{\mathbf{e}}^T\dot{\mathbf{e}}/2\dot{\sigma}^2)}$$

where $\mathbf{Y}$ is a column vector with i^{th} component y_i, and $\dot{\mathbf{e}}$ is a column vector with i^{th} component $\dot{e}_i$.

Nearly normally distributed errors are, in fact, encountered quite frequently in practice. This is because measurement errors are often the sum of a number of random errors from unknown sources, and, by the central limit theorem, the sum of these errors is approximately normally distributed whatever the distribution of the individual errors that make up the sum.

In practice, the estimated values of the parameters $\hat{\boldsymbol{\theta}}$ will not equal the true values $\dot{\boldsymbol{\theta}}$ because of the random errors, $\dot{e}_i$, in the data. Since $\hat{\boldsymbol{\theta}}$ is a random variable, however, it may be possible to indicate with some specific

[1]Contribution of the National Bureau of Standards and not subject to copyright in the United States.

[2]This research supported by ARO contract DAAG 29-84-K-0140

probability $1-\alpha$ in what region about $\hat{\boldsymbol{\theta}}$ we might reasonably expect $\dot{\boldsymbol{\theta}}$ to be. Such regions are known as $100\cdot(1-\alpha)\%$ confidence regions. A joint confidence region about all of the parameters is defined using a function

$$CR_\alpha : \mathbf{Y} \rightarrow \text{a region in } R^p$$

which satisfies

$$Pr[\ \dot{\boldsymbol{\theta}} \in CR_\alpha(\mathbf{Y})\ ] = 1-\alpha.$$

Similarly, a confidence interval about an individual parameter $\hat{\theta}_j$ is defined using a function

$$CI_{j,\alpha} : \mathbf{Y} \rightarrow \text{an interval in } R$$

which satisfies

$$Pr[\ \dot{\theta}_j \in CI_{j,\alpha}(\mathbf{Y})\ ] = 1-\alpha.$$

The above definitions state that, before the data are sampled, the probability that the confidence regions and confidence intervals to be constructed will contain the true parameter values is $1-\alpha$. Thus, if we repeatedly draw samples and construct confidence regions and intervals about the least squares estimates for each sample, then in the long run $100\cdot(1-\alpha)\%$ of these confidence regions and intervals should contain the true values. Procedures that, for all functions $f(\mathbf{x}_i;\boldsymbol{\theta})$ and confidence levels $1-\alpha$, are statistically guaranteed asymptotically to contain the true value $100\cdot(1-\alpha)\%$ of the time are called exact; all other procedures are called approximate.

Various methods have been proposed for calculating confidence regions and intervals for parameter estimation by nonlinear least squares. These include several variants of the linearization method, as well as the likelihood and lack-of-fit methods. [See e.g. Bard (1974), Gallant (1976), Draper and Smith (1981).] We review all these methods briefly in Section 2. They all are equivalent, and exact, for linear models. For nonlinear models, only the lack-of-fit method for computing confidence regions is exact; the other methods for computing confidence regions and all the methods for computing confidence intervals are approximate. The linearization regions and intervals appear to be the most approximate for nonlinear models, but they also are far less expensive to compute than the likelihood or lack-of-fit regions and intervals, and are the predominant methods implemented in production software. Some nonlinear least squares packages, including NL2SOL [Dennis, Gay, and Welsch (1981)], include three variants of the linearization method, which differ only in that the variance-covariance matrix of the estimated parameters is approximated in three different ways, namely

$$\hat{\mathbf{V}}_a = s^2\left(\mathbf{J}(\hat{\boldsymbol{\theta}})^T\mathbf{J}(\hat{\boldsymbol{\theta}})\right)^{-1},$$

$$\hat{\mathbf{V}}_b = s^2\,\mathbf{H}(\hat{\boldsymbol{\theta}})^{-1},$$

or

$$\hat{\mathbf{V}}_c = s^2\,\mathbf{H}(\hat{\boldsymbol{\theta}})^{-1}\left(\mathbf{J}(\hat{\boldsymbol{\theta}})^T\mathbf{J}(\hat{\boldsymbol{\theta}})\right)\mathbf{H}(\hat{\boldsymbol{\theta}})^{-1},$$

where $s^2 = S(\hat{\boldsymbol{\theta}})/(n-p)$ is the estimated residual variance; $\mathbf{J}(\hat{\boldsymbol{\theta}})$ is the Jacobian of $f(\mathbf{x}_i;\boldsymbol{\theta})$, $i=1,\ldots,n$, at $\hat{\boldsymbol{\theta}}$; and $\mathbf{H}(\hat{\boldsymbol{\theta}})$ is the Hessian of $S(\boldsymbol{\theta})$ at $\hat{\boldsymbol{\theta}}$.

Sections 3-6 of this paper describe and analyze a Monte Carlo study that compares all of these methods for computing confidence regions and intervals on 20 nonlinear models. The study is used to empirically observe how often the true parameter values are contained in the confidence regions and confidence intervals constructed using a given method. The actual percent of the nominally $100\cdot(1-\alpha)\%$ confidence regions and intervals which are found to contain the true value is known as the observed coverage. The observed coverage will generally depend on the method used to construct the confidence regions and confidence intervals, on the nominal confidence level, $1-\alpha$, on the degree of nonlinearity of the function, $f(\mathbf{x}_i;\boldsymbol{\theta})$, and to a small extent, on the number of replications in the simulation. If the experiment used to generate the data is repeated a large number of times under the same conditions, and if CR_α and $CI_{j,\alpha}$ are exact and the model is correct, then the observed coverage will approach the nominal coverage. When CR_α and $CI_{j,\alpha}$ are only approximate, the observed coverage will not necessarily approach the nominal coverage, although one would hope that the difference between the observed and nominal coverage for a reasonable approximate method would be small for most functions.

No similar study of this magnitude appears to have been reported previously. The properties of confidence regions and confidence intervals computed using the linearization, likelihood, and lack-of-fit methods have been analyzed by several authors, including Jennrich (1959), Beale (1960), Guttman and Meeter (1965), Gallant (1976), Duncan (1978), and Bates and Watts (1980). While the literature includes numerous warnings regarding the possible inaccuracy of the approximate methods, it contains little empirical data to illustrate the size of the discrepancies between observed and nominal coverage that might be expected. In those studies which do contain empirical data on confidence regions and intervals, the largest reported differences between the observed and nominal coverage is only 9% for a 95% confidence region computed using the linearization method, and is even smaller for the likelihood method [Gallant (1976)]. In many practical applications, potential differences of 9% might not be cause for concern. Evidence of much larger differences, however, would indicate the need for improved methods. Our results provide such evidence.

Our Monte Carlo study has several purposes. First we wish to determine whether the observed coverage of the linearization method is significantly affected by how the variance-covariance matrix is computed. Second, we wish to determine whether the approximate confidence regions and confidence intervals constructed using the linearization and likelihood methods, and the approximate confidence intervals constructed using the lack-of-fit method have observed coverage significantly different from nominal. In particular, we want to know whether the frequently used linearization method is significantly better or worse than the more expensive likelihood and lack-of-fit methods. Section 3 describes how we designed our study to answer these questions. The results are presented and discussed in Section 4. We have also investigated how effective the diagnostics of Bates and Watts (1980) are in predicting when the confidence regions produced by the linearization and likelihood methods should be reliable; this part of the study is the subject of Section 5.

Our study is oriented toward nonlinear least squares software developers who need assurance that the methods they implement are reasonable for a wide variety of problems. We make only the customary assumptions that th model is correct and that the errors are normally distributed. We do not assume that we can change the representation of the parameters, e.g., by reparameterizing θ as $\log(\theta)$, in order to reduce the difference between the observed and nominal coverage, because reparameterization is not a technique that can be routinely implemented by software developers who have no control over the functions analyzed. Readers interested in using reparameterization to improve their results are refered to Ratkowsky (1983).

The conclusions we draw from this study are presented in Section 6. The first conclusion is that among the variants of the linearization method, the one using $\hat{\mathbf{V}}_a$ is the best choice because it is the cheapest, and is always at least as reliable as the other two variants and sometimes more reliable. The second conclusion is that even the best linearization method can be very poor; confidence regions with observed coverage as low as 12.4% for a nominal 95% region, and confidence intervals with observed coverage as low as 75.0% for a nominal 95% interval are reported. In contrast, for each of the datasets tested, the confidence regions and confidence intervals constructed using the likelihood method and lack-of-fit methods are quite close to nominal. Finally, our study indicates that the diagnostics of Bates and Watts (1980) appear quite successful at predicting when linearization confidence regions will be poor. Our recommendations as to how nonlinear least squares software should calculate confidence regions and intervals, in light of these conclusions, also are given in Section 6.

2. Background

This section briefly discusses methods for constructing confidence regions and confidence intervals. First, we give a very quick survey of confidence regions and confidence intervals for linear least squares. Next, we describe the two different ways function nonlinearity can affect the solution locus. Then, we review the linearization, likelihood, and lack-of-fit methods for constructing confidence regions and confidence intervals when the model is nonlinear. For a more complete discussion, see Bard (1974), Gallant (1976), Draper and Smith (1981), or Donaldson (1985).

Linear least squares

When $f(\mathbf{x}_i;\boldsymbol{\theta})$ is linear in the parameters $\boldsymbol{\theta}$, then $f(\mathbf{x}_i;\boldsymbol{\theta}) = \mathbf{x}_i\,\boldsymbol{\theta}$. Consequently, the Jacobian of $\mathbf{F}(\boldsymbol{\theta})$ is $\mathbf{X}$, an n by p matrix with i^{th} row $\mathbf{x}_i$. If we assume that $\mathbf{X}$ is of full rank, then $\mathbf{X}^T\mathbf{X}$ is nonsingular, and the linear least squares estimators can be expressed in closed form by

$$\hat{\boldsymbol{\theta}} = (\mathbf{X}^T\mathbf{X})^{-1}\,\mathbf{X}^T\mathbf{Y}\ .$$

When $\dot{\epsilon} \sim N(\mathbf{0},\dot{\sigma}^2\,\mathbf{I})$, a $100{\cdot}(1-\alpha)\%$ confidence region about $\hat{\boldsymbol{\theta}}$ contains those values $\tilde{\boldsymbol{\theta}}$ for which

$$S(\tilde{\boldsymbol{\theta}}) - S(\hat{\boldsymbol{\theta}}) \le s^2\,p\,F_{p,n-p,1-\alpha}\,. \tag{2.1}$$

Equation (2.1) is equivalent to

$$(\boldsymbol{\theta}-\hat{\boldsymbol{\theta}})^T\,\mathbf{X}^T\mathbf{X}\,(\boldsymbol{\theta}-\hat{\boldsymbol{\theta}}) \le s^2\,p\,F_{p,n-p,1-\alpha} \tag{2.2}$$

for all linear models, which shows that the shape of the confidence regions about $\hat{\boldsymbol{\theta}}$ is ellipsoidal for all linear models.

A $100{\cdot}(1-\alpha)\%$ confidence interval about $\hat{\theta}_j$ contains those values $\tilde{\theta}_j$ for which

$$|\tilde{\theta}_j - \hat{\theta}_j| \le s\,\sqrt{(\mathbf{X}^T\mathbf{X})_{jj}^{-1}}\;t_{n-p,1-\alpha/2} \tag{2.3}$$

where $(\mathbf{X}^T\mathbf{X})_{jj}^{-1}$ is the $(j,j)^{th}$ element of the inverse of $\mathbf{X}^T\mathbf{X}$. The limits of this confidence interval can be shown to be those values θ_j which

$$\text{maximize } (\theta_j - \hat{\theta}_j)^2 \text{ subject to} \tag{2.4}$$

$$S(\boldsymbol{\theta}) - S(\hat{\boldsymbol{\theta}}) = s^2\left(t_{n-p,1-\alpha/2}\right)^2 = s^2\,F_{1,n-p,1-\alpha}\ .$$

Nonlinearity and the Solution Locus

The solution locus, or estimation space, of $f(\mathbf{x}_i;\boldsymbol{\theta})$, $i=1,\ldots,n$, consists of all points with coordinates expressible as

$$\left(f(\mathbf{x}_1;\boldsymbol{\theta}), f(\mathbf{x}_2;\boldsymbol{\theta}), \ldots, f(\mathbf{x}_n;\boldsymbol{\theta})\right)$$

where the $\mathbf{x}_i$, $i=1,\ldots,n$, are the fixed values of the predictor variables, and $\boldsymbol{\theta}$ is allowed to vary over all possible values of the p unknown parameters. The solution locus is planar if there exists a reparameterization of $f(\mathbf{x}_i;\boldsymbol{\theta})$ that makes the function linear in the p parameters. Otherwise, the solution locus is curved.

A coordinate grid on the solution locus can be formed by tracing the paths obtained when each parameter is individually allowed to vary while all other parameters are held fixed. The coordinate grid is curvilinear whenever the function $f(\mathbf{x}_i;\boldsymbol{\theta})$ is nonlinear in one or more of its parameters. It is linear only when the function itself is linear.

Curvature of the solution locus is called "intrinsic" curvature [Beale (1960); Bates and Watts (1980)]. Curvature of the coordinate grid is called "parameter-effects" or simply "parameter" curvature [Bates and Watts (1980)]. Intrinsic curvature is not affected by reparameterization. Parameter-effects curvature is. Linear functions have zero parameter-effects curvature and zero intrinsic curvature. Nonlinear functions always have nonzero parameter-effects curvature, and can have either zero or nonzero intrinsic curvature, i.e., a planar or curved solution locus, respectively.

Nonlinear Least Squares

When the function is nonlinear, the least squares estimators of the parameters cannot in general be expressed in closed form, and must instead be computed by iterative techniques. Construction of exact confidence regions and confidence intervals also is much more difficult, and so approximate methods are frequently used. The leading methods, linearization, likelihood, and lack-of-fit, are described briefly below.

Linearization methods. Linearization methods for constructing confidence regions and confidence intervals assume that the nonlinear function can be adequately approximated by an affine, or linear, approximation to the function at the solution. That is, this method assumes that the solution locus is planar, and that the coordinate grid is linear throughout the area to be covered by the confidence regions and confidence intervals. Under this assumption, linear least squares theory tells us that the confidence region about $\hat{\boldsymbol{\theta}}$ consists of those values $\tilde{\boldsymbol{\theta}}$ for which

$$(\tilde{\boldsymbol{\theta}}-\hat{\boldsymbol{\theta}})^T\hat{\mathbf{V}}^{-1}(\tilde{\boldsymbol{\theta}}-\hat{\boldsymbol{\theta}}) \le p\,F_{p,n-p,1-\alpha}$$

while a confidence interval about $\hat{\theta}_j$, $j=1,\ldots,p$, consists of those values $\tilde{\theta}_j$ for which

$$|\tilde{\theta}_j - \hat{\theta}_j| \le \hat{V}_{jj}^{1/2}\, t_{n-p,1-\alpha/2},$$

where $\hat{V}$ is the estimated variance-covariance matrix of the parameters, and $\hat{V}_{jj}$ is the $(j,j)^{th}$ element of $\hat{V}$.

Three approximations to $\hat{V}$ are frequently used. These are

$$\hat{V}_a = s^2\left(J(\hat{\theta})^T J(\hat{\theta})\right)^{-1}, \qquad (A)$$

$$\hat{V}_b = s^2\, H(\hat{\theta})^{-1}, \qquad (B)$$

and

$$\hat{V}_c = s^2\, H(\hat{\theta})^{-1}\left(J(\hat{\theta})^T J(\hat{\theta})\right) H(\hat{\theta})^{-1}, \qquad (C)$$

where $J(\theta)$ is the Jacobian of $F(\theta)$ at $\hat{\theta}$; $H(\hat{\theta})$ is the Hessian of $S(\theta)$ at $\hat{\theta}$; and s^2 is the residual variance, $s^2 = S(\hat{\theta})/n-p$. Approximation (A) is the most common approximation to $\hat{V}$, and is the direct analog from linear least squares theory. Approximation (B) can be obtained using maximum likelihood theory, and can be viewed as using observed rather than expected information in forming the variance-covariance matrix. Approximation (C) is obtained by using a quadratic model of $S(\theta)$. [For a more detailed discussion of these variants, see Bard (1974) or Donaldson (1985).] When certain regularity conditions are met [Jennrich (1959)], these approximations to $\hat{V}$ asymptotically will approach the true variance-covariance matrix of the model. Note also that these approximation differ only when

$$\sum_{i=1}^{n} r_i(\theta)\, \frac{\partial^2 f(x_i;\theta)}{\partial\theta_j \partial\theta_k}$$

is nonzero. In particular, for linear functions, each of these representations of $\hat{V}$ is equal to

$$s^2\left(J(\hat{\theta})^T J(\hat{\theta})\right)^{-1} = s^2\,(X^T X)^{-1} .$$

Linearization methods have the advantage that their resulting confidence regions and intervals are simple and inexpensive to construct, and that they produce bounded, convex confidence regions. In addition, the information needed to construct confidence regions and intervals using this method can be parsimoniously summarized by the p by p matrix $\hat{V}$, and is well understood by users familiar with linear least squares. Because the linearization methods assume that both the intrinsic curvature and the parameter-effects curvature of $f(x_i;\theta)$ are zero, however, we expect that the linearization methods could sometimes produce observed coverages very far from the expected nominal coverage. The results of our Monte Carlo study show this to be true.

Likelihood method. The likelihood method is another approximate method for producing confidence regions and confidence intervals. The likelihood method confidence region about $\hat{\theta}$ consists of those values $\tilde{\theta}$ for which

$$S(\tilde{\theta}) - S(\hat{\theta}) \le s^2\, p\, F_{p,n-p,1-\alpha} .$$

This is analogous to equation (2.1) for confidence regions for the parameters of a linear function, although when $f(x_i;\theta)$ is nonlinear in the parameters the resulting confidence region is no longer ellipsoidal. The likelihood method confidence interval about $\hat{\theta}_j$ is the interval bounded by the points which

$$\text{maximize } (\theta_j - \hat{\theta}_j)^2 \text{ subject to}$$

$$S(\theta) - S(\hat{\theta}) \le s^2\, F_{1,n-p,1-\alpha} .$$

This confidence interval is the projection onto the appropriate parameter axis of the above region, and is analogous to equation (2.4) for confidence intervals in the case of linear least squares.

When the solution locus is planar, the confidence regions (but not the confidence intervals) constructed using the likelihood method are exact. In addition, likelihood method confidence regions and intervals have the desirable property that they are constructed from contours of constant likelihood, and that the regions and intervals are not affected by reparameterization of the function $f(x_i;\theta)$. Thus we might expect the likelihood method to produce confidence regions and confidence intervals with observed coverage closer to nominal than those produced using the linearization methods. However, the likelihood method has several practical disadvantages. Both the confidence regions and confidence intervals produced using the likelihood method can be disjoint and unbounded because the contours of a nonlinear function can be disjoint and unbounded. The method also is very expensive to use, and, when the data arrays are large, it can be awkward to publish the information necessary to reconstruct the confidence region because this information is not succinctly summarized as it is in the case of the linearization method.

Lack-of-fit method. The lack-of-fit method can be used to produce exact joint confidence regions for all p of the parameters, and to produce approximate confidence intervals and confidence regions for subsets of the parameters. An exact $100\cdot(1-\alpha)\%$ confidence region consists of all values $\tilde{\theta}$ such that

$$\frac{R(\tilde{\theta})^T P(\tilde{\theta}) R(\tilde{\theta})}{R(\tilde{\theta})^T \left(I - P(\tilde{\theta})\right) R(\tilde{\theta})} \le \frac{p}{n-p}\, F_{p,n-p,1-\alpha}$$

where

$$P = J(\tilde{\theta})\left(J(\tilde{\theta})^T J(\tilde{\theta})^T\right)^{-1} J(\tilde{\theta})^T .$$

Note that the lack-of-fit method does not require that the least squares solution be found prior to constructing the confidence region. Similarly, a confidence interval for the jth parameter consists of those values $\tilde{\theta}_j$ for which there exists values of θ_k, $k=1,\ldots,j-1,j+1,\ldots,p$, such that for these p parameter values, $\tilde{\theta}$,

$$\frac{S^L(\hat{\theta}_{J_{k\ne j}(\tilde{\theta})}) - S^L(\hat{\theta}_{J(\tilde{\theta})})}{S^L(\hat{\theta}_{J(\tilde{\theta})})/(n-p)} \le F_{1,n-p,1-\alpha}$$

where $S^L(\hat{\theta}_{J_{k\ne j}(\tilde{\theta})})$ is the residual sum of squares obtained when $R(\tilde{\theta})$ is *linearly* fit to all the columns of $J(\tilde{\theta})$ excluding the jth, and $S^L(\hat{\theta}_{J(\tilde{\theta})})$ is the residual sum of squares obtained when $R(\tilde{\theta})$ is *linearly* fit to $J(\tilde{\theta})$. This interval is exact if $f(x_i;\theta)$ is linear in θ_k, $k=1,\ldots,j-1,j+1,\ldots,p$; otherwise it is approximate.

The lack-of-fit method is even more expensive to use than the likelihood method, and, as is the case for the likelihood method, the information needed to construct the confidence regions cannot be succinctly summarize for publication. Also, the confidence regions and confidence intervals constructed using the lack-of-fit method are guaranteed to contain every minimum, maximum, and/or saddle point of the likelihood surface. This makes the lack-of-fit method structurally undesirable.

3. The Monte Carlo Study

This section briefly describes how our Monte Carlo study was constructed. Full details are provided by Donaldson (1985).

The Monte Carlo method uses the computer to simulate the results of repeating an experiment many times in order to obtain a large sample from which the statistical properties of a system can be examined. For each simulation, we first generated the errors and response variables. The errors, $\dot{e}$, were produced using the Marsaglia and Tsang pseudo-normal random number algorithm (1984) as implemented by James Blue and David Kahanar of the National Bureau of Standards Scientific Computing Division. The response variable, $\mathbf{Y}$, was then constructed so that its ith component is

$$y_i = f(\mathbf{x}_i;\dot{\boldsymbol{\theta}}) + \dot{e}_i \ .$$

Then the least squares estimate, $\hat{\boldsymbol{\theta}}$, was calculated using NL2SOL, an unconstrained quasi-Newton code for nonlinear least squares [Dennis, Gay, and Welsch (1981)]. Starting values for NL2SOL were set to the true values of the parameters, $\dot{\boldsymbol{\theta}}$, and the stopping criteria for the convergence tests based on the relative change in the parameters and in the sum of squares both were set to 10^{-5}.

Finally, for each confidence region or interval method and each derivative configuration being analyzed, we recorded whether the true values of the parameters were contained within the confidence regions and confidence intervals for this realization of the data. Determining whether the true parameter values lay within the confidence regions and confidence intervals about the least squares estimates fortunately did not require that we construct the full confidence regions and confidence intervals for each confidence level and method. Instead, we simply calculated the smallest confidence level, $1-\omega$, such that a $100\cdot(1-\omega)\%$ confidence region or confidence interval constructed using the method being analyzed will contain the true parameter values. When $\omega > \alpha$, the true value did not lie in the $100\cdot(1-\alpha)\%$ confidence region or confidence interval; when $\omega \leq \alpha$, it did. The values $1-\omega$ were obtained using the hypothesis tests corresponding to the formulas for confidence regions and intervals given in Section 2, and the appropriate cumulative distribution functions; the procedures are described in detail in Donaldson (1985). The cumulative distribution functions were obtained from the STARPAC subprogram library [Donaldson and Tryon (1983)].

The observed coverage, γ_α, for the particular nominal confidence level, method and system under analysis is the percentage of the total number of realizations of the data, N, for which $\omega \leq \alpha$. When N is large, the standard deviation of γ_α can be approximated using the normal approximation to the binomial distribution. In this study we used $N=500$, so the maximum standard deviation of the observed coverage at any coverage level is approximately 2.2%.

Note that substituting a new realization of the data for one which could not be completely analyzed because either (a) the nonlinear least squares algorithm did not converge, or (b) the test statistics could not be computed for every method being analyzed, is a form of censoring which will bias the observed coverages obtained. In our analysis, we adjusted the value of $\dot{\sigma}$ for each dataset so that every realization could be completely analyzed, and therefore the results reported in this paper are not derived from censored data.

We computed the observed coverage for four nominal confidence levels, 0.50, 0.75, 0.95, and 0.99. In this paper we only include our data for the level 0.95, although we comment briefly in Section 4 on our results at the other levels. Data for the full study are given in Donaldson (1985).

The references for the datasets used in our Monte Carlo study are given in Appendix A and described in detail in Donaldson (1985). With only two exceptions, the functions and data which comprise our datasets have been taken from Ratkowsky (1983), Himmelblau (1970), Guttman and Meeter (1965), and Duncan (1978). The standard deviation of the errors of some of the datasets has been adjusted in order to allow us to successfully analyze each realization of the data for each dataset. The two datasets not taken from the published literature are identified as 8ACA and 9AAG. Dataset 8ACA was created especially for this study by generalizing function 3 to a larger number of parameters. Dataset 9AAG involves a microwave absorption line function taken from a consulting session at the National Bureau of Standards in Boulder, Colorado.

The number of parameters in the 20 datasets analyzed range from 2 to 8 and the ratio of the number of parameters to the number of observations range from 2/42 to 3/5. While these datasets are often troublesome, they are mostly real world problems that have not been made artificially difficult.

Each dataset was analyzed twice to allow us to examine the effect of increasing the standard deviation of the errors. In the first analysis, $\dot{e} \sim N(\mathbf{0},\dot{\sigma}^2\,\mathbf{I})$; in the second analysis, $\dot{e} \sim N\big(\mathbf{0},(\eta\,\dot{\sigma})^2\,\mathbf{I}\big)$, where η is approximately the largest number $\leq$ 10 for which every realization of the data could be successfully analyzed. The methods analyzed in the second analysis were the same as in the first except that variants B and C of the linearization method were excluded from the second analysis because, when $\eta > 1.0$, we were frequently unable to compute the required test statistics using these two variants.

Computation of the linearization method and the lack-of-fit method requires that certain derivatives be available. The Jacobian of $\mathbf{F}(\boldsymbol{\theta})$ is used by both the linearization and lack-of-fit methods. Variants B and C of the linearization method use the Hessian of $S(\boldsymbol{\theta})$ as well. In practice, analytic derivatives often are not available. Therefore, in our study each method was implemented and analyzed using three different derivative configurations. These configurations are (1) the Jacobian and Hessian both approximated by finite-differences, (2) the Jacobian computed analytically and the Hessian computed by finite-differences, and (3) both the Jacobian and the Hessian computed analytically. For derivative configurations (1) and (2), the variance-covariance matrix needed by the linearization method was returned directly from NL2SOL. For configuration (3), it was constructed outside of NL2SOL. For details on the formulas used to compute the finite-difference derivative approximations, see Donaldson (1985).

We ran our Monte Carlo study in single precision on a 60 bit word length computer. All subroutines extracted from other sources were used without modification except for NL2SOL, which was changed for this study in two important ways. First we disabled the two tests within NL2SOL used to detect near singularity. Second, we used the STARPAC front end to NL2SOL. With this front end, the finite difference approximation to the Jacobian is computed with the optimal derivative step sizes selected using the algorithm developed by Schnabel (1981), thus maximizing the number of correct digits in each element of the finite difference Jacobian.

4. Results and Observations

This section presents the results of our Monte Carlo study of the lack-of-fit method, the likelihood method, and the three variants of the linearization method. The section is divided into a discussion of confidence regions and confidence intervals. For each, we also make a number of observations about the results. The conclusions we draw from our analysis are discussed in the next chapter.

The material in this chapter includes a number of figures. These are printed at the end of the paper.

Confidence Regions

Results. The results for nominally 95% confidence regions constructed using each of the methods analyzed in this study with $\dot{e} \sim N(\mathbf{0}, \dot{\sigma}^2 \mathbf{I})$ are graphically displayed in Figure 1. For each dataset, the observed coverage is plotted against the method and derivative configuration used to obtain it.

The three derivative configurations are labeled DC1, DC2, and DC3 in these and the following figures and tables, as well as in Appendix B, Here DC1 denotes use of finite difference approximations for both the Jacobian and the Hessian, DC2 denotes use of analytic Jacobian and finite difference Hessian, and DC3 denotes use of analytic Jacobian and Hessian. Since the computations required to calculate the lack-of-fit method results and the likelihood method results using derivative configurations DC2 and DC3 are exactly the same, these results are displayed together.

Figure 2 shows the analogous results for $\dot{e} \sim N(\mathbf{0}, (\eta\, \dot{\sigma})^2 \mathbf{I})$. As noted in Section 3, variants B and C of the linearization method are excluded from the analysis displayed in Figure 2 because computational difficulties were encountered for these variants when the variance of the errors was increased.

A conservative 95% confidence interval about the nominal confidence level is indicated on each plot by a pair of horizontal lines which represent the values $100 \cdot (1-\alpha) \pm 4.4$, where 4.4 is two times the maximum observed coverage at any coverage level. This confidence interval provides a quick means of determining whether any of the observed coverages for each method are significantly different from the nominal confidence level at the 5% level. When the method used to construct the confidence regions and confidence intervals is exact, we expect that the observed coverage for 95% of all possible datasets will lie within this confidence interval.

Observations. Figures 1 and 2 show that the lack-of-fit and likelihood method confidence regions are quite reliable, and that the results are not affected by use of finite difference derivatives. In all our tests, they produced observed coverages which seldom vary from nominal by an amount that is significant at the 5% level. In fact, for these datasets, there is only one instance (dataset 3AAA, $\dot{e} \sim N(\mathbf{0}, (\eta\, \dot{\sigma})^2 \mathbf{I})$) where the difference between the nominal and observed coverages produced using these two methods is greater than 5%, and in this instance, the observed coverage is greater than nominal, not less.

The three variants of the linearization method, on the other hand, frequently produced far less reliable confidence regions, although, as discussed below, the results still do not appear to be affected by the use of finite-difference derivatives. The difference between the nominal and observed coverages obtained using the linearization methods often are considerably more than 20%, which is a difference that many if not most users would find unacceptable.

By comparing Figure 1 to Figure 2, it is apparent that increasing the variance of the errors does, in fact, increase the differences between observed and nominal coverage for all methods. Our tests at confidence levels 0.50, 0.75, and 0.99, which are not reported in detail here, also showed that the spread between the observed and nominal coverage obtained using the linearization method increases as the nominal confidence level is increased.

The large differences for some datasets between the observed coverage of confidence regions constructed using the likelihood method and those obtained using the linearization method may be explained by the difference in the shape of the two regions. The likelihood method confidence region corresponds to the boundary and interior of a contour of the sum of squares surface, i.e., a contour of constant likelihood, whereas the linearization method confidence regions are always ellipsoidal. We plotted these contours for various datasets, and the difference sometimes were very large. Examples for datasets 3AAA and 14AAG are given in Donaldson (1985).

Figure 1 also indicates that the observed coverage obtained using variants A, B, and C of the linearization method are nearly identical. The results of two-sided paired-sample t-tests indicate that there is no statistically significant differences at the 5% level between the observed coverages obtained using any of the variants of the linearization method with any of the derivative configurations. The same results were obtained for our tests at the 0.50, 0.75, and 0.99 confidence levels.

Confidence Intervals

Results. Figures 3 and 4 provide information for confidence intervals which is analogous to that shown in figures 1 and 2 for confidence regions. The observed coverages plotted are the *smallest* of the p confidence interval coverages obtained for each dataset. Figure 3 displays the observed confidence interval results for nominally 95% confidence levels, when $\dot{e} \sim N(\mathbf{0}, \dot{\sigma}^2 \mathbf{I})$; figure 4 shows the results when $\dot{e} \sim N(\mathbf{0}, (\eta\, \dot{\sigma})^2 \mathbf{I})$, excluding linearization method variants B and C as was done for the linearization method confidence regions.

Observations. Figure 3 shows that for confidence intervals, the best results are obtained using the lack-of-fit and likelihood methods, and the worst results are obtained using the linearization method, as was the case for confidence regions. The lack-of-fit and likelihood methods produce confidence intervals which seldom vary from nominal by an amount that is significant at the 5% level, and never are less than nominal by more than 5.0%. Again, use of finite difference Jacobians does not appear to affect the results for these two methods.

The three variants of the linearization method, on the other hand, frequently produce far less reliable confidence intervals than the lack-of-fit and likelihood methods. Disturbing differences between observed and nominal coverages occur when each of the variants of the linearization method is used to construct confidence intervals. The observed coverage for a nominally 95% confidence interval is as low as 75.0%, 44.0%, and 10.8% for variants A, B, and C, respectively. For most of the datasets tested in our study, however, the span between observed and nominal coverage produced by the three variants of the linearization method is considerably less for confidence intervals than for linearization method confidence regions constructed about the parameters of the same dataset. This is especially true when derivative configurations DC2 and DC3 are used.

One reason why linearization method confidence intervals have better coverage than linearization method confidence regions is that, when the parameter estimates are correlated with each other, a number of points may be included in the linearization method confidence intervals but not in the confidence regions. Note, however, that if a confidence interval was computed for the linear combination of the parameters given by the eigenvector corresponding to the minor axis of the linearization method confidence region ellipsoid, then the linearization method confidence interval observed coverage should approximately equal that of the linearization method confidence region. In our Monte Carlo study, we actually computed the linearization method confidence interval observed coverage for this linear combination of the parameters. In every case, the observed coverage we obtained for the confidence interval about this linear combination was approximately equal to that of the linearization method confidence region observed coverage.

The use of finite differences to approximate both the Jacobian and the Hessian appears to significantly degrade the confidence interval results for linearization variants B and C. Figure 3 shows that, while there is no striking difference in the results obtained using the three variants of the linearization method with derivative configurations DC2 and DC3, variants B and C degrade significantly more than variant A when using DC1, i.e., finite difference Jacobian and Hessian. A two-sided paired-sample t-test was used to determine whether, for a given derivative configuration, the observed coverages obtained using the different linearization method variants are statistically different at the 5% significance level. The results indicate that when derivative configuration DC2 and DC3 are used, the differences in the results obtained using variants A, B, and C are seldom statistically significant at the 5% level, but that when the Jacobian and Hessian are approximated using finite differences (derivative configuration DC1) then the differences in results are often significant.

Comparing Figures 3 and 4 shows that as the variance of the errors is increased, the differences between observed and nominal coverage also are increased, as was the case for the confidence region results. However, this increase is not as pronounced for confidence intervals as for confidence regions. The results at confidence levels 0.50, 0.75, 0.95, and 0.99 also showed that as the nominal confidence level approaches 100%, the spread between observed and nominal coverages obtained using the linearization method is increased.

5. Diagnostic tools

The preceding section demonstrates a pressing need for diagnostics to warn users when the commonly used linearization method confidence region will not have adequate coverage. In addition, it would be useful to have a warning to indicate when the approximate likelihood method may be inadequate. Bates and Watts (1980) have proposed measures of nonlinearity that provide such diagnostics.

According to Bates and Watts, when their relative measure of parameter effects curvature is small compared to the critical value $(F_{p,n-p,0.05})^{-1/2}$, then the linear coordinate grid assumption is valid over the region of interest, and therefore the linearization method confidence region should be adequate. Similarly, when their relative measure of intrinsic curvature is small compared to the same critical value, then the assumption that the solution locus is planar is valid over the region of interest, and therefore the likelihood method confidence region should be adequate.

In Figure 5 we plot the 20 confidence region observed coverages obtained using linearization method variant A with analytic derivatives (derivative configuration DC3) and $\dot{\varepsilon} \sim N\big(\mathbf{0},(\eta\, \dot{\sigma})^2\, \mathbf{I}\big)$ against the Bates and Watts relative measure of parameter effects curvature. Likewise, in figure 6 we plot the corresponding 20 likelihood method confidence region observed coverages against the Bates and Watts relative measure of intrinsic curvature. The relative curvature measures were computed at the true parameter values using the true variance of the errors. In these plots, we have scaled the measures of parameter effects curvature and intrinsic curvature by dividing the measure by the appropriate critical value. Thus, in both of these plots, a scaled curvature measure less than 1 indicates the relative measure was less than the critical value, while a value greater than 1 indicates the curvature exceeded the critical value.

It is clear from figure 5 that the Bates and Watts parameter effects curvature measure is strongly correlated with the observed coverage obtained using the linearization method. In fact, for our data, as the parameter effects curvature increases, the observed coverage for the linearization method confidence regions decreases nearly monotonically and linearly as the logarithm of the scaled parameter effects curvature. Furthermore, in all datasets where the parameter effects curvature is less than the critical value, the observed confidence region is very close to nominal, while in all cases where the parameter effects curvature is greater than ten times the critical value, the observed coverage is unsatisfactorily low. Datasets with parameter effects curvature between one and ten times the critical value had observed confidence region coverage between 83.2% and 91.6%. From these results, it appears that the Bates and Watts parameter effects curvature is a reliable, if perhaps stringent, indicator of when the linearization method will produce reliable confidence regions.

Figure 6 shows that all but one of the 20 datasets tested in this study have intrinsic curvature which is less than the critical value, which means that each of these datasets is nearly planar. For nearly planar datasets we expected good observed coverage from the likelihood method, and, as figure 6 shows, that is what we got. Since none of our datasets have high intrinsic curvature, however, we do not know how the likelihood method will perform when the solution locus is not nearly planar. We cannot assume that the accurate results obtained in our study using the likelihood method will necessarily carry over to datasets with large intrinsic curvature.

Cook, Tsai and Wei (1984) provide an example which has scaled parameter effects curvature of 934.5 and scaled intrinsic curvature of 8.4. Both the parameter effects curvature and intrinsic curvature of this dataset exceed any curvature measure we observed in the 20 datasets in our study. For this dataset, we computed observed confidence region coverages of 19.0% and 95.0% using the linearization method and likelihood methods, respectively. While the linearization method confidence region observed coverage is very far from nominal as we would expect based on the parameter effects curvature of this model, the likelihood method confidence region observed coverage is not. We cannot conclude anything from this one observation. It is clear, however, that additional analysis of datasets with high intrinsic curvature would be useful to further assess the effect of a non-planar solution locus on the likelihood method.

6. Conclusions

Based on our computational study, we can draw conclusions about : i) the comparison between the three variants of the linearization method; ii) the reliability of linearization methods for calculating confidence regions and confidence intervals; and iii) the reliability of the likelihood and lack-of-fit methods for calculating confidence regions and confidence intervals.

When using the linearization method to construct confidence regions and intervals, our Monte Carlo study has shown no clearcut difference in the observed coverage of one variant as compared to another. In our tests, the only statistically significant difference among the results produced by the three linearization variants was in constructing confidence intervals with finite difference Jacobians and Hessians; here variant A was superior to variants B and C. We found no empirical evidence that one should prefer variants B or C, even though they may be appealing from a theoretical point of view. Therefore we conclude that variant A of the linearization method, which is computed using

$$\hat{\mathbf{V}}_a = s^2 \left(\mathbf{J}(\hat{\boldsymbol{\theta}})^T \mathbf{J}(\hat{\boldsymbol{\theta}})\right)^{-1}, \tag{6.1}$$

is the best variant to use for constructing both confidence regions and confidence intervals, because it is simpler, less expensive, and more numerically stable to compute than variants B or C, which use

$$\hat{\mathbf{V}}_b = s^2 \, \mathbf{H}(\hat{\boldsymbol{\theta}})^{-1} \tag{6.2}$$

and

$$\hat{\mathbf{V}}_c = s^2 \, \mathbf{H}(\hat{\boldsymbol{\theta}})^{-1} \left(\mathbf{J}(\hat{\boldsymbol{\theta}})^T \mathbf{J}(\hat{\boldsymbol{\theta}})\right) \mathbf{H}(\hat{\boldsymbol{\theta}})^{-1}, \tag{6.3}$$

respectively. Variant A is simpler and less expensive because it only requires the Jacobian of the model function at the solution and not the additional second order terms that are also required to form the Hessian. It is more stable because it can be formed by inverting the upper triangular factor R of the QR factorization of the Jacobian rather than by calculating the inverse of the Hessian; the former calculation can be expected to lose roughly half as many digits as the latter in finite precision arithmetic.

The linearization method is not always an adequate method for approximating confidence regions and confidence intervals for the parameters of a nonlinear model, however. The results presented in the preceding section show just how poor the linearization method can be in some cases. Although there are many examples where the linearization method's observed coverage differs from nominal by only a very small amount, there are also many cases where the observed coverage is far lower than the nominal. In our tests, the best linearization method variant, A, produced observed coverages as low as 12.4% for nominal 95% confidence regions and 75.0% for nominal 95% confidence intervals.

Users will continue to use the linearization method, however, because it is readily available in software packages and provides a concise representation of the information needed to construct confidence regions and intervals. The erratic results obtained in our study when using the linearization method lead us to conclude that users of nonlinear least squares software must be helped to cautiously assess the results they obtain using the linearization method. The results of the preceding section show that the diagnostic tools proposed by Bates and Watts (1980) are very successful in indicating cases where the linearization method confidence regions are likely to be unreliable. In these cases, more reliable methods, such as the likelihood or lack-of-fit methods, are required to produce accurate confidence regions or intervals.

Our study shows that the lack-of-fit and likelihood methods both produce observed coverages acceptably close to nominal in every test case. Although the difficulties and expense associated with using these two methods to compute confidence regions make it unlikely that they will ever routinely replace the commonly used linearization method for this purpose, they appear to be a reliable alternative that should be considered when diagnostics show that linearization confidence regions are unreliable. It is not as difficult and expensive to construct confidence intervals using the lack-of-fit or likelihood methods, and we believe that producers of nonlinear least squares software should consider this possibility. (Constructing these intervals requires the solution of a series of nonlinearly constrained optimization problems; it may be necessary to construct special purpose software to solve these problems as efficiently as possible.) Performing hypothesis tests using the likelihood or lack-of-fit methods is computationally simple for both confidence regions and intervals, so we recommend that one of these two methods be employed for hypothesis tests whenever possible.

Users may prefer the likelihood method to the lack-of-fit method even though it is approximate and the lack-of-fit method is exact, because the likelihood method has more desirable structural characteristics than the lack-of-fit method. Our study provides no empirical evidence that the results produced by the likelihood method are inferior to those produced by the lack-of-fit method. This does not guarantee that similar results will be obtained on other datasets, however. In particular, the results of the diagnostic test proposed by Bates and Watts showed that all our datasets have low intrinsic curvature, which is precisely the situation when likelihood methods are expected to be very reliable. The additional dataset we analyzed with high intrinsic curvature also produced likelihood method confidence region observed coverage close to nominal. Additional analysis is required to determine whether the likelihood method is reliable for datasets with high intrinsic curvature, and to determine whether the Bates and Watts measure of intrinsic curvature is a useful tool for indicating when the likelihood method confidence regions are likely to be unreliable.

In addition to diagnostics, it appears that there is a need for new methods for estimating confidence regions that are both reliable and easy to report. We are especially interested in investigating two methods that would result in conservative elliptical confidence regions. The first method is to find the minimal magnification of the (95%) linearization confidence region that encloses the (95%) likelihood or lack-of-fit confidence region. This would require the solution of a constrained optimization problem with one nonlinear equality constraint. The second method is to find the smallest volume ellipse that encloses the desired likelihood or lack-of-fit confidence region. This would require the solution of a semi-infinite programming poblem, i.e. an optimization problem with an infinite set of constraints.

7. Summary

We have presented the results of a Monte Carlo study comparing the linearization, likelihood and lack-of-fit methods for constructing confidence regions and confidence intervals. Our results indicate that the linearization method should be constructed using the simplest approximation to the variance-covariance matrix, (6.1), as it is simpler, less expensive, more numerically stable, and at least as accurate as the other two linearization variants, which are constructed using (6.2) and (6.3). We have also given considerable evidence that confidence regions, and to some extent confidence intervals, constructed using the linearization method can be essentially meaningless.

Our study shows that the likelihood and lack-of-fit methods, on the other hand, produced consistently good results for the datasets tested. However, because the likelihood method is approximate it is not clear that the good results we obtained with it will necessarily be characteristic of all datasets. Also, because of the undesirable structural characteristics of the lack-of-fit method, it is unlikely to be used routinely, although in cases where accuracy is of extreme importance, it may be a useful tool to have.

Because of the uncertainty associated with the linearization and likelihood methods, we also have briefly examined how the Bates and Watts curvature measures relate to the confidence region observed coverages we obtained in this study. Our results show that the Bates and Watts parameter effects curvature appears to provide excellent indication of when the linearization method may produce less than satisfactory results. Our results are not as conclusive, however, about the relation between intrinsic curvature and likelihood method coverage since the solution locus for all of our datasets were nearly planar.

References

Bard, Y. (1974), *Nonlinear Parameter Estimation,* Academic Press, New York.

Bates, D. M., Watts, D. G. (1980), "Relative curvature measures of nonlinearity," *Journal of the Royal Statistical Society (B),* 42(1), pp. 1-25.

Beale, E. M. L. (1960), "Confidence regions in non-linear estimation," *Journal of the Royal Statistical Society (B),* 22(1), pp. 41-76.

Cook, R. D., Tsai, C. L., Wei, B. C. (1984), "Bias in nonlinear regression," Mathematics Research Center Technical Summary Report No. 2645, University of Wisconsin, Madison, Wisconsin.

Dennis, J. E. Jr., Gay, D. M., Welsch, R. E. (1981), "Algorithm 573: NL2SOL - An Adaptive Nonlinear Least Squares Algorithm [E4]," *ACM Transactions on Mathematical Software,* 7(3), pp. 369-383.

Donaldson, J. R. (1985), "Computational experience with confidence regions and confidence intervals for nonlinear least squares," MS Thesis, Department of Computer Science, University of Colorado, Boulder, Colorado.

Donaldson, J. R., Tryon, P. V. (1983), "Nonlinear least squares regression using STARPAC," National Bureau of Standards Technical Note 1068-2, U.S. Government Printing Office, Washington, D. C.

Draper, N. R., Smith, H. (1981), *Applied Regression Analysis (Second Edition),* John Wiley and Sons, New York.

Duncan, G. T. (1978), "An empirical study of jackknife-constructed confidence regions in nonlinear regression," *Technometrics,* 20(2), pp. 123-129.

Gallant, R. A. (1976), "Confidence regions for the parameters of a nonlinear regression model," Institute of Statistics Mimeograph Series No. 1077, North Carolina State University, Raleigh, North Carolina.

Guttman, I., Meeter, D. A. (1965), "On Beale's measures of non-linearity," *Technometrics,* 7(4), pp. 623-637.

Himmelblau, D. M. (1970), *Process Analysis by Statistical Methods,* John Wiley and Sons, Inc., New York.

Jennrich, R. I. (1968), "Asymptotic properties of non-linear least squares estimators," *The Annals of Mathematical Statistics,* 40(2), pp. 633-643.

Marsaglia, G., Tsang, W. W. (1984), "A fast, easily implemented method for sampling from decreasing or symmetric unimodal density functions," *SIAM Journal on Scientific and Statistical Computing,* 5(2), pp. 349-359.

Ratkowsky, D. A. (1983), *Nonlinear Regression Modeling,* Marcel Dekker, Inc., New York.

Schnabel, R. B. (1982), "Finite difference derivatives - theory and practice," Internal Report.

Appendix

Dataset	Id.	p/n	Reference
1	2AAA	2/12	Guttman and Meeter (1965) model η_2 , page 628
2	3AAA	2/12	Guttman and Meeter (1965) model η_3 , page 628
3	4AAA	2/24	Duncan (1978) model III , page 127
4	5AAF	4/18	Himmelblau (1970) model 6.2-3 , page 183
5	6AAA	3/13	Himmelblau (1970) model 6.2-4 , page 188
6	8ACA	4/24	None
7	9AAG	8/25	Inghold Hertel Microwave Absorption Line Function (personal communication)
8	11AAB	4/9	Ratkowsky (1983) model 4.4 , page 62
9	12AAB	4/9	Ratkowsky (1983) model 4.14 , page 77
10	14ACG	3/10	Ratkowsky (1983) model 3.5 , page 51 and 58
11	14ABG	3/21	Ratkowsky (1983) model 3.5 , page 51 and 58
12	14AAG	3/42	Ratkowsky (1983) model 3.5 , page 51 and 58
13	15AAA	3/16	Ratkowsky (1983) model 6.11 , page 120 and 58
14	16AAF	5/27	Ratkowsky (1983) model 6.12 , page 122, 123 and 125
15	17AAA	2/42	Ratkowsky (1983) model 3.4 , page 50 and 58
16	18AAA	3/9	Ratkowsky (1983) model 4.1 , page 61 and 88
17	19AAA	3/9	Ratkowsky (1983) model 4.2 , page 61 and 88
18	20AAG	4/9	Ratkowsky (1983) model 4.3 , page 62 and 88
19	21AAA	4/9	Ratkowsky (1983) model 4.5 , page 63 and 88
20	22AAB	3/5	Ratkowsky (1983) model 5.1 , page 93 and 102

Figures

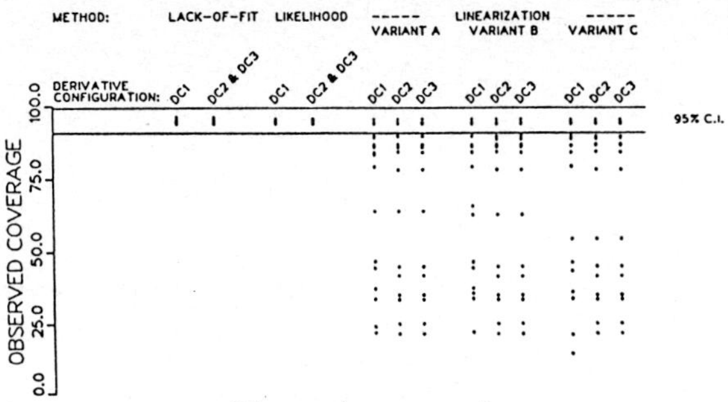

Observed coverage for
nominally 95% confidence regions with $\dot{e} \sim N(\mathbf{0}, \dot{\sigma}^2 \mathbf{I})$
versus
method by derivative configuration

Figure 1

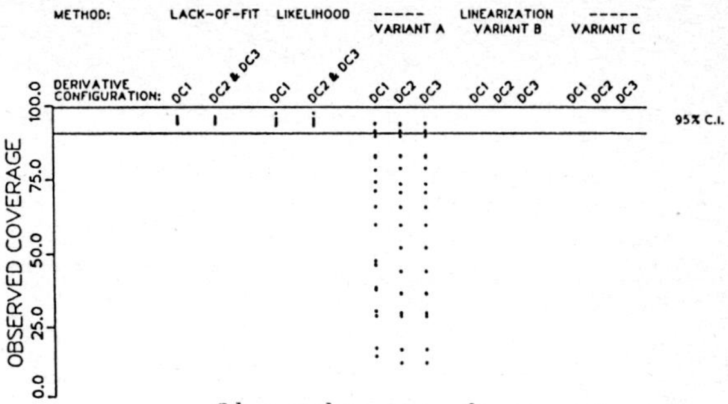

Observed coverage for
nominally 95% confidence regions with $\dot{e} \sim N\left(\mathbf{0}, (\eta\ \dot{\sigma})^2 \mathbf{I}\right)$
versus
method by derivative configuration

Figure 2

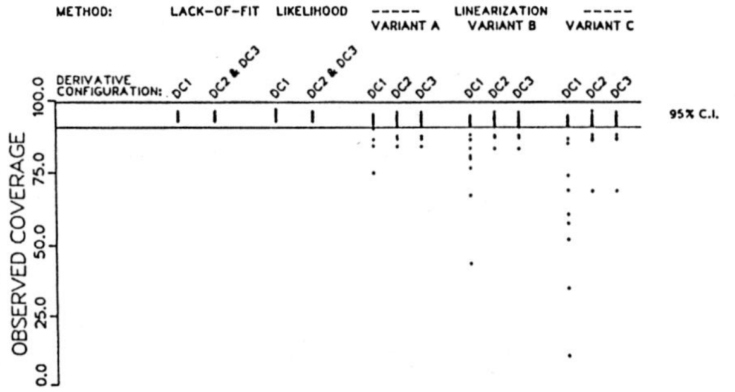

Observed coverage for
nominally 95% confidence intervals with $\dot{e} \sim N(\mathbf{0}, \dot{\sigma}^2 \mathbf{I})$
versus
method by derivative configuration

Figure 3

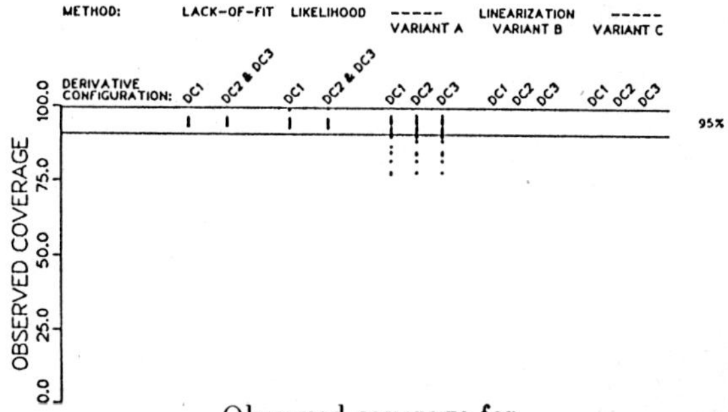

Observed coverage for
nominally 95% confidence intervals with $\dot{e} \sim N\left(\mathbf{0}, (\eta\ \dot{\sigma})^2 \mathbf{I}\right)$
versus
method by derivative configuration

Figure 4

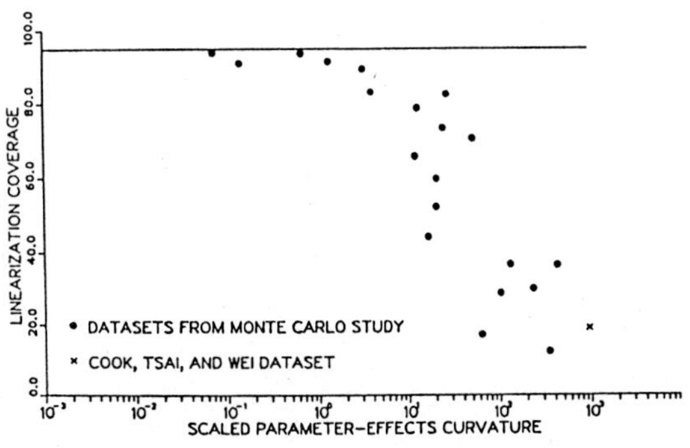

Linearization method confidence region observed coverage
versus
parameter effects curvature scaled by $(F_{p,n-p,0.05})^{-1/2}$

Figure 5

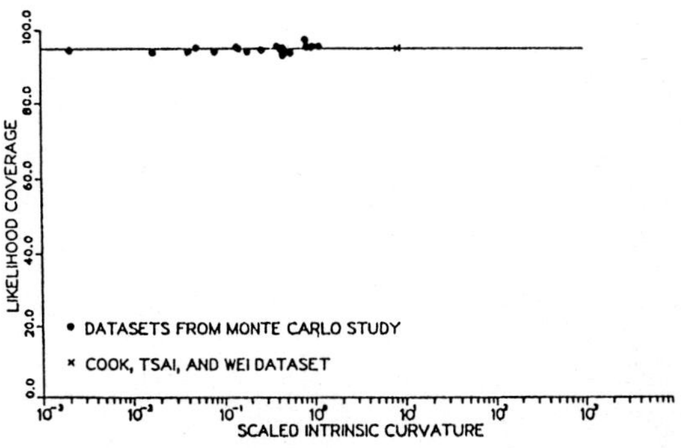

Likelihood method confidence region observed coverage
versus
intrinsic curvature scaled by $(F_{p,n-p,0.05})^{-1/2}$

Figure 6

COMPUTER SCIENCE AND STATISTICS:
The Interface, D.M. Allen (ed.)

CURVATURES FOR PARAMETER SUBSETS IN NONLINEAR REGRESSION

R. Dennis Cook, University of Minnesota; Miriam L. Goldberg, University of Wisconsin

The relative curvature measures of nonlinearity proposed by Bates and Watts (1980) are extended to an arbitrary subset of the parameters in a normal, nonlinear regression model. In particular, the subset curvatures proposed indicate the validity of linearization-based approximate confidence intervals for single parameters. The derivation produces the original Bates-Watts measures directly from the likelihood function. When the intrinsic curvature is negligible, the parameter-effects curvature array contains all information necessary to construct curvature measures for parameter subsets.

Key Words: Confidence regions, Curvature measures, Least squares, Likelihood.

1. INTRODUCTION

Confidence regions for parameters of a normal nonlinear regression model are commonly constructed by using linear regression methods, replacing the solution locus with the tangent plane at the maximum likelihood estimate. Such tangent plane regions are generally easier to construct than corresponding likelihood regions. More importantly, the elliptical contours of tangent plane regions are relatively easy to characterize and understand, particularly for one- or two-dimensional parameter subsets which are often of interest. Likelihood regions, on the other hand, are not influenced by parameter-effects nonlinearity and, therefore, generally have true coverage closer to the nominal level than do tangent plane regions. Under suitable regularity conditions and with a sufficiently large sample size, tangent plane and likelihood regions will be in good agreement, but in any particular problem the strength of this agreement is usually uncertain.

Bates and Watts (1980) propose measures of intrinsic and parameter-effects curvature for assessing the adequacy of the tangent plane approximation: Relatively small values for both the maximum intrinsic curvature Γ^{η} and the maximum parameter-effects curvature Γ^{τ} indicate that the tangent plane approximation is reasonable, while relatively large values for either Γ^{η} or Γ^{τ} indicate that this approximation is questionable. These ideas are extended and refined by Bates and Watts (1981), and Hamilton, Bates and Watts (1982). For a review of related literature, see Bates and Watts (1980) and Ratkowsky (1983). Programs for calculating Γ^{η} and Γ^{τ} are given by Bates, Hamilton and Watts (1983).

The material in Bates and Watts (1980) represents an important step forward, but their method for assessing the adequacy of the tangent plane approximation applies only to tangent plane regions for the full parameter vector. This method is not appropriate for assessing the adequacy of tangent plane regions for a subset of parameters, as indicated by Cook and Witmer (1984) and Linssen (1980). It is fairly easy to construct examples where Γ^{τ} is relatively large and yet there is good agreement between the tangent plane and likelihood regions for a subset of the parameters. One such example is given in Section 2 which is a brief review of the tangent plane approximation and the Bates-Watts methodology. We are often interested in confidence regions for subsets, particularly for individual parameters. Thus, the inability of the Bates-Watts methodology to assess the adequacy of subset regions reflects an important gap in our understanding and ability to deal with nonlinear models.

In Section 3 we develop measures for assessing the agreement between tangent plane and likelihood regions for an arbitrary subset of parameters from a nonlinear regression model. The measures require the same building blocks as needed for the construction of Γ^{τ}, and reduce to Γ^{τ} when the full parameter vector is considered. Computationally, these measures require little more effort than Γ^{τ} itself. Section 4 contains several examples and our concluding comments are given in Section 5. In the remainder of this section, we establish notation and briefly review relevant background information.

A nonlinear regression model can be represented in the form

$$y_i = f(x_i, \theta) + \varepsilon_i \ , \ i=1,\dots,n \quad (1)$$

where y_i is the i-th response, x_i is a vector of known variables, θ is a px1 vector of unknown parameters, the response function f is a known, scalar-valued function that is twice continuously differentiable in θ, and the errors ε_i are independent and identically distributed normal random variables with mean 0 and variance σ^2.

The maximum likelihood (ML) estimator $\hat{\theta}$ of θ can be obtained by minimizing the residual sum of squares

$$RSS(\theta) = \sum_{i=1}^{n} (y_i - f(x_i, \theta))^2 \quad (2)$$

Kennedy and Gentle (1980) discuss methods for obtaining $\hat{\theta}$. For our purposes we assume that $\hat{\theta}$ is available.

For notational convenience, let $f_i(\theta) = f(x_i,\theta)$ and let V denote the nxp matrix with elements $f_i^r = \partial f_i/\partial\theta_r$, i=1,...,n r=1,...,p.

Here and in what follows all derivatives are evaluated at $\hat{\theta}$ unless explicitly indicated otherwise.

Various quadratic approximations to be used in the following sections involve the pxp matrices W_i , i=1,...,n, with elements $f_i^{rs} = \partial^2 f_i/\partial\theta_r\partial\theta_s$, r,s=1,...,p. These matrices can be written conveniently in an nxpxp array W (Bates and Watts, 1980). The ab-th "column" of W is the ab-th second derivative vector W_{ab} with elements f_i^{ab} , i=1,...,n, while the i-th face W_i of W is the pxp matrix consisting of the i-th elements of the second derivative vectors W_{ab}.

2. CURVATURES AND THE TANGENT PLANE APPROXIMATION

Let $F(\theta)$ denote the nx1 vector with elements $f_i(\theta)$. The standard elliptical confidence region for θ based on replacing $F(\theta)$ with the tangent plane at $\hat{\theta}$ can be written as

$$\{\theta: \phi^T V^T V \phi \leq s^2 G\} \quad (3)$$

where $\phi = (\phi_a) = \theta - \hat{\theta}$, $s^2 = RSS(\hat{\theta})/(n-p)$, $G = pF_{\alpha}(p, n-p)$ and $F_{\alpha}(v_1, v_2)$ is the upper α probability point of an F distribution with v_1 and v_2 degrees of freedom.

To assess the adequacy of the region in (3), we need the standard quadratic expansion of F about $\hat{\theta}$:

$$F(\theta) = F(\hat{\theta}) + V\phi + \frac{1}{2}\phi^T W \phi \quad (4)$$

Multiplication involving three-dimensional arrays is defined as in Bates and Watts (1980) so that $\phi^T W \phi$ is an nx1 vector with elements $\phi^T W_i \phi$, i=1,...,n. Generally, if F is quadratic over a sufficiently large neighborhood of $\hat{\theta}$ and the quadratic term of (4) is sufficiently small relative to the linear term, the tangent plane region (3) should be reasonable; otherwise, this approximation may be in doubt. Bates and Watts (1980, 1981) implement this idea by first decomposing each column of W into its projections onto the column and null spaces of V: $W_{ab} = P_V W_{ab} + (I-P_V) W_{ab} = W^{\tau}_{ab} + W^{\eta}_{ab}$, where P_V is the orthogonal projection operator for the column space of V. With this decomposition, the quadratic expansion (4) becomes

$$F(\theta) \cong F(\hat{\theta}) + V\phi + \frac{1}{2}\phi^T W^{\tau}\phi + \frac{1}{2}\phi^T W^{\eta}\phi \qquad (5)$$

where W^{τ} and W^{η} are the nxpxp arrays whose columns are W^{τ}_{ab} and W^{η}_{ab}, respectively.

Next, the adequacy of the tangent plane region is assessed by using the maximum parameter-effects curvature

$$\Gamma^{\tau} = \max \frac{||\phi^T W^{\tau}\phi||}{||V\phi||^2}\sqrt{ps} \qquad (6)$$

and the maximum intrinsic curvature

$$\Gamma^{\eta} = \max \frac{||\phi^T W^{\eta}\phi||}{||V\phi||^2}\sqrt{ps} \qquad (7)$$

where the maximum is taken over all ϕ in R^p. These curvatures as well as the decomposition of $\phi^T W\phi$ displayed in (5), reflect different characteristics of the nonlinearity of the model. The intrinsic curvature Γ^{η} is invariant under reparameterizations and is thus a measure of the intrinsic nonlinearity of the solution locus. In contrast, Γ^{τ} depends on the parameterization: different parameterizations can result in substantially different values of Γ^{τ}. If both Γ^{τ} and Γ^{η} are sufficiently small, the tangent plane region (3) should be adequate.

More specifically, for a tangent plane region of the form (3), Bates and Watts (1980) suggest that the linear approximation should be adequate if Γ^{η} and Γ^{τ} are both small compared to the guide $c = 1/\sqrt{F_{\alpha}(p, n-p)}$. When Γ^{η} or Γ^{τ} is greater than c, the linear approximation and the circular approximation that is the basis of the curvature measures both break down within the tangent plane region. Thus, Ratkowsky (1983) proposes that c/2 be used as a cutoff level, beyond which the tangent plane region is presumed inadequate.

To demonstrate that the Bates-Watts methodology can fail for subsets of θ, we consider the Fieller-Creasy problem in which the ratio of the means of two normal populations is of interest. The corresponding nonlinear model can be written as

$$f(x_i, \theta) = \theta_1 x_i + \theta_1\theta_2(1-x_i) \qquad (8)$$

where x_i is an indicator variable that takes the values 1 and 0 for populations 1 and 2, respectively. For convenience we assume equal sample sizes for the two populations $n_1=n_2=n/2$ and, without loss of generality, we assume that σ^2 is known.

The model given in (8) is intrinsically linear so that $\Gamma^{\eta} = 0$. Further, Cook and Witmer (1984) show that

$$\Gamma^{\tau} = \frac{\sqrt{2}\sigma\{(\hat{\theta}_2^2 + 1)^{1/2} + |\hat{\theta}_2|\}}{|\hat{\theta}_1|\sqrt{n}} \qquad (9)$$

In this case the Bates-Watts (1980) guide for judging the adequacy of the tangent plane approximation is $c = (\chi(\alpha;2))^{-1/2}$ where $\chi(\alpha;v)$ is the upper α probability point of the chi-squared distribution with v degrees of freedom. However, it is clear that standard methods can be used to form exact confidence intervals for θ_1, the mean of the first population, regardless of the value of Γ^{τ}. In other words, the tangent plane and likelihood regions for θ_1 are identical for all Γ^{τ}.

A similar phenomenon occurs in connection with θ_2. Let $r = \sigma^2\chi(\alpha;1)/n\hat{\theta}_1^2$. Assuming that r<1, Cook and Witmer (1984) show that the $1-\alpha$ likelihood region for θ_2 can be written as

$$[\hat{\theta}_2 \pm \{r(1-r) + r\hat{\theta}_2^2\}^{1/2}]/(1-r) \qquad (10)$$

The level associated with this region is exact. The corresponding tangent plane region is

$$\hat{\theta}_2 \pm (r + r\hat{\theta}_2^2)^{1/2} \qquad (11)$$

Clearly, (10) and (11) will be close only if r is sufficiently small. For any fixed value of r, however, Γ^τ may be large or small depending on the value of $\hat{\theta}_2$ so that again the Bates-Watts criterion fails to reflect accurately the agreement between the tangent plane and likelihood regions for a parameter subset. We will return to this example at the end of the next section.

3. SUBSETS

Let $L(\theta,\sigma^2)$ denote the log likelihood for model (1), and partition $\theta^T=(\theta_1^T, \theta_2^T)$ where θ_i is a p_i x 1 vector, i=1,2. The standard likelihood region for θ_2 can be written in the form (Cox and Hinkley, 1974, p. 343).

$$\{\theta_2 : 2[L(\hat{\theta},\hat{\sigma}^2) - L(g(\theta_2), \theta_2, \tilde{\sigma}^2(\theta_2))] \le \rho\} \qquad (12)$$

where ρ, a selected positive constant, is used to set the nominal level and $(g^T(\theta_2), \tilde{\sigma}^2(\theta_2))$ represents the vector-valued function that maximizes $L(\theta_1, \theta_2, \sigma^2)$ for each value of θ_2. Evaluating (12), the likelihood region for θ_2 can be written equivalently as

$$\{\theta_2 : n\cdot\log[\sum_{i=1}^{n} (y_i - f_i(g(\theta_2), \theta_2))^2/n\hat{\sigma}^2] \le \rho\} \qquad (13)$$

Clearly, the form of this region is governed by the vector-valued function $h(\theta_2) = F(g(\theta_2), \theta_2)$. If h is essentially linear over a sufficiently large neighborhood of $\hat{\theta}_2$, the contours of (13) will be elliptical and we can expect (13) and the corresponding tangent plane region to agree; otherwise these regions will tend to be dissimilar. To determine when these regions are in substantial agreement, we investigate the behavior of h by using the method described in Section 2, except that F is replaced by h which, in combination with $Y = (y_i)$, contains essential information on θ_2. Thus, in exact analogy with the Bates-Watts development, we will produce expressions for the curvature of the solution locus submanifold defined by h. Where necessary for clarity, we refer to this as "subset curvature". Similarly, "subset parameter-effects", and "subset intrinsic" refer to the decomposition of the subset curvature into components in the submanifold tangent plane and its orthogonal complement.

Let $\alpha^T(\theta_2) = (\alpha_i(\theta_2)) = (g^T(\theta_2), \theta_2^T)$, let Δ_1 denote the $p x p_2$ matrix with elements $\partial\alpha_i/\partial\theta_{2j}$, i=1,2,...,p, j=1,2,...,p_2, and let Δ_2 denote the $p x p_2 x p_2$ array with i-th face Δ_{2i}, i=1,2,...,p; the elements of Δ_{2i} are $\partial^2\alpha_i/\partial\theta_{2j}\partial\theta_{2k}$, j,k=1,...,$p_2$. We assume, of course, that g is a twice continuously differentiable function of θ_2. With these definitions the straightforward quadratic approximation of $h(\theta_2)$ about $\hat{\theta}_2$ can be written as

$$h(\theta_2) \simeq F(\hat{\theta}) + V\Delta_1\phi_2 \qquad (14a)$$

$$+ \tfrac{1}{2}\phi_2^T\Delta_1^T W\Delta_1\phi_2 \qquad (14b) \qquad (14)$$

$$+ \tfrac{1}{2}V(\phi_2^T\Delta_2\phi_2) \qquad (14c)$$

where $\phi_2 = \theta_2 - \hat{\theta}_2$.

3.1 Refining Equation (14).

For the quadratic expansion in (14) to be useful, we need to develop explicit forms for Δ_1 and Δ_2 to produce a reexpression of (14) that displays the (subset) parameter-effects and intrinsic components of h at $\hat{\theta}_2$. To avoid interruption, the details of this development have been relegated to the Appendix. Here we discuss the final form.

The final form of (14) is based on the assumption that the intrinsic curvature of F at $\hat{\theta}$ is negligible. That assumption is somewhat restrictive but it is valid in the important class of problems where the parameters of interest are nonlinear functions of the location parameters in a linear model. In any event, we judge the practical advantages of allowing for substantial intrinsic curvatures to be minimal since experience has shown (See Bates and Watts 1980, and Ratkowsky 1983) that they are typically small. Of course, Γ^n can and should be evaluated in practice so that this assumption can be checked.

In the remainder of this paper we use C(M) and C'(M) to indicate the column and null spaces, respectively, of the matrix M; the corresponding orthogonal projection operators will be denoted by P_M and P_M', respectively.

In their development of the intrinsic and parameter-effects curvatures for the full parameter vector, Bates and Watts (1980) found it convenient and revealing to work in transformed coordinates. Similarly, the quadratic expansion (14) is most easily understood in terms of these same transformed coordinates: Let V = UR denote the unique QR-factorization of V where R is upper triangular and the columns of the nxp matrix U form an orthonormal basis for C(V). Next, partition R as

$$R = \begin{matrix} R_{11} & R_{12} \\ 0 & R_{22} \end{matrix} \qquad (15)$$

where R_{ii} is $p_i \times p_i$, i=1,2. Transformed coordinates $\tilde{\phi}$ can now be defined as $\tilde{\phi}^T = (\tilde{\phi}_1^T, \tilde{\phi}_2^T) = \phi^T R^T$ so that

$$\tilde{\phi}_1 = R_{11}\phi_1 + R_{12}\,\phi_2 \qquad (16)$$

and

$$\tilde{\phi}_2 = R_{22}\,\phi_2 \qquad (17)$$

In the following any quantity with a tilde added above indicates evaluation in the ϕ coordinates. Thus, for example, $\tilde{V} = U$ and $\tilde{W} = R^{-T}\, W\, R^{-1}$. Partition the i-th face $\tilde{W}_i$ of $\tilde{W}$ as

$$\tilde{W}_i = \begin{pmatrix} \tilde{W}_{i11} & \tilde{W}_{i12} \\ \tilde{W}_{i21} & \tilde{W}_{i22} \end{pmatrix}, \quad i=1,\ldots,n \qquad (18)$$

where the dimension of $\tilde{W}_{ijj}$ is $p_j \times p_j$, j=1,2. Next, define $\tilde{W}_{22}$ to be the nxp_2xp_2 subarray of $\tilde{W}$ with i-th face $\tilde{W}_{i22}$ and similarly define $\tilde{W}_{12}$ to be the nxp_1xp_2 subarray of $\tilde{W}$ with i-th face $\tilde{W}_{i12}$, i=1,...,n. Finally, partition $V = (V_1, V_2)$ and $U = (U_1, U_2)$ where U_i and V_i are $n \times p_i$ matrices.

With this structure, the quadratic expansion of h can be reexpressed informatively as

$$\begin{aligned} h(\theta_2) \approx\ & F(\hat{\theta}) + U_2\tilde{\phi}_2 && (19a) \\ & + \tfrac{1}{2}\tilde{\phi}_2^T[P_{U_2}][\tilde{W}_{22}]\tilde{\phi}_2 && (19b) \quad (19) \\ & - U_1[\tilde{\phi}_2^T U_2^T][\tilde{W}_{12}]\tilde{\phi}_2 && (19c) \end{aligned}$$

where the brackets [·][·] indicate column (sample space) multiplication as defined in Bates and Watts (1980), and discussed briefly in the Appendix. Term (19a) describes the plane tangent to h at $\hat{\theta}_2$. Since $C(U_2) = C(P_V' V_2)$, this plane is simply the affine subspace $F(\hat{\theta}) + C(P_V' V_2)$. This is the same as the subspace obtained when using the tangent plane approximation to form a confidence region for θ_2. In other words, the confidence contour based on the tangent plane approximation will coincide with those based on substituting the linear approximation of h into (13), as expected.

Term (19b) contains the projections of the columns of $\tilde{W}_{22}$ onto the plane tangent to h at $\hat{\theta}_2$. Thus, this term reflects the (subset) parameter-effects curvature of h in the direction $\tilde{\phi}_2$. The maximum parameter-effects curvature Γ_s^{τ} for the subset θ_2 can now be defined as

$$\Gamma_s^{\tau}(\theta_2) = \max||d^T[P_{U_2}][\tilde{W}_{22}]d||\sqrt{p_2}s \qquad (20)$$

where the maximum is taken over all d in $D = \{d: d\in R^2, ||d|| = 1\}$. Since $\tilde{\phi}_2$ is a linear transformation of ϕ_2 as described in (17), $\Gamma_s^{\tau}(\theta_2)$ will be the same in both coordinate systems.

To further understand (20), partition the i-th face A_i of the pxpxp unscaled parameter-effects curvature array $A = [U^T][\tilde{W}]$ as

$$A_i = \begin{pmatrix} A_{i11} & A_{i12} \\ A_{i21} & A_{i22} \end{pmatrix} \qquad (21)$$

where the dimension of A_{ijj} is $p_j x p_j$, j=1,2, i=1,...,p. Next, let A_{22} denote the $p_2 x p_2 x p_2$ subarray of A with faces A_{i22}, $i=p_1+1,\ldots,p$. Then

$$[P_{U_2}][\tilde{W}_{22}] = [U_2][A_{22}]$$

and

$$\Gamma_s^{\tau}(\theta_2) = \max_D||d^T A_{22} d||\ \sqrt{p_2}s \qquad (22)$$

In this form it is clear that the maximum parameter-effects curvature for the subset problem depends only on the behavior of the $\tilde{\phi}_2$ parameter-curves. The elements of A_{22} can be used to understand the behavior of these parameter-curves in terms of arcing, "compansion", fanning and torsion, as described in Bates and Watts (1981).

Term (19c) is clearly in $C(V_1)$ and is thus orthogonal to the subspace tangent plane. This term then reflects the intrinsic curvature of h at $\hat{\theta}_2$ so that the maximum intrinsic curvature can be defined as

$$\Gamma_s^{\eta}(\theta_2) = \max_D||[d^T U_2^T][\tilde{W}_{12}]d||2\sqrt{p_2}s \qquad (23)$$

Note that (23) contains the extra factor 2, corresponding to the absence of the factor 1/2 in (19c).

This curvature can also be expressed in terms of a subarray of A. Let A_{12} denote the $p_2 x p_1 x p_2$ subarray of A that has faces A_{i12}, $i=p_1+1,\ldots,p$. Then $A_{12} = [U_2^T][\tilde{W}_{12}]$ and

$$\Gamma_s^{\eta}(\theta_2) = \max||[d^T][A_{12}]d||2\sqrt{p_2}s$$

$$= \max||\sum_{j=p_1+1}^{p} d_j A_{j12} d||2\sqrt{p_2}s \qquad (24)$$

where d_j is the $(j-p_1)$-th element of d. Interestingly, the intrinsic curvature for the subset problem depends only on fanning and torsion components of A; compansion and arcing play no role in the determination of Γ_s^{η}. The fanning and torsion terms of A depend in part on how the columns of V are ordered. Since we have assumed that the last p_2 columns of V correspond to θ_2, it is the fanning and torsion with respect to this ordering that are important.

If both Γ_s^{η} and Γ_s^{τ} are sufficiently small, the likelihood and tangent plane confidence regions for θ_2 will be similar; otherwise we can expect these regions to be dissimilar. Following Bates and Watts (1980), $c = (F_{\alpha}(p_2, n-p))^{-\frac{1}{2}}$ can be used as a rough guide for judging the size of these curvatures. As noted earlier, our experience indicates that curvatures must be substantially less than c to insure close agreement between tangent plane and likelihood regions. This will be illustrated in sections 3.3 and 4.

Finally, we combine the intrinsic and parameter-effects components of (19) to define the total curvature $\Gamma_s(\theta_2)$ of h at $\hat{\theta}_2$ as

$$\Gamma_s(\theta_2) = \sqrt{p_2}s \max_D\{||d^T A_{22} d||^2 + 4||[d^T][A_{12}]d||^2\}^{1/2} \qquad (25)$$

As will be demonstrated in the next subsection, the total subset curvature Γ_s may be more relevant than both Γ_s^η and Γ_s^τ. For example, it is possible to have $\Gamma_s^\eta < c$ and $\Gamma_s^\tau < c$ while $\Gamma_s > c$. In such situations Γ_s^τ and Γ_s^η may incorrectly indicate that the tangent plane approximation is adequate, while Γ_s correctly indicates otherwise.

When the full parameter θ is of interest, we have $\theta_2 = \theta$ and $p_2 = p$. In this case, the subset intrinsic curvature (24) is zero, A_{22} is the Bates-Watts parameter-effects array, and both (22) and (25) represent the maximum parameter-effects curvature for θ. Thus, our derivation based on the likelihood reproduces the primary quantity developed by Bates and Watts (1980).

The main conclusion of this section is that the unscaled parameter-effects curvature array A for the full parameter contains all necessary information for evaluating the adequacy of tangent plane confidence regions for certain subsets of θ. For example, if the last parameter θ_p is of interest then $\Gamma_s^\tau(\theta_p)$ is simply $s|a_{ppp}|$ where a_{ijk} is the (j,k)-th element of the i-th face of A. Similarly,

$$\Gamma_s^\eta(\theta_p) = 2s\left(\sum_{i=1}^{p-1} a_{pip}^2\right)^{1/2} \qquad (26)$$

Thus, compansion and fanning are the only effects that are relevant to an assessment of the agreement between likelihood and tangent plane confidence regions for a single parameter.

3.2 Computation

Recall that the developments of this section are based on the assumption that the last p_2 columns of V correspond to the parameters of interest. This assumption is necessary to maintain the collective identity of θ_2 as indicated in (17). This implies that the ordering of the columns of V is critical and consequently θ_p is the only single parameter for which curvatures can be constructed from a given parameter-effects array A. The A-array for other orderings can be constructed by permuting the columns of V and beginning again, of course.

Alternatively, a computationally more efficient method for obtaining the A-array in a rotated coordinate system can be constructed as follows. Let $\phi_z = Z\phi$ where Z is a selected pxp permutation matrix. In what follows, the subscript z added to any quantity indicate evaluation in the coordinates ϕ_z. Clearly, $V_z = VZ^T = URZ^T$. Let U^{*T} be an orthogonal matrix such that $R^* = U^*RZ^T$ is upper triangular. Since the QR-factorization of V_z is unique, it follows that $V_z = U_zR_z$ where $U_z = UU^*$ and $R_z = R^*$. Using this structure it is not difficult to verify that

$$A_z = [U^*][U^*AU^{*T}] \qquad (27)$$

Thus, to find A_z, the parameter-effects curvature array for the rotated coordinates ϕ_z, we need only the pxp matrix U^* to diagonalize RZ^T. A single call to LINPACH (1979) routine SCHEX produces R^*, $[U^{*T}][A]$ and the information necessary to construct U^*.

3.3 Fieller-Creasy Again

To apply Γ_s^η and Γ_s^τ in the Fieller-Creasy problem when θ_2 is the subset of interest, we require only the 2x2x2 parameter-effects curvature array A for

$$V = \left(x+\hat{\theta}_2(b-x),\ \hat{\theta}_1(b-x)\right)$$

where x is the nx1 vector with elements x_i as defined following (8) and b is an nx1 vector of one's. The faces A_i of A are (Cook and Witmer, 1984)

$$A_1 = \frac{\hat{\theta}_2\sqrt{2}}{\hat{\theta}_1\{n(1+\hat{\theta}_2)\}^{1/2}} \begin{bmatrix} 0 & 1 \\ 1 & -2\hat{\theta}_2 \end{bmatrix} \tag{28}$$

and

$$A_2 = A_1/\hat{\theta}_2 \tag{29}$$

Reading directly from this array we have

$$\Gamma_s^\tau(\theta_2) = \sigma|a_{222}| = \frac{2^{3/2}\sigma}{\sqrt{n}|\hat{\theta}_1|} \cdot \frac{|\hat{\theta}_2|}{(1+\hat{\theta}_2^2)^{1/2}} \tag{30}$$

and

$$\Gamma_s^\eta(\theta_2) = 2\sigma|a_{212}| = \frac{2^{3/2}\sigma}{\sqrt{n}|\hat{\theta}_1|} \cdot \frac{1}{(1+\hat{\theta}_2^2)^{1/2}} \tag{31}$$

Recall that we are assuming σ to be known in this example so that the guide for assessing the magnitudes of Γ_s^η and Γ_s^τ is $c=(\chi(\alpha;1))^{-1/2}$.

From (30) we see that $\Gamma_s^T(\theta_2)$ will be zero only if $\hat{\theta}_2=0$; in this case

$\Gamma_s^\eta(\theta_2)=2^{3/2}\sigma/\sqrt{n}|\hat{\theta}_1| < c$ or, equivalently,

$r = 2\sigma^2\chi(\alpha;1)/n\hat{\theta}_1^2 < 1/4$ is necessary for the subset intrinsic curvature to be less than the guide. Further $r < 1/4$ is a sufficient - although not necessary - condition for both $\Gamma_s^\eta(\theta_2)$ and $\Gamma_s^\tau(\theta_2)$ to be less than c when $\hat{\theta}_2$ is arbitrary.

Next, using (25) it follows that the total subset curvature is simply

$$\Gamma_s(\theta_2) = 2^{3/2}\sigma/\sqrt{n}|\hat{\theta}_1| \tag{32}$$

and thus $\Gamma_s(\theta_2) < c$ if and only if $r < 1/4$. When $r > 1$ the likelihood region for θ_2 will be either the complement of an interval or else the entire real line; otherwise this region will be the interval given in (10). In this example, the total subset curvature recovers the critical quantity r as introduced in section 2, and the condition $\Gamma_s < c$ insures that the tangent plane interval (11) will in fact be approximating a likelihood interval rather than some dissimilar region. This condition also provides for an added measure of agreement between these intervals since it is equivalent to $r < 1/4$ rather than simply $r < 1$.

Applying (22) and (24) when θ_1 is the subset of interest gives $\Gamma_s^\eta(\theta_1) = \Gamma_s^\tau(\theta_1)=0$, as expected. Notice that this conclusion cannot be obtained by inspecting the A array given in (28) and (29). As mentioned previously, different subsets in general require different orderings for the columns of V and thus different coordinates. This is the case here.

Finally, we consider the special case characterized by $(\hat{\theta}_1, \hat{\theta}_2) = (3,0)$ and $r = .428$. These conditions correspond to $n = 2\sigma^2$. From (9), $\Gamma^\tau=.33 < .41 = \chi^{-1/2}(.05;2)$. From Figure 1 (Cook and Witmer 1984), we see that the likelihood region, whose level is exact in this case, does not seem to be adequately approximated by the tangent plane region for small values of θ_1.

Further insight into this problem can be gained by inspecting marginal regions for θ_1 and θ_2. Generally, marginal regions for subsets can be obtained by projecting all points in the joint region onto the appropriate subspaces. The projections of the regions in Figure 1 onto the θ_1 axis show that the likelihood and tangent plane intervals for θ_1 will be identical, as expected. The projections onto the θ_2 axis show that the resulting 98.6 percent likelihood interval will be about 60 percent longer than the corresponding tangent plane interval! This dissimilarity is clearly indicated by $\Gamma_s^\eta(\theta_2) = .67 > .41 = \chi^{-1/2}(.014;1)$.

Our experience leads to the following heuristic characterization of the problem described in the previous paragraph. Consider a p_2-dimensional subset θ_2 with guide $c_2=(F_\alpha(p_2,n-p))^{-1/2}$ and partition

$\theta_2^T = (\theta_{21}^T, \theta_{22}^T)$ where θ_{2i} is p_{2i}x1, i=1,2. The guide corresponding to the confidence region for θ_{2i} obtained by projecting the selected $1-\alpha$ region for θ_2 is simply $c_{2i} = c_2(p_{2i}/p_2)^{1/2}$, i=1,2. When the subset curvatures for θ_{21} are large relative to c_{21} and the subset curvatures for θ_{22} are near zero, it can happen that the curvatures for θ_2 are moderate. In such cases the curvatures for θ_2 can provide a misleading indication that the tangent plane and likelihood regions for θ_2 are in acceptable agreement. As hinted above, this problem might be overcome by requiring that all subsets θ_{21} of θ_2 have curvatures less than the respective guides c_{21}. When $\theta_2 = \theta$ this added requirement seems to represent a useful fine tuning of the basic Bates-Watts methodology.

4. ILLUSTRATIONS

In this section we present several numerical examples to illustrate selected results of the previous sections.

For the first example we use the Michaelis-Menton model

$$f_i = \theta_1 x_i/(\theta_2+x_i) \qquad (33)$$

in combination with the 12 observations reported in Bates and Watts (1980). Figure 2 gives 87 percent tangent plane (broken contour) and likelihood (solid contour) confidence regions for (θ_1, θ_2). Here and in the following examples the levels of displayed bivariate confidence regions are chosen so that the corresponding univariate marginal regions have a nominal 95 percent coverage rate. It seems clear from Figure 2 that the tangent plane region for (θ_1, θ_2) is not an adequate approximation of the likelihood region, although interpreting the Bates-Watts guide directly as the cutoff value would lead to the opposite conclusion, since $\Gamma^\tau = .598 < c = .635$. The subset curvatures for θ_1 and θ_2 are listed in Table 1; the corresponding guide is c = .449. Again, the curvatures are less than the guide while the marginal likelihood regions do not seem to be well represented by the corresponding tangent plane regions. This reenforces our previous remark that curvatures must be substantially less than c to insure close agreement. With this interpretation we see that all curvatures successfully indicate the dissimilarity between the various likelihood and tangent plane regions in Figure 2.

Figure 3 gives 88% likelihood and tangent plane regions for (θ_1, θ_2) obtained by using model (33) and the 7 observations reported by Michaelis and Menton (1913). For these data $\Gamma^\tau = .079$. This value and the subset curvatures reported in Table 1 are relatively small, indicating reasonable agreement between the regions displayed in Figure 3.

For our next example we use the exponential model

$$f_i = \theta_1(1-\exp(\theta_2 x_i)) \qquad (34)$$

in combination with the 6 observations reported in Draper and Smith (1981, p. 522., data set 3). In this case $\Gamma^\tau = 1.92$ clearly indicates the dissimilarity between the 88 percent regions for (θ_1, θ_2) shown in Figure 4. However, the 95% marginal regions for θ_2 are in close agreement, while the agreement between the marginal regions for θ_1 seems less than adequate. These conclusions are clearly indicated by the subset curvatures $\Gamma_s(\theta_2) = .069$ and $\Gamma_s(\theta_1) = .314$ which may be judged relative to the guide c = .360.

For the three-parameter asymptotic regression model

$$f_i = \theta_1 + \theta_2\exp(\theta_3 x_i) \qquad (35)$$

and the 27 observations reported in Ratkowsky (1983, p. 101, data set 1), we obtain $\Gamma^\tau = 1.53$. The corresponding guide is

$c = [F_{.05}(3,24)]^{-1/2} = .58$. This suggests that

the 95 percent likelihood region for $\theta^T = (\theta_1,\theta_2,\theta_3)$ cannot be adequately approximated by the corresponding tangent plane region. The subset curvatures for selected subsets of θ are listed in Table 1. From these curvatures alone we would reach the following conclusions: 1) The likelihood and tangent plane regions for θ_2 are in very close agreement. 2) The marginal regions for θ_1 and θ_3 will be noticeably different, but the agreement is probably adequate for most purposes. 3) The usual 95 percent tangent plane regions for (θ_1,θ_3) and (θ_2,θ_3) should be used for only very rough analyses, although lower level regions may be acceptable replacements for the corresponding likelihood regions. These conclusions are supported by the 86 percent regions for (θ_2,θ_3) and (θ_1,θ_3) shown in Figures 5 and 6, respectively.

For our final example we again use the asymptotic regression model (35), this time in combination with the 9 observations reported by Hunt (1970). Subset curvatures for 4 parameter subsets are listed in Table 1. The subset curvature for θ_3 is small, indicating good agreement between the corresponding likelihood and tangent plane regions. The subset curvatures for the remaining subsets, particularly (θ_2,θ_3), are large.

The 87 percent likelihood and tangent plane confidence regions for (θ_2,θ_3) are given in Figure 7. The large total curvature, $\Gamma_s(\theta_2,\theta_3) = 36.4$, correctly indicates that use of the tangent plane region as an approximation of the disjoint likelihood region would be a disaster for this pair of parameters. In fairness, however, it should be recalled that the approximations used to derive the subset curvatures are local so that $\Gamma_s(\theta_1,\theta_2)$ is responding primarily to the disagreement between the tangent plane region and the portion of the likelihood region that contains $\hat{\theta}$. Similar comments apply when only θ_2 is of interest.

From Figure 7, there is reasonable agreement between the tangent plane and likelihood regions for θ_3, as indicated by the small curvature $\Gamma_a(\theta_3) = .095$. It can be argued justifiably, however, that this correct indication from the curvature is largely fortuitous since the curvatures do not recognize the contribution of the smaller piece of the likelihood region for (θ_2,θ_3) to the likelihood region for θ_3. Under this argument, the subset curvature measure for θ_3 has failed to indicate the dissimilarity between the tangent plane region for θ_3 and the likelihood region $(-.0191,0)$ obtained by using only the larger subregion that contains $\hat{\theta}$.

The reason that the curvatures give some inappropriate indications in this final example is that both the linear and quadratic approximations to the model function fail. This failure is evident from a very low R^2 from the regression used by Goldberg, Bates and Watts (1983) to obtain numerical curvatures, and from related measures of "lack of quadraticity" explored by the present authors. In cases where the quadratic approximation to the model function is poor, curvature measures based on that approximation may not be meaningful.

Nevertheless, these subset curvature measures represent an important advance in our understanding of nonlinear models, and provide useful information about the adequacy of the linear approximation when the quadratic approximation is appropriate. Further work is needed on methods of identifying cases where the quadratic approximation may fail.

5. CONCLUSIONS

The subset curvatures developed in this paper appear to be reliable indicators of the adequacy of tangent plane confidence regions for most nonlinear models. In particular, the curvature for a single parameter is a useful tool for assessing the agreement between standard large sample confidence intervals and corresponding marginal likelihood regions. This ability to deal with subsets greatly extends the usefulness of the Bates-Watts methodology.

Because the original Bates-Watts framework applies only to the complete parameter vector, guidelines developed in that framework can be misleading when the adequacy of the linear approximation is very different for different subsets. To ensure good agreement between the tangent plane and likelihood regions, the maximum curvature must be considerably smaller than the Bates-Watts guide. However, this criterion can be too stringent for certain parameter subsets if the whole-parameter curvature Γ^{τ} is used. By contrast, the subset curvature describes the shape of the likelihood region in the parameter subspace of interest. Thus, the subset curvature is more directly relevant to the tangent plane adequacy question and, based on the examples described above, is evidently more accurate.

The practical usefulness of the methods described here depends, in part, on their ease of implementation. The subset curvatures for any selected subset can be computed directly from the Bates-Watts parameter-effects curvature array. This array can be obtained either analytically (Bates and Watts, 1980) or numerically by using the procedure given in Goldberg, Bates and Watts (1983).

The usefulness of the subset curvatures depends also on the restriction that the intrinsic curvature of F at $\hat{\theta}$ is small. This restriction is not of great practical importance since it has been found to hold in most cases. Nevertheless, a unified approach which incorporates the intrinsic curvature component might offer further insight in some situations.

Another area for further research is the development of measures that indicate when the subset curvatures themselves may be unreliable due to the failure the second-order approximation to the model function. While the possibility of such failure is of concern, the class of models adequately described by a quadratic function is considerably larger than the class for which the linear approximation alone is adequate.

ACKNOWLEDGEMENTS

Data computations for this work were performed on the University of Wisconsin Statistics Department's research computer. The authors thank Douglas Bates for access to software and data libraries used for some examples presented, and for suggestions on computational methods.

APPENDIX

Derivation of Equation (19)

To develop equation (19) from equation (14), we first require explicit expressions for Δ_1 and Δ_2.

A.1. Δ_1 and Δ_2

Let $\ddot{L}$ and $\dddot{L}$ denote the pxp matrix and pxpxp array of second and third partial derivatives of the log likelihood L with respect to the elements of θ, respectively. Let g_a denote the a-th component of g as defined following (12) and partition $\ddot{L}$ as

$$\ddot{L} = \begin{pmatrix} L_{11} & L_{12} \\ L_{21} & L_{22} \end{pmatrix}$$

where L_{jj} is $p_j \times p_j$, j=1,2.

Since g maximizes $L(\theta_1,\theta_2)$ for each fixed value of θ_2 we clearly have

$$\left.\frac{\partial L(g(\theta_2),\theta_2)}{\partial g_a}\right|_{g=g(\theta_2)} = 0 \qquad (A.1)$$

for $a=1,2,\ldots,p_1$ and all θ_2. This identity will be used as the basis for obtaining Δ_1 and Δ_2.

Differentiating both sides of (A.1) with respect to θ_2 and evaluating at $\hat{\theta}_2$ gives

$$(L_{11}, L_{12})\Delta_1 = 0$$

Since the submatrix consisting of the last p_2 rows of Δ_1 is an identity matrix it follows that

$$\Delta_1 = \begin{pmatrix} -L_{11}^{-1} L_{12} \\ I \end{pmatrix} \qquad (A.2)$$

Let $e_i = y_i - f_i(\hat{\theta})$. The the first term of

$$\ddot{L} = (\sum_{i=1}^{n} e_i W_i - V^T V)/\sigma^2$$

represents intrinsic curvature of F at $\hat{\theta}$. Since this curvature is assumed to be negligible, $\ddot{L} = -V^T V/\sigma^2$ and therefore

$$\Delta_1 = \begin{pmatrix} -(V_1^T V_1)^{-1} V_1^T V_2 \\ I \end{pmatrix} = \begin{pmatrix} -R_{11}^{-1} R_{12} \\ I \end{pmatrix} \qquad (A.3)$$

where $V = (V_1, V_2)$ and R_{ij} is defined in (15).

An expression for Δ_2 can be obtained similarly by taking second partial derivatives of (A.1) with respect to θ_{2r} and θ_{2s}, $r,s=1,2,\ldots,p_2$. This yields

$$\sum_{b=1}^{p_1} L_{ab} \frac{\partial^2 \alpha_b}{\partial\theta_{2r}\partial\theta_{2s}}$$

$$= -\sum_{b=1}^{p}\sum_{c=1}^{p} L_{abc} \frac{\partial\alpha_c}{\partial\theta_{2r}} \frac{\partial\alpha_b}{\partial\theta_{2s}} \qquad (A.4)$$

where L_{ab}, L_{abc} and α_b denote the indicated elements of L, L and $\alpha^T = (g^T(\theta_2), \theta_2^T)$, respectively, and $a=1,2,\ldots,p_1$. The component $\partial^2\alpha_b/\partial\theta_{2r}\partial\theta_{2s}$ is the (r,s)-th element of the b-th face Δ_{2b} of Δ_2. Since $\Delta_{2b} = 0$ for $b=p_1+1,\ldots,p$ the summation on the left of (A.4) need only range from 1 to p_1. Notice also that $\partial\alpha_c/\partial\theta_{2r}$ is simply the (c,r)-th element of Δ_1. Expressing (A.4) in matrix notation and solving for Δ_2 gives

$$\Delta_2 = -\begin{bmatrix} L_{11}^{-1} & 0 \\ 0 & 0 \end{bmatrix} \left[\Delta_1^T \overset{\cdots}{L} \Delta_1\right] \qquad (A.5)$$

Here and in what follows brackets [ ][ ] indicate column multiplication as defined in Bates and Watts (1980). (Generally, if A is an axb matrix and B is a bxcxd array then the elements of the i-th face C_i, $i=1,\ldots,a$, of the axcxd array $C = [A][B]$ are $A_i^T B_{jk}$, $j=1,2,\ldots,c$, $k=1,2,\ldots,d$, where A_i^T is the i-th row of A and B_{jk} is the jk-th column of B.) As before we will take $L_{11} = -V_1^T V_1/\sigma^2$.

To further evaluate Δ_2, we require the pxpxp array $\overset{\cdots}{L}$. Straightforward algebra will verify that

$$L_{abc} = \frac{1}{\sigma^2}\sum_{i=1}^{n}$$

$$(e_i f_i^{abc} - f_i^a f_i^{bc} - f_i^b f_i^{ac} - f_i^c f_i^{ab})$$

Using this representation it is easily verified that the a-th face $\overset{\cdots}{L}_a$ of $\overset{\cdots}{L}$ is

$$\overset{\cdots}{L}_a = -\frac{1}{\sigma^2}\{[b_a^T V^T][W] + V^T K_a + K_a^T V\} \qquad (A.6)$$

where b_a is the a-th standard basis vector for R^p and $K_a = b_a^T W$ is the nxp matrix with W_{ac} as the c-th column. Finally, it follows from (A.6) that

$$Z^T \overset{\cdots}{L} Z = -\frac{1}{\sigma^2}\{Z^T[V^T][W]Z + 2[Z^T V^T][W]Z\} \qquad (A.7)$$

where Z is an arbitrary px1 vector. This form will be useful in later developments.

A.2 Tangent plane, Term (19a)

It follows immediately from (A.3) that

$$V\Delta_1 = P'_{V_1} V_2 = U_2 R_{22} \qquad (A.8)$$

where U_2 is defined following (18). Thus, the relevant tangent plane is the affine subspace $F(\hat{\theta}) + C(P'_{V_1} V_2)$. Transforming term (14a) according to (16) and (17) immediately gives term (19a).

A.3 Parameter-Effects, Term (19b)

From the form of Δ_2 given by (A.3), it is clear that term (14c) is in $C(V_1)$ and is thus orthogonal to the θ_2-subspace tangent plane. The parameter-effects component of (14) must therefore come from term (14b).

The three-dimensional array W in (14b) can be decomposed into the sum of three arrays with orthogonal columns,

$$W = [P_V - P_{V_1}][W] + [P_{V_1}][W] + [P'_V][W] \qquad (A.9)$$

The first term in this decomposition contains the projections of the columns of W onto $C(P'_{V_1} V_2)$ and thus it represents parameter-effects curvature for the subset problem. The second and third terms are intrinsic components for h and F, respectively. Since the intrinsic curvature of F at $\hat{\theta}$ is assumed to be negligible, the third term of (A.9) is set to zero. Addend (14b) can now be reexpressed as

$$\frac{1}{2} \phi_2^T \Delta_1^T W \Delta_1 \phi_2$$

$$= \frac{1}{2} \phi_2^T \Delta_1^T [P_V - P_{V_1}][W]\Delta_1 \phi_2 \qquad (A.10a)$$

$$+ \frac{1}{2} \phi_2^T \Delta_1^T [P_{V_1}][W]\Delta_1 \phi_2 \qquad (A.10b)$$

$$(A.10)$$

From (18) and (A.3) it follows that

$$\tilde{W}_{22} = R_{22}^{-T} \Delta_1^T W \Delta_1 R_{22}^{-1}$$

Using this in combination with (17) and (A.8) to transform the coordinates in term (A.10a) gives term (19b).

A.4 Intrinsic Curvature, Term (19c)

In the expansion of h given in (14), we still have the sum of terms (14c) and (A.10b) to deal with. We first consider (14c).

Using (A.5) and (A.7) with $Z = V_1 \phi_2$ we have

$$\frac{1}{2} V(\phi_2^T \Delta_2 \phi_2) = \frac{\rho^2}{2} M (\phi_2^T \Delta_1^T \dddot{L} \Delta_1 \phi_2)$$

$$= -\frac{1}{2} M \{\phi_2^T \Delta_1^T [V^T][W]\Delta_1 \phi_2\} \qquad (A.11)$$

$$= - M [\phi_2^T \Delta_1^T V^T][W]\Delta_1 \phi_2$$

where $M = \left(V_1 (V_1^T V_1)^{-1}, 0\right)$. The first term of (A.11) is exactly the negative of term (A.10b) so that in an obvious notation

$$(14c) + (A.10b) = - M[\phi_2^T \Delta_1^T V^T][W]\Delta_1 \phi_2$$

$$= - [\phi_2^T \Delta_1^T V^T][MW\Delta_1] \phi_2 \qquad (A.12)$$

From (18) and the definition of $\tilde{W}$, it can be shown that

$$MW\Delta_1 = U_1 \tilde{W}_{12} R_{22}$$

Finally, using this relationship, (A.8) and (17) to transform the coordinates in (A.12) we obtain term (19c).

TABLE 1

Subset Curvatures

Model/Data	Parameter Subset	Γ_s^{τ}	Γ_s^{η}	Γ_s
(33)	θ_1	.330	.183	.377
Bates & Watts	θ_2	.393	.089	.403
(33)	θ_1	.014	.025	.029
Michaelis & Menton	θ_2	.050	.019	.053
(34)	θ_1	.277	.148	.314
Draper & Smith	θ_2	.053	.044	.069
	θ_1	.165	.180	.244
(35)	θ_2	.003	.059	.059
Ratkowsky	θ_3	.153	.132	.203
	(θ_1,θ_3)	1.07	.088	1.07
	(θ_2,θ_3)	.518	.000	.518
	θ_1	1.75	.190	1.76
(35)	θ_2	1.80	.256	1.82
Hunt	θ_3	.018	.094	.095
	(θ_2,θ_3)	36.4	.000	36.4

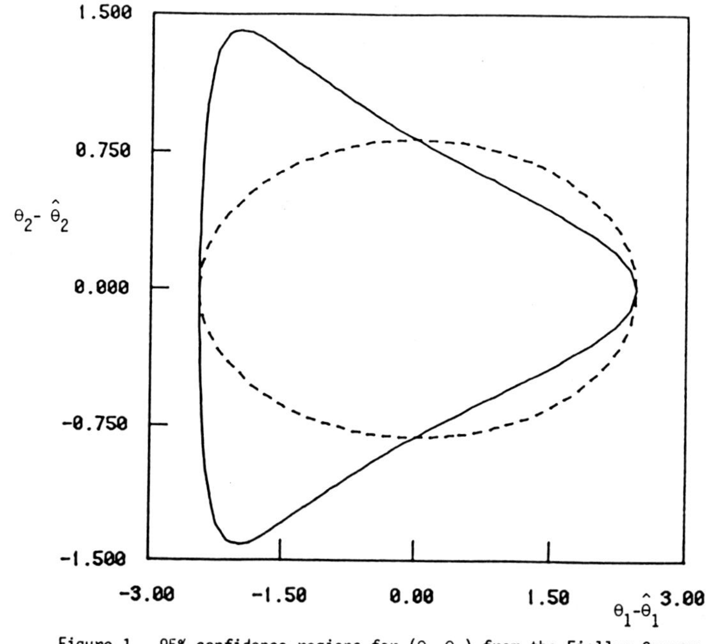

Figure 1. 95% confidence regions for (θ_1,θ_2) from the Fieller-Creasy model (8); $(\hat{\theta}_1,\hat{\theta}_2) = (3,0)$. Likelihood region ——— . Tangent plane region ---- .

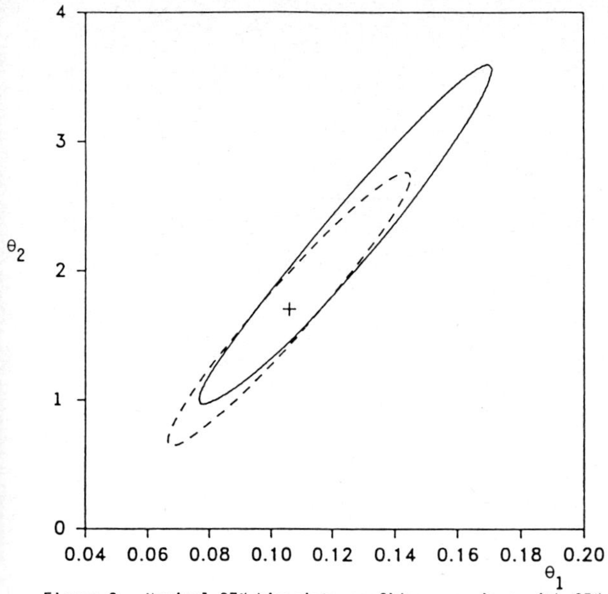

Figure 2. Nominal 87% bivariate confidence regions with 95% marginal regions for (θ_1,θ_2) from model (33) and the Bates-Watts (1980) data. Likelihood —— . Tangent plane ---- .

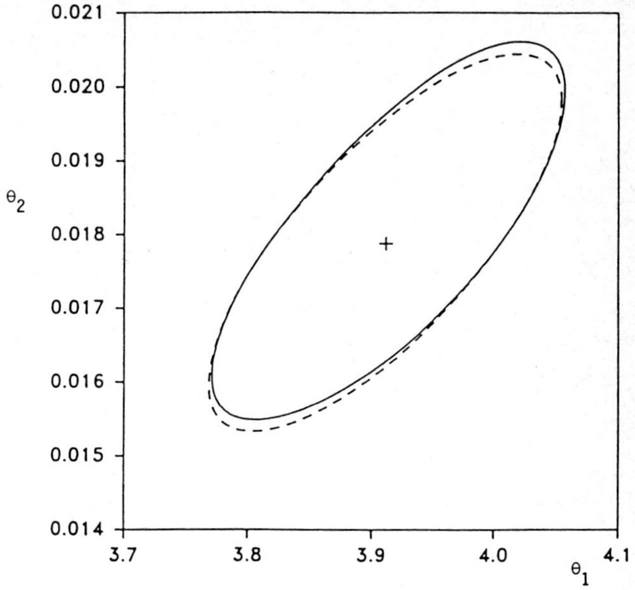

Figure 3. Nominal 88% bivariate regions with 95% marginal regions for (θ_1,θ_2) from model (33) and the Michaelis-Menton (1913) data. Likelihood ——. Tangent plane ----.

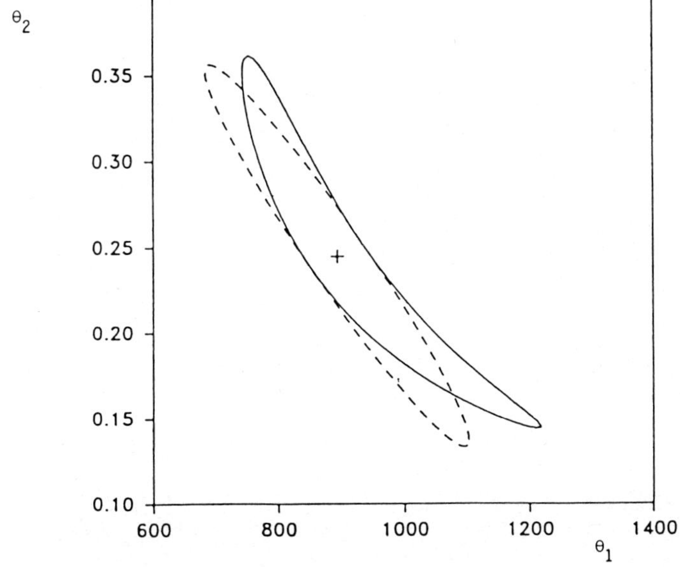

Figure 4. Nominal 88% bivariate regions with 95% marginal regions for (θ_1,θ_2) from model (34) and the Draper-Smith (1981) data. Likelihood —— . Exact ---- .

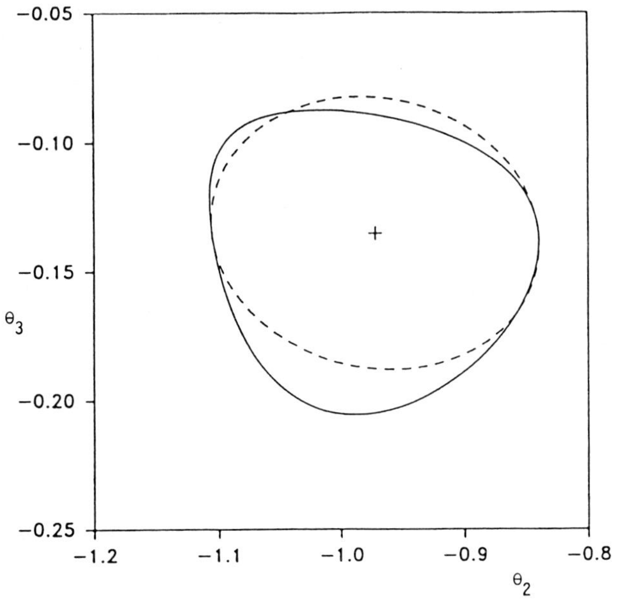

Figure 5. Nominal 86% bivariate regions with 95% marginal regions for (θ_2, θ_3) from model (35) and the Ratkowsky (1983) data. Likelihood ——— . Tangent plane ---- .

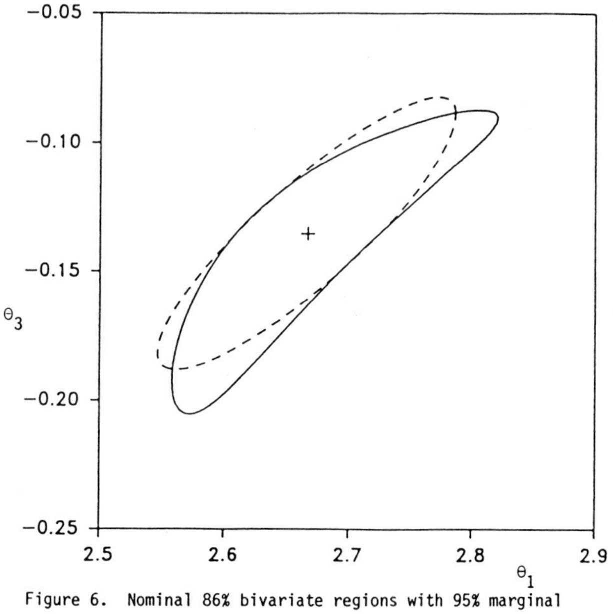

Figure 6. Nominal 86% bivariate regions with 95% marginal regions for (θ_1, θ_3) from model (35) and the Ratkowsky (1983) data. Likelihood ——— . Exact ---- .

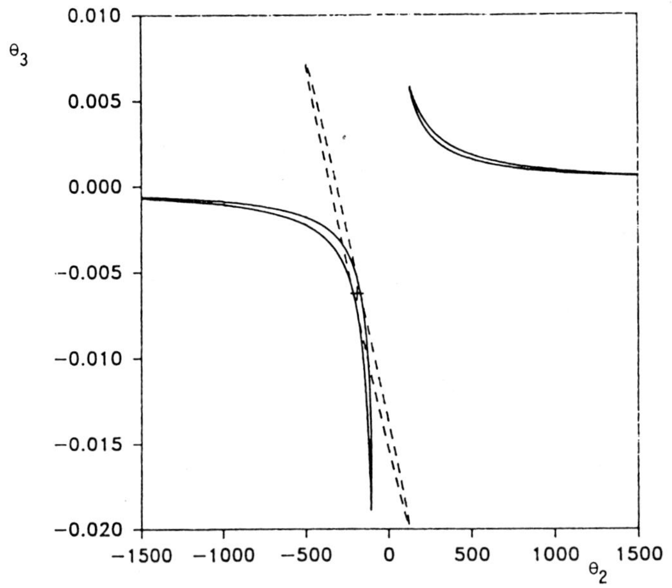

Figure 7. Nominal 87% bivariate regions with 95% marginal regions for (θ_2, θ_3) from model (35) and the Hunt (1970) data. Likelihood ——— . Exact ---- .

REFERENCES

1. Bates, D. M. and Watts, D. G. (1980), "Relative Curvature Measures of Nonlinearity," Journal of the Royal Statistical Society, Ser. B, 42, 1-25

2. Bates, D. M. and Watts, D. G. (1981), "Parameter Transformations for Improved Approximate Confidence Regions in Nonlinear Least Squares", The Annals of Statistics, 9, 1152-1167.

3. Bates, D. M., Hamilton, D. C. and Watts, D. G. (1983), "Calculation of Intrinsic and Parameter-Effects Curvatures for Nonlinear Regression Models", Communications in Statistics, Part B--Simulation and Computation, 12, 469-477

4. Cook, R. D. and Witmer, J. A. (1984), "A Note on Parameter-effects Curvature", Technical Report 439, School of Statistics, University of Minnesota.

5. Cox, R. D. and Hinkley, D. V., (1974), Theoretical Statistics, London: Chapman and Hall.

6. Dongarra, J. J., Moler, C. B., Bunch, J. R., and Stewart, G. W. (1979), LINPACK Users' Guide, Philadelphia: SIAM.

7. Draper, N. R. and Smith, H. (1981), Applied Regression Analysis, New York: John Wiley.

8. Goldberg, M. L., Bates, D. M. and Watts, D. G. (1984), "Curvature Measures Avoiding Second Derivatives", Technical Report No. 739, Department of Statistics, University of Wisconsin.

9. Hamilton, D. C., Watts, D. G. and Bates, D. M. (1982), "Accounting for Intrinsic Nonlinearity in Nonlinear Regression Parameter Inference Regions", The Annals of Statistics, 10, 386-393.

10. Hunt, W. F. (1970), "The Influence of Leaf Death on the Rate of Accumulation of Green Herbage During Pasture", Journal of Applied Ecology, 7, 41-50.

11. Kennedy, W. J. and Gentle, J. E. (1980), "Statistical Computing, New York: Marcel Dekker.

12. Linssen, H. N. (1980), "Discussion of the Paper by Dr. Bates and Professor Watts", Journal of the Royal Statistical Society B, 42, p. 21.

13. Michaelis, L. and Menton, M. L. (1913), "Kinetik der Invertinwirkung", Biochemische Zeitschrift, 49, 333.

14. Ratkowsky, D. A. (1983). Nonlinear Regression Modeling. Marcel Dekker, Inc.: New York.

The Metadata of Computational Processes

Organizer: *Gordon Sande*

Invited Presentations:

Knowledge Acquisition in REX and Student
William A. Gale

Managing Data Analysis Through Save-States
Paul J. Cowley and Mark A. Whiting

COMPUTER SCIENCE AND STATISTICS:
The Interface, D.M. Allen (ed.)

Knowledge Acquisition in REX and Student

William A. Gale

AT&T Bell Laboratories
Murray Hill, New Jersey 07974

ABSTRACT

REX (Regression EXpert) demonstrated the feasibility of building data analysis consultation programs using expert system techniques. However, experience with REX development showed the need for automated assistance in building, maintaining, and extending knowledge bases for other specific data analytic tasks. Symptoms of this need were difficulty maintaining consistency across examples, need for the statistician to learn an obscure language, and difficulty of specialization.

Programming by examples is a natural approach in the statistics domain, because working examples is necessary in any case. Such an approach would address the problems noted in the development of REX. Three fundamental steps in the development of a practical programmed-by-example system are the acquisition of the first example, acquisition of an additional consistent example, and the integration of an inconsistent example.

By restricting the domain within which knowledge can be acquired to data analysis, it has been possible to design practical solutions to these three steps. The first phase of Student, a system designed to learn data analysis strategies from examples, has been implemented. It acquires the first example in any data analysis area, and incorporates many features required for handling problems of additional consistent and inconsistent knowledge.

1. Background

REX is a consultation program in an area of statistics, regression analysis, built using expert system techniques. Its performance was described in [Pregibon and Gale, 1984]. It had an active life as a demonstration system, running about weekly for a year. It demonstrated the feasibility of using expert system techniques to build a consultant in data analysis. However, as detailed in the next section, the knowledge acquisition process for REX left a lot to be desired.

Regression analysis is one technique of a broader category of *data analysis* techniques. Other techniques include spectrum analysis, analysis of variance, and cluster analysis, for example. A statistician doing data analysis operates on a *data set* or *example*. A general goal of the analysis is to meaningfully summarize the salient features of the data set. The standard form of summary is a *statistical model*, typically with parameters estimated from the data set. By using plots and numerical tests, the statistician detects incompatibilities between the model and the data set, which are ameliorated by some *action*, such as transforming a variable, changing the model, or changing the method of estimating parameters.

In mimicking this process, REX checks for problems using tests, and recommends actions to the client after verifying that a proposed action will solve the problem found. It offers to show the client plots whenever it detects a problem or recommends an action.

In building REX, the statistical knowledge it contains has come to be called a "strategy" for regression analysis. The term seems appropriate as the nature of the knowledge includes

what problems to look for,
when to look for them,
how to look for them,
how to decide if they are real, and
what to do if they are.

There is very little statistical literature relevant to strategy, and indeed, REX, as an environment for developing and testing strategy has opened up a new area of research.

2. A Critique of Knowledge Acquisition in REX

Developing a strategy for use in REX was a labor-intensive process. Two phases can be distinguished. In the first phase the statistician responsible for the strategy, Daryl Pregibon, chose a half dozen regression examples that clearly showed some common problems. He then analyzed them using interactive statistical software with an automatic trace. After analyzing the group of examples, he studied the traces and abstracted a description of what he was doing. We coded this as a strategy for REX and tried it on a few more examples. He revised the strategy completely at this point, and the second phase began.

In the second and longer phase, one of us selected one additional regression example and ran REX interactively on the chosen example. Typically the strategy would not handle the example (since the example was selected knowing what would stretch REX), and we modified the strategy so that the example would be handled. This process was iterated through about three dozen more examples.

Based on this experience, and on a feeling that it was typical of other techniques, we do not believe it is possible to construct a data analysis strategy without working through many examples. The range of the decisions needed to construct a strategy is extreme, and there is no literature simplifying the task. Therefore the only available defense of a strategy is to demonstrate performance, which requires working many examples more than those used to construct the system. Our experience also leads us to believe that it is easy to generalize from data analysis examples -- relatively few examples are needed to exhibit the required distinctions.

However, the way in which we worked examples for REX was far from ideal. The first difficulty with our method was assuring ourselves that a strategy modified to work one additional example still worked all previous examples. We could by brute force run REX in batch mode on all previous examples and see if the performance was the same. Usually we reasoned that most of the previous examples could not be affected, and checked the few that might be affected by hand. Naturally, the more examples worked, the more severe this problem became. The necessity to check consistency in batch mode for a system designed to be interactive reduced the flexibility of the strategy developed.

Secondly, the method used was the epitome of the currently standard two-person development of expert systems. I wrote the inference engine used while Daryl was responsible for the strategy developed. Whenever Daryl wanted to do something he hadn't done before, we had to huddle, as Daryl was learning a language he would only use to construct one program. In a department with twenty professional statisticians and one person intimately familiar with the inference engine, it was not clear how many additional data analysis techniques could be handled by this two person approach.

Thirdly, it would be difficult to modify the strategy in REX. Modifiability is important first because a growing literature on strategy can be expected to suggest desirable changes. It is important secondly because strategies need to be specialized to the needs of a particular group. Statistics is a discipline that is applied in other, "ground", domains. Current knowledge representation and language generation techniques are not adequate to producing a tool that will speak physics with physicists and psychology with psychologists. An alternative to one broad tool is a tool that is readily specialized. However, the first two problems would make this difficult: to specialize the program a local statistician would have to learn a language used by no other program in the world, and the modifications made might inadvertently destroy some capabilities of the strategy.

One valuable insight gained from building REX is an abstract view of its strategy that we believe can be transferred to other data analysis techniques. A practical data analysis consists of an attempt to use a simple technique that is well understood (by statisticians!). However, its use is subject to a number of assumptions which may or not hold in a particular data set. When an assumption is violated, either the data must be modified to fit the simple technique, or a more advanced technique must be used. In other words, it has been possible to view data analysis as a diagnosis problem (although not all statisticians agree!) This view is "meta-knowledge" about data analysis which has been built into Student, as described below.

3. Requirements for Learning By Example

The necessity of working examples to construct a data analysis strategy suggests examining the possibility of acquiring strategies directly through some process of working examples. The previous discussion suggests that the process would need to assist the user in establishing consistency across all examples worked, and should not require the statistician to learn an obscure language.

I am suggesting that progress in knowledge acquisition is possible through restriction of the domain of knowledge to acquire. An issue for this approach is whether the restricted domain is broad enough to be worth the difficulty of constructing a special tool. For data analysis, I believe the answer is yes. A human statistician is typically expert in one or a few types of data analysis, while a dozen data analysis techniques would cover the bulk of data sets analyzed [Snee, 1980]. One might ultimately distinguish a few dozen data analysis techniques. Therefore, many statisticians will be needed to construct a reasonably comprehensive data analysis expert system.

A program by example system is enticing for other reasons. First, it would be useful for the study of statistical strategies to collect multiple strategies for the same type of data analysis. Combination of knowledge from multiple experts is an open problem in expert system construction. I view collection of a body of strategies from multiple experts as a necessary precursor to serious study of this problem. Second, a statistician at a specific location could specialize the system by working examples typical of local practice. The value of specialization was discussed in the previous section.

A few systems previously developed come to mind in considering construction of an expert system by working examples. Teiresias [Davis, 1979] is the chief example of a program designed for interactive transfer of expertise to an expert system. The mode of using Teiresias was to be that of selecting of an example, letting the system run until it made a mistake, eliciting the key piece of knowledge to avoid the mistake, and adding the new knowledge. The system therefore operates by acquiring an additional piece of knowledge presumed consistent with that previously acquired. In addition to adding consistent knowledge, however, there are two other major problems that need to be solved for a practical learning by example system.

First, the system must support the acquisition of a *first* example or rule. In a production system, the first rules acquired are typically different from later rules, because the system uses a core of rules to encode control information. A subject matter expert will not be able to provide control information.

Second, the system must support deliberate changes to the knowledge base over time. We need to directly determine the consistency of new examples with previous examples, not just assume it. We do not want to take a "debugging" attitude, but one of showing what is right the first time.

On the other hand, there are some systems that support programming by example, although none of them are for construction of expert systems. Tinker [Lieberman, 1983], PHD [Attardi and Simi, 1983], SBA [Zloof and De Jong, 1977], and a system by Bierman and Krishnaswamy [1976] are examples. Attardi and Simi review several of these systems, which are designed for office automation programming. Tinker appears to be the closest to our ideas for Student.

In using Tinker, the programmer selects a concrete typical example of data for the procedure. He then performs the procedure step by step. The system is therefore able to learn how to do the first example. As more examples are supplied, the program required for them is compared with the already constructed program. If the two differ, the user is queried for a predicate that will distinguish the two cases. Therefore, the user ultimately provides one example for each branch of the final program.

Tinker seems to assume that the user knows how each example should be worked; there is no means to change the program by deleting an example already worked. The way a particular data analysis should be done is not cut and dried, and indeed, the statistician is typically learning about a particular example while doing the analysis. I have built into Student some means of modeling what the statistician has learned, or may have learned, to capitalize on this opportunity for knowledge acquisition. I do not yet know how effective this will be.

On the other hand, Tinker is tackling a harder problem in that it hopes to support Lisp programming of any procedure. Lieberman demonstrates its level of success in this by creating a simple editor. It is an encouraging demonstration. Tinker's use of menus, pointing, and question answering are suggestive techniques.

4. Preliminary Experience with Student

Student is a system designed to allow a statistician working alone to build an expert consultation system in a data analysis area. A first phase has been implemented. The first phase is designed to acquire the first example in any data analysis area.

Student can be operated in two modes -- consultation mode and learning mode. In consultation mode, it will work functionally in a manner similar to REX, suggesting acceptable ways to analyze a given data set. Since it is general to the extent of data analysis, it would handle a much wider range of problems than REX does, given the requisite strategic information.

Student is able to acquire the first example because it is limited to data analysis, and is not a general purpose tool for learning arbitrary things by example. In particular, the meta-knowledge about what a practical data analysis is, inferred from building REX, is built into Student. This meta-knowledge is represented as a network of eleven types of frames, as shown in the following table.

- input variables
  - data types
- assumption testing
  - plot
    - generic plot
  - test
    - generic test
- action
  - question discriminator
  - predicate discriminator
- words

Each type of frame has its own set of slots, which represent the things that must be known in order to carry out a consultation. When a slot has not been filled, the system knows that it doesn't know that information. It can then do something to acquire the information, which is usually just to ask the statistician.

Student manages two major data structures. One, the strategy, has just been discussed. The other is a second network of frames that represents a trace of the analysis of the current example. It is built of three types of frames: entry points, decisions, and actions. The trace can branch at each decision point, if the user gives more than one response (at different times) to a question posed by Student. A decision frame records all the responses to a given question, and book keeping information to uniquely express the set of answers effective at a given point in the trace. The action frames represent each side effect action taken by the program. The entry points are created each time an assumption testing frame is begun in the strategy. They allow the user to return to the same exact context in which the frame was begun at any time. The user can then reach any decision previously made by stepping through decisions to be left standing.

These two data structures support phase 1 and have been designed with an eye towards work on phase 2 (acquiring an additional consistent example) and phase 3 (acquiring an inconsistent example). The remaining paragraphs in this section discuss how consistency and inconsistency are expected to be handled.

The analyses demonstrated by the statistician are assumed to be acceptable analyses of the examples (as judged by a statistician). A major focus of design in Student has been to assure that as a data analysis strategy evolves, all previous analyses remain acceptable analyses (as judged by Student's strategy). This is the basic test of consistency. Points at which consistency is not obvious have been found to fall in four categories: provably consistent, mechanically consistent, mechanically checkable, and provably inconsistent. A provably consistent change results when pre-specifiable data is sufficient to prove consistency. A mechanically consistent change results when information needs to be gathered by reexamining previous examples, but the result must be a consistent strategy. A mechanically checkable change requires reexamination of prior examples in order to show consistency, and the review may establish inconsistency. A provably inconsistent change results when pre-specifiable data is sufficient to prove inconsistency.

Treatment of inconsistent changes rests on how the trace of the latest example is related to the accumulated strategy. Each example worked produces a trace with all the information gathered from the statistician. Each trace represents an example worked in the context of the strategy accumulated to that point, and the strategy changes called for by the trace are guaranteed to be consistent with that accumulated strategy. Therefore, an ordered set of traces is a kind of "source code" from which it is possible to "compile" an integrated strategy consistent with all the examples represented in the traces.

A provably inconsistent change will conflict with parts of the the traces of some prior examples. Those parts will have to be reworked manually, and it is a service to isolate them for attention. The remaining parts of the traces can be retained, assuming that the actions based on the (incorrectly derived) data, although incorrect for the example, were correct for the data. The result will be a tree of partial strategies, each branch representing an inconsistent difference between two strategies. Each node represents a strategy which can be derived by integrating the ordered set of traces from the root to the node.

5. Summary

REX is a working demonstration of the feasibility of expert systems for data analysis. It has several strengths: a convenient user interface, ability to solve standard textbook regression problems, and a modest ability to explain the reasons for its suggestions. However, it also has limitations, mainly in supporting strategy acquisition, modification, and specialization.

Student has been designed to build upon REX's strengths while overcoming its limitations. Student will allow statisticians to construct or extend knowledge based consultation systems by working examples and answering questions. This will provide easier and faster construction of better consultation systems in data analysis.

The proposition that Student explores is that by restricting the domain within which knowledge can be acquired, significant assistance in knowledge acquisition is possible. The control information needed to structure the first example can be provided. The information necessary to prove whether a change of knowledge is consistent can be specified and collected. Support for changing inconsistently with some previous examples can be provided.

6. References

Attardi, G. and Simi, M. (1983). "Extending the Power of Programming by Examples." Appearing on pp. 3-26 in *Integrated Interactive Computing Systems*, ed. Degano and Sandewall, North-Holland Publishing Company, Amsterdam.

Biermann, A.W. and Krishnaswamy, R. (1976). "Constructing Programs from Example Computations." *IEEE Transactions on Software Engineering SE-2*, 3, 141-153.

Davis, R. (1979). "Interactive Transfer of Expertise: Acquisition of New Inference Rules." *Artificial Intelligence* 12, 121-157.

Lieberman, H. (1983). "Designing Interactive Systems from the User's Viewpoint." Appearing on pp. 45-59 in *Integrated Interactive Computing Systems*, ed. Degano and Sandewall, North-Holland Publishing Company, Amsterdam.

Pregibon, D. and Gale, W.A. (1984). "REX: An Expert System for Regression Analysis." *Proceedings COMPSTAT 84*, Prague, Czechoslovakia.

Snee, R. D., (1980) "Preparing Statisticians for Careers in Industry" *The American Statistician*, May 1980, 34: 65-75.

Zloof, M.M. and DeJong, S.P. (1977) "The System for Business Administration (SBA): Programming Language." *CACM 20*, 6, 385-396.

COMPUTER SCIENCE AND STATISTICS:
The Interface, D.M. Allen (ed.)
Elsevier Science Publishers B.V. (North-Holland), 1986

MANAGING DATA ANALYSIS THROUGH SAVE-STATES *

Paula J. Cowley and Mark A. Whiting

Pacific Northwest Laboratory
Richland, WA 99352

Data analysis management is a methodology intended to increase the productivity of the data analyst. A primary entity for data analysis management is the "save-state", a collection of metadata and data that captures a state of the analysis. The analyst may create a save-state to designate a milestone of the analysis. The save-state may be used later to return to that milestone by restoring the conditions of the analysis that existed at the time the save-state was created. Scientist at Pacific Northwest Laboratory (PNL) have developed a prototype data analysis management system. In that system, a save-state includes pointers to the data sets and command procedures active at the time the save-state was created, active plot descriptions and other graphics parameters, and comments supplied by the analyst. Associated with each save-state is a record of the sequence of commands or operations used to accomplish the transition from the previous (parent) save-state. Metadata also describes the overall relationships between the save-states that have been created during the analysis.

NATURE OF THE DATA ANALYSIS PROCESS

For the past several years, a team of computer scientists and statisticians working on the Analysis of Large Data Sets Project (ALDS) at the Pacific Northwest Laboratory (PNL) has been investigating the nature of the data analysis process [1,2,3,4]. There were several motivations for this work. We wanted to understand the analysis process better. We wanted to provide better tools. We hoped we could learn more about how an expert data analyst worked in order to help less experienced analysts.

When we examined the data analysis process, we were able to identify several properties of it. The process tends to be iterative with similar operations applied repeatedly for different data sets and subsets. The process tended to be exploratory. The analyst has some basic ideas of how to approach the analysis at the onset, but the direction the analysis takes often results from the knowledge gained from previous points in the analysis. Because of this, data analysis is best pursued interactively. It is very difficult to write the complete script for the analysis before it begins.

The analysis process can result in many dead ends. Because of its exploratory nature, the analyst may try several approaches that simply don't work. This fact is often not apparent when the final results of the analysis are presented, since only the successful results are presented. However, it is useful to keep track of these dead ends. They can be useful in showing that the analysis was rigorous and complete and that reasonable alternatives were investigated.

The analyst may have several alternatives to explore at various points in the analysis. Since only one alternative can be dealt with at a time, it is useful to be able to return to previous points in the analyses in order to try another alternative.

The process is characterized by fits and starts, dead ends, and decision points with many options to explore. Although we can think of the process as proceeding linearly through time from beginning to end, the process really has more structure to it than that. Rather than representing the process as a straight line, the process is better characterized as a tree where the nodes of the tree represent significant points in the analysis, which we call "save-states," and the lines between the nodes represent the steps in the analysis that took place to create the child node from the parent node. Figure 1 shows how such a tree can be depicted graphically. From any point in the analysis designated as a save-state, the analyst can proceed until a significant point in

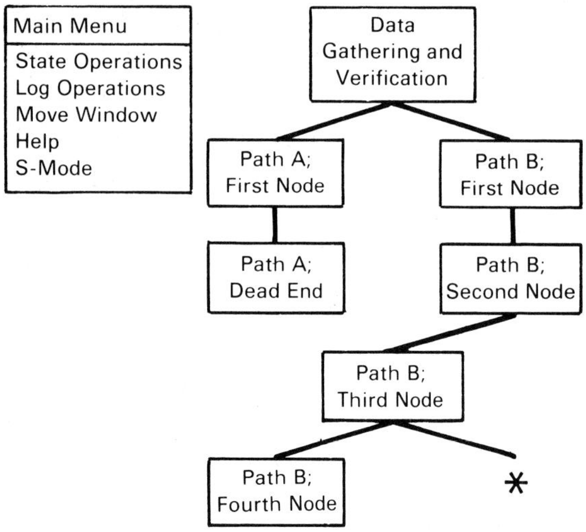

FIGURE 1. The Data Analysis Management Display

the analysis is reached. This point can be defined as a new node of the tree. The analysis can proceed on from that point or the analyst can return to a previous node in the analysis and begin another path. By allowing the analyst to go back to previous nodes and proceeding from that point, a tree can be created. This graphical representation also depicts where the analyst is currently working. The star at the end of a line segment indicates that the analyst is currently proceeding down the path indicated by the line segment and the analyst may create a new node at any time. The new node will replace the star in the graph.

COMMON TOOLS FOR DATA ANALYSIS

There are a number of tools available as aids in performing computer-based data analysis. Although these tools have improved steadily over the past several years, there is very little to help the analyst manage the process. The sections below discuss the desirable characteristics of data analysis tools and, with the exception of the section on statistical functions, the areas in which they lack capabilities for helping the analyst manage data analysis.

Statistical Functions. Most statistical analysis packages are built around a library of statistical functions that can be applied to data sets. However, no package can anticipate (or afford to develop and maintain) all functions the analyst may want to apply. Some systems such as AT&T Bell Laboratories' S System [5] have been designed to be extensible so that the analyst can add new functions as they are identified and can run functions available in languages such as FORTRAN within the environment provided by S.

At PNL, we have always attempted to build systems which can utilize existing data analysis packages. The work we have done in data analysis management was built using S as a base.

Data Management. Most systems for data analysis have facilities for data management that allow the analyst to organize, store, and retrieve data. All provide capabilities for data to be brought into the system for analysis and for data and results to be displayed and printed as output. Some systems have better facilities for organizing data than others. Some support more complicated data structures than others. Some allow the analyst to provide meaningful names rather than simply assigning column numbers to data variables and leaving it to the analyst to keep track of which column contains which variable.

It is often useful to store data that is derived during the course of the analysis. Some of this derived data may consist of intermediate results that can be useful in later phases of the analysis. The process of storing and recalling derived data should be easy to perform.

Both raw data and derived data need metadata to describe characteristics of the data itself, such as the data's source, its units, how it was calculated (if derived), what missing value code(s) are used, why it was generated, and what its role is in the analysis. It is important to be able to associate the metadata with the data and make it easily accessible in a meaningful way to the analyst.

It is important to provide the analyst with a way of keeping track of the data sets. Some of this can be provided through the metadata and through good naming conventions, but current packages provide no facilities for associating data sets with particular stages of the analysis. The analyst has no automatic way of knowing when the data set was analyzed or where it is used in later stages of the analysis.

It is useful if the analyst can record the context in which a data set was created. Only the analyst can provide this context by describing such things as why the operations that created the data set were performed, how the operations were useful, the relevance of the operations, why the data is being preserved, and what insights were gained. It is not only useful to associate this context with the data itself but also with the portion of the analysis process in which the data set was created or used.

We see a need to provide the analyst with tools for recording this type of information. The most common mode is for the analyst to type in the information through a keyboard -- perhaps using an available text editor. Another way to capture this information is by using an audio tape recorder. The analyst can dictate insights and comments and store them so they can be played back later. Our tape recorder is computer-controlled so the recorder can automatically advance to the segment of tape containing the comments relevant to a particular save-state. The system should be designed so the analyst can use the mode of annotation with which he/she is most comfortable.

Graphics. Graphics is recognized as an essential tool for data analysis. It is currently used during all phases of data analysis including data checking and validation, data exploration, and data confirmation and presentation. However, it is often difficult to regenerate a given graph. In order to do so, the data sets must be available exactly as they were before and the conditions under which the graphics were generated must be the same. Sometimes it is difficult to even recall when during the course of the analysis the graph was produced.

Logs. Many statistical analysis packages will record the course of the analysis in a log (also called a diary or journal). The analyst can turn the log on and off as desired. When the log is turned on, all the commands entered by the user at a terminal are also written to a file. The log can provide a history of the course of the analysis, including all useful commands, non-useful commands, and mistakes. The analyst can also insert comments into the log as additional documentation. Some systems permit the analyst to have results (output) added to the log.

Even with comments inserted by the analyst, logs can often be unintelligible without detailed, time-consuming study. While they record the actions in the order in which they transpired, the data analysis process is not strictly linear in time. As mentioned earlier, the process can be depicted as a tree. One of the advantages of such a graphical depiction of the course of the analysis is that segments of the logs can be associated with particular nodes in the tree. The log segment that is associated with each node is the set of commands that caused the node to be created from its parent. This technique gives structure to the log.

Procedural Capabilities. As mentioned before, the data analysis process is iterative. The same operations are often applied to several data sets or subsets. Analysts routinely create macros (or procedures) consisting of sets of commands that are saved and stored as entities. These procedures are often parameterized so they can operate as needed against various data sets. Analysts often build macros from the log. The log is edited to remove errors and superfluous commands and then tested. It is refined and debugged. When the analyst is satisfied, the macro can be stored for later use.

In S, macros are stored in structures similar to those used for data. The analyst can differentiate between macros and data because the names of macro data structures are prefixed with "mac." Just as data sets should be associated with portions of the analysis, it is useful to associate macros with portions of the analysis in order to identify where they were created and where they were applied.

THE SAVE-STATE

We have developed a new methodology to aid the analyst in managing the data analysis process. The primary entity for managing data analysis is the "save-state," a collection of metadata and data that captures significant information about the state of the analysis at a certain point in the analysis process. The analyst may create a save-state at any time during the analysis. Save-state may be created for any number of reasons:

- The analyst may wish to designate a milestone in the analysis because a significant insight was gained at that point in the analysis.

- A decision point was reached in the analysis and several different alternatives can be explored from this point in the analysis.

- A dead end was reached that is worthy of being preserved for documentation purposes.

- A more significant alternative needs to be explored but the portion of work is incomplete and the analyst must return to it later.

Once a save-state exists, the analyst can "restore" that save-state in order to resume analysis from the point at which the save-state was created. The effect is as if the analyst had moved back in time to the point at which the save-state was created.

Information associated with the save-state includes the name of the save-state, the data and time the save-state was created and last accessed, the name of the analyst who created the save-state, the states of various icons (described below) that are part of the save-state, a list of the data sets and macros associated with the save-state, a list of plots associated with the save-state, written comments entered at the keyboard, and information that points to verbal comments saved on cassette tape. This information is sufficient to give the analyst a quick overview of what the save-state contains and why it was created. The analyst can "scan" the save-state to view this type of information without having to restore the save-state and incurring the overhead of moving data sets around.

Besides storing the save-states themselves, information is stored that allows the relationships between the various save-states to be graphically depicted. In order to do this, the system stores an internal name for the save-state; its title; a set of indices that depict the parent, child, and sibling relations between the save-states; a flag that indicates whether the save-state has actually been deleted and only a place marker is preserved; and a flag that indicates whether the save-state is the currently active state, the last scanned state, or is an incomplete state waiting to be created.

Also associated with each save-state is the segment of the log that contains the set of commands that describe the transition between the state and its parent state.

ADAM

In order to provide easy access to the tree of data analysis save-states and take advantage of its natural structure, our prototype data analysis management system, ADAM, is graphics based. The prototype has been implemented on a Digital Equipment Corporation VAX 11/780. The high resolution graphics display device is a Ramtek 9400. The audio cassette deck used for recording and playing dictation is a Yamaha K-700 cassette deck. As mentioned earlier, we are using AT&T Bell Laboratories' S statistical analysis system. S is running under Eunice, a UNIX derivative that allows VMS to be run as the base operating system while still providing UNIX functionality.

The tree of save-states is always present on the high resolution color graphics device whenever the analyst is performing data analysis management functions (see Figure 1). This is the same device that is used to display graphics during data analysis. The analyst interacts with ADAM through a series of menus. Priority windows [6] are used. Both the menus and the windows are based on the principle of successive disclosure. The analyst controls the level of detail displayed at any time. The analyst can select more or less detailed menus or graphical displays of the save-states and log segments as desired.

We have defined three classes of functions that can be performed using the menus. There are (1) functions that are performed on save-states, (2) functions that are performed on segments of the log, and (3) utility functions. The utility functions include RETURN, which allows the analyst to move to a higher level menu; HELP, which provides help on the menu currently displayed; MOVE WINDOW, which allows the various windows on the screen to be moved from place to place; and S-MODE which allows the analyst to exit the data analysis management mode and return to the S statistical analysis package to perform further analysis.

The functions that can be performed on save-states include SCAN, RESTORE, MODIFY, SHOW NETWORK, ERASE NETWORK, CREATE, and DELETE. Each of these functions is discussed in more detail below.

The SCAN function provides the analyst with an overview of the save-state being scanned. Information from the save-state is displayed in a window that

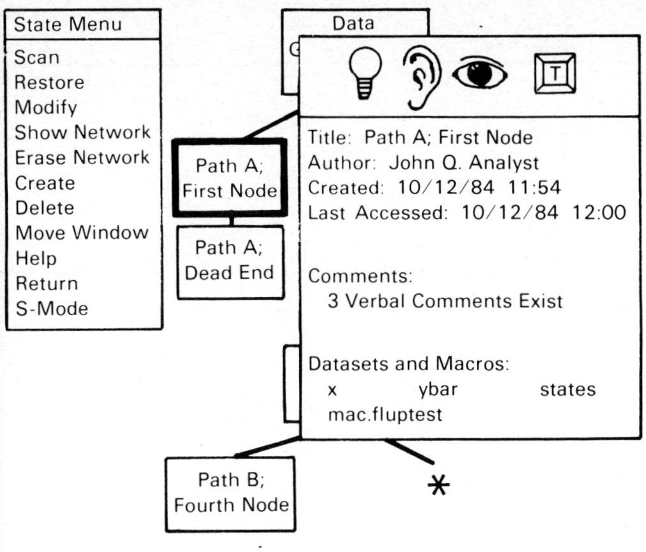

FIGURE 2. Scanning a Save-State

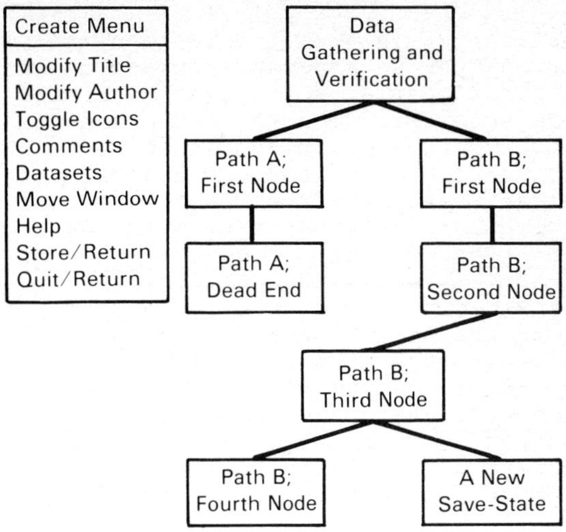

FIGURE 3. Creating a New Save-State

overlays the graph of the save-states. The SCAN display is shown in Figure 2. On the color graphics device, the save-state being scanned is shown in a color different from the color used for the other save-states. In the figure, the save-state being scanned is shown with a thickened line. The icons indicate special characteristics of the save-state. The light bulb icon indicates that special insight was gained at this point in the analysis. The ear icon indicates that the analyst has dictated some ideas on cassette tape which can be played on a computer-controlled audio cassette deck. The eye icon indicates that some graphics are associated with this save-state and can be recreated if desired. The keyboard key icon indicates that the analyst has keyed in some documentation which can be displayed in a window if desired. The SCAN function can be performed with very little overhead. No data sets are accessed or moved except the small data structure that contains information on the save-state.

The RESTORE function allows the analyst to move back to a previous point in the analysis at which the save-state was created. Whenever the analyst restores a save-state and returns to the statistical system to do more analysis, the evolution of a new save-state has begun. The RESTORE function requires that data sets currently in the working data base are replaced by the data sets belonging to the restored save-state.

The MODIFY function allows the analyst to directly modify the save-state. The analyst can modify the author or title of the save-state, turn the icons on and off, modify documentation associated with the save-state, and modify the list of data sets and macros associated with the save-state (although this does not change their contents).

Although the data analysis process is most often depicted as a tree, we recognize that the process is not strictly a tree. It is really better characterized as a network. The process becomes a network whenever the analyst includes a data set in a save-state that the save-state did not inherit from its parent (e. g., a data set is imported from another save-state not in the current analysis path). However, we recognized that continually depicting the network would make the display so confusing that it would be very difficult to get a good overview of what was going on. We created the SHOW NETWORK and ERASE NETWORK functions to allow the analyst to see the underlying network structure when desired and to remove it in order to restore the uncluttered tree representation. When the network is displayed, arrows are drawn from the appropriate non-ancestral save-states to the save-state currently being scanned or restored.

The CREATE function can be invoked as desired whenever the analyst feels that a significant point in the analysis has been reached. The options of the CREATE function are shown in Figure 3. When the analyst creates a new save-state, the analyst is prompted for a title. The analyst's name was provided as the author's name at the beginning of the ADAM session. Both the title and the author can be modified if desired. The analyst can turn icons on and off, add verbal and/or written comments, and include or exclude datasets and macros during the creation process. When the creation process is complete, the analyst can either choose the option to store the newly created save-state and return to the higher-level menu or quit and return to the higher-level menu without creating the save-state. The analyst can move back and forth between the statistical analysis system and ADAM without creating save-states.

When a new save-state is created, the star (asterisk) on the tree that marks the current point in the data analysis process is replaced with a box representing the save-state. The newly created save-state becomes the current save-state and any further processing will proceed from that point in the process. If the analyst restores another save-state, processing will proceed from that point instead.

The DELETE function can be used to mark save-states as deleted. The analyst cannot restore, scan, or modify a deleted save-state. The deleted save-state appears on the display as a circle without a title in it.

The log functions are SCAN LOG, SCAN PLOTS, EDIT LOG, and CREATE MACRO. The analyst can perform any of these functions on any log segment. When the analyst chooses SCAN LOG or SCAN PLOTS, a window is opened on the graphics device and the information is displayed as typified in Figure 4. There may be more information available than will fit in the window. The analyst can scroll between portions of information. If the analyst wants to edit the log or create macros, ADAM will invoke a standard text editor so the analyst can edit the log segment of interest.

FUTURE DIRECTIONS

ADAM was designed by a group of computer scientists and statisticians with the needs and desires of the statisticians in mind. Our next step is to test ADAM under the conditions of a real analysis. There are a number of questions we are seeking to answer. We want to know how well the concept actually works in practice. Our experience with ADAM will form the basis for the next generation data analysis management system.

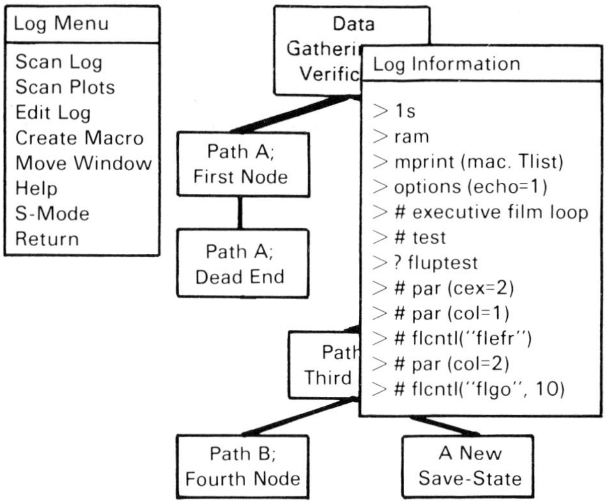

FIGURE 4. Scanning the Log

We are concerned about how we can clean up the log in order to make it more intelligible and still maintain in it what is necessary and sufficient to replicate graphics and restore the save-state. Our current DELETE command only marks a save-state as deleted. We need to determine how to truly delete save-states and what the implications of these deletes are with respect to other save-states which share the same data sets. We already know that a delete of a state with no children is different from a delete of a state that has several children. We want to investigate whether comments recorded on cassette tape are really useful and how their usefulness compares to comments that are typed into a file by the analyst.

The environment provided by machines designed for artificial intelligence work show great promise for both the programmer and the analyst. We are investigating whether these machines can provide a better environment in which to do both data analysis and data analysis management.

* This work was supported by the Applie Mathematical Sciences Group, Scientific Computing Staff, U. S. Department of Energy, under contract DE-AC06-76RLO 1830.

REFERENCES

[1] Thomas, J. J., "A User Interaction Model for Manipulation of Large Data Sets," Proceedings of the 14th Symposium on the Interface, (Springer Verlag, New York, 1982) pp. 118-128.

[2] Thomas, J. J. and Hall, D. L., "ALDS Project: Motivation, Statistical Database Management Issues, Perspectives, and Directions," Proceedings of the Second International Workshop on Statistical Database Management, Los Altos, California, September 27-29, 1983, pp. 82-88.

[3] Nicholson, W. L., Carr, D. B., Cowley, P. J., and Whiting, M. A., "The Role of Environments in Managing Data Analysis," American Statistical Association 1984 Proceedings of the Statistical Computing Section, (American Statistical Association, 1984).

[4] Carr, D. B., Cowley, P. J., Whiting, M. A., and Nicholson, W. L., "Organizational Tools for Data Analysis Environments" American Statistical Association 1984 Proceedings of the Statistical Computing Section, (American Statistical Association, 1984).

[5] Becker, R. A. and Chambers, J. M, S: A Language and System for Data Analysis, (Bell Laboratories, 1981).

[6] Littlefield, R. J., "Priority Windows: A Device Independent, Vector Oriented Approach," Computer Graphics, Vol. 18, No. 2 (1984), pp. 187-193.

Teaching of Statistical Computing

Organizer: *Richard M. Heiberger*

Invited Presentations:

Implications of Advances in Computing for Graduate Study in Statistics

William F. Eddy, Albyn C. Jones, Robert E. Kass and Mark J. Schervish

A Numerical Analysis Approach to the Teaching of Statistical Computing

Sallie Keller McNulty

Animating Statistical Algorithms (Abstract Only)

Marc H. Brown

Discussion on Teaching of Statistical Computing

Richard M. Heiberger

COMPUTER SCIENCE AND STATISTICS:
The Interface, D.M. Allen (ed.)

IMPLICATIONS OF ADVANCES IN COMPUTING FOR GRADUATE STUDY IN STATISTICS

William F. Eddy, Albyn C. Jones, Robert E. Kass, and Mark J. Schervish

Department of Statistics,
Carnegie-Mellon University
Pittsburgh, PA USA

1. INTRODUCTION

The ability of statisticians to perform calculations, both numerical and non-numerical has changed radically in the last few decades and the pace of change continues to increase. In providing graduate students with appropriately modern training, statistics departments must respond by modernizing both computing environments and curricula. These are intertwined, the course serving needs created by the environment, and the environment determining some choices among topics in the course. This paper will describe the current environment at Carnegie-Mellon and the content of a course that we believe should be taken by all Ph. D. students in statistics. We make further introductory remarks, then present the resource description in Section 2 and the course description in Section 3.

1.1. The Past at Carnegie-Mellon

Thirty years ago statisticians did their computation on desk calculators. As recently as 10 years ago, the CMU Statistics Department relied on the campus computing center's IBM 360/67. Course work was primarily theoretical, using pencil and paper exercises and no computing. At about that time the university made a strong commitment to the wide-spread use of interactive computing for educational purposes. By 1980 CMU was acquiring about one DEC 2060 and 100 terminals per year for the central computing facility, and had acquired software such as BMDP, SPSS, MINITAB, IMSL, DISSPLA, and TELLAGRAF. These facilities are used for coursework for both undergrad and grad students. The system can support about 500 simultaneous users.

In 1981, the department began acquiring its own computer terminals; in 1982, we purchased our first VAX. By the time this appears in print, the CMU Statistics Department will have its own local area network with at least six personal computers and ten workstations (including some color). Our VAX has an attached array processor and we provide our own I/O facilites (including a pen plotter and a graphics laser printer).

We are part of a very large local area network with more than 250 nodes, which is scheduled to become an order of magnitude larger in the next 18 months. In less than five years we have gone from total dependence on a large central computer facility to our own independent operation based on a substantial number of interconnected machines. Our situation has changed dramatically and will continue to change; it is our job to adapt our graduate programs to the new environment.

1.2. Intelligent Consumption

Computer hardware and computer software have become an integral part of our daily activities. We find it necessary to devote substantial effort to keeping abreast of developments in both arenas so that our environment continues to improve. We think it is wise to transfer some of what we learn to our students, as they, too, will soon make such decisions wherever they might be.

At the same time, because we do not yet have essentially unlimited computational resources, we have to be constantly aware of the limitations of our environment, in terms of both numerical accuracy and also computational efficiency. Again, we think it is good to transfer this awareness to our students. Our motive in this case is partly selfish; graduate students can have a negative impact on our shared environment if they do not appreciate the various tradeoffs amongst the resources available.

1.3. Curriculum

Computing is an integral part of the curriculum at all levels of study in statisitics at CMU. Virtually all courses other than probability theory and the theory of inference make moderate to heavy use of our computing facilities. We summarize computational activity within our program according to level of study.

1. Undergraduate: Introductory courses for routine elementary data analysis; Special topics courses such as: (i) Statistical Software Packages and (ii) Elements of Statistical Computation.

2. Masters Degree: Data analysis in the various statistical methodology courses; Special topics courses.

3. Doctoral Degree: Advanced Data Analysis coursework; Statisitical Computing coursework; Advanced topics and seminars.

4. Specialist in Computation: Software design; theoretical work on algorithms; numerical analysis.

2. RESOURCES

We list some of the hardware and software resources available, and then discuss the approach taken at CMU.

2.1. Hardware

2.1.1. Microcomputers

The IBM/PC (and its variants) and Apple/Macintosh are fairly powerful machines, particularly when compared with what was available in a central facility a decade or two ago. Random access memory used to be a scarce resource; now, personal computers may have half a megabyte of storage or more. Some statistics departments rely heavily on them, and many students will eventually be doing much of their work on these machines.

A substantial increase in the value of personal computers occurs when they are linked together in a local area network. We say more on this point below; we will also briefly discuss software for microcomputers.

2.1.2. Workstations

A workstation is a high resolution graphics terminal connected to a dedicated host computer. A workstation offers an improved environment for most computing tasks, including data analysis and software development. Muliple windows allow one to perform a variety of tasks nearly simultaneously. For example, a data analyst can look at a dataset plotted in several different projections at the same time, or can look at plots of several datasets side by side, or can compose text in one window while displaying plots in another.

Like personal computers, workstations are more valuable when connected in a local area network. The disadvantage of a timesharing system is that with many potential users, the system is often overloaded. Adding a network of workstations to the system allows the individual users access to a system that is essentially independent of the number of simultaneous users. Our experience with workstations is very positive and we find the communication capabilities imparted by a network to be essential.

2.1.3. Printers, Plotters, and Terminals

Printers and plotters are necessary, and local production of good quality text and figures is convenient. Laser printers are very nice even for non-production documents, but for routine hard-copy output a line printer and an inexpensive plotter will suffice. Graphics terminals, however, vary substantially in providing the capabilities that are essential for some research. Since it is likely that prices will continue to come down, and the use of graphical methods of statistical analysis will continue to increase, it is a distinct advantage for students to become familiar with locally programmable graphics terminals (whether slow, like the DEC GIGI, or fast and powerful, like the Tektronix 4215). Driven by host computers that are available for general computing as well, these devices can be less expensive alternatives to stand-alone workstations.

2.1.4. Parallel Processing

We have recently added an array processor (attached processor) to our hardware stable. It has roughly the power of an IBM 3083, but only costs about $25,000. We don't yet have enough experience with it to make useful statements about its role in training students, but we feel that there is much potential gain from parallel computation for statistics.

2.1.5. Networks

Networks come in various flavors. We have access to several national networks such as Bitnet and Telenet, as well as an extensive Local Area Network (LAN); the best guess is that we have about 250 machines on our LAN but some of them are located in Cleveland, New York City, Poughkeepsie, and elsewhere so the term local is somewhat abused.

In 1982 CMU and IBM signed an agreement to develop a prototype personal computing network. The goal is to provide all students, faculty and professional staff with access to personal computing workstations integrated into a network which will provide access to data-bases such as the library card catalog, communications via mail and bulletin boards, and software. With the development of effective tools for non-numeric data processing (eg: text processing, graphics, etc), even departments in the liberal and fine arts are rapidly expanding their use, and incorporating computing into their curriculum.

The CMU distributed network will have the following features:

1. Independent access: Access to a personal computer workstation and its performance is not affected by the number of simultaneous users on the network.
2. Flexible access: Users can enter the system from any suitably equipped site, for example a suitably equipped workstation at home.
3. Multiple windows: Users will be able to maintain several contexts simultaneously, moving easily from one task to another.
4. Communications: Users will be able to communicate with each other through the network. There are a mail facilities, file transfer capabilities, and central database access.
5. Multi-media capabilities: The system will be able to generate, transmit, and store video information, including both static and dynamic images. There are plans for audio capabilities as well.
6. Expandibility: Currently the system has about 50 workstations. Within 2 years the system will expand to thousands of workstations.
7. Cost-effectiveness: The prices of personal computers are declining relative to computing power much more rapidly than the prices of large scale time-sharing computers.

The planned environment has four system elements:

1. Personal computer workstations: 32-bit processors capable of executing 1 *M*illion instructions per second, having 1 *M*egabyte of memory with a 1 *M*illion pixel display of bit-map graphics (a 3*M* machine). The machine wil have no disk drives (to keep the price in the range $3000-$6000).

2. File servers with local disk storage and other special facilities such as laser printers, optical scanners, etc.

3. A communications network linking the workstations to file servers and the central facilities.

4. Central computing facilities for large scale online storage, large scale computation, and other specialized services.

Since the cost of personal computing facilities is decreasing more rapidly than that of large systems, this approach appears to be the least expensive way to provide access to computing facilities for the campus community.

2.1.6. Computing in Our Department

We have 15 FTE faculty, about 30 graduate students, and 6 administrative/secretarial staff (The staff are an integral part of our environment and, in fact, are the only ones with guaranteed access to our VAX.) Our main processor is a VAX 11/750 with 4MB of memory, 900MB of disk storage, a magnetic tape drive, 24 terminal lines, 3 distinct network interfaces and a floating-point accelerator. Our terminals are connected to it through a large central switching facility which provides terminals the opportunity to connect to any of a number of other computers on the campus (and, equally, provides other terminals the opportunity to connect to our computer). In addition, we have 5 IBM PC/XT personal computers, 2 Apple/Macintosh personal computers, 2 SUN 2/120 workstations, and (by the time this appears in print) 7 VAXstation II workstations (each with 3MB of memory, a 30MB disk, and the power of a VAX 11/780) and a Tektronics 4125 color workstation.

2.2. Software

There are several categories of software that are relevant.

1. Operating Systems: UNIX is clearly becoming the most widely-used operating system for mini- and microcomputers. Students should get some experience with it. On the other hand, detailed knowledge of operating systems is rarely of great use to statisticians. (One exception is when one has to handle large arrays with virtual memory operating systems.)

2. Statistical Packages: It is, of course, essential for students to get experience with the most common statistical packages, such as BMDP and SAS, and it is also helpful for them to use the newer, extensible programs designed for interactive data analysis, such as S and ISP.

3. Graphics: Life with a graphics terminal is easiest when there is a good library of graphics subroutines, including, if possible, some for performing transformation and rotation locally. It does not seem especially desirable for most students to program in a low-level language.

4. Subroutine Libraries: Among the most important tools for the research worker and practitioner is the subroutine library. Gaining an ability to understand computational aspects of a problem at a depth sufficient for writing good programs that make use of high quality subroutines, such as those in IMSL and LINPACK, should be a central goal of computing education for graduate students in statistics.

5. Symbolic Computing: Statistical problems are being solved with the aid of symbol manipulators, such as MACSYMA. (See Gong, 1983.) Like faculty, students will benefit by having a manipulation package available.

6. Data Base Management: Although data base management systems are not often appreciated as contributing to statistical aspects of solutions to problems, their great utility makes experience with them valuable for students who will subsequently work with large data sets.

7. Microcomputer Software: We have examined several statistical packages with mostly discouraging results. A detailed review of one reasonably good package is given by Schervish (1985). After leaving CMU, some of our students will work primarily on microcomputers, and it is worthwhile to give them the opportunity to learn about software for micros while they are here.

8. Text processing: Faculty and students alike make use of SCRIBE for document production ranging from course handouts to articles and dissertations. In conjunction with computer communications facilities, this promises to alter the way many of us conduct our research. For example, this article was a collaboration of four authors who communicated primarily by computer mail, including passing drafts and revisions back and forth.

3. CURRICULUM FOR PH. D. STUDENTS

Computing has become a basic tool for both the theory and the practice of statistics, much as measure theoretic probability is a basic tool for mathematical statistics, and should have a similar place in the curriculum. Students, even those who are not planning to specialize in statistical computing, need to be aware of the theory and practice of computing.

We outline here a one-semester course in statistical computing. One of us (Eddy) has taught a similar course several times, and a related course was taught by two of us together (Eddy and Kass). At Carnegie-Mellon, this course is presently integrated into a two-semester course in Data Analysis. The integration, however, is quite rough -- for the most part we deal first with computing and then with data analysis. There are some nice opportunities for taking advantage of the complementary nature of these two areas of statistics, but many of the topics are basic elements of computing and so must be taught first on their own.

Clearly some topics must be left out of a one-semester course. Our choices reflect not only judgments about the relative importance of various topics, but also the existence of related courses in the department. Some topics that are sometimes mentioned as important in a course on statistical computing fit better into other parts of our curriculum. For example, the addition of routines to an extensible software package is a topic that can be covered in the statistical software packages course. This is an undergraduate course in our curriculum, though several graduate students usually attend it. For alternative suggestions see Bates (1983) and Kennedy (1983).

3.1. Fundamental Topics in Computer Science

3.1.1. Computer Organization and Hardware

We feel that it is important to have an appreciation for the organizational structure of a computer and suspect that this will become somewhat more important as various kinds of concurrent computation become more commonplace. We therefore discuss the most basic elements of architecture, describing the central processing unit, memory, and input-output devices. It can be worthwhile to discuss busses and microprogramming. We usually talk about the architecture of a particular machine in some detail, and it makes sense to discuss the machine that students will use most heavily. Currently, that machine is the DEC VAX11/750; in our next iteration we may also discuss the VAXstation II. Various kinds of parallel architectures could be included here but we prefer to postpone that material until we actually discuss concurrent processing in detail.

3.1.2. Data Representation

A thorough knowledge of internal data representation is a prerequisite to understanding of several other topics, such as arithmetic, error analysis, random number generation, and hashing. Obviously this knowledge is also critical to program debugging. We feel it is essential that students understand the representations used not only on their machine but also on a variety of other machines. We cover fixed point numbers, floating point numbers (including the IEEE P754 Floating point standard), character data (BCD, EBCDIC, ASCII), and bit strings.

3.1.3. Computer Arithmetic

Basic to understanding of numerical analysis is understanding of computer arithmetic and rounding error. Students should be aware of the basic operations which are available, how they are performed, and the types of errors that can occur, such as overflow, underflow, and rounding. For example, it is well known to the computing community (though often not to students) that computation of a sum of squares by the so-called "desk-calculator" algorithm

$$\Sigma x_i^2 - n\bar{x}^2$$

is numerically unstable. Students should learn of the better methods, and why they are superior. (See Chan, et al., 1983.) We also introduce the techniques of error analysis, including backwards error analysis and stochastic error analysis.

Students need to understand that different machines use different data representations, and different technologies for rounding. They should appreciate the effect of these differences on the accuracy of computer arithmetic. They should also know of the IEEE standard for floating point computations, and understand its advantages, and they should be cognizant of programming methods that achieve the effects of extended precision.

3.1.4. Data Structures

Students who have programmed in Fortran or Pascal will know what an array is, but typically they have no experience, or even awareness of other data structures, the use of pointers, and related algorithms. We introduce students to a variety of useful data structures including: Linear lists and linked lists, arrays, graphs, trees, and hash tables. At the same time we cover a variety of related algorithms, such as: insertion and deletion of data items from these structures, balancing trees, garbage collection, etc.

3.1.5. Basic Algorithms

In addition to algorithms relating to data structures, there are basic algorithms and theoretical issues that students should be aware of. Our list includes: Iteration (most students already know this), recursion (the divide and conquer strategy; FFT, linear-time medians), sorting, searching, and NP-Completeness (e.g., the Traveling Salesman problem).

3.2. Numerical Techniques

3.2.1. Linear Algebra

It is essential that students understand how the computations for least squares linear regression are, or should be, performed. They need to understand the computational details of Gaussian elimination and the Cholesky decomposition of X^TX. They need to understand the orthogonal decomposition techniques: Householder rotations, the QR decomposition, and the singular value decomposition. In our program these topics are covered briefly during a first-year graduate course in mathematical methods for statistics, but it is worth reviewing and elaborating here. Students should also understand what is gained and what is lost when the computations are performed on X^TX.

There is a variety of other topics we cover in less detail, including eigenvector-eigenvalue methods such as the symmetric QR method, condition numbers and computational accuracy, ANOVA calculations for orthogonal designs and conjugate gradient techniques for non-orthogonal designs.

3.2.2. Optimization

Among computational problems of applied statistics, optimization is ubiquitous, the most common application being maximum likelihood estimation. An excellent recent text on nonlinear optimization that we have used is Dennis and Schnabel (1983); see Kass (1985) for an extended review. High-quality Fortran programs are also available, e.g., in MINPACK and IMSL.

We believe that students need to understand both the theoretical and computational issues in optimization. We consider Newton and Newton-like methods, and the simplex method (Nelder and Mead, 1965) for general minimization problems, and the Gauss-Newton method and derivative-free least-squares (e.g., Ralston and Jennrich, 1978) for nonlinear least-squares. We expect the students to learn the basic analysis including the convergence and rates of convergence arguments; they should also understand the stopping rules. In addition, we discuss some ideas for dealing with constraints. We devote most of our efforts to Newton's method and its variants.

In our teaching experience we have found it quite worthwhile to go over various Newton-like methods in the one-dimensional case, and require the students to write programs implementing each of the techniques discussed. It is important for students to understand the motivation for the use of Newton-like methods, as well as to gain some idea of the options available and their possible pitfalls. Furthermore, the use of difference quotients in place of derivatives opens the door to the study of secant methods, which form the basis of most available high-quality general-purpose code. There is clever and sometimes elegant mathematics involved, and this class of methods has received substantial attention in recent years. An aspect of secant methods that is extremely important for statistics is that the approximations to the Hessian deteriorate. We make sure the students appreciate that the final Hessian in the output of most quasi-Newton programs should not be trusted.

We also spend some time on linear programming and constrained optimization problems. We focus on Lagrange multipliers and the Kuhn-Tucker conditions. In addition, we discuss some of the discrete optimization problems, a topic that we believe will become increasingly important for statistics.

3.2.3. Approximation of Functions

Approximation of functions is already familiar to statistics students in the form of L_p approximation. The choices p=1 and 2 are most familiar, but for many computational purposes, $p=\infty$ is more appropriate. Orthogonal polynomials are also familiar, but discussion of their use in computing will present them in a manner that many students will not have seen. This is a crucial part of their mathematical training, as well, so it should not be skipped. Also fundamental is an introduction to the theory of rational function approximations; in addition, we include discussion of interpolation. Although trigonometric approximation and the Fast Fourier Transform may be covered in other courses, its importance makes inclusion of it here highly desirable. Some discussion of approximation by splines is also useful.

3.2.4. Quadrature

Students should be aware of the popular quadrature methods for evaluating definite integrals. It is important to discuss both one-step use of quadrature formulas (e.g., for Gaussian or Newton-Cotes quadrature), having well-known error bounds, and also "adaptive quadrature" methods, such as the IMSL routine DCADRE. The stopping rule for adaptive quadrature is crucial, since one can introduce errors into the solution by using stopping rules based only on the change in approximations achieved at successive iterations. (See Bohrer and Schervish, 1981, and Schervish, 1984, for examples.) The errors become particularly troublesome in multivariate integrals. Students should also be exposed to the Monte Carlo methods of integration, including importance sampling.

3.3. Computer Intensive Methods

3.3.1. Graphics

Much interesting recent research in statistical computing involves graphics. There is certainly room in a course such as this for extensive discussion of statistical graphics, including methods such as projection pursuit, but we have not yet emphasized this area within our version of the course. At the least, students need some awareness of the ongoing efforts and the existing methods for displaying multivariate data as described, for example, in Gnanadesikan (1977) and Chambers et al. (1983).

3.3.2. Random Numbers and Simulation

Simulation methods are often used to evaluate the accuracy of asymptotic aproximations; in some cases where analytical results are not available a simulation is the only available technique. Since random numbers from a given distribution may be generated from a sequence of uniformly distributed random numbers, the basic problem is the generation of uniformly distributed random numbers. Standard methods include the linear congruential method, the feedback shift register method, and combination methods such as the idea of MacLaren and Marsaglia (1965) use one sequence of random numbers to shuffle another. Once a sequence of uniformly distributed random numbers has been generated, observations from arbitrary distributions may be derived by various techniques, including use of the inverse distribution function and acceptance-rejection methods.

The usual goal of a Monte Carlo experiment is to estimate the mean or some other functional of the sampling distribution of a statistic. Various techniques for variance reduction are used, including: increased sample size, use of antithetic variables, and stratification.

Monte Carlo style techniques have other applications to statistical practice, including the evaluation of high dimensional integrals, evaluation of posterior distributions, and bootstrapping. Students must gain a solid understanding of the basic elements of this central topic in statistical computing.

3.4. Concurrent Processing

We believe that the most dramatic change in computing in the next decade is going to be the evolution of the various very high-speed computers. Our students need some appreciation of this, and we discuss concurrent computation in several parts of the course. Our detailed introduction includes description of various architectures (see, e.g., Schwartz, 1983), interprocessor communication networks, and a little material on numerical analysis (see, e.g., Schendel, 1984). We expect that the next iteration of our course will include some actual hands-on work with our array processor.

3.5. Writing Software

Bates (1983) reports that completion of a term project of writing, testing, and documenting a piece of statistical software gives students a valuable sense of the requirements of producing good software. We prefer to have students devote their time to learning the large amount of material we cover, but we share with Bates the desire to impart an appreciation of some of the concepts of software engineering, such as top-down and modular design and structured programming languages, and the variety of useful tools for software writing, including the subroutine packages such as IMSL and LINPACK, interactive languages such as APL, and matrix manipulation languages such as those in SAS or S. Thus, we integrate these topics into the course where we can, but do not devote much time to software writing per se.

REFERENCES

[1] Bates, D. (1983). Teaching statistical computing. *Proceedings of the Statistical Computing Section of the American Statistical Association*, 63-64.

[2] Bohrer, R.E. and M.J. Schervish (1981). An error-bounded algorithm for normal probabilities of rectangular regions. *Technometrics* **23**, 297-300.

[3] Chambers, J.M., W.S. Cleveland, B. Kleiner, and P.A. Tukey (1983). *Graphical Methods for Data Analysis*. Wadsworth, Monterey.

[4] Chan, T.F., Golub, G.H., and LeVeque, R.J. (1983). Algorithms for computing the sample variance: Analysis and recommendations. *Am. Statist.*,**37**, 242-247.

[5] Dennis, J.E. and R.B. Schnabel (1983). *Numerical Methods for Unconstrained Optimization and Nonlinear Equations*. Prentice-Hall, Englewood Cliffs, NJ.

[6] Gong, G. (1983). Letting MACSYMA help. *Computer Science and Statistics*, fifteenth symposium. Springer-Verlag, New York.

[7] Gnanadesikan, R. (1977). *Methods for Statistical Data Analysis of Multivariate Observations*. Wiley, New York.

[8] Kass, R.E. (1985). Review of *Numerical Methods for Unconstrained Optimization* by J.E. Dennis and R.B. Schnabel. *Journal of the American Statistical Association*, **80**, 247-248.

[9] Kennedy, W.J. (1983). A curriculum in statistical computing. *Proceedings of the Statistical Computing Section of the American Statistical Association*, 65-66.

[10] MacLaren, M.D. and G. Marsaglia (1965). Uniform random number generators. *Journal of the ACM*,**12**, 83-89.

[11] Nelder, J.A. and R. Mead (1965). A simplex method for function minimization. *Computer J.*, **7**, 308-313.

[12] Ralston, M.L. and R.I. Jennrich (1978). Dud, a derivative-free algorithm for nonlinear least squares. *Technometrics*, **20**, 7-14.

[13] Schendel, U. (1984). *Introduction to Numerical Methods for Parallel Computers*. Wiley, New York.

[14] Schervish, M.J. (1984). Multivariate normal probabilities with error bound. *Applied Statistics* **33**, 81-94 (correction forthcoming).

[15] Schervish, M.J. (1985). Review of the statistical package SYSTAT. *Am. Statist.* **39**, 67-70.

[16] Schwartz, J.T. (1983). Design alternatives for ultraperformance parallel computers. Technical Report, Computer Science Department, Courant Institute of Mathematical Sciences.

COMPUTER SCIENCE AND STATISTICS:
The Interface, D.M. Allen (ed.)

A NUMERICAL ANALYSIS APPROACH TO THE TEACHING OF STATISTICAL COMPUTING

Sallie Keller McNulty

University of North Carolina at Greensboro
Greensboro, North Carolina

The growing field of statistical computing has created the need for students to obtain a more formal education in the subject. This gives rise to the following questions. Where does statistical computing fit into the education of statistics majors? Is there some common statistical computing body of knowledge these students should receive? How machine oriented should this training be? These topics are addressed from the perspectives of both undergraduate and graduate study in statistics. Is it our goal to teach students studying statistical computing a skill or the theory behind that skill? The answer to this question may be based on the level of education and the background required of the student before entering a statistical computing course.

1. INTRODUCTION

This section of the conference is about the teaching of statistical computing. Is statistical computing sufficiently important to be included in a statistics program? Rather than give my own, perhaps biased, opinion of the importance and nontrivial nature of statistical computing, I quote M.G. Kendall (1972).

"... bright ideas do not fructify unless we can bring them to bear on numerical material, and for many of our outstanding problems, as we shall see, the computer is necessary."

"... the statistician requires a full mathematical armory to bring his solving process to the point where the machine can take over if required."

Statistical computing, unlike other areas of specialization within the discipline of statistics, has an ambiguous connotation. A popular notion about someone trained in statistical computing is that they are simply very clever in manipulating statistical software packages. This is neither the goal nor the outcome of a statistical computing education.

One way to remove this ambiguous connotation is for those of us in the field to establish what major topics should be included in statistical computing courses. It is clear in what course a student will learn about stratified random sampling and ratio estimators. It is not evident in what course, if any, a student will learn about random number generation, sweep operators, and numerical stability.

This paper outlines topics that ought to be included in statistical computing courses. Statistical computing training for both graduate and undergraduate students is discussed. Suggestions are made regarding where these courses fit into the statistics curriculum and how machine oriented they should be. It is hoped that a result of the papers presented in this section of the conference will be to stimulate discussion among those of us involved in statistical computing about the issues mentioned above.

2. STATISTICAL COMPUTING TOPICS

Two interesting committee reports about the training of statisticians have appeared recently in The American Statistician . The first appeared in May 1980 and was directed at the training of statisticians for employment in industry. The second appeared in May 1982 and dealt with the training of the statistician for the federal government. As might be expected, there is considerable overlap in the recommendations given in these reports. Computing skills and a knowledge of statistical computing was indicated to be important by both reports. The specific recommendations in these areas fell into four categories.

1. Knowledge of a scientific programming language.

2. Experience with several of the most popular statistical software packages.

3. Experience with the construction and maintenance of large data base files.

4. Instruction in proper numerical analysis techniques for statistical computations.

Most statisticians would concur with Kennedy (1982) that Items 1 and 2 should be a required part of every statistics student's education. Kennedy also points out that the experience of Item 3 is frequently attained through involvement in consulting. For students specializing in statistical computing, a special effort should be made to acquire this experience in data base management. To fulfill Item 4 the student would need to complete one or two statistical computing courses.

There appears to be mixed feelings within the statistical computing community as to whether a statistical computing course should be a requirement or an elective for the statistics major. In any case, statistics majors should gain an awareness of what general topics are considered to be in the field of statistical computing from their overall statistics education. One purpose of the text Statistical Computing by Kennedy and Gentle (1980) was to present, in one place, material that is central to the area of statistical computing. A brief outline of the topics included in their book is as follows.

1. Introduction to the history and literature of statistical computing.

2. Computer hardware operating characteristics.

3. Computer software and programming considerations for package design.

4. Floating-point arithmetic and an introduction to error analysis.

5. Random number generation, testing, and an introduction to general simulation methodology.

6. Approximating probabilities, percentiles and other special functions.

7. Numerical methods in linear algebra with emphasis on methods most useful in statistics.

8. Linear least squares computations including model building and solutions under constraints.

9. Nonlinear least squares computations for unconstrained and constrained problems.

10. Computational methods for alternatives to least squares --- robust methods.

A partial overlap with the material listed here can be found in Computational Methods for Data Analysis by Chambers (1977). An additional topic included in Chambers's text is graphical procedures. Another interesting book on the subject of statistical computing is Statistical Computation by Maindonald (1984) which deals extensively with Topics 7 to 10 in the outline. It is appropriate, in this author's opinion, to include all of the topics listed above as well as some graphical procedures in the battery of statistical computing courses which is offered.

3. UNDERGRADUATE PROGRAM

A distinction has not yet been made in this paper between undergraduate and graduate education in statistical computing. In general, the difference between undergraduate and graduate study in any area of specialization is usually the amount and depth of the material covered. The basic content of the material remains largely the same. There is no reason for statistical computing to be handled differently.

At present, there are several recurring themes in undergraduate statistical computing courses. These are data structures, data base management, and the use of statistical packages. This may be due to the lack of appropriate prerequisites for a statistical computing course such as calculus and undergraduate mathematical statistics, thus making it difficult to consider many of the topics listed in Section 2. Data structures and data base management are some of the ACM (Association for Computing Machinery) curriculum recomendations for computer science. Thus, students could probably acquire expertise in these areas by taking a course(s) to be found among the university's computer science offerings. If statistical package experience other than what is obtained in the required statistics courses is needed, then perhaps a specific statistics package course should be offered. To avoid unnecessary confusion with respect to the field of statistical computing, it is suggested, by this author, that courses of the nature just discussed be titled something other than statistical computing.

With the prerequisites of calculus, probability theory, and some computer programming, a first course in statistical computing for the undergraduate statistics major could include Topics 1 to 6 from Section 2 and some graphical procedures. This set of material does not require a sophisticated background in either linear algebra or linear models. It would be very easy for such a course to turn into a general numerical analysis class. When teaching statistical computing, care must be taken to emphasize which numerical methods are important to the statistician and why they are important. A second course in statistical computing for the undergraduate student is probably not necessary. The student may benefit more from an additional mathematics class or exposure to another area of specialization within statistics.

4. GRADUATE PROGRAM

To study statistical computing at the graduate level, prerequisite knowledge of a scientific programming language, statistical theory, and statistical methods is needed. For a first semester graduate course in statistical computing, Kennedy (1982) recommends covering all the material listed in Section 2. This results in breadth but not depth of coverage. Kennedy suggests an advanced selected topics course be offered for those students interested in specializing in statistical computing. The success of the statistical computing program at Iowa State University indicates that Kennedy's plan works well in both exposing the statistics graduate student to statistical computing and in preparing students to carry out statistical computing research.

An alternative to the program described by Kennedy (1982) might be to offer two <u>nonsequential</u> graduate statistical computing courses. One would cover Topics 1 to 6 and a second would cover Topics 7 to 10 from Section 2. This would provide a more in depth coverage of the subjects. Courses of this nature could also allow time for inclusion of extra statistical computing material of special interest to the professor. Although greater depth of coverage and a choice of topics may better suit the needs of the student who completes only one course in statistical computing, this program may fall short of meeting all the needs of students wishing to specialize in statistical computing.

5. MACHINE ORIENTATION

As is indicated by the title of this paper, this author believes the numerical analysis aspect of statistical computing should be emphasized more than the computer programming aspect. It is, however, important for students to understand the constraints imposed on the numerical methods by the computer. This knowledge is the key to improving their computer programming skills.

The computer can be effectively used to reinforce the algorithm construction and the numerical analysis necessary to bring a statistical concept from its theoretical form to its computer approximation. Too much computing tends to pull the emphasis of the statistical computing course away from the statistical and numerical analysis issues and towards computer science and computer programming problems.

To some degree, the amount of computing will depend on the background of the students in the class. Also, more computing would tend to be included in an undergraduate statistical computing course than in a graduate course. Direct exposure to some of the statistical libraries would be beneficial to the undergraduate student. The graduate student will usually become familiar with these libraries through other statistics classes or in their cousulting experience.

It is more important that a student who has studied statistical computing be able to determine if the statistical needs are being statisfied by a given algorithm than to be a top notch scientific programmer. The statistical computing student should, however, gain enough background in a statistical computing course to communicate effectively with the computer scientist about stabilty of algorithms and programming considerations which optimize computer resources.

Perhaps due to the increased interest in general computational methods, numerical linear algebra courses and database management courses are more readily available in mathematics and/or computer science departments. Also, students are exposed sooner and in more depth to the statistical software packages in the standard statistics core classes. Students completing the courses mentioned above learn to handle canned routines and learn to do some scientific programming. Using these courses as prerequistes to the statistical computing course(s) would make it possible to concentrate on numerical methods in statistics with less emphasis on both pure numerical analysis and computer programming. The software packages cannot possible keep up with all the new statistical methodology or incorporate every possible twist in the more common methods. Consequently, it is desirable to educate students in statistical computing in such a way that when confronted with a statistical analysis problem the will not be constrained to those methods that are available in existing computer software.

6. CONCLUDING REMARKS

Minton (1983) discusses the establishment of statistics as a discipline. The criterion given in Minton's paper for the visibilty and recognition of a discipline can also be applied to visibilty and recognition of an area of specialization within a discipline. The criteria are

1. A theory and body of literature;

2. A significant number of professionals working in the field;

3. More than a few professional journals regularly publishing new advances in the subject;

4. A significant market demand for its services.

The field of statistical computing is moving towards fulfilling all of these criteria. To expedite this effort it would be helpful to define more clearly the statistical computing body of knowledge. This can be accomplished through our statistical computing course offerings and through the exposure we give students to statistical computing in their other statistics classes.

ACKNOWLEDGEMENTS

The author would like to thank Dr. William J. Kennedy and Dr. Robert W. Jernigan for their suggestions on the topic and their review of this paper.

REFERENCES

[1] Chambers, J. M., Computational Methods for Data Analysis (Wiley, New York, 1977).

[2] Eldridge, M. D. et. al., Preparing Statisticians for Careers in the Federal Government: Report of the ASA Section on Statistical Education Committee on Training of Statisticians for Government, The American Statisitcian, 34 (1982) 69-81.

[3] Kendall, M. G., The History and Future of Statisitics, Statistical Papers in Honor of George W. Snedecor (Iowa State Press, Ames, Iowa, 1972).

[4] Kennedy, W. J., The Statistical Computing Portion of a Graduate Education in Statistics, Teaching of Statistics and Statistical Consulting (Academic Press, New York, 1982).

[5] Kennedy, W. J. and Gentle, J. E., Statistical Computing (Marcel Dekker, New York, 1980).

[6] Maindonald, J. H., Statistical Computation (Wiley, New York, 1984).

[7] Minton P. D., The Visibility of Statisitics as a Discipline, The American Statistician, 37 (1983) 284-289.

[8] Snee, R. D. et. al., Preparing Statisticians for Careers in Industry: Report of the ASA Section on Statistical Education Committee on Training of Statisticians for Industry, The American Statistician, 34 (1980) 65-75.

COMPUTER SCIENCE AND STATISTICS:
The Interface, D.M. Allen (ed.)

ANIMATING STATISTICAL ALGORITHMS

Marc H. Brown

Department of Computer Science
Brown University
Providence, RI 02912

High-performance graphics-based workstations have made possible a quantum leap forward in the quality of tools available for teaching and studying statistical algorithms. For example, the Department of Computer Science at Brown University has a specially designed auditorium/lecture-hall containing 60 such workstations, interconnected by a high-bandwidth resource-sharing local area network (LAN). Rather than using a chalkboard or viewgraphs, instructors are able to use dynamic simulations of algorithms being taught. Students are able to interact with these real-time animations in order to gain better insight into their operational characteristics. Students are transformed from passive listeners to active participants in the learning process.

In this talk, we will describe the software environment we have developed for animating algorithms. Typically, animations contain multiple *views* of the data. As the algorithm progresses, all of the views are updated simultaneously. Users are able to stop the animation at any time, control the speed of the animation and even whether it should run in reverse, single-step and set breakpoints using entities meaningful to the algorithm being animated. In addition, multiple algorithms may be run in parallel in order to better compare and contrast them. We will also give examples from the host of computer science and statistics algorithms that we have animaged, and show a videotape of some animations.

COMPUTER SCIENCE AND STATISTICS:
The Interface, D.M. Allen (ed.)

Discussion on Teaching of Statistical Computing

Richard M. Heiberger

Temple University
Department of Statistics
Philadelphia, PA 19122

This discussion comments primarily on software design topics other than the **numerical analysis** issues covered by the other speakers. It includes a short **discussion of my** attempt to illustrate by counterexample the dictum that Householder **reflection calculations** should be based on the numerically optimum reflection angle.

The speakers were consistent in their emphasis on the fundamental area of numerical analysis. The major addition I have to the numerical analysis theme is a recommendation for the new book by John R. Rice [1] as a major reference for everyone's library and as an excellent text for a numerical analysis course. Rice discusses the derivation of algorithms, programming of algorithms, and use of published software. Graphs, examples, subroutines, and problem sets are in abundance. Pathological cases are carefully treated. There are several chapters on design and use of program libraries. The book includes the ACM index of all algorithms published from 1960-1980 in 17 major journals and a detailed index to the IMSL Library Subroutines.

In my course I also place a strong emphasis on issues of design of programming systems and packages. Not only do I discuss individual algorithms, I also place them in the context of a package. Therefore I discuss communication among subroutines, design of overlay structures, and design of user-friendly user interfaces. I stress the importance of adhering strictly to professional programming standards to make long-term maintenance of a system possible. One of my class projects is an assignment to write a simple subroutine and attach it to an existing package to take advantage of the user control language and data management facilities developed for the package. I have used MINITAB [2] and SAS [3] for this purpose.

I also find it helpful to explore the boundaries of a problem. For example, while discussing the Householder reflection, I decided to verify that the sign of the reflection really made the important difference to numerical stability that is claimed for it. It does, of course, but it was initially difficult to construct a case where choosing the wrong sign caused cancellation of significant digits.

I found two conditions were necessary for an example to display numerical difficulties. The two defining vectors, the ones that are to be reflected onto a constant times the direction of the other, must be nearly linear dependent and the computations must use single precision accumulation of inner products. Only with that combination of conditions was I able to use the non-optimal sign to create a "reflection" matrix that did not reflect the two vectors onto each other. Neither near dependence nor single precision accumulation by itself was enough to make the wrong sign give trouble. Only when instability was present in both the data and in the computational process was there an incorrect calculation. This example reinforces two conclusions. One, proper computational paranoia, such as always automatically using double precision accumulation, can protect you from some potential problems. Two, well-posed problems with stable data can lead to correct computations even if there are instabilities in the algorithm.

[1] Rice, John R., Numerical Methods, Software, and Analysis: IMSL(r) Reference Edition (McGraw Hill, New York, 1983)
[2] Ryan, Thomas A., Jr., Brian L Joiner, and Barbara F Ryan, MINITAB Reference Manual (MINITAB Project, University Park, PA, 1982)
[3] SAS Programmers Guide (SAS Institute, Inc., Cary, NC, 1982)

Repeated Measures Data with Missing Observations

Organizer: *Alan B. Forsythe*

Invited Presentations:

A Monte Carlo Study of Parallelism Tests for Complete and Incomplete Growth Curve Data

Neil C. Schwertman, Sallysue Stein, William Flynn, and Kathryn L. Schenk

An Algorithmic Approach for the Fitting of a General Mixed ANOVA Model Appropriate in Longitudinal Settings

Daniel O. Stram, Nan M. Laird, and James H. Ware

COMPUTER SCIENCE AND STATISTICS:
The Interface, D.M. Allen (ed.)
Elsevier Science Publishers B.V. (North-Holland), 1986

A MONTE CARLO STUDY OF PARALLELISM TESTS FOR COMPLETE AND INCOMPLETE GROWTH CURVE DATA

Neil C. Schwertman, Sallysue Stein, William Flynn, Kathryn L. Schenk

California State University, Chico, CA, IBM Corporation, Boca Raton, FL,
& Data Management Computer Systems, Auburn, CA

Monte Carlo simulations using a broad spectrum of dispersion structures are used to compare for significance level and power tests for the parallelism of the response curves for both complete and incomplete data.
The methods used are the split-plot, Hotelling's T-square, analysis of the estimated regression coefficients for each subject, successive differences, and estimation of missing data. For complete data the split-plot analysis using the Geisser-Greenhouse correction and Hotelling's T-square on the estimated regression coefficients for each subject were best. For incomplete data the split-plot analysis using the Geisser-Greenhouse correction from the smoothed dispersion matrix was most satisfactory.

COMPUTER SCIENCE AND STATISTICS:
The Interface, D.M. Allen (ed.)

AN ALGORITHMIC APPROACH FOR THE FITTING OF A GENERAL MIXED ANOVA MODEL APPROPRIATE IN LONGITUDINAL SETTINGS.

Daniel O. Stram, Nan M. Laird and James H. Ware

Department of Biostatistics,
Harvard School of Public Health.
667 Huntington Avenue,
Boston, MA 02134

The utility of the EM algorithm in fitting mixed ANOVA models is discussed. Issues addressed range from practical programming considerations to the suitability of the EM technique for the inclusion of empirical or investigator Bayesian prior information into the estimates of fixed effects and variance components. The class of ANOVA models considered are appropriate in many longitudinal problems including growth curve and repeated measures analysis with arbitrary patterns of missing data. An example of growth curve modeling is used as an illustration of the estimation techniques and model specification issues -- and for the purposes of comparing the approach with a simpler 'two-stage' analysis.

1. INTRODUCTION.

This paper discusses the use of the EM algorithm for fitting a subclass of mixed (fixed and random effects) linear models to longitudinal data. The class of models considered includes growth curves as important special cases. We illustrate growth curve modeling with an example taken from an energy conservation study, which serves to illustrate the general principles of the longitudinal mixed model approach. Also discussed is the rate of convergence of the EM algorithm in this variance component setting. Simple approaches to speeding the convergence of the algorithm are described and illustrated.

2. THE CLASS OF LONGITUDINAL MODELS.

The class of models considered here, which we term 'longitudinal random effects' models (Laird and Ware 1982) may be written, as a representation for n_i different responses for the ith subject, in the form

$$Y_i = X_i\alpha + Z_i\beta_i + \varepsilon_i \quad (i=1,\ldots,m) \qquad (1)$$

Here X_i and Z_i are known design matrices (of order $n_i$xp and $n_i$xq respectively), α is a px1 unknown vector of fixed effects, β_i is the qx1 vector of random effects for the ith subject, which we assume to be multivariate normally distributed as N(0,D) independently of ε_i and β_j for $i \neq j$. The 'intra-subject' error term ε_i is assumed to be normal, $N(0,\sigma^2 I)$. The parameters of the model which are to be estimated are then the vector of fixed effects, α, and the variance components, namely σ^2 and the (q+1)q/2 distinct elements of D. In addition one often considers the estimation of the random effects, β_i, themselves for the purposes of residual analysis and assessing the influence of outliers. The LRE class of models is characterized by the nesting of the random effects within subject. The model imposes a specific form on the covariance structure of the distribution of the y_i. That is, the model for the independent y_i vectors is multivariate normal with means $X_i\alpha$ and covariance matrix $\Sigma_i = \sigma^2 I + Z_i D Z_i'$.

Growth curves can be considered as a special subclass of these models characterized by a linear relationship between the columns of the X_i and Z_i matrices which we may write as

$$X_i = Z_i A_i,$$

with A_i a known matrix.

3. AN EXAMPLE OF GROWTH CURVE DATA.

3.1 The Princeton 'Modular Retrofit Experiment'.

The data used as an illustration here are from an experiment in energy conservation conducted by Princeton University's Center for Energy and Environmental Studies (Dutt et. al. 1982). In the late 1970's the Center organized a study which sought to measure the impact of two levels of conservation activities on energy utilization in preexisting single family New Jersey housing (Dutt et. al. 1982). The levels of so-called energy 'retrofit' activity were:

1. 'House-doctor'

2. 'Major-retrofit'

The House-doctor level involved a single day visit by personnel trained in making relatively inexpensive repairs to ventilation, heating, and insulation systems. The major-retrofit level included the house-doctor treatment and the addition of attic and wall insulation. To test the efficacy of these two retrofit regimens a total of 138 New Jersey houses heated with natural gas were enrolled in the study known as the 'Modular Retrofit Experiment' (MRE) and were randomly assigned to one of the treatment groups -- control, where no actions were performed by the study, house-doctor, and major-retrofit. With the cooperation of participating gas utilities utility billing data (usually collected on a one-month billing cycle) was obtained for one year prior to the retrofit and house-doctor activity (pre-intervention). Post-intervention data were obtained by collecting meter reading data for an additional year following the retrofit period. In the subsequent paragraphs we consider two approaches to these data -- a two-stage model and a unified longitudinal random effects model.

3.2 Two-stage analysis of the MRE data.

A two-stage analysis of the MRE data can be performed in the following fashion. Let y_{ijk}, $j=0,1$, $k=1,\dots,n_{ij}$, $i=1,\dots,138$ be the average daily natural gas consumptions by the ith house for the kth meter reading in the jth period (j=0 for pre-period and j=1 for post). For each house in each period (pre and post) models of the following form were fit, using least squares.

$$y_{ijk} = a_{ij} + b_{ij}HDD_{ijk} + \varepsilon_{ijk} \qquad (2)$$

Here a_{ij} is the heating-insensitive or 'base-level' consumption for house i in period j, b_{ij} is the weather sensitive 'heating-slope' and HDD_{ijk} is the average daily heating degree-days observed for meter reading period k for house i in period j. Once $\hat{a}_{ij}$ and $\hat{b}_{ij}$ were obtained by least squares the effects of the levels of retrofit activity on the intercept and heating slope over the experiment as a whole were assessed by calculating the differences

$$\Delta\hat{a}_{ij} = \hat{a}_{i2} - \hat{a}_{i1}$$

and

$$\Delta\hat{b}_{ij} = \hat{b}_{i2} - \hat{b}_{i1}$$

and fitting two separate univariate ANOVA models to these data of form

$$\Delta\hat{a}_{ij} = \mu_o + \mu_1 H_i + \mu_2 R_i + \varepsilon_i$$

and

$$\Delta\hat{b}_{ij} = \gamma_o + \gamma_1 H_i + \gamma_2 R_i + \varepsilon_i \qquad (3)$$

where H_i and R_i are dummy variables indicating membership in the house-doctor and major retrofit groups respectively.

3.3 The longitudinal random effects approach to the MRE data.

An alternative to the two-stage analysis is the application of a growth curve analysis which fits a single overall model for all the consumption data in the experiment. The way of writing such a LRE model which seems most analogous to the two-stage model is as

$$y_i = Z_i \overbrace{\begin{bmatrix} a_i \\ \Delta a_i \\ b_i \\ \Delta b_i \end{bmatrix}}^{\beta_i} + X_i \overbrace{\begin{bmatrix} a \\ \mu_o \\ \mu_1 \\ \mu_2 \\ b \\ \gamma_o \\ q_1 \\ \gamma_2 \end{bmatrix}}^{\alpha} + \varepsilon_i \qquad (4)$$

where Z_i is of form

$$\begin{bmatrix} \mathbf{1}_{i,o} & \mathbf{0}_{i,o} & \mathbf{HDD}_{i,o} & \mathbf{0}_{i,o} \\ \mathbf{1}_{i,1} & \mathbf{1}_{i,1} & \mathbf{HDD}_{i,1} & \mathbf{HDD}_{i,1} \end{bmatrix}$$

and X_i equals Z_iA_i with A_i equal to

$$\begin{bmatrix} 1 & 0 & 0 & 0 & 0 & 0 & 0 & 0 \\ 0 & 1 & H_i & R_i & 0 & 0 & 0 & 0 \\ 0 & 0 & 0 & 0 & 1 & 0 & 0 & 0 \\ 0 & 0 & 0 & 0 & 0 & 1 & H_i & R_i \end{bmatrix}.$$

Here y_i consists of all the $n_{i,o}+n_{i,1}$ consumption readings for house i, $\mathbf{1}_{i,j}$ is an n_{ij}x1 vector of 1s, $\mathbf{0}_{i,j}$, is a n_{ij}x1 vector of 0s and $\mathbf{HDD}_{i,j}$ is an n_{ij}x1 vector of average daily heating degree days for the meter reading periods in period j (j=0,1) for house i, and H_i and R_i are dummy variables indicating membership in the two treatment groups. The random effects, α, in model (4) are a_i, Δa_i, b_i, and Δb_i, which are the individual house pre-period base-levels, change in base-level, from pre- to post-periods, pre-period heating-slope and change in heating-slope, respectively. The fixed effects are a, an overall mean of pre-period base-level consumption, μ_o, μ_1, μ_2, which are the overall means of changes in the base-level for the control, house-doctor, and major-retrofit groups, respectively. The remaining fixed effect parameters for heating-slope, b, γ_o, γ_1, and γ_2, are analogous to the parameters a, μ_o, μ_1, and μ_2 for base-level.

3.4 Which analysis is preferable, LRE or two-stage?

The longitudinal random effects estimates of the treatment responses, μ_0, μ_1, and μ_2, and γ_0, γ_1, and γ_2, may be thought of as optimally weighted versions of estimates of the same parameters in the second stage ANOVAs of model (3). In particular if every house in the MRE experiment were to have the same Z_i matrix for the random effects (that is, the same number of meter-reads and the same heating degree-days in each meter-reading period) then the two-stage and the LRE analyses would be essentially equivalent. In the MRE study this was not the case. The number of meter reading periods where data were observed for each house varied, as did the timing, meaning that the dates of the beginning and end of each period and hence the heating degree-days differed for each house. While the ideal number of readings was twenty-four for each house, corresponding to two one-year intervals, about 16% of these data were missing.

The two-stage analysis makes no allowances for different variances in the intra-subject parameter estimates (model 2) either due to missing data or to differences in heating degree-days across houses. For example, a house with missing data in the summer will have a less reliable estimate of Δa_i than one having a full complement of data. Although the two-stage analysis does not take this information into account, the random assignment of houses to the treatments insures that such data can be regarded as ancillary to the experiment, as long as the causes of missing data are independent of y_i . The size of the test of hypothesis concerning the effect of the treatment should be correct for the two-stage analysis. The test will, however, have less power than for the mixed model approach -- if the longitudinal random effects model is correct.

Note that the assumptions underlying the LRE model (1) are more restrictive than for the two-stage analysis. In particular the intra-subject error variances, σ^2, as we have specified the model here, are assumed to be the same for all the houses. For the Modular Retrofit Experiment this is a dubious assumption. For example, the R-squares for the individual house heating degree-day models range from above 0.98 at the highest down to 0.80 for the lowest. This tends to imply considerable differences in the intra-subject error variances from subject to subject. Therefore an important research question raised by the application of the LRE model to this data set is just how sensitive the size and power of hypothesis tests based on these models will be to this particular source of model misspecification.

The two-stage analysis, on the other hand, is immune to differences in the intra-subject error variances because of the random assignment of subjects to treatment. The assumptions required for the two-stage analysis is that the unconditional distribution of the error terms in model (3) are homoscedastic and Gaussian. Thus information collected in the course of fitting the individual subject models may be regard as ancillary to the experiment and may be ignored without biasing the unconditional size of hypothesis test based on the two-stage approach.

4. ALGORITHMIC APPROACH.

4.1 The EM algorithm.

Once it is decided that the longitudinal random effects model is appropriate for these data we are faced with the problem of estimating the variance components in the model, namely σ^2 and the elements of D. We follow Laird and Ware (1982) in employing the EM algorithm -- with certain modifications for speeding convergence, for the iterative estimation of these parameters. We choose the EM algorithm for the following reasons. First, when used for maximum likelihood estimation the EM is known to always increase the likelihood at each stage of the iterations (see Dempster Laird and Rubin, 1977). Second for the LRE model (1) it has a very simple interpretation and implementation in terms of the unobservable random effects, β_i. Third, forms of prior information such as an aprior distribution on the fixed effects and certain types of prior estimates of the variance components can be included directly into the EM algorithm.

4.2 Maximum likelihood and restricted maximum likelihood estimation.

To use the EM algorithm in the mixed model setting, we assume that the individual subject random effects are missing data. If the β_is were all known then the likelihood equations for the variance components, D and σ^2 would be

$$\hat{D} = \sum_{i=1}^{m} \beta_i\beta_i'/m \qquad (5)$$

and

$$\hat{\sigma}^2 = \{\sum_{i=1}^{m} (y_i - Z_i\beta_i - X_i\hat{\alpha})(y_i - Z_i\beta_i - X_i\hat{\alpha})'\}/N \qquad (6)$$

with

$$\hat{\alpha} = \{\sum_{i=1}^{m} (X_i'\Sigma_i^{-1}X_i)\}^{-1}X_i'\Sigma_i^{-1}(y_i - Z_i\beta_i) \qquad (7)$$

$$\text{and } N = \sum_{i=1}^{m} n_i .$$

Laird and Ware (1982) show that iteratively replacing the right hand side of equations (5) and (6) with their expected values, given the data y_i and the parameter estimates $D=D^{(\omega-1)}$ and $\sigma^2=\sigma^{2(\omega-1)}$, from the previous iteration, and then recalculating, is an EM algorithm algorithm as studied in Dempster Laird and Rubin (1977). Laird and Ware discuss two different approaches towards performing the estimation -- that is, of calculating (iteratively) the expected values of $\hat{D}$ and $\hat{\sigma}^2$. The first is to assume that the fixed effects, α, are fixed but unknown parameters to be estimated, in which case the EM procedure yields maximum likelihood estimates (MLE). The second approach they call empirical Bayes. They impose an improper prior distribution on α as normal with mean zero and an (infinite) covariance matrix V, defined so that $V^{-1} = 0$. In this case the EM procedure gives estimates of D and σ^2 which are equivalent to the restricted maximum likelihood estimates (REML) discussed by Patterson and Thompson (1971).

4.2 Computing Formulae.

For maximum likelihood estimation the iterations become:

$$\hat{\alpha}^{(\omega)} = \Big(\sum_{i=1}^{m} X_i'X_i\Big)^{-1} \sum_{i=1}^{m} X_i(y_i-Z_iE(\beta_i|y))$$

$$\hat{D}^{(\omega)} = (1/m)\sum_{i=1}^{m} [E(\beta_i|y)E(\beta_i|y)'+V(\beta_i|y)]$$

$$\hat{\sigma}^{2(\omega)} = (1/N)\sum_{i=1}^{m} [(y_i-X_i\alpha)(y_i-X_i\alpha)' -2(y_i-X_i\alpha)'Z_iE(b_i|y) + E(\beta_i|y)'Z_i'Z_iE(\beta_i|y)+tr(Z_i'Z_iE(\beta_i|y)E(\beta_i|y)') + tr(Z_i'Z_iV(\beta_i|y))]$$

Where $E(\beta_i|y) = (Z_i'Z_i+\sigma^2D^{-1})^{-1}Z_i'(y_i-X_i\alpha)$

and $V(\beta_i|y) = \sigma^2(Z_i'Z_i+\sigma^2D^{-1})^{-1}$. (8)

For notational convenience the iteration number $(\omega-1)$ has been suppressed in the right hand side of these expressions.

For REML estimation the computing formulae (from Cook 1982) are a bit more complicated, they are:

$$\hat{D}^{(\omega)} = (1/m)\sum_{i=1}^{m} E(\beta_i|y)E(\beta_i|y)'+V(\beta_i|y)$$

$$\begin{aligned}\hat{\sigma}^{2(\omega)} = N^{-1}\Big[&\sum_{i=1}^{m} y_i'y_i-2E(\alpha|y)\sum_{i=1}^{m} X_i'y_i -2\sum_{i=1}^{m} E(\beta_i|y)'Z_i'y_i \\ &+ E(\alpha|y)'\Big(\sum_{i=1}^{m} X_i'X_i\Big)E(\alpha|y) + tr\Big(V(\alpha|y)\sum_{i=1}^{m} X_i'X_i\Big) \\ &+ \sum_{i=1}^{m} E(\beta_i|y)'Z_i'Z_i\, E(\beta_i|y) + tr\Big(\sum_{i=1}^{m} Z_i'Z_i\, V(\beta_i|y)\Big) \\ &+ 2\, E(\alpha|y)'\sum_{i=1}^{m} X_i'Z_i\, E(\beta_i|y) \\ &+ 2\sum_{i=1}^{m} tr(X_i'Z_i\, COV(b_i,\alpha|y))\Big]\end{aligned}$$

where

$$E(\alpha|y) = H^{-1}\sum_{i=1}^{m} X_i'y_i - H^{-1}\sum_{i=1}^{m} (X_i'Z_i(Z_i'Z_i+\sigma^2D^{-1})^{-1} Z_i'y_i$$

$$V(\alpha|y) = \sigma^2\Big[\sum_{i=1}^{m} X_i'X_i - \sum_{i=1}^{m} X_i'Z_i(Z_i'Z_i+\sigma^2D^{-1})^{-1} Z_i'X_i\Big]^{-1}$$

$$E(\beta_i|y) = (Z_i'Z_i+\sigma^2D^{-1})^{-1}(Z_i'y_i- Z_i' X_iE(\alpha|y))$$

$$V(\beta_i|y) = \sigma^2\ [(Z_i'Z_i+\sigma^2D^{-1})^{-1}+ G_iH^{-1}G_i']$$

$$COV(b_i,\alpha|y) = -\sigma^2\ (Z_i'Z_i+\sigma^2D^{-1})^{-1}\ Z_i'X_iH^{-1}$$

and where

$$H = X'X - X'Z\ F^{-1}\ Z'X$$

$$F=\begin{bmatrix}(Z_1'Z_1+\sigma^2D^{-1}) & 0 & \cdots & & 0\\ 0 & (Z_2'Z_2+\sigma^2D^{-1}) & 0 & \cdots & 0\\ \vdots & & \ddots & & \vdots\\ & & & & 0\\ 0 & \cdots & & 0 & (Z_m'Z_m+\sigma^2D^{-1})\end{bmatrix}$$

and

$$G_i = (Z_i'Z_i+\sigma^2D^{-1})^{-1}\ Z_i'X_i$$

4.4 The EM algorithm's speed of convergence.

One common criticism of the use of the EM algorithm in many settings, not just variance component estimation, is that it can be extremely slow to converge — often even when other methods such as Newton-Raphson or Fisher' scoring converge rapidly. The reason that this can be the case is that the EM algorithm is a first order successive substitution method -- and thus will exhibit linear convergence at the end of the iterations. To see this let θ be the vector of parameters to be estimated by EM. For the longitudinal random effects model θ consists of σ^2 and the $q(q+1)/2$ distinct components of D. The EM algorithm at the ωth iteration consists of the successive substitution step

$$\theta^{(\omega)} = g(\theta^{(\omega-1)}) \qquad (11)$$

where g represents the entire EM step -- ie the updating formulae (8) or (9) and (10). Using the first term of a Taylor series expansion of g we can write

$$\theta^{(\omega+1)}-\theta^{(\omega)} = g(\theta^{(\omega)})-g(\theta^{(\omega-1)}) \doteq J^{(\omega-1)}(\theta^{(\omega)}-\theta^{(\omega-1)})$$

where J is the matrix of partial derivatives

$$\frac{\partial}{\partial\theta} g(\theta)$$

evaluated at $\theta^{(\omega-1)}$. Assuming suitable differentiability conditions hold, as $\theta^{(\omega)}$ approaches $g(\theta^{\infty})$ -- that is, as the EM algorithm converges, $J^{(\omega)}$ will converge to J^{∞}, and for ω large enough we will have

$$\theta^{(\omega+1)} - \theta^{(\omega)} \doteq J^{\infty}(\theta^{(\omega)} - \theta^{(\omega-1)})$$

to any desired degree of precision. Further iterations produce differences in the parameter estimates iteratively as

$$\theta^{(k+\omega+1)} - \theta^{(k+\omega)} \doteq (J^{\infty})^{k}(\theta^{(\omega)} - \theta^{(\omega-1)}). \qquad (12)$$

But this implies (see Gerald 1970 p 182, for example) that the left hand side of the preceding equation will approach an eigenvector associated with λ, the largest eigenvalue of J^{∞} (so long as λ is distinct). We see, therefore, that the limiting rate of convergence of the EM will be determined by the size of λ, which can be shown to be real and between zero and one (see Dempster Laird and Rubin, 1977). If λ is near one then the EM algorithm will be extremely slow in converging since the step sizes will be small. On the other hand if λ is near zero the algorithm will be rapid (though still linearly convergent) in the final stages.

4.5 Speeding up the EM algorithm.

The methods we discuss here are applicable for accelerating the convergence of any linearly convergent successive substitution algorithm. They can be considered to be multivariate forms of the Aitken acceleration method (Gerald 1970). The basic idea is to employ either an estimate of J^{∞} or of λ to change the convergence behavior of the EM algorithm from linear to quadratic.

It is useful to monitor the convergence of the EM algorithm by estimating λ is the course of the iterations. One reasonable estimate of λ might be

$$\hat{\lambda} = 1/s \sum_{i=i}^{s} (\theta_i^{(\omega)}-\theta_i^{(\omega-1)}) / (\theta_i^{(\omega-1)}-\theta_i^{(\omega-2)}) \qquad (13)$$

where s is the number of components of θ. This is the mean of the ratios of the differences of the individual parameter estimates obtained in the most recent two iterations. From equation (12) it is clear that as ω approaches ∞ this will converge to λ. If all of the parameter changes are approximately proportional, that is, if

$$(\theta^{(\omega)} - \theta^{(\omega-1)}) \doteq \hat{\lambda}(\theta^{(\omega-1)} - \theta^{(\omega-2)}) \quad \text{for } i=1,\ldots,s,$$

and if $\hat{\lambda}$ is between zero and 1, then it is appropriate to use λ to speed convergence. From equation (12) we can write

$$\theta^{\infty} - \theta^{(\omega-1)} = \theta^{(\omega)} - \theta^{(\omega-1)} + \theta^{(\omega+1)} - \theta^{(\omega)} + \ldots$$

$$\doteq \sum_{i=1}^{\infty} \lambda^{k}(\theta^{(\omega)}-\theta^{(\omega-1)}) = 1/(1-\lambda)\,(\theta^{(\omega)}-\theta^{(\omega-1)}).$$

Thus we can estimate

$$\hat{\theta}^{\infty} = \theta^{(\omega-1)} + 1/(1-\lambda)\,(\theta^{(\omega)}-\theta^{(\omega-1)}). \qquad (14)$$

This estimate, $\hat{\theta}^{\infty}$, could then be used instead of $\theta^{(\omega)}$ in further iterations. Of course it would be advisable to check that $\hat{\theta}^{\infty}$ actually increases the likelihood over $\theta^{(\omega)}$, just to be sure. This is essentially the same thing as applying a univariate Aitkens acceleration to each of the parameters being estimated.

Figure 1. Shows plots of several of the variance component estimates, against iteration number, calculated for the MRE data. For illustrative purposes in this plot, extremely poor initial values for D and σ^2 were purposely used here. After the iterations had been run six times we calculated $\hat{\lambda}$ and the RMSE of the summands of (13) as equal to 0.2204 and 0.0265 respectively. Since $\hat{\lambda}$ was relatively close to zero with a small heterogeneity over the s components of θ, we expect at this point in the iterations that the EM will converge readily, as seen in Figure 1. At this point in the iterations we can apply equation (14) with $\omega = 6$ although it probably is unnecessary to do so since λ is so small.

Another approach towards speeding up the algorithm is to estimate J^{∞} rather than λ and use a multivariate generalization of the Aitken acceleration procedure. Since $J^{(\omega)}$ generally must approach J^{∞} before $\theta^{(\omega)} - \theta^{(\omega-1)}$ converges to an eigenvector of J^{∞} we see that J^{∞} can often be estimated earlier than λ.

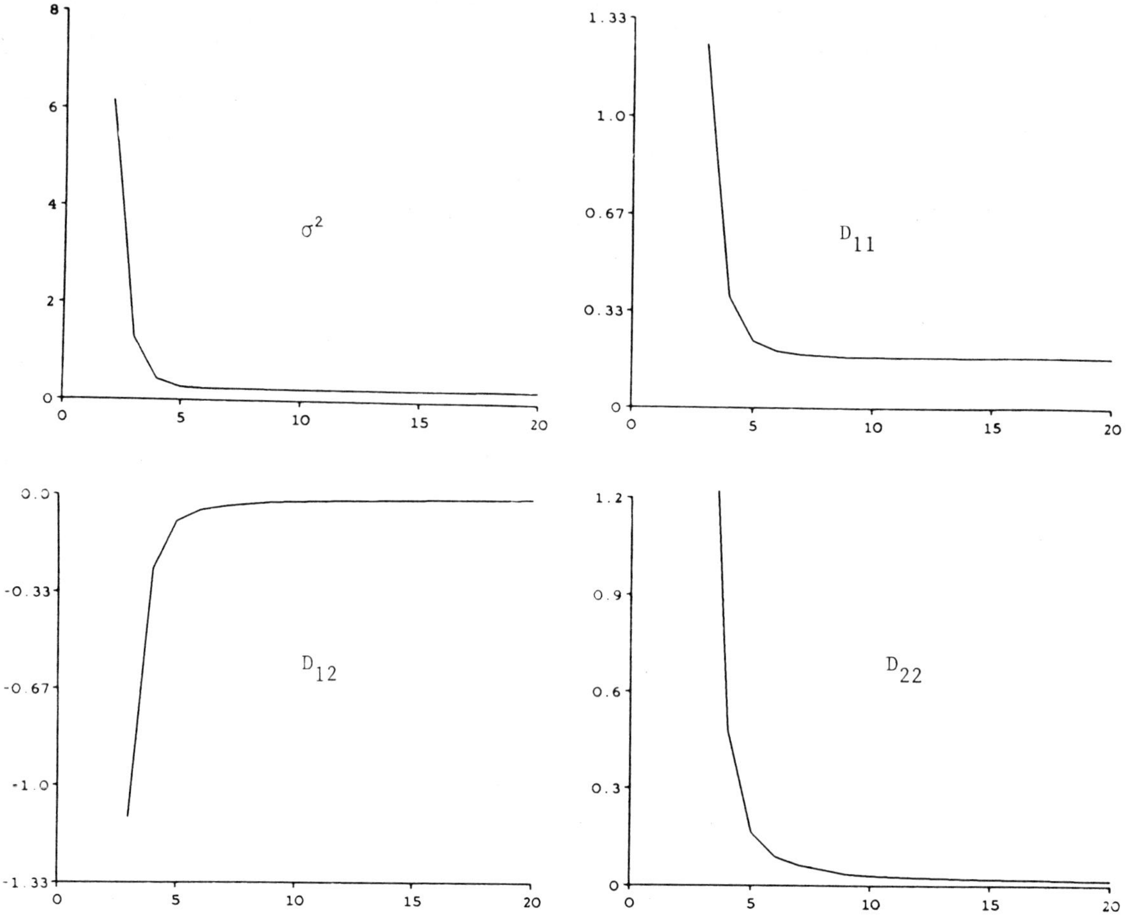

Figure 1 : Plots of Several Variance Component Estimates Against Iteration Number for the MRE Data.

This method essentially amounts to employing a Newton step to help solve the likelihood equations as written in expression (11). If we again write $\theta^{\infty} - \theta^{(\omega-1)}$ as $\theta^{(\omega)} - \theta^{(\omega-1)} + \theta^{(\omega+1)} - \theta^{(\omega)} + \ldots$, and if we approximate $\theta^{(k+\omega+1)} - \theta^{(k+\omega)}$ as

$$J^{(k+\omega-1)} \; (\theta^{(k+\omega)} - \theta^{(k+\omega-1)}$$

we have as l approaches ∞

$$\theta^{\infty} - \theta^{(\omega-1)} \doteq \{ \sum_{k=1}^{\infty} (J^{\infty})^{k} \} \; (\theta^{(\omega)} - \theta^{(\omega-1)}).$$

Since, from Dempster Laird Rubin (1977), J^{∞} has all its eigenvalues between zero and one the power series converges and is equal to $(I-J^{\infty})^{-1}$.

Thus by approximating J^{∞} we can try speeding up the algorithm, estimating

$$\hat{\theta}^{\infty} = \theta^{(\omega-1)} + (I-J^{\infty})^{-1} \; (\theta^{(\omega)} - \theta^{(\omega-1)}). \quad (15)$$

Then (after checking that $\hat{\hat{\theta}}^{\infty}$ indeed increases the likelihood over $\theta^{(\omega)}$) we can substitute $\hat{\theta}^{\infty}$ for $\theta^{(\omega)}$ in further iterations.

The question which remains is: How does one estimate J^{∞}? One could of course estimate J^{∞} by $J^{(\omega)}$. For maximum likelihood estimation, it is not too hard to give explicit formulae for $J^{(\omega)}$, either by directly differentiating the updating formulae presented in Eq (8) of Section 4.2, or by using methods discussed by Louis (1982). These calculations, however, would seem to get unbearably messy for REML estimation. It is, nevertheless, not generally necessary to know the form of $J^{(\omega)}$, to attempt the speedup. We can instead approximate J from the past history of the iterations themselves. Thus for $\omega > s$ we can approximate $J^{(\omega)}$ as

$$J = \Theta_{S}^{\omega} \{\Theta_{S}^{\omega-1}\}^{-1} \quad (16)$$

where Θ_{S}^{ω} is an sxs matrix of form

$$[\theta^{(\omega)} - \theta^{(\omega-1)} | \theta^{(\omega-1)} - \theta^{(\omega-2)} | \ldots | \theta^{(\omega-s+1)} - \theta^{(\omega-s)}].$$

As ω approaches ∞ this procedure becomes numerically unstable because

$$(\theta^{(\omega-1)} - \theta^{(\omega-2)}) \doteq \lambda \; (\theta^{(\omega-2)} - \theta^{(\omega-3)})$$

and so the inverse of $\Theta_{S}^{\omega-1}$ no longer exists. Of course when this occurs we can simply switch to the 'λ-method' to accomplish the same thing.

While for the MRE data the EM iterations converged quite readily it is not hard to find examples of slow convergence. Figure 2 gives plots of estimates arising from a growth curve problem where convergence was extremely slow. We notice that the first few iterations (starting from fairly poor initial values) produced large step sizes but in the later iterations the algorithm was very reluctant in approaching its final values. Even after more than one-hundred iterations the variance component estimates continued to change in the third decimal place from step to step. After six iterations of the EM on these data we estimated J using Eq (16) as

0.7607	2.7226	1.3997	-4.1169
-0.0178	1.7790	0.3019	-1.3840
-0.4552	-6.0342	-2.5391	8.0242
-0.1122	-0.5491	-0.5458	1.4118

We find that the largest eigenvalue of this matrix equals 0.899 which corresponds well with the slow convergence of the estimates observed in Figure 2. However at iteration 6 the use of the 'λ-method' seemed inappropriate since the summands in Eq (13), namely

$$(\theta_i^{(6)} - \theta_i^{(5)}) \;/\; (\theta_i^{(5)} - \theta_i^{(4)})$$

varied greatly, from 2.75 to 0.09, indicating that $(\theta_i^{(5)} - \theta_i^{(4)})$ was nowhere near an eigenvector of J^{∞}. Nevertheless good results for these data were obtained by the use of the multivariate Aitken's acceleration method (15) when this procedure was applied at the 6th, 12th, and 18th iterations. The results are shown in Figure 2 as the line on the plots which begins at iteration 7.

Our recommendation for exploiting these extremely simple procedures for accelerating convergence is to attempt to use Aitken's acceleration method, Eq (15), first, but, if $\theta_S^{\omega-1}$ is too illconditioned to invert, to switch to the λ-method, Eq (14) where the largest eigenvalue, λ, is estimated from Eq (13). In passing we note that the computational burden of these techniques is far less than that of performing an EM step and thus should always be considered as a convergence accelerator, or in fact in any linearly convergent iterative algorithm.

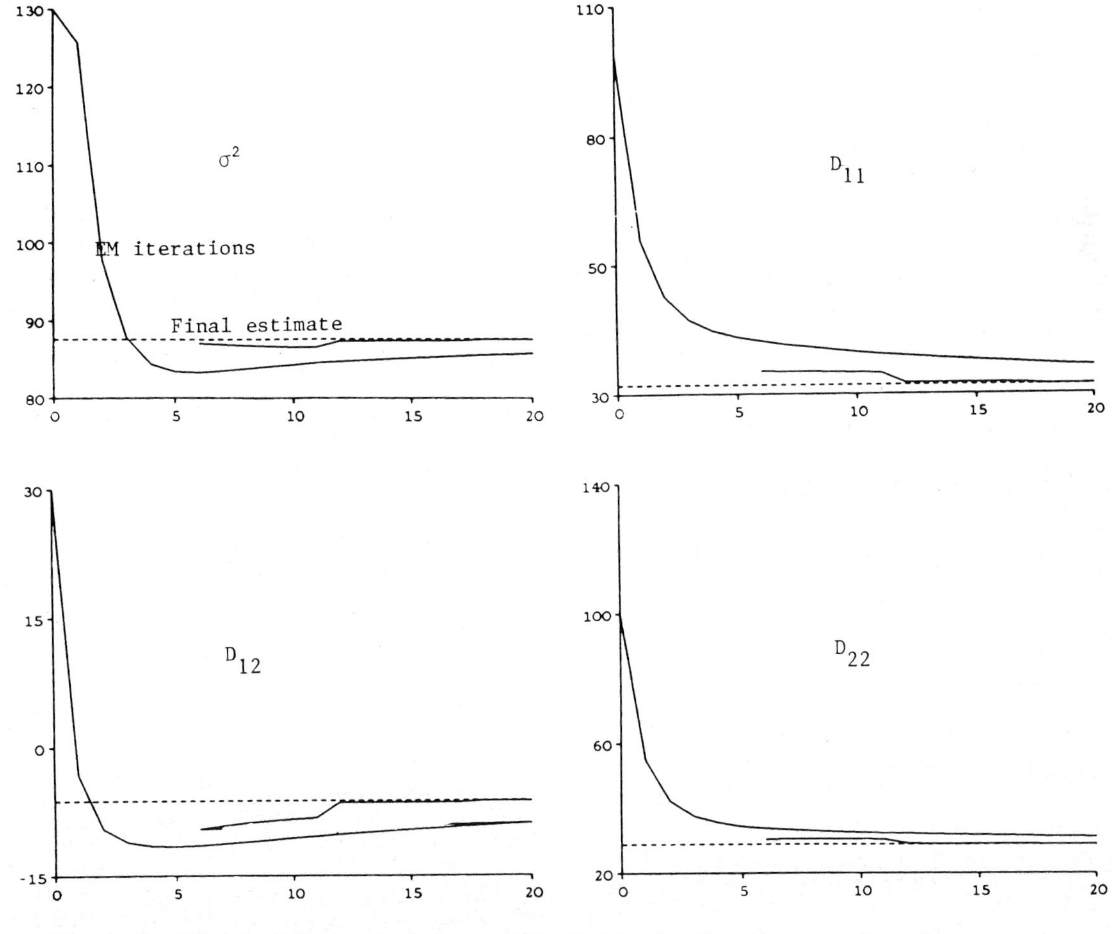

Figure 2 : Plots of Variance Component Estimates for Growth Curve Example. Also Shown are the Results of Aitken's Acceleration Procedure.

4.6 Incorporation of 'prior-information' on the variance components.

In Section 4.2 we noted that the EM algorithm is well suited to estimation when an improper prior distribution is placed on the 'fixed effects', α, for empirical Bayes estimation, which we note is equivalent to REML estimation. The EM algorithm is also suited to the incorporation of certain types of prior information on the components of D.

Suppose that we have a prior estimate D_p of D and further suppose that we think of D_p as having resulted from observing n_p independent 'unobservables', β_i, for $i=-n_p+1,\ldots,0$. (The negative index indicating the prior nature of the knowledge of the β_i). Under this, admittedly artificial, assumption it can easily be shown that the EM step for maximizing the combined likelihood of the observed y_i $(i>0)$ and β_i $(i \leq 0)$ data is to simply let

$$\hat{D}^{(\omega)*} = \frac{m\hat{D}^{(\omega)} + n_p\hat{D}_p}{m+n_p}$$

Here $\hat{D}^{(\omega)}$ is the usual EM estimate, as given in Section 4.3, at the ωth iteration, but calculated using $\hat{D}^{(\omega-1)*}$ as the estimate at the previous iteration.

Of course it would be quite unusual if a meaningful estimate of D was available before the start of the experiment, much less that the estimate had been derived by measuring unobservables. Nevertheless this procedure may still have utility in certain cases. Helms (1985, paper read this session) reports a number of instances when (using Fisher's scoring to find MLE estimates) the values of D and σ^2 which solved the likelihood equations were outside the parameter space. That is, $\hat{D}$ had one or more negative eigenvalues. When using the EM algorithm in such circumstances the eigenvalues of $\hat{D}$ will not actually be permitted to become negative, the estimate, $\hat{D}$, will instead head towards a point on the boundary of the parameter space as a limit which is never entirely obtained. In this case it would seem entirely justifiable to pull back $\hat{D}$ from the boundary in a specified direction, perhaps towards the identity matrix. Thinking about this procedure in terms of the employment of a 'prior' estimate of D means that we can characterize our final estimate in terms of the strength of the prior information used, that is, the size of n_p, to produce the final estimate. This process would seem to roughly correspond to the ridge regression approach towards least squares fitting.

5. RESULTS FOR THE PRINCETON DATA, TWO-STAGE VS LONGITUDINAL RANDOM EFFECTS.

The results for fitting, using REML estimation, the longitudinal random effects model discussed in Section 3.3 to the MRE data are shown below.

$$\hat{\begin{bmatrix} a \\ \mu_o \\ \mu_1 \\ \mu_2 \\ b \\ \gamma_o \\ \gamma_1 \\ \gamma_2 \end{bmatrix}} = \begin{bmatrix} 1.512 \\ -0.081 \\ -0.152 \\ -0.173 \\ 0.228 \\ -0.026 \\ -0.013 \\ -0.034 \end{bmatrix} \quad \begin{matrix} \text{CCF/DAY} \\ \\ \\ \\ \text{CCF/HDD} \\ \\ \\ \\ \end{matrix}$$

$$\hat{\sigma}^2 = 0.2418$$

and

$$\hat{D} = \begin{bmatrix} 0.1724 & & & \\ -0.0180 & 0.0106 & & \\ 0.0136 & -0.0005 & 0.0047 & \\ -0.0010 & 0.0007 & -0.0010 & 0.0006 \end{bmatrix}.$$

The estimate of the variance covariance matrix of the fixed effects is:

$$\begin{bmatrix} 15.28 & & & & & & & \\ -4.12 & 24.03 & & & & & & \\ -0.15 & -22.24 & 36.62 & & & & & \\ 0.09 & -22.34 & 22.30 & 42.53 & & & & \\ 0.77 & 0.18 & 0.01 & -0.01 & 0.28 & & & \\ 0.18 & -0.95 & 0.85 & 0.86 & -0.02 & 0.09 & & \\ 0.00 & 0.85 & -1.41 & -0.86 & 0.00 & -0.09 & 0.15 & \\ -0.01 & 0.86 & -0.86 & -1.63 & 0.00 & -0.09 & 0.09 & 0.17 \end{bmatrix} \times 10^{-5}$$

Table 1 compares the results obtained using the longitudinal random effects methods with those from the two-stage analysis.

We note that the LRE analysis gives greater statistical significance to the changes in heating slope and less to the changes in base level than does the two-stage analysis. This seems to reflect the fact that missing data were more common in the summer months of the study than in the winter, since in general missing a summer datapoint has more effect on the variability of the base-level parameter than on the heating slope. The group with the largest proportion of missing data was the house doctor group in the post-period. It is this group's estimate of change in heating slope (over that of the controls) for which the conclusions of the two analyses differ most markedly.

6. CONCLUSIONS AND REMARKS.

One reason for offering the Modular Retrofit Experiment data as an illustrative example for the longitudinal random effects model is that it raises several interesting model specification issues. For example, the assumption that the error variance, σ^2, is the same for all subjects is likely inappropriate for these data. Moreover, the form of the intra-subject models used here is less than

TABLE 1

ESTIMATED PARAMETER CHANGES IN TREATMENT GROUPS OVER CONTROLS FOR THE MRE DATA

LRE ANALYSIS

	Heating-slope	Base-level
House-doctor	-0.011 (-2.898)	-0.138 (-2.274)
Major-retrofit	-0.033 (-7.968)	-0.161 (-2.474)

TWO-STAGE ANALYSIS

	Heating-slope	Base-level
House-doctor	-0.008 (-1.21)	-0.176 (-3.42)
Major-retrofit	-0.030 (-4.57)	-0.212 (-3.79)

t-statistics are shown in parentheses.

optimal as well. In modeling the individual houses here, all heating degree-days were calculated at the arbitrarily fixed temperature setting of 60°F. A more physically meaningful model for the individual houses involves the estimation of the heating degree-day reference temperature, as in Dutt et al (1982), for each house in each of the pre- and post-periods. That is, the model should take into account the possibility of between-subject variation in thermostat settings or other physical factors which affect the reference temperature at which a house's gas furnace turns on as temperature decreases. Such reference temperature estimation, however, produces an intra-subject model which is intrinsically nonlinear in its parameters. The incorporation of nonlinear intra-subject models into the LRE setting must be regarded as an area open for further research. Until the significance of these departures in model specification are further investigated -- or until the LRE model is further extended, the two-stage analysis of these data would seem to be the most trustworthy. Nevertheless, the comparisons between the two-stage results and those for the LRE model are very intriguing.

One common complaint about the EM algorithm, when compared to gradient methods like Fisher's scoring, is that at the end of the iterations we are left without the usual information matrix estimate of the variance covariance matrix of the parameter estimates. We note here, however, that in the variance components problem, this criticism is inappropriate if one is primarily interested in the estimates of the fixed effect parameters. When using ML estimation the expected information estimate of the asymptotic variance of the fixed effects gives

$$\text{Asm Var } (\hat{\alpha}) = [\sum_{i=1}^{m} X_i'(\hat{\sigma}^2 I + Z_i\hat{D}Z_i')^{-1}X_i]^{-1}.$$

Thus when computing the asymptotic variance covariance matrix of α we do not include any information concerning the variability of our estimates of D or σ^2. This estimate of the variance of $\hat{\alpha}$, or any linear combination of $\hat{\alpha}$, can be computed once, at the end of the iterations. While Fisher's scoring, unlike the EM, automatically gives information about the variability of D and σ^2, at the end of the iterations, it does not give any way to make use of this information in refining the estimates of the variance of α, which is the issue most often of interest. The fact that for ML estimation an information matrix for the variance components is available using Fisher's scoring, but not from the EM algorithm, does not alone seem to be important enough to govern the choice between algorithms, at least in most common applications of the LRE model.

REFERENCES

[1] Cook, N.R. (1982) , A General Linear Model Approach to Longitudinal Data Analysis, Ph.D. Thesis, Department of Biostatistics, Harvard School of Public Health.

[2] Dempster, A.P., Laird, N.M., and Rubin, D.B. (1977). 'Maximum Likelihood from Incomplete Data via the EM Algorithm,' Journal of the Royal Statistical Society B, 39, 1-38.

[3] Dutt, G.S., Lavine, M.L., Levi, B.G., and Socolow, R.H. (1982), ' The Modular Retrofit Experiment: Exploring the House Doctor Concept', Princeton University Center for Energy and Environmental Studies Report No. 130, Princeton, New Jersey 08544.

[4] Gerald, C.F. (1970), Applied Numerical Analysis, Addison-Wesley.

[5] Helms, R. (1985). 'Algorithms and Software for the Analysis of Incomplete/Mistimed Longitudinal Data,'. Presented at the Seventeenth Symposium on the Interface of Computer Science and Statistics, Lexington KY.

[6] Laird, N.M. and Ware, J.H. (1982), 'Random Effects Models for Longitudinal Data,' Biometrics, 38, 963-974.

[7] Louis, T.A. (1982), 'Finding the Observed Information Matrix When Using the EM Algorithm,' Journal of the Royal Statistical Society B, 44, 226-233.

ACKNOWLEDGMENTS

Daniel Stram's work was supported in part by NHLBI grant number HL07427. The authors would like to thank Nancy Cook and Tom Louis for their contributions in this area.

Performance of Statisticians with Statistical Software

Organizer: *John C. Nash*

Invited Presentations:

Measuring the Performance of Statisticians with Statistical Software

John C. Nash

COMPUTER SCIENCE AND STATISTICS:
The Interface, D.M. Allen (ed.)

MEASURING THE PERFORMANCE OF STATISTICIANS WITH STATISTICAL SOFTWARE

A discussion workshop moderated by

John C. Nash
Faculty of Administration
University of Ottawa
Ottawa, Ontario, K1N 9B5
Canada

The intended purpose of this workshop was to bring to light ideas relating to the effectiveness of use of statistical software. That is, the fact that a particular piece of statistical software is capable of performing a given task is to be considered within the perspective of the ease and efficiency with which a user can avail himself/herself of this functionality. The discussions reported herein focussed on benchmarking and the desire of users to be able to deal with categorical variables which have an underlying ordering, as well as some of the mundane but important details of statistical computing.

LIST OF PARTICIPANTS

Commins, Bill (National Science Foundation);
Dumas, Bonnie P. (Westvaco);
Dvorin, Marian (U. of Maryland);
Easley, Diane (Cameron Iron Works);
Hertsgaard, Doris (North Dakota State U.);
Jennings, Dennis (U. of Illinois);
Kao, Tzu-Chey (U. of Wisconsin - Oshkosh);
Kolesar, Bob (Engelhaard);
Lane, Peter (Rothamsted Experimental Station);
Lee, Shui T. (NIOSH);
Ling, Robert (Northwestern U.);
Mehra, Munish (U. of Kentucky);
Morgan, Blaine (U. of Tennessee);
Nash, John C. (moderator, U. of Ottawa);
Nelson, Elizabeth (Internal Revenue Service);
Robinson, David (Henderson State U.);
Sacher, Richard S.(RPI);
Schuenemeyer, Jack (U. of Delaware);
Scott, Del T. (Brigham Young U.);
Simon, Steve (Bowling Green State U.);
Simpson, Pippa (U. of Kentucky);
Tung, Sarah (U. of Delaware);
Walstenholme, Dave (Imperial College);
Wang, Chyan-Ji (U. of Kentucky);
Wang, Lung-Chu (U. of Kentucky);

INTRODUCTION

This workshop was organized in an attempt to bring together statisticians and designers of statistical software so that an exchange of ideas might result in the future development of statistical software well-suited to particular classes of users and procedures for assessing this suitability. A previous workshop entitled "Which tools to use in statistical analysis? Choices of hardware and software" [9] was held in Ottawa on November 8, 1984 as a prelude to the present discussion. It took as its perspective the choices open to the user in attempting to solve a particular statistical problem. In the present workshop it was intended that the emphasis would shift slightly to give an overview of the process by which software might be assessed and selected in order to develop measures of "performance" of the statistician with the statistical software. Recognizing that this goal is ambitious, it was gratifying to note the willingness of conference participants to cooperate in developing ideas in this area.

The workshop was moderated by J.C.Nash who wrote notes directly on overhead slides which were then drafted into this report with the help of some of the participants (identified in the Acknowledgements) The report is structured as a dialog, though the editor has taken some liberties in expanding the original notes to clarify the ideas. Due to time constraints in preparing copy for publication, some references and statements remain incomplete and are marked as such by "??".

DISCUSSION

Tung: Can we focus on the following 2 ideas:

1. the development of benchmark problems and data sets;

2. the criteria for assessing how well the software has handled these?

General group: - agreement to this suggestion

- suggestion that linear regression benchmarks be the first target problem.

Simon: There are two benchmark data sets, the Longley dataset which is routinely applied by software reviewers and the Wampler/Lauchli dataset.

Nash: There is also the Wampler polynomial data sets.

a) Longley [6]

Scheunemeyer: This data has 7 independent variables (plus the constant) for 16 time periods. The dependent variables relate to employment. The independent variables are highly collinear and there is a scaling problem.

Kolesar: The Longley set is good for testing for these difficulties.

Scheunemeyer: Even extra years of data do not improve the collinearity to an appreciable extent.

b) Wampler polynomial least squares [10]

Nash: This data offers several problems with increasing collinearity, though they are not parametrized [8].

c) Wampler/Lauchli [4, 11]

Simon: The Wampler/Lauchli set is a parametrized set whose regression coefficients can be shown analytically to be a column of 1's. By steadily lowering a parameter epsilon towards zero, you make the columns of the X matrix (excluding the intercept) increasingly collinear. The advantage of this set is that you can examine a package's performance with both moderate and extreme examples of ill-conditioning. Longley gives a single extreme that may or may not be representative of the data sets one is likely to encounter. One criticism of the W/L set is that it is artificially generated (see Lesage and Simon, [4]).

Lane: Such data sets are mostly useful for testing diagnostics.

Several: It is important to decide what to do when collinearity is diagnosed.

Nash: In forecasting/prediction applications we may not need the coefficients so that a minimum length least squares solution [8, p.17] may be useful. This is equivalent to some principal component solutions.

Scheunemeyer: The ridge regression methods may also be reasonable alternatives for both forecasts and parameters.

Simon: There are software design questions related to this discussion. Most packages use either of the following choices:

1. try to give the "best possible" answers for any data set (with a warning given to the user when needed);

2. refuse to analyze any data set with extreme ill-conditioning;

Lee: Choice 3 is choice (2) with remedial actions suggested by the program.

Several: Choice (1) makes it too easy for users to continue BUT also more options for informed user.

Nash: It is not widely recognized that elimination methods ("sweeping") may not flag rank-deficiencies. The typical pivot tests are sufficient but not necessary conditions. There are some examples of matrices which appear well behaved but are quite close to being singular, for example, the Moler matrix, [8, p.210].

Kolesar: We need to distinguish special cases versus general packages. This could involve different control parameters for the software.

Consensus: 1. A "long pause" is needed when collinearity is detected, with questions posed to a user which require that he/she understand the consequences of proceeding.

2. Software should suggest remedial action to overcome the collinearity.

Several: What about judging software?

Ling: There are questions of numerical versus statistical accuracy.

Wang: Numerical accuracy is a function of algorithm and precision available.

Simon: IF the algorithm and arithmetic are correctly programmed.

Ling: NOT true, due to hardware and data dependencies.

(Editor: This difference was not totally resolved.)

Several: Statistical accuracy -- is beta-hat "near" center of the distribution of possible estimates given minor perturbations in the data? This is related to the numerical accuracy. [Readers should also note the paper by G.W.Stewart elsewhere in these proceedings.]

Nash: Are there any parametrized data sets which allow the difficulties discussed to be tested AND perturbations to be introduced?

Simon: We have some work in progress [5].

PROBLEM: Marathon run times (courtesy Roland Thomas, Carleton U.)

Problem type: estimation of distributional form / parameters and testing of various hypotheses.

Originators: Roland Thomas, John Nash

Features: The times of all marathon races run in Canada under an arbitrary 3 hour limit were recorded for several years for both men and women. It is desired to fit the distribution of times for each sex separately to a mixture model that might represent competetive and recreational runners within each separate sex group. The underlying hypothesis is that, though the "average" difference between the sexes is of the order of 30 minutes, the difference between "competetive" times for males and females is much less, perhaps closer to the actual record differences, which are of the order of 15 minutes. Since several years of data are available, one may also hope to observe the changes in performance levels of both sexes.

There are particular aspects of the problem which make it quite difficult:

1) the data set is quite large (5000+ observations each year)

2) there is no a priori model which may be suggested other than the two population mixture

3) the sample is censored by the arbitrary 3 hour limit, which eliminates a larger proportion of one sex than the other. Furthermore, we have no way to include runners who do not finish. (Several participants wanted to know if there was any indication of the number of starters.)

Commins: Higher participation rates by women might explain relative improvement in performance by women.

Suggestion by Innis Sande (Statistics Canada) conveyed to participants via moderator: Need a "marathon" effect to account for differences in terrain and weather.

Moderator briefly presented R.Thomas approach:

1) plot the cumulative distribution of times for each race/sex

2) try a mixture of normal distributions for 2 groups (elite, recreational)

3) use a (random) subset of the data for preliminary determination of the distributional parameters to save computing time.

Sacher: Is it really necessary to consider a random subset, since a convenience subset (systematic sampling) would probably suffice for the purpose?

PROBLEM: Data handling and presentation

Problem type: data manipulation, tabulation and graphing

Originator: Judd Hampton, Agriculture Canada

Features: This problem involves the handling of a relatively large number of variables on an ongoing basis, and the preparation of tables and graphs based on this data on both a regular and ad hoc basis.

Background/client group: Marketing and Economics Branch, Agriculture Canada, produces quarterly Market Commentaries for Grains and Oilseeds, Dairy, Livestock, Horticulture and Special Crops, Poultry and (consumer) Food sectors. These commentaries report the situation and outlook for each sector and are used by producers, the financial community, government at different levels, agribusiness and consumer agencies. Note that the output requires accented characters.(Many popular software packages such as Lotus 1-2-3 cannot easily be modified to allow graphical or printed output with such characters.)

Problem: The Statistical Analysis Group of M&E Branch has the task of producing detailed tables and graphs for the commentaries which are used at the annual (Agricultural) Outlook conference. This involves approximately 1500 camera-ready graphs, which must contain accurate, up-to-date information presented clearly in both official languages in accordance with strict editorial standards. The publication deadlines are tight, and often are close to the release date of the source information.

Current approach: Originally, hand-drawn graphs and typed tables were used. In the early 1970s, Hewlett-Packard desk-top computers were introduced (9820 and 9830 series) with plotters for graphical output. The quantity of output was such that one plotter actually wore out the potentiometer slide wire (the only case the HP technicians had heard of in which this happened). Now HP 9845 series machines are used. Software was created to maintain and update databases containing monthly and annual time series and to print current and past data and five-year averages. A Multiwriter letter quality printer produces the final tables for photoreduction. HP flatbed plotters are still used to produce high quality plots. In some cases the multiple pen capability is used not for colour but for different pen widths. HP software has generally not proved adequate to meet production standards, and software produced in-house is prepared as needed.

Kolesar: For regular, i.e. routine, use, special purpose programs are likely to be the best choice.

Consensus: the decision to use a special program should be governed by a decision rule

E(no. of uses) * E(saving/use) <

Cost of preparing program

Lane: A session at the Prague COMPSTAT meeting discussed such problems [3, see also 1].

(Note: there was actually more discussion of this problem, but the major points raised are covered here.)

PROBLEM: Contigency table with ordered categories

Originator: Roland Thomas (Carleton U.)

Features: A cross-tabulation shows a 2 state response against 3 categories of one predictor and 5 of another -- a 2 by 3 by 5 table. One or more predictors have categories which have an ordering e.g. they represent the ranges of a numerical variable which are observed. How can such data be analyzed efficiently while using the ordinality present in the predictors?

Scheunemeyer: Try assigning weights to the different categories...

Lee: Then use a logistic model on the assigned weighting.

Lane: Chapter ?? in McCullagh and Nelder [7] discusses this model. They parametrize the proportion of response as one cumulates through the (ordered) categories.

Lee: One has to choose particular break-points in a variable to give appropriate categories.

Scheunemeyer: We are trying to force ordered categories into a continuous case.

Dumas: SAS GSK command can be used to perform weighted least squares. Another approach is GENCAT (Landis, 1974??). Brown (BMDP) is writing a new code to ...??

Commins: If we get a "good" fit without using the ordering, should we continue our analysis?

Jennings: Scaled model has one parameter per variable, while the independence model has one per level of variable. If the contingency table indicates independence, we may wish to continue analysis.

Simon: Conover [2, pp. 232-234, 335-338 and problems 3 and 4 on p.386] mentions using ranks in a contingency table with an ordinal category. This approach relies heavily on average ranks.

At this point, despite fairly active

discussion, the moderator had to bring the session to a close.

ACKNOWLEDGEMENTS

Stephen Simon and Peter Lane contributed written comments on a draft of this report which were extremely helpful in resolving ambiguities in the text. David Allen arranged that the conference room seating was appropriate to a workshop situation. The Ottawa Chapter of the American Statistical Association, in particular Elaine Hoskins, President, organized a similar workshop at which the format and background material for the present activity were developed.

REFERENCES

[1] Bellm, Th.H.J., and Verbeek, A., Standard packages versus tailor made software: some experiences in statistical production, Statistical Software Newsletter, 10 (2) (September 1984) 68-74.

[2] Conover, W.J., Practical nonparametric statistics, John Wiley: New York (1980).

[3] Havranek, T., Lane, P., Molenaar, I., Nelder, J.A. (chairman), Tiit, E-M., Verbeek, A. and Victor, N., Standard packages versus tailor made software, a panel discussion at COMPSTAT'84 in Prague, Statistical Software Newsletter, 10 (2) (September 1984) 56-67.

[4] Lesage, J.P. and Simon, S.D., Numerical accuracy of statistical algorithms for microcomputers, American Statistical Association, Proceedings of the Statistical Computing Section (1984) 53-58.

[5] Lesage, J.P. and Simon, S.D., The impact of centering and scaling on the numerical accuracy of regression algorithms, submitted to the ASM conference on Mini and Microcomputers and their Applications, to be held June 3-5, 1985 in Montreal.

[6] Longley, J.W., An appraisal of least squares programs for the electronic computer from the point of view of the user, J. Amer. Statistical Assoc. 62 (1967) 819-831.

[7] McCullagh, P. and Nelder, J.A., Generalized linear models, Chapman and Hall, London (1983).

[8] Nash, J.C., Compact numerical methods for computers: linear algebra and function minimisation, Adam Hilger: Bristol and Halsted Press: New York (1979).

[8] Nash, J.C., Accuracy of least squares computer programs: another reminder:comment, Amer. J. Agricultural Economics 61 (4) (November 1979) 703-709.

[9] Nash, J.C.(ed.), Which tools to use in statistical analysis -- choices of hardware and software, Notes of a workshop. American Statistical Association Ottawa Chapter (February 1985)

[10] Wampler, R.H., A report on the accuracy of some widely used least squares computer programs, J. Amer. Statistical Assoc., 65 (1970) 549-565.

[11] Wampler, R.H., Test procedures and problems for least squares algorithms, Journal of Econometrics 12 (1980) 3-22.

APPENDIX

The Wampler/Lauchli dataset

Y_i = (n - 1 + epsilon), for i=1

= epsilon, for i=2,...,n-1

= (n - 1 - epsilon), for i=n

X_{ij} = 1, for i=1, j=1,...,n

= 1, for j=1, i=1,...,n

= epsilon, for i=j=2,...,n-1

= 0 otherwise

(X is a bordered diagonal matrix)

Statistical Workstations

Organizer: *Thomas J. Boardman and Gary Anderson*

Invited Presentations:

Essential Ingredients for a Statistical Workstation

Thomas J. Boardman

Statistical Software, Graphics and Future Workstations for Data Analysis

Richard A. Becker, John M. Chambers, and Allan R. Wilks

COMPUTER SCIENCE AND STATISTICS:
The Interface, D.M. Allen (ed.)

ESSENTIAL INGREDIENTS FOR A STATISTICAL WORKSTATION

Thomas J. Boardman

Department of Statistics
Colorado State University
Fort Collins, Colorado

In the future engineers, scientists, and other professionals will perform many of their work assignments on computer workstations. In part, the renewed interest in statistical methods as one tool for helping industry and government improves the quality of goods and services, justifies the need for statistical components in the workstation. Some design objectives for workstations are discussed in order to lead into a discussion of the necessary hardware and software ingredients for workstations. One scenario is proposed by describing how the statistical functionalities on a workstation might appear to the user if the hardware has a bit mapped screen similar to the Apple Macintosh. Finally several challenges for the future are described which offer encouragement for improvements in statistical software in the future.

1. INTRODUCTION

My intention is to challenge your thinking about how people may use statistical methods in the future, perhaps even the way people will first learn about statistical methods. You might ask why should workstations have statistical components? Let us start back at the beginning and discuss the increased interest in computing. I met Bruce Woolbert, of Hewlett Packard's Personal Computer Division, at the Pharmaceutical Manufacturers Association Biostatistics Subsection Annual Meeting, held in San Francisco in October 1984. He and I had been asked to address the conference. Hewlett Packard authorized a firm to do market research for them. Bruce Woolbert reported some of the results during his presentation. One statistic he reported is that one in thirteen office professionals is currently using computers in his/her job function. He went on to say that we are increasingly seeing new uses for personal computers. People are finding there are ways that they can use computers that they had not even considered in the past. For example, networking of computing systems will be much more popular in the future. More about this topic later. In fashion at the moment are ideas for using computers in new ways such as computer-aided design, computer-aided engineering, computer-aided manufacturing, and computer-aided office. All of these reflect the market's movement toward integrated systems.

Computers are used from the bottom up. By that we mean that computers are now used all the way from secondary education through college. Their availability in education certainly has an effect on what we are doing in our course work in higher education. Within the last couple of years I have seen considerable changes in my own department in terms of the quality and types of computing that we are doing in our course work.

Computers are also used in industry and in many of our homes. Furthermore I suspect that computers will be in a great number of the graduate students' homes after they complete their studies. So we can say that computers are for you and me and the kids. The pressure on adults from youngsters using computers will force adults to think about how computing actually needs to be done. Bruce Woolbert made an estimate based on the same research that Hewlett Packard had commissioned. He reported that by 1995, 65% of ALL office workers will be using computers. Are the adults ready for that magnitude of commitment?

Interest in statistical methods has been generated by the renewed interest in quality and productivity. Competition from Japan and other countries has awakened U.S. industry to the fact that statistical methods can help improve processes. Of course statistical methods are only a part of the quality improvement efforts and processes do not involve just goods. Some estimates show that in excess of 85% of all employees are actually in the service area. There are many opportunities for improving quality and therefore productivity in the service area.

The new emphasis on quality is affecting the way management deals with their employees. There is a new awareness of the employees' roles: to know their job, and to get their job done more effectively. The annual National Quality Month is one indication from Congress and the President of the importance of this area. Other activities such as the American Statistical Association's Committee on Quality and Productivity, are of course concerned with smaller audiences but still show some commitment from ASA.

The software business is booming. The "Directory of Software for Quality Assurance/Quality Control" in the March 1985 issue of Quality Progress, listed 118 packages. Almost all of them have some statistical components. Quite a few of the packages are strictly statistical packages such as SPSS, MINITAB, etc. What does this mean? There is a renewed interest in statistical computing. The proliferation of statistical software is important because it is another sign of the beginning of the understanding that statisticians and, more importantly, statistical methods can really help. W. Edwards Deming says that American management has to change. Even though it is only a small part of the transformation process, the use of statistical methods is nevertheless part of the process. Management is faced with making meaningful decisions in the face of uncertainty and variation. Using the scientific method to get meaningful information upon which to base some of their decisions is beginning to be recognized as a valid approach. Statisticians and statistical methods can help managers make decisions in a scientific manner.

2. WORKSTATIONS

Why should the statistical computations be implemented in a workstation environment? There are a couple of key points here. The order is not important. One is the proliferation of microcomputers. I do not have recent estimates of the number of microcomputers at CSU but I suspect that it is probably upwards of a thousand at this point. In the spring of 1984 the estimate was in the neighborhood of 400 with new orders at about 80 a month. Unfortunately our statistics department is not expanding in microcomputers as rapidly. Nevertheless the growth is dramatic.

Another reason why a workstation environment makes sense is that people who use computers have more than one task to do. Although they tend to be focused around one speciality, computer users find themselves using word processors, statistical packages, and wanting to do lots of different tasks at a computer. The idea behind a workstation is to put together all of the tools necessary to help a user perform any number of tasks. Thus it is an important design concept to make workstations simple for workers to use. Workstations are being looked at as an effective way to get the job done. Computers are not a substitute for good hard thinking or good creative work. Dr. Deming discusses what he calls "instant putting solutions"; that is, any solution to a problem that is easy--not necessarily cheap but easy to do. Some look at computers as being an effective way to make better quality products and to increase people's quality and productivity. Deming is convinced computers will not replace good and creative thinking. Workstations should be viewed by management as one potential tool for improving the quality and productivity of the workers. Of course, the cost versus benefit of using any tool must be evaluated. As someone said, if the only tool you have is a hammer, it is surprising how many problems look like nails.

Another justification for workstations involves the concept of networking of computing resources. The networking concept involves more than just sharing computer peripherals such as printers and plotters. Ideally networks of computing devices will free the user from having to make decisions about which computer offers the proper environment for today's tasks. A network system should provide simple ways for users to communicate with many computers without having to know many different protocols. Consider my own situation. Currently I am working on the IBM XT in the Stat. Lab., I have a Macintosh at home, I use the CSU CDC CYBER mainframe computer for many statistical applications, I am on the Engineering College collection of VAX's and I recently tried SAS on the Vet Hospital's Data General. It is mind boggling to try to remember all the different protocols to get on all of these machines. One potential advantage for a network environment at CSU is that interfacing to the various computers could be much simpler. You would not have to remember anything except the protocol for the one machine you prefer to use. The computer network would interface to all the others. If one machine needs a caret C or whatever, the network systems could remember that and take care of it for you.

Finally workstation environments will abound because the suppliers of these systems will convince us through advertisement that we cannot do without their systems. This reason may actually dominate all the others. Why? Because software vendors are going to make a lot of money on workstation software. Vendors are just beginning to push the concept of integrated packages. The workstation environment is a step beyond several integrated packages. In this environment almost all tasks which we would like to do on a computer are "integrated" together.

What are some of the essential components of a workstation? Consider the following three categories: the design objectives for a workstation the necessary hardware ingredients, and the necessary software ingredients. At a conference sponsored by SIGNUM of ACM in March 1984 I heard a presentation by John K. Wooten of the Computing Division of Los Alamos National Laboratory. His talk touched on the first two areas above. Blending my experience as a project investigator and consultant for Hewlett Packard with recent visits to AT&T Bell Laboratories, discussion with those at previous Interface Conferences, reading articles on the topic and considerable thinking, I have prepared the following lists under the three categories mentioned above.

Consider first the design objectives which an organization should have when considering how a workstation should ideally be used.

Design Objectives for a Workstation Environment

To be most effective, workstations should be used throughout an organization.

The interface must be user friendly.

Certainly job specific software will be needed and must be available shortly after introduction to the worker.

Good response time is essential.

The hardware and software must be compatible with other equipment already in place.

The hardware and software must be expandable and upgradeable as new developments come on line.

The user must be able to program in one or more languages but the user should not have to program to use the equipment.

There must be software for office automation such as: word, text, and composition processors; file organizer; information retrieval system interface; electronic mail; inventory control modules; data communication links; data base management systems; graphics presentations; ledger analysis packages; etc.

Since many different data bases exist in an organization the system must be able to access them.

Through network environments or whatever, one should be able to share resources such as peripherals.

The list of hardware ingredients which follows may be lacking. I do not claim any particular wisdom here. Then too if we wait a week or two the list will probably change.

Some Hardware Ingredients For A Workstation

Full Bit-Mapped Screen of sufficient size to be read more than a foot or two away

Good Resolution, color graphics both on the screen and a graphics output device (may be at a remote site)

A simple keyboard

A cursor control device such as a mouse

Multiwindow screen capability

Considerable ram perhaps 1 to 1.5 megabytes

Considerable local storage, 10-20 megabytes

Inexpensive printer close by and peripherals such as a laser printer, hard disk with large storage, plotters connected to your phone for all forms of communications

The next list is the software characteristics which should be designed into a workstation system. Although each of the items could be described in great detail this will not be done for two reasons. First, since most readers of this paper will have a general idea of what is meant by each of the characteristics, the author does not intend to create an argument on semantics. Secondly, all of our definitions will be likely to change as we view new approaches to software development. Therefore this list is merely included to suggest the general characteristics which should be considered.

Software Characteristics For A Workstation System

The operation must appear to the user to be FRIENDLY.

The system should appear to the user to Do Harder Tasks Simply.

The system ought to remember what has been done before using what is often referred to as Intrip System.

To the extent that it is needed Help Features should be available.

There ought to be effective ways to Allow the Sophisticated User to Move Quickly Inside the System.

The system should provide for Repeatable Work with minimal user specification.

As the science of Artificial Intelligence develops the workstation system should incorporate some of the better features.

The workstation system should provide for Multitasking both in the CPU and on the display.

The user should be able to develop User Specified Procedures/Routines which can be called up in the future.

3. SCENARIO OF RESEARCH ON A WORKSTATION

Consider for a moment how an engineer or a scientist might use a workstation environment to solve a problem. The notion to keep in mind is that the tasks which I am describing can be performed at one station. The researcher is sitting at his/her desk. The researcher has been confronted with a problem. The first thing you might want to think about is to formulate the initial concepts associated with a possible solution, organize and develop ideas, and save those things for future use. (See Exhibit 1 for a list of tasks and workstation tools to be used.)

You might use a word and text processor and an idea processor. You will need a file organizer to save all the ideas for the next steps.

Before you chance reinventing the proverbial wheel you might want to perform a literature search using one of the several available information retrieval systems. Once the search is complete the results will be saved. At this point you should be ready to formulate the proposed research objectives and prepare a draft including the budget and other financial implications. The tools involved in this step are word and text processors, a financial modelor, and a spread sheet package.

The draft is submitted via electronic mail for peer evaluation followed by a possible revision. Once approval has been obtained it is necessary to check on the availability of the equipment and supplies to be used in the experiment. If this information is not immediately at hand one could use the inventory control and order processing components of the workstation. Indeed, since others may wish to use the equipment, the requirements should be noted through the network environment so others will not make claim on the equipment.

Using an experiment design package the researcher is assisted in making final decisions about which factors to use, the levels of those factors, and the type of design to be run. The hardware is interfaced with the appropriate instrumentation, the order of the experimental design is randomized and the experiment is performed.

We should mention here that at several of these steps we do not expect immediate response from the system. For example, the time involved to complete all experimental runs may be several days or weeks. The important thing to remember though is that we can expect that the user at his workstation will be receiving information, when appropriate, on the progress of the experimentation.

The data as received are stored in a data bases management system and verification procedures are performed continuously. At various stages the meta data are input to the data bases management system. Meta data are essentially all the non-numerical information associated with the data base that you would like to remember. Everything you might record in a lab book which normally gets lost when you input the results to the computer can be saved as meta data.

The researcher completes the various data manipulation operations such as handling missing values, transformations, sorting, merging, etc. most likely in a data base management system. At this point you are ready for appropriate statistical analyses including exploratory analyses on the data. There could be many steps involved here. The process should be iterative and augmented with statistical graphics as well.

After completing the analyses the researcher will need to prepare some graphics for presentation of the results. These can be done in the statistical graphics package or perhaps in a graphics presentation package specifically designed for high resolution graphics. A ledger analysis or spread sheet package can be used for summarizing the final accounting for the report. We complete the written report on a word and text processor, develop slides for presentation of the results, and give the oral report to management throughout the corporation. The presentation may be a real time "dog and pony" show on the CRT screen to the various managers and colleagues who need to know the results.

Finally the researcher saves all of the results in a file organizer for future reference. Subsequently the researcher reads his mail and discovers a new project awaiting him. Or perhaps the previous project needs to be studied under new conditions. The point is, of course, that the workstation environment can perform a myriad of tasks--all accomplished at one location. Note also that only a few of the tasks involve statistical operations. The workstation environment must allow a complex array of tasks to be performed. From the user's point of view the operation should appear to be blended together. The resources used in one task should be available to other portions without great effort on the part of the user.

4. ADDITIONAL FEATURES FOR THE STATISTICAL COMPONENTS

There are a few specific additional features which should be part of the statistical components of a workstation. These are in addition to those software characteristics discussed in section 2. The software must be user friendly both for the beginner and the experienced user. Many will experience their first use of statistical methods in a workstation environment. It is therefore important that their experience with statistical analyses be friendly. By this we mean that at whatever level of complexity, the operation of the workstation should appear to be straightforward.

Of course we expect that the statistical components should offer comprehensive and complete solutions for the task selected. The software should be powerful. The statistical analyses should cover a wide range of types of situations. And of course we expect that the results should be correct statistically and numerically.

Three special operations are quite important for the statistical components in a workstation environment. First, the system should allow the user to branch back up through the path of the analysis and choose another route. The system must remember what has been done before and allow the user to try new routines. Secondly, the system should offer repeatable sessions in which the user can request similar paths through the analysis with perhaps a different selection of variables and/or subsets. And third, the system should allow the user to customize his or her own steps through the data analysis. The sequence of operations and decisions which are made could be given a procedure name and requested subsequently by that name.

Finally it is imperative that the statistical and other components incorporate graphics into every segment of the routines. In particular the statistical components should have graphics which are fine-tuned to the analyses and integrated into all components of the software. Furthermore, the user interface should most likely be graphical in nature with pull-down menus, pop-out windows, etc. In "The Visual Mind and the Macintosh", Benzon [1] describes why he believes the visual mind is now recognized as being so important in user interfacing. While his article focuses on the Apple Macintosh computer, most of his remarks would also apply to other operating systems and software as well. Indeed those vendors and software developers who do not make use of the left side/right side characteristics of the brain are missing an important way to interface with the user.

5. ONE POSSIBLE SCREEN IMPLEMENTATION

Let us consider how the user interface for the statistical components might be implemented in an integrated workstation. The hardware will have to include a high resolution screen. A color screen would be helpful but is not essential. We will need to control the cursor with either a mouse or some other type of controller. The mouse is my preference at this time.

There are several characteristics of the operating environment to be mentioned. The user should not have to remember a lot of commands to start the system. The start-up sequence is often a frustrating exercise for most novice users. Of course it can be frustrating even to the experienced use if he/she must remember the sequence for several different machines. Typically, smaller machines are easier to use, but this is not always true.

We want to have simple, easy-to-understand displays. In some of the packages we have been evaluating at CSU the initial display is very difficult to understand. The operations ought to be simple enough so that the user does not need a manual. We like pull-down menus which lead to pop-out menus. We have discovered that most users like to have the ability to fill in the answer blanks on a screen. One must have a fairly sophisticated computer to be able to move cursor around, fill in answers, and/or check off various options. Up to this point the argument has been that you need to provide a command language interface for the more sophisticated users. After all, the story goes, these users will want to move around rapidly in this software. The paper by Velleman and Lekowitz in these Proceedings, however, suggests that even sophisticated users can use a mouse interface more efficiently than a command language approach. More research needs to be conducted on this topic but the preliminary results are encouraging.

The ability to view and operate on multi-windows on the screen is essential. The windows will naturally overlap. Thus many different events can be shown on the screen at the same time. We should be able to page through the windows. A data window will more than likely contain more information than can reasonably be displayed on the screen. Paging is essential.

The system ought to support multi-processing which is visible on the screen. For example, the results of an analysis might be displayed in one window while the user cycles through the data in another window, and a scattergram is created in another window. In addition we might expect background processing to occur while other operations are displayed on the screen.

Other features include a help operation without tears. On some systems once you enter the help sequence the system essentially sets aside the current operations and branches to some other part of the program. You may have to recyle through the entire operation again to get back to where you were. Another feature which has already been mentioned would allow one to back up the steps in the analysis, make changes, and start down another path. Finally we require user defined routines. The system should allow us to specify a particular sequence of events and identify that sequence as ours.

During the oral presentation of this topic, the author mentioned that because he has been associated with a project with Hewlett Packard that is not complete at this point and his wife is a consultant for IBM, he decided to "show" one possible implementation on a Macintosh computer- the computer they have at home. All of the characteristics such as pull-down menus and pop-out menus were described during the oral presentation. Another concept which was described is an event window at the bottom of the screen. In this window events are displayed such as the time, date, elapsed time for certain events, busy signal for disk I/O, and status of operations such as multiprocessing of computations with nonlinear regression. In addition, the author described several other features such as how one might scroll through various windows, enlarge or shrink windows, and telescope or magnify portions of a window.

6. CHALLENGES FOR THE FUTURE

The concern has been raised that many people in the future may first learn about statistical methods on a workstation. Their first exposure to comprehensive sets of statistical tools may be when they sit down at their workstation. The idea is a little frightening. Assuming that people will first encounter statistical methods this way, it means that the quality of the statistical software is paramount. The software developers have a more important responsibility in workstation environments. And statisticians have a responsibility to make sure that the software developers produce quality products. Some might question why statisticians should be involved; after all, computer scientists will more than likely be doing the software development. Sure, the computer scientist will design the systems and they ought to do so. However, statisticians should help the computer scientist in the following areas: in defining the depth, breadth, and completeness of the statistical coverage; determining the algorithms to be used for the computation; reviewing the user interface with regard to at least the terminology used; supplying the test data sets; and evaluating the overall performance of the software. We must realize that there will be new forms of statistical software in the future. One can speculate that the way computer packages now interact with the user will be considered "old fashioned" in three years or less.

Statistical methods in the future will be changing as well. Large data sets will be more prevalent. Rapid arrival, on-line data collection will be commonplace. New types of data analyses to accommodate large multivariate data sets will be needed. We will no longer be satisfied with simply giving our clients analyses of variance tables. They will need and expect far more from the statistics packages.

Many believe that there will be a dramatic emphasis on the use of good and insightful statistical graphics. Certainly the hardware can display the graphics. It will be up to the developers to integrate statistical graphics throughout the routines. The American Statistical Association will shortly have a new section of Statistcal Graphics. The statistics profession obviously feels graphics are important. Statisticians have a chance to "show" users of our methodology that statistics really can help. Statistical graphics may be our best tool.

The supercomputers with various forms of parallel processors may indeed change the type of problems we consider to be statistical in nature. This topic should receive more attention at the future Interface Conferences.

Finally, one subject for the future which may affect how users do statistical methods is Artificial Intelligence (AI). After asking several people for their definitions of AI and receiving somewhat different answers from each, someone finally said that a program which could recognize information never specifically programmed and draw inferences and conclusions would have AI features. While the concept may be plausible, the reality may make us wonder. Is it possible to encode the knowledge systems of a brilliant statistician such as John Tukey into software so that the user will have the benefit of Tukey's help on the user's problem (smart system)? And can the system carry the process further so that even though the smart system has not seen the user's problem before it will lead the user through decision-making processes? WOW!

7. CONCLUSIONS

Small computers will be increasingly involved in all aspects of our lives. Our children will begin learning how to use computers in elementary school and can reasonably expect to use them throughout their lives. Employees will use computers and computer technology on the job. Indeed many employers may install identical computers in their employees' homes so that they can follow up on good ideas even while they are at home. Whether supplied by their employer or not, most homes in the future will use computers for a wide range of tasks. We can expect that many tasks have not even been envisioned today. The computer revolution may not have even arrived. Perhaps we are only at the dawn of the revolution. We do not fully appreciate the place which computers will have in societies of the future.

There is every indication that statistical methods will be even more important in the future. The renewed emphasis on improving quality and productivity is helping. Because everything we do can be thought of as a process which needs continuous improvement, recognition of the proper use of statistical methods should increase greatly. These are great times for statisticians. Statistical software will continue to proliferate and change. Some feel that the software developers may help to lead statistical methods into the 21st century.

Workstations are but one result of high technology which should affect our lives in a positive way. The statistical components incorporated in these workstations would be impressive. Statisticians need to get involved to make sure that this happens.

REFERENCES:

[1] Benzon, W., The Visual Mind and the Macintosh, Byte, January (1985), 113-130.

Exhibit 1
Scenario of Research Done on a Workstation

Step	Tasks	Workstation Tools
1.	Formulate initial concept, organize & develop ideas, & save for future	Word & Text Processor (WTP), Idea Processor (IP) & File Organizer (FO)

2.	Perform literature search & save	Information Retrieval Systems and FO
3.	Formulate proposed research objectives & prepare draft, including financial implications & budget	WTP, Financial Modeler & Spread-sheet package
4.	Submit to peers for evaluation, criti-cism & revise	Electronic Mail & WTO
5.	Check on availabil-ity of equipment to be used	Inventory Control & Order Processing
6.	Decide on appropri-ate experimental design, etc.	Exper. Design Routines
7.	Interface the instrumentation, randomize the runs, run exp., & store results in data base	Data Comm. Linkage Rand. and Data Base Management systems (DBMS)
8.	Perform any number of data verification procedures	DBMS
9.	Input Meta data to data base	DBMS
10.	Complete data mani-operations such as MV's, transform sorting, merging, etc.	DBMS or Stat. Library
11.	Perform appropriate stat. analysis in-cluding EDA. Note: Many steps involved here	SL which should include many Stat. Graphics (SG)
12.	Prepare graphics for presentation results	SG & Graphics Presen-tation Package
13.	Prepare final accounting summary of cost vs benefits	Ledger analysis package & spread-sheet package
14.	Complete written report on results	Word & text processor
15.	Develop slides for oral presentation of results	Graphics Presentation
16.	Give oral report to management through-out corporation	Real Time Oral "Dog & Pony Show" on CRT
17.	Save all results for future	File Organizer
18.	Begin new project or read mail to discover what the boss wants next	Electronic Mail & then go to step 1

COMPUTER SCIENCE AND STATISTICS:
The Interface, D.M. Allen (ed.)

Statistical Software, Graphics and Future Workstations for Data Analysis

Richard A. Becker

John M. Chambers

Allan R. Wilks

AT&T Bell Laboratories
Murray Hill, New Jersey 07974

ABSTRACT

The personal workstation is rapidly emerging as a powerful tool for conducting data analysis, particularly in contrast to either the large mainframe or the small personal computer. This talk describes some user experiences in working with a variety of workstations and in providing data analysis software for them, especially for graphical display of data. The discussion includes the present state and desirable future evolution of workstations from the viewpoint of statistical applications.

1. A Computer for Every Data Analyst

Recent trends in computer technology have caused drastic changes in the price of hardware. At present, a workstation computer, may be purchased for approximately $15,000, and the trend in price is still definitely downward: soon such machines will be priced near $5,000. With prices at this level, it will not be long before any serious data analyst will be able to afford a personal workstation.

When we speak of a workstation, we mean something quite different from the type of machine currently called a "personal computer". The "personal computer" is generally characterized by slow processor speed, limited internal storage capacity, and small amounts of external storage. At present, most of these machines are based upon processors with either 8-bit or 16-bit architectures; this naturally limits the amount of memory that the machine can address. The current personal machines are also limited in software. Although there are numerous small programs for these machines, large, integrated systems are not normally present. Typically, they have very primitive operating systems, and because of hardware constraints, the operator is often forced to load and unload programs manually as the need arises.

The personal workstation is very different, indeed. The workstation is a "real computer", not unlike the super-mini machines designed for scientific processing. A workstation typically has a 32-bit processor, at least 1 megabyte of main memory, and disk storage capacity in the tens of megabytes. Also, workstations normally have modern operating systems and sophisticated software. They are capable of running the application systems and user programs written for mainframe time-shared computers, without the competition for resources and the usage costs associated with time-sharing. Workstation users do not suffer due to inadequacies of hardware or software. In fact, the workstation opens new opportunities for the development of an environment which emphasizes the human interface.

Of course, it is appropriate to ask "Why not a time-shared mainframe?" The answer to this is that the workstation gives the analyst control over the computing resources necessary for the job. The price of a mainframe typically means that it is controlled by a group that may not be responsive to the need for modern data analytic software. Also, since the processor is shared, work for other users may interfere with the processing power needed for data analysis.

The Human Interface

There are several characteristics of the workstation which have a major impact on its human interface. These are high-resolution display devices, provision for user control of multiple processes, interaction, and networking. The display is often a bitmap raster-scan device, with resolution of approximately 100 pixels per inch. This relatively high resolution enables the display to produce approximations to typeset documents, with various fonts and point sizes. It also can produce quite satisfactory graphical displays and, since the local processor can change the bitmap rapidly, the display can give the appearance of continuous motion.

One of the major uses of such a display device involves the creation of "windows" on the screen. Each window acts as an independent connection to the processor, much as multiple timesharing terminals could be running on a large, shared machine. Any particular window can run a process that is tailored to a specific task, such as producing graphical displays, text editing, or document display.

The user has control over the workstation not only through a conventional keyboard, but also through a visual interface. Dynamic interaction with the display is carried out through a mouse (or touch screen, tablet, light pen, etc.) that enables the user to point at the display, draw, and make menu selections. The combination of pointing device and good, fast graphics makes menu-style interfaces to application software much more attractive.

Because certain peripheral computer facilities are expensive or infrequently used, workstations use local area networks to provide access to them. At one extreme, workstations can be used as very intelligent terminals to current mainframe computers.

Statistical Computing in the New Environment

How will a workstation environment affect statistical computing? Major impacts will be made by:

- Local Power
- Graphics
- Dynamic Displays
- Multiple Windows
- Interaction
- Networking

2. Local Power

The availability of large amounts of essentially free computing power is likely to change the way that data analysis is done. Once a workstation is available, it costs nothing to have it computing. Therefore more processor intensive data analytic techniques are likely to be attempted. Techniques like the bootstrap, in which thousands of similar analyses may be carried out in order to find confidence limits, are likely to be more common. In general, simulation techniques are much more likely to be used.

However, perhaps the more important contribution of the local processing power, is that it will encourage the analyst to consider more analyses. Data analysis is a process that does not have one fixed answer; often it is important to come up with several different views of the same data. When the analyst is able to look at the data from many different viewpoints, without having to incur large processing costs or to share his machine with others (and suffer degraded performance), the quality of the analysis is likely to improve.

3. Graphics

No longer will computer graphics be limited to those able to afford expensive equipment. With the advent of workstations, *every* user will have graphic capabilities. Graphical techniques have long been known as powerful aids to data analysis. The human mind is far superior to any computer software in the area of pattern recognition. When shown a scatter plot, a human data analyst can recognize curvature, clustering, and a host of other interesting characteristics of the data displayed. The combination of interactive graphic displays with an interactive computing environment will provide a synergistic effect, leading again to better data analysis.

Perhaps a less obvious benefit of graphics will be the ability of using graphical symbols to aid user interaction. Just as international road signs use pictures to guide automobile drivers, so will computers be able to use non-verbal graphic images, known as icons, to guide data analysts.

4. Dynamic Displays

Static graphical displays have always been available to people who want to look at data. Many of the displays common today were invented in past centuries. Much of the research into new methods of displaying data involves dynamically changing pictures. This can involve, for example, movie-like sequences of views of a point-cloud. A good example of such research is described in PRIM9 (Fisherkeller, 1974), ORION (Friedman, 1982), and PRIMH (Donoho, 1982).

At AT&T Bell Laboratories, we have experimented with a number of these dynamic displays. Most of this research was done on a Teletype 5620 Dot-Mapped Display terminal (which is basically a diskless workstation). We have rotating point clouds, a straightedge display that moves under control of the mouse, dynamic display of identifiers on a scatterplot, and a more advanced technique for multivariate data, known as "brushing" (see Becker and Cleveland, 1984).

Dynamic displays can also be used for the presentation of several distinct but related pictures in alternation, the process called *alternagraphics* by Tukey (1982). Given a multi-plane graphics color display terminal with a color map (such as the Advanced Electronics Design Model 512), it is possible to rapidly cycle through pre-computed scenes. Such displays are not slowed down by their complexity, but have a limited number of views and precomputation overhead. We have used this technique to show rotation of 3-dimensional surfaces with perspective generated by stereo glasses. We have also looked at the behavior of smoothers as a locality parameter was varied.

Another use of local processing power in conjunction with dynamic displays is in fast-changing displays. For example, it should be possible to plot the data in a univariate regression problem and to interactively move, delete, or add points to the plot and to see the regression line continuously updated. We could also choose power transformations for the x- and y- variables on a scatter plot by observing the picture as the transformation powers were changed under control of a graphical input device.

5. Multiple Windows

Since workstations allow the user to control separate activities from separate windows, a number of difficulties of current statistical software melt away. For example, it becomes easy to allow the user to interact with the statistical software in one window (either through a keyboard or menus), to see graphical results in another window, and to get on-line assistance at the same time in another window. The size, shape, and position of the windows can reflect the users wishes, and they can be rearranged at any time.

6. Interaction

Since workstations normally provide hardware and software facilities for user interaction, there is much flexibility in the face that statistical software presents to the user. Dynamically changing menus can be provided; menus can "pop-up" on the display until the user makes a selection, and then can disappear; icons can be used for non-verbal interaction. Multiple windows allow users to explore on-line documentation or pursue any other background computing they like, without interrupting or removing from the display the current interaction.

7. Networking

Networking is one method for providing a number of workstations with shared resources. However, networking facilities will do much more for the statistician. Workstation networks are often configured as in Exhibit 1.

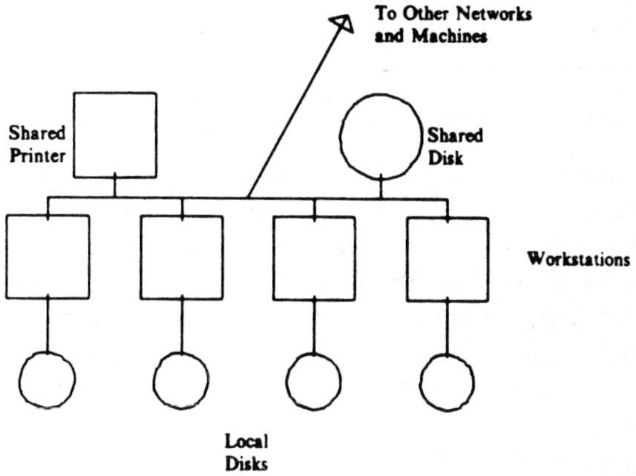

Data transmissions around the local network are typically very fast, often several megabits per second. At these speeds, users can share data, documentation and software without experiencing any loss in apparent performance because a particular file is actually at a remote location. As the figure suggests, relatively expensive and infrequently used resources (fast printers, hardcopy plotters, very large disks, special processors) can be connected to the network and used by all the workstations, without seriously slowing down access to them. The ability to connect the local network to other networks and to other types of computing environment is particularly important. Users need not sever their links to the conventional mainframe computing world, from which much of the data for analysis will continue to come.

The personal effects of the networking environment are at least as important as the technical effects. Electronic communication, both local and remote, is one of the most fundamental changes being made by the current computer revolution. Workstations linked by local and remote networks give the user full access to this communication. For example, the UNIX[1] system provides both one-to-one communication (e.g., *mail* and *write* commands) and broadcast communication (*news* and *netnews*). The style of communication stimulated by these facilities is qualitatively different from traditional paper communication, emphasizing rapid response and brief documents. In many ways it is more *communication* in contrast to the *publication* mode typical of paper documents. The publication process itself is also mightily changed in the workstation environment: the hardware and user interface facilities allow authors to interact with the editing, design and production process much more directly.

Modern Software for Data Analysis

Once workstation hardware is available, it becomes necessary to think of appropriate software for the new environment. Of course, it will not be sufficient to use old batch software inherited from the 1960s, or to think of the display screen as a fast line printer. In addition, it must be remembered that hardware evolution will continue, and hence the software must be adaptable to tomorrow's hardware.

Most people think of statistical computations such as regression or transformations when thinking of statistical computing. However, there is much more involved than that. The software must be able to store and retrieve data, work with a wide variety of data structures, and provide interactive graphics on various graphical devices (which, like workstations, have proliferated rapidly and are continually undergoing change).

S is a system which is meant to fulfill the needs for modern data analysis software. It runs under a number of versions of the UNIX operating system on a variety of hardware. S is described in a recent book by Becker and Chambers (1984).

1. UNIX is a Trademark of AT&T Bell Laboratories.

The primary goal for S is to allow users to perform good data analysis. Judging from the experience of some thousands of users, S satisfies this goal quite well. However, in order for people to be able to use S to analyze data, they must have access to S. Hence it is desirable to have S readily available, on inexpensive but appropriate hardware.

Luckily, the general trend in computer hardware is for more power at less cost, and the current selection of professional workstations is the manifestation of this trend. Modern workstations combine computing power, large amounts of addressable memory, and quick and consistent response time, and often come with the UNIX operating system. Many of these workstations also have provision for bitmap graphic displays. These machines not only provide an excellent environment for S, but they also have the potential for providing better understanding of data through dynamic graphic displays. These new UNIX-based workstations are a desirable environment for S because of their low price, good graphics (bitmap, dynamic), and responsiveness. We now have experimented with S on the following workstations: Sun, Hewlett-Packard 9000, AT&T 3B2, and Wicat. The machines run a variety of UNIX systems, including AT&T System V and Berkeley 4.2BSD.

We have had experience in porting S to the following machines and variants of the UNIX system:

Hardware	Operating System
HP Series 200 (MC68000)	HP-UX (System III)
SUN 100 (MC68010)	4.2BSD
3B2-300 (WE32000)	System V
Wicat 150 (MC68000)	7th Edition, System V
Perkin-Elmer 32/30	7th Edition
HP Series 500 (HP 32-bit chip)	HP-UX (System III)
Apollo	AEGIS
IBM 370	UNIX/370
DEC VAX 11/780	32V
DEC PDP 11/70, 11/45	7th Edition
Pyramid	System V, 4.2BSD
Ridge	4.2BSD
DEC VAX 11/780, 750	BSD 4.1, 4.1c, 4.2

DEC VAX 11/780	System III, V
DEC VAX 11/780, 750	8th Edition
AT&T 3B20S	System V

When we first wrote S for the UNIX system, one of the major decisions we made was the basic choice of programming language. Because of the large amount of FORTRAN computational code already available, we decided to use that language. However, we decided that the primitive operations of the S system should be implemented in C. C provides the natural linkage with the underlying UNIX operating system calls.

Conclusions

The statistical computing arena is undergoing a quiet revolution. In the near future, increased computing power, good graphics and new modes of human interaction will be available to a greatly increased population of *potential* users of statistical systems. Such users will benefit, and indeed require, high-quality on-line help in using statistical software. Fortunately, the personal workstation is well suited to provide such help. Its resources are essentially free to the user, encouraging the approach that as much effort as needed should be spent by the computer in presenting data dynamically and in supporting interaction with the user.

The statistician will also find many new opportunities in such an environment. The computer power should greatly increase the use of simulation as a routine tool, whenever the behavior of a model or estimate needs to be studied. In the choice of theoretical work in statistics, as well, the statistician with a real concern for the healthy practice of data analysis will find new challenges in providing support for this new user population. For example, graphical presentation of data, diagnostics of value to the non-professional analyst and more advanced techniques such as expert systems are all exciting possibilities in the new environment.

References

Richard A. Becker and John M. Chambers, *S: An Interactive Environment for Data Analysis and Graphics*, Wadsworth, Belmont, CA, 1984.

Richard A. Becker and John M. Chambers, "Design of the S System for Data Analysis", *Communications of the ACM*, May 1984.

Richard A. Becker and William S. Cleveland, "Brushing a Scatterplot Matrix: High Interaction Graphical Methods for Analyzing Multidimensional Data", submitted for publication, 1984.

David L. Donoho, Peter J. Huber, Ernesto Ramos, and H. Mathis Thoma (1982), "Kinematic display of multivariate data", *Proceedings of the Third Annual Conference and Exposition of the National Computer Graphics Association*, **1**, 393-400.

Mary Ann Fisherkeller, Jerome H. Friedman, and John W. Tukey (1974), "PRIM-9: An interactive multidimensional data display and analysis system", Publication 1408, Stanford Linear Accelerator Center.

Jerome H. Friedman and Werner Stuetzle (1983), "Hardware for Kinematic Statistical Graphics", pp. 163-169, *Comp. Sci. & Stat.: Proc. of the 15th Symp. on the Interface*, James Gentle, ed., North Holland.

John W. Tukey (1982), "Another look at the future," *Comp. Sci. & Stat.: Proc. of the 14th Symp. on the Interface*, Springer-Verlag, New York.

Numerical Methods

Organizer: *William Kennedy*

Invited Presentations:

Methods for Multidimensional Scaling
Douglas B. Clarkson and James E. Gentle

Collinearity, Scaling, and Rounding Error
G. W. Stewart

Bivariate Density Estimation and Automated Stick-Pin Maps
Michael E. Tarter and William Freeman

COMPUTER SCIENCE AND STATISTICS:
The Interface, D.M. Allen (ed.)

Methods for Multidimensional Scaling

Douglas B. Clarkson and James E. Gentle

IMSL, Inc.

Multidimensional scaling is an often used technique with many similarities to factor analysis. This paper discusses and compares several models for multidimensional scaling, and gives some generalizations of some of these models. It proposes new (to multidimensional scaling) fitting criteria, and compares the results obtained by their use. Some solutions to problems encountered in the optimization algorithms are discussed. Finally, some statistical implications of multidimensional scaling models are given.

1. Introduction

In the general multidimensional scaling (MDS) problem the data consists of one or more dissimilarity matrices, where a dissimilarity is some measure of distance, and the matrices give, in some sense, the distances between the objects (or stimuli) considered. An easy example of such a matrix is the mileage distances between cities often found on road maps. Here, the distances between cities is the dissimilarity measure, and the MDS problem is to locate the cites in a two (or three) dimensional space based upon these distances. As a second, more complicated example, consider the purely fictitious data in Table 1. In this example, the stimuli represent 7 stores and the dissimilarity is a ranking of the distances in each row of the distance matrix. Rather than using the actual distances, the ranks of the distances are used as the dissimilarity measure. From row 1 of the table, one can see that store 3 is closest to store 1, store 2 is second closest to store 1, store 5 is third closest, etc.

Table 1

Store Distances

Store	Store Ranks 1	2	3	4	5	6	7
1	0	2	1	5	3	4	6
2	4	0	3	1	2	5	6
3	1	2	0	3	4	5	6
4	3	1	4	0	2	5	6
5	4	1	5	2	0	3	6
6	2	3	4	5	1	0	6
7	1	4	2	6	5	3	0

The data is an example of ordinal dissimilarity data. In the general MDS problem, the data can be categorical, ordinal, interval, or ratio. Also, since the ranks in one row have no direct relationship to the ranks in a second row, each row represents a different stratum (or conditionality group) in the sense that dissimilarities in the two rows cannot be compared. In the general MDS problem, more than one dissimilarity matrix can be observed, and a stratum can be a row of a dissimilarity matrix, an entire dissimilarity matrix, or all of the data. These strata correspond to what is called row conditional, matrix conditional, or unconditional data, respectively.

The idea in multidimensional scaling is to locate objects in a τ-dimensional Euclidean space in such a manner that the agreement between the observed dissimilarties and the distances predicted by the location of the objects in the space is in some sense optimal. In this example, the usual Euclidean distance given by

$$\delta_{ij}^2 = \sum_{k=1}^{\tau} (X_{ik} - X_{jk})^2$$

is used, where X_{ik} is the coordinate of the i-th object in the k-th of τ dimensions in the Euclidean space. (The matrix consisting of the X_{ik} is called the configuration matrix.)

Generally, a criterion function of some form is used to obtain an optimal solution. In this case, the criterion function is given by:

$$q = \frac{\sum_{i,j=1}^{n} (\delta_{ij}^* - \delta_{ij})^2}{\sum_{i,j=1}^{n} (\delta_{ij}^*)^2}$$

where n is the number of stimuli, $\delta^*_{..}$ denotes the optimal dissimilarities, called disparities (see below), and $\delta_{..}$ is the predicted distance given above.

The criterion function is optimized with respect to both the configuration, through δ, and the disparities δ^*. If the data is ratio or interval, the disparities δ^* are the observed distances and there is no optimization with respect to δ^*. In ordinal data the disparities are the predicted dissimilarities, where the prediction is made via a monotonic regression of the ranks of the observed data on the predicted distances δ within each stratum. Finally, in categorical data, the disparities δ^* are the average of all predicted distances δ which have the same observed dissimilarity within a stratum.

The numerator in the above expression is the least squares criterion. The denominator is a normalizing factor which prevents the solution from becoming degenerate in ordinal (or categorical) data (a different criterion might be used in ratio and interval data). The denominator is required here because the optimization is with respect to both δ and δ^*. If the denominator were not present, q could be made as small as desired simply by simultaneously scaling both δ and δ^*.

As an example of the monotonic regression used in ordinal data, consider the following table:

	Ranks for Store 7					
Store	1	3	6	2	5	4
Rank	1	2	3	4	5	6
δ	.69	.65	.44	1.10	1.07	1.23
δ^*	.59	.59	.59	1.08	1.08	1.23

In this table the original rankings of the dissimilarities for each store compared to the ranks for store 7 are given in the second row. (The data is presented in its original rank order.) Using the estimated configuration, the distances in the third row are computed. In computing the disparities, note that in the third row .65 is less than .69, .44 is less than .65, and 1.07 is less than 1.10. Since the disparities must be in the same rank order as the observed dissimilarities in the second row, the monotonic regression averages the elements in the third row as required in order to preserve the originally observed ordering in the disparities δ^* given in the fourth row.

When the criterion function is optimized, the resulting configuration is given in Table 2. A plot of these results is given in Figure 1, with a plot of the store locations which gave rise to the rankings in Table 1 presented in Figure 2. In comparing these figures note that the scale is meaningless since no distances were actually observed. The differences in the location of the stores in these two figures come about because of the lack of uniqueness of the estimated configuration in ordinal (or categorical) data. (For that matter, note that the store locations given in Figure 1 are not unique, as an infinite number of such plots could have given rise to the same rankings in Table 1, even after eliminating variation due to reflections and rotation.)

Table 2

The Configuration X

Stimulus	X_{i1}	X_{i2}
1	0.37	-0.12
2	-0.24	-0.86
3	0.58	-0.96
4	-1.07	-0.65
5	-1.00	0.78
6	-0.15	1.24
7	1.51	0.57

Figure 1

The Estimated Configuration

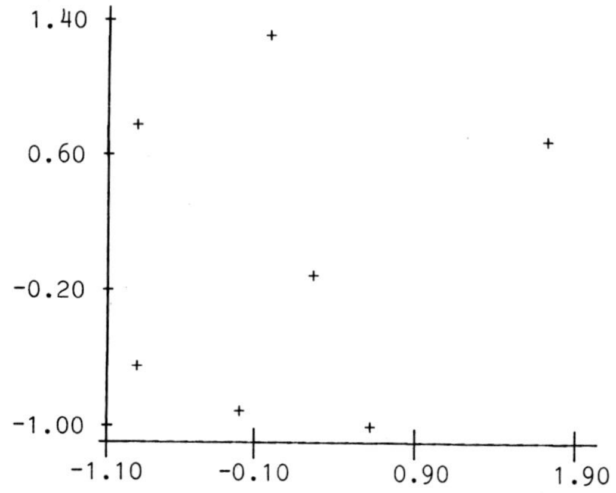

The fact that the estimated configuration in Figure 2 is not unique (even after allowing for changes in sign and for rotations) can be seen as follows: If the fit is perfect, then the numerator is 0.0 and the denominator has no effect. One can then change the configuration in such a manner that rankings in the ordering of the distances are unchanged. The monotonic regression will then change the disparities so that they are exactly equal to the new distances, and the numerator in the criterion function remains 0.0.

Figure 2

The Actual Store Locations

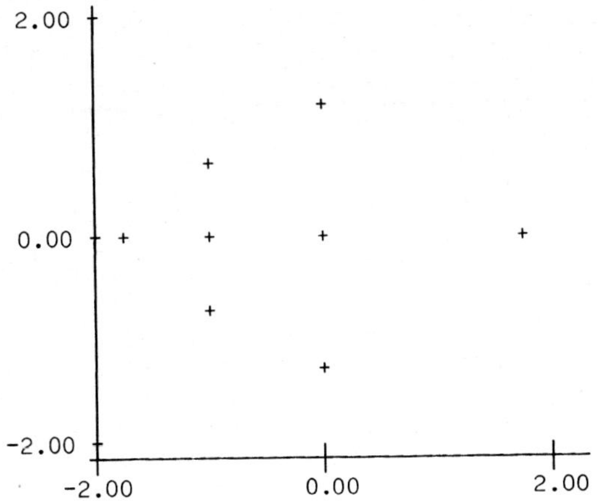

In section 2 the general criterion function (and thus, the model) used by subroutine MSCAL in the IMSL library is described, along with possible generalizations. (Subroutine MSCAL will be released in the next edition of the libraries.) Section 3 describes the methods used for fitting the model, while section four gives a more complicated example.

2. The General Criterion Function

The general criterion function in subroutine MSCAL is given as follows:

$$q = \sum_{h} \omega_h \times \sum_{i,j} \left| f(\delta^{*}_{ijm}) - a_h - b_h \times f(\delta_{ijm}) \right|^{p}$$

where ω_h depends upon the $f(\delta^{*}_{ijm})$ in the h-th stratum, h indexes the strata, f is a transformation discussed below, a_h and b_h are constants to be estimated in some models, m indexes the subjects and will depend upon i, j, and h according to the stratification used, and p allows for L_p estimates other than least squares to be used in the criterion function. (The most likely values for p are 2.0 for least squares and 1.0 for least absolute value.)

Null and Sarle (1982) also suggested a criterion function involving p-th power estimates for use with ratio and interval data. MSCAL allows categorical and ordinal data as well as ratio and interval.

The function f in the criterion function allows the user to make various assumptions about the distribution of the observed dissimilarities. This is clearly most important in ratio or interval data, but it also has effects in ordinal and categorical data, primarily through the weights ω_h. Since least squares and maximum likelihood estimates are equivalent (in one stratum) when the distribution of the transformed random variables are normal, the function f may be used as a transformation to normality. This is equivalent to using f to obtain homogeneous variances within each stratum.

Choices for f in MSCAL are:

$$f(x) = \begin{array}{l} x^2 \\ x \\ \log(x). \end{array}$$

If one believes that squared distances have constant variance (and are approximately normally distributed), then $f(x)=x^2$ should be used. Similarly, $f(x) = x$, or $f(x) = \log(x)$ should be used if these transformations yield constant variance.

The squared transformation is the transformation used in the ALSCAL program of Takane, Young, and DeLeeuw (1977), while distances are used in MULTISCALE (Ramsey, (1983)), and KYST (Kruskal, Young, and Seery (1973)), among others, and log distances are allowed in MULTISCALE.

The Distance Models

The models for the distances δ_{ijm} are equivalent to those used in ALSCAL. They are given as follows:

The Euclidean model:

$$\delta^2_{ijm} = \sum_{k=1}^{\tau} (X_{ik} - X_{jk})^2$$

The individual differences model:

$$\delta^2_{ijm} = \sum_{k=1}^{\tau} W_{mk} \times (X_{ik} - X_{jk})^2$$

where W_{mk} is the weight on the k-th dimension for the m-th subject.

The stimulus weighted model:

$$\delta^2_{ijm} = \sum_{k=1}^{\tau} S_{ik} \times (X_{ik} - X_{jk})^2$$

where S_{ik} is the weight on the k-th dimension for the i-th stimulus.

The stimulus weighted individual differences model:

$$\delta^2_{ijm} = \sum_{k=1}^{\tau} W_{mk} \times S_{ik} \times (X_{ik} - X_{jk})^2$$

Other distance models are possible. For example, instead of a weighted space, one could allow a rotation of each individual's coordinate axis. This yields the IDIOSCALE model of Carroll and Chang (1970). Additionally, one could allow for asymmetric models via the skew symmetric matrices of Weeks and Bentler (1982). Future refinements of the MSCAL subroutine may allow for such refinements.

The Strata Weights

In metric scaling, strata weights are used to weight the observations within a stratum. In this case, weights which are inversely proportional to the variances are preferred because such weights lead to normal distribution theory maximum likelihood estimates. Thus, in metric scaling, one would use

$$\omega_h^{-1} = \sum |f(\delta^*_{ijm}) - f(\delta_{ijm})|^p / n_h$$

where ω_h is the weight in the h-th stratum, the sum is over all observations in the stratum, and n_h is the number of observation in the stratum.

In nonmetric scaling, because the criterion function is minimized with respect to both δ and δ^*, the criterion function is degenerate unless strata weights are used as a normalizing factor. An optimum criterion value of zero could always be obtained without this normalization. In most multidimensional scaling programs, normalization is provided by the use of one of two possible weights proposed by Kruskal (1964). These weights are given by:

$$\omega_h = \sum |f(\delta^*_{ijm})|^p$$

or

$$\omega_h = \sum |f(\delta^*_{ijm}) - \bar{f}(\delta^*_{...})|^p$$

where the sum is over the observations in the h-th stratum, and where $\bar{f}(\delta^*_{...})$ is the average of the disparities in this stratum.

3. Fitting the model

Initial estimates of all parameters are obtained via the same methods which are employed in the ALSCAL program of Young, Takane, and DeLeeuw (1977). For the configuration this amounts to obtaining the average of the product moments matrices (double centering the dissimilarities), computing the τ largest eigenvectors of this matrix, and multiplying by the square root of the matrix of eigenvalues. When subject weights are required, the method of Schonemann as modified by Young, Takane, and Lewyckyj (1978) is used. Finally, when stimulus weights are required, a multiple regression method in conjunction with the method of Schonemann is employed.

After the initial estimates are obtained, a modified Gauss-Newton algorithm is used to obtain estimates of most parameters. In the multidimensional scaling models discussed here, this amounts to iteratively reweighted least squares. To speed convergence, the initial iterations are performed on subsets of the parameters, while in the final iterations all parameters but the disparities are optimized simultaneously. In all iterations, optimal values for the disparities are computed via a secant based method discussed later.

All parameters appearing in the general criterion function do not have to be used in the multidimensional scaling. Thus, with some exceptions, the presence of the subject weights W, the stimulus weights S, the scaling factor b_h, and the additive constant a_h, is optional. Moreover, any parameter matrix (including the configuration matrix X) can be fixed in the optimization procedure. (The disparities are fixed by declaring the data to be interval or ratio data.)

The initial iterations proceed as follows:

1. In nonmetric scaling, the disparities estimates δ^* are computed within each strata assuming that all other parameters are fixed. The estimates of a_h and b_h within each stratum are also computed at this time.

2. The optimal configuration estimates (X) are computed.

3. The optimal subject weights estimates (W) are computed (one subject at a time).

4. The optimal stimulus weights (S) are computed.

When the maximum change in any parameter is less than a user specified constant (100.0×EPS), the iterative method changes. In the iterations at this point, steps 2, 3, and 4 above are combined so that optimal estimates of X, W, and S are obtained simultaneously. (Note that in metric scaling, the Hessian for all parameters is computed. The inverse of this matrix is commonly used as an estimate of the variancecovariance matrix of the parameters. Some additional uses of this matrix are discussed later.)

Convergence is said to have occurred when the change in any parameter from one iteration to the next is less than a user specified constant EPS.

The L_p Gauss-Newton Algorithm

As stated earlier, a modified Gauss-Newton algorithm is used in the estimation of all parameters but the disparities (and the parameters a_h and b_h). This algorithm, discussed by Merle and Späth (1974), uses iteratively reweighted least squares on the criterion function. In discussing this algorithm, first rewrite the criterion function as follows:

$$q = \sum_{i,j,h} \frac{\omega_h \times |f(\delta^*_{ijm}) - a_h - b_h \times f(\delta_{ijm})|^2}{|f(\delta^*_{ijm}) - a_h - b_h \times f(\delta_{ijm})|^{2-p}}$$

Least squares is then used on a linearization of the parameters in δ_{ijm} to obtain the estimates. In this least squares estimation, it is assumed that ω_h and the denominator of q are fixed. (I.e., for each observation, ω_h and the denominator of q are combined to yield an observation weight which is fixed with respect to the iteration.) The only problem occurs when $p<2$ and the denominator is zero, at which time a division by zero would occur. In this situation, the denominator is set to 0.001, and the calculations then proceed as usual.

Estimating the Disparities

I. Ordinal data

As was discussed earlier, in least squares MDS monotonic regression is used in the computation of the disparities in ordinal data. Because p-th power estimates are computed, these methods cannot be used (when p is not 2.0), because they would not yield optimal estimates. A severely modified monotonic regression must be used instead. Within each stratum the criterion function is given by:

$$q = \frac{\sum |y_{(k)} - f(\delta_{(k)})|^p}{\sum |y_{(k)}|^p}$$

where $y_{(k)} = f(\delta^*_{(k)})$, and k is the rank of the observed dissimilarity in its stratum . (k is enclosed in parenthesis to emphasize this ranking.) In this equation, the $y_{(k)}$ are all parameters, while in this phase of the optimization, it is assumed that the $\delta_{(k)}$ are fixed. The monotonic assumption in ordinal data requires that the $y_{(1)} \leq y_{(2)} \leq y_{(3)} \ldots \leq y_{(s)}$, where s is the number of observations in the stratum.

Using Lagrange multipliers the criterion function within each stratum is transformed to:

$$q = \sum |y_{(k)} - b \times f(\delta_{(k)})|^p - \lambda(\sum |y_{(k)}|^p - c)$$

where b is the scaling parameter b_h for this stratum (a_h is not used), and λ is the Lagrange multiplier.

Within each stratum the criterion function involves parameters $y_{(k)}$, b, and λ in this phase of the optimization. Because of the monotonic restrictions on the $y_{(k)}$, it is not easily possible to use the usual Newton-Raphson techniques on all parameters simultaneously. Because of λ and the second term in the criterion function, simple modification of the usual monotone regression techniques may not be employed. The following algorithm, while sometimes slow to converge, seems to yield the optimal estimates (Kuhn-Tucker theory guarantees that the estimates are optimal if convergence occurs.):

1. Set λ and b to 0.0.
2. Estimate $y_{(k)}$ for the criterion q holding λ and b fixed.
3. Estimate b.
4. Estimate λ.
5. If the change in any parameter from one iteration to the next is greater than EPS, go back to step 2.

In step 2 a secant algorithm is used to compute each isotonic parameter $y_{(k)}$ based upon the observations $f(\delta_{(k)})$ associated with the parameter. The $y_{(k)}$ are made monotone by restricting all y's which would otherwise violate the monotonic restriction to be equal. This has the effect of increasing the number of observations which are used in the L_p location estimate of the restricted parameters. For example, the monotonicity restrictions may require that the ranked transformed disparities $y_{(2)}$ through $y_{(7)}$ be equal. One would then compute the estimate of these 6 parameters as the L_p location estimate of the transformed distances $f(\delta_{(2)})$ through $f(\delta_{(7)})$.

In step 3, a secant algorithm for fixed $y_{(k)}$ and λ is used. The computation of λ in step 4 is direct, and is obtained by setting the derivative of q with respect to each $y_{(k)}$ equal to zero, and then summing over all possible $y_{(k)}$.

The algorithm seems to converge for all values of p in the interval [1,2]. Convergence is slowest for p near 1 and is fastest at p=2.

II. Categorical data

In categorical data the p-th power estimate of location is used on the transformed distances as the disparity estimate for all observations with the same observed dissimilarity within each stratum. A secant algorithm is used to compute the p-th power location estimate.

Case Analysis

Because the Hessian is computed in full in metric scaling, a case analysis (also called a residual analysis) can be performed. Clearly, one quantity of interest for each observed dissimilarity is its residual. Some measure of influence may also be of interest, as well the observation weight and its standardized residual. These statistics may be computed as follows:

1. Compute the observation weight and residual in the usual manner.

2. Compute the influences as follows: Let g_{ijm} denote the row vector of weighted partial derivatives of the $a_h + b_h \times f(\delta_{ijm})$ with respect to the parameters a_h, b_h, and all parameters in δ_{ijm}, and let G denote the matrix of these partial derivatives. Compute the influences (or leverages) as the diagonal elements of the matrix $G(G'G)^{-1}G'$.

3. The studentized residual is given as:

$$r = e/SQRT(MSE \times (1-h))$$

where e is the residual, h is the leverage, and MSE is the (weighted) mean square error estimate computed via the criterion function and adjusted for the number of parameters estimated.

4. An Example

As a second example of ordinal row conditional dissimilarity data, consider the matrix in Table 3 in which nine wines are judged with respect to their dissimilarity by one of nine people. The data consists of nine such matrices, one for each of the nine judges Each person ranked the dissimilarity of the remaining eight wines with the row wine. Thus the ij element in Table 3 gives the ranked dissimilarity of the j-th to the i-th wine, where ranking is within each row. Thus, in row 1, wine 2 is judged most similar to wine 1 in this table, while wine 8 is judged least similar. The study was blind in the sense that no individual knew the name of the wine being tasted.

Table 3

Wine Tasting Data

Wine	1	2	3	4	5	6	7	8	9
1	0	1	2	5	6	3	7	8	4
2	1	0	2	4	3	8	5	7	6
3	1	8	0	4	7	5	3	2	6
4	7	2	5	0	1	3	6	4	8
5	2	8	4	6	0	1	7	5	3
6	6	7	5	3	1	0	4	8	2
7	4	8	1	5	2	7	0	3	6
8	7	2	8	5	6	1	4	0	3
9	3	1	4	8	7	5	2	6	0

When a multidimensional scaling analysis of the data is performed using least squares as the criterion function, and the individual differences model for the distances, the resulting criterion function is given as follows for each of 1, 2, 3, and 4 dimensions.

τ	Criterion
1	12.599
2	4.857
3	3.425
4	1.287

Because there is a large decrease in the criterion function from 1 to 2 dimensions, and then a leveling off at 2, 3, and 4 dimensions, two dimensions are retained.

Figure 3

The Derived Wine Configuration

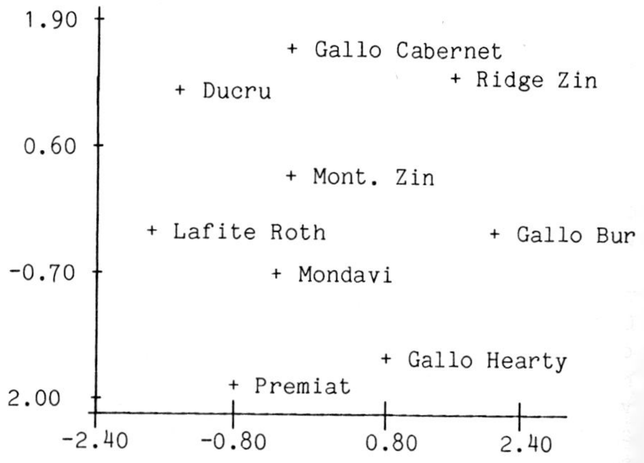

Figure 4

The Judges' Weights

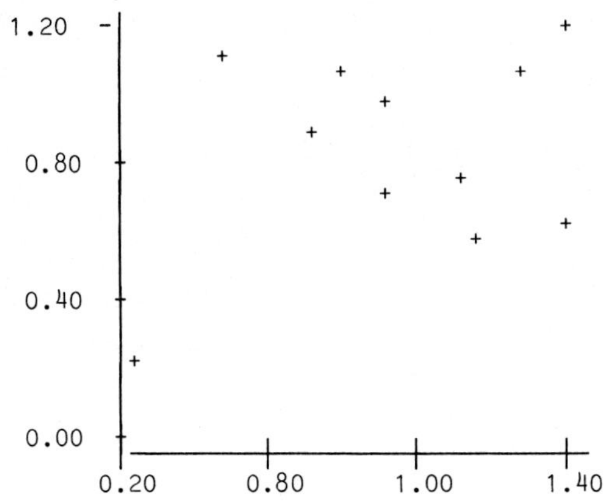

A plot of the configuration in this two dimensional solution is given in Figure 3, in which the scaled wine location is the leftmost letter in the wines name. A plot of the subject weights for each of these two dimensions is given in Figure 4. In Figure 4 note that subject 3 gives almost no weight to dimension 1 and gives comparatively little weight to the second dimension. This outcome can be explained by the fact that subject 3 had a bad head cold during the judging. It is encouraging that the multidimensional scaling seems to be picking up this fact.

Interpretation of the stimulus configuration is difficult. The fact that the Gallo Hearty Burgundy and the Gallo Burgundy are closely related is encouraging because of their close proximity on the plot. Also encouraging is the fact that the two wines made primarily from zinfandel grapes are also close together on the plot. Still, the meaning of each of the two dimensions is difficult to interpret, especially for one who prefers drinking, to learning about, wine.

5. Discussion

Because the asymptotic theory in multidimensional scaling is complicated by the fact that the number of parameters increases in most models with the number of subjects (see Ramsey, 1978), the validity of all asymptotic results in samples of even moderate size is questionable. One should also question the validity of the estimated variances and covariances and any residual analysis. Still, some estimate of a variance is better than none in most cases, and a residual analysis in metric data should yield some information.

The meaning of a residual analysis in nonmetric data is not well understood, however. In such data, because of the monotonic regression, the residuals may not be meaningful. Since the leverages also depend upon the residuals, (and in any event do not include information in the disparity derivatives) they may not be meaningful, either.

A residual analysis in L_p estimation also needs to be investigated more fully. Indeed, the validity and estimates of parameter variances is required when p is not 2. Estimates of leverages also needs investigation.

The fact that the estimates are not unique in the nonmetric scaling models, even after allowing for sign changes and rotations, is disconcerting. This lack of uniqueness comes about because of the disparity estimation. In classical nonmetric scaling, ordinal data becomes pseudo continuous via the monotonic regression. (After the monotonic regression, the disparities are analyzed as if they were continuous.) It seems that a better method would start from the premise that the data are ranks, and compute the configuration estimates directly from the premise. In this regard, the MAXSCAL algorithm of Takane and Carroll (1981) shows promise. This algorithm uses the information in the ranks in the same way that the Cox proportional hazards model can be thought of using it, through the marginal likelihood.

References

1. Carroll, J.D., and Chang, J.J., (1970), Analysis of individual differences in multidimensional scaling via an N-way generalization of "Eckart-Young" decomposition, Psychometrika, **35**, 283-319.

2. Kruskal, J.B., (1964), Multidimensional scaling by optimizing goodness of fit to a nonmetric hypothesis, Psychometrika, **29**, 1-27.

3. Kruskal J.B., Young, F.W., and Seery, J.B., (1977), How to use KYST, a very flexible program to do multidimensional scaling and unfolding, Unpublished manuscript, Bell Telephone Laboratories, Murray Hill, New Jersey.

4. Merle, G., and Späth, H., (1974), Computational experiences with discrete L_p approximation, Computing, **12**, 315-321.

5. Null, Cynthia H., and Sarle, Warren S., (1982), Multidimensional Scaling by Least Squares, in Proceedings of the Seventh Annual SAS Users Group International Conference, SAS Institute Inc., Cary, North Carolina.

6. Ramsey, J.O., (1978), Confidence regions for multidimensional scaling analysis, Psychometrika, **43**, 145-160.

7. Ramsey, J. O.,(1983), Multiscale II Manual, Unpublished manuscript, McGill University, Montreal, Quebec, Canada.

8. Takane, Yoshio, and Carroll, J. Douglas, (1981), Nonmetric maximum likelihood multidimensional scaling from directional ranking of similarities, Psychometrika, **46**, 389-405.

9. Takane, Yoshio, Young, Forrest W., and DeLeeuw, Jan, (1977), Nonmetric individual differences multidimensional scaling: an alternating least squares method with optimal scaling properties, Psychometrika, **42**, 7-67.

10. Weeks, David G., and Bentler, P.M., (1982), Restricted multidimensional scaling models for asymmetric proximities, Psychometrika, **47**, 201-208.

11. Young, F.W., Takane, Y., and Lewyckyj, R., (1978), Three notes on ALSCAL, Psychometrika, **43**, 433-435.

COMPUTER SCIENCE AND STATISTICS:
The Interface, D.M. Allen (ed.)

COLLINEARITY, SCALING, AND ROUNDING ERROR

G. W. Stewart

Department of Computer Science
Institute for Physical Science and Technology
University of Maryland
College Park

1. Introduction

In this paper we shall be concerned with the effects of near collinearity in the linear model

$$y = Xb + e \ , \qquad (1.1)$$

where X is an $n \times p$ matrix of rank p. The qualification "near" is important, for the case where X is exactly collinear, that is where $\mathrm{rank}(X) < p$, is well understood, at least mathematically. Here the theory of estimation tells us that the model (1.1) does not contain enough information to estimate the vector of regression coefficients. The cure is usually to supply additional information in the form of identifiability constraints on b, or more rarely, when the collinearity results from missing data, to supply additional observations to the model. Design matrices are the most important source of exactly collinear models, and the associated theory usually provides a clue to the appropriate fix.

Near collinearities, on the other hand, arise from various sources, and their detection and treatment present a number of research problems that have not yet been satisfactorily resolved. In this paper we shall be concerned with their detection. In principle this problem may be solved by deciding what deleterious effects of collinearity one wishes to avoid and then computing a measure of these effects for the problem at hand. If the effects are are acceptably small, one can continue with the analysis. If not, one must take special action.

The chief ill effects of near collinearities are that they inflate the variance of the least squares estimate of b and that they magnify the effects of errors in the regression variables. In this paper we shall be concerned with how collinearity interacts with errors in the variables. Because this is an interface conference, I will start from my own field and consider errors arising from rounding during the computation of the regression coefficients. At the end of the paper, I will speculate on the problem in general.

In the next section, we will introduce the condition number of the regression matrix X and indicate why it may be considered a measure of collinearity. The condition number shares with other measures of collinearity the property that it changes when the columns of the regression matrix are scaled. The chief technical problem in this paper is to find an appropriate scaling. In §3 we shall present an argument for scaling so that the columns of the regression matrix have the same norms (unit column scaling) and show how the results of rounding error analysis vitiate the argument. In §4 we shall present a different argument, based on rounding error analysis, that also supports unit column scaling. The paper concludes with some general observations.

2. The Condition Number

The condition number of a square matrix was first introduced by Alan Turning in 1948 to measure the sensitivity of the solution of systems of linear equations to perturbations in their coefficients. A related condition number for the solution of linear least squares problems was introduced by Golub and Wilkinson in 1966. For the regression matrix of (1.1) the condition number is defined as

$$\kappa(X) = \| X \| \, \| X^{\dagger} \| \,, \qquad (2.1)$$

where

$$X^{\dagger} = (X^{T}X)^{-1}X^{T} \qquad (2.2)$$

is the pseudo-inverse of X. Here $\| \cdot \|$ is the Euclidean norm of a vector or the spectral norm of a matrix; i.e.

$$\| X \| =_{df} \min_{\| b \| = 1} \| Xb \| \,. \qquad (3.3)$$

For the properties of these norms as well as proofs of the statements to follow in this section, see (Golub and Van Loan, 1983).

The connection of the condition number with collinearity can be made clear by observing that the condition number remains unchanged when X is multiplied by a scalar. Consequently, we may assume without loss of generality that $\| X \| = 1$. With this scaling the reciprocal of the condition number has the following characterizations.

1. κ^{-1} = the smallest singular value of X.
2. $\kappa^{-1} = \min_{\| b \| = 1} \| Xb \|$.
3. $\kappa^{-1} = \min\{ \| E \| : \mathrm{rank}(X + E) < p \}$.

Let us discuss each of these characterizations in turn.

Although the first characterization is phrased as a numerical analyst might put it, it can easily be recast in language that a statistician would appreciate. The singular values of X are the square roots of the eigenvalues of $X^{T}X$. Hence if κ^{-1} is small, $X^{T}X$ has a small eigenvalue, and the inverse cross-product matrix $(X^{T}X)^{-1}$ has a large eigenvalue and hence is itself large. Since the largest element of a positive definite matrix occurs on its diagonal, at least one diagonal of the inverse cross product matrix is large. These diagonals are called variance inflation factors, and the connection between collinearity and large variance inflation factors has often been remarked in the statistical literature.

The second characterization says that if κ^{-1} is small, then there is a vector b for which Xb, is small. In other words, X has an approximate null vector — a sure sign of near collinearity.

The third characterization expresses the relation between condition and collinearity in a very natural manner. Specifically, if κ^{-1} is small then a small perturbation of X is exactly collinear.

Although the condition number is closely related to the notion of near collinearity, it was originally introduced to measure the sensitivity of least squares coefficients to perturbations in the least squares matrix; that is, the sensitivity of regression coefficients to errors in the variables. The principle result is the following. In the model (1.1) let

$$\hat{b} = X^{\dagger}y \qquad (2.3)$$

be the estimated regression coefficients. Let

$$\bar{X} = X + E \qquad (2.4)$$

be a perturbation of X and let

$$\bar{b} = \bar{X}^{\dagger}y \qquad (2.5)$$

be the corresponding estimated regression coefficients. Then

$$\frac{\| \bar{b} - \hat{b} \|}{\| \hat{b} \|} \dot{\leq} \qquad (2.6)$$

$$\left[\kappa(X) + \kappa^{2}(X) \frac{\| \hat{e} \|}{\| X \| \, \| \hat{b} \|} \right] \frac{\| E \|}{\| X \|} \,,$$

where $\hat{e}$ is the residual vector $y - X\hat{b}$. A dot has been placed over the inequality sign to indicate that term in $\| E \|^{2}$ and higher powers have been ignored.

The left hand side of (2.6) represents a relative error in b; as it approaches unity, at least one of the components of b must loose all accuracy. Likewise, the factor $\|E\| / \|X\|$ represents a relative error in X due to the perturbation E. The factor in the brackets is always greater than one and grows with κ. Thus, if κ is large, the regression coefficients can be expected to be sensitive to errors in the variables.

Although the condition number provides a great deal of insight into the nature of collinearity and especially into its interaction with errors in the variables, it is not much used by statisticians. There are two reasons for this. The first is that the right hand side of (2.6) is usually an overestimate of the actual error. This is not surprising, since the bound was derived by numerical analysts, who typically encounter the very small errors caused by rounding on a computer and can therefore afford to use a loose bound. On the other hand, the errors in the variables of a regression problem may be comparatively large, and a loose bound may cause the analyst to give up on a tractable problem.

The second reason, which is the one we shall be concerned with in this paper, is that the condition number is not invariant under scaling of the columns of the matrix. To see how this comes about let us partition X in the form

$$X = (X_* \; x) \qquad (2.7)$$

and define

$$X_\alpha = (X_* \; \alpha x)\,; \qquad (2.8)$$

that is, X_α is X with its last column scaled by the factor α.

Now as α approaches zero,

$$\lim_{\alpha \to 0} \|X_\alpha\| = \|X_*\| > 0\,. \qquad (2.9)$$

On the other hand

$$X_\alpha^\dagger = \begin{bmatrix} X_*^{(\dagger)} \\ \alpha^{-1} x^{(\dagger)} \end{bmatrix}, \qquad (2.10)$$

where $X_*^{(\dagger)}$ consists of the first $p-1$ rows of $X^\dagger$ and $x^{(\dagger)}$ is the last row. It follows that

$$\lim_{\alpha \to 0} \|X_\alpha^\dagger\| = \infty \qquad (2.11)$$

and hence

$$\lim_{\alpha \to 0} X_\alpha^\dagger = \infty\,. \qquad (2.12)$$

From (2.12) we see that by scaling a column of X in a suitable manner, we can make the condition number as large as we like. One feels instinctively that there is something phony about this inflation of the condition number, and on this account the phenomenon has been dubbed artificial ill conditioning. But calling names does not solve problems, and there remains the question of what scaling is correct. We shall now turn to this problem.

3. A Facile Argument

There is one scaling which is widely recommended in regression analysis: scale the columns of X so that they have norm one. If column means have been subtracted from X, this scaling makes the cross-product matrix X^TX a correlation matrix; hence the name correlation scaling is sometimes found in the literature. However, we do not wish to confine our analysis to models with a constant term, and we will instead refer to the strategy as ***unit column scaling***.

Where rounding error is concerned, there is an easy argument in favor of unit column scaling. It is based on two observations.

1. Unit column scaling approximately minimizes the condition number.
2. Unit column scaling ameliorates the effects of rounding errors on computed solutions.

The first observation is due to van der Sluice (1969). The second is widespread throughout the statistical literature (see for example (Draper and Smith, 1981, p.264)). Together they place unit column scaling in the enviable position of minimizing error bounds such as (2.6) while at the same time minimizing the effects of rounding error. On could hardly ask for more.

Unfortunately for this argument, the second observation above is false. Provided that no exponent exceptions occur in the calculation of the regression coefficients, the effects of rounding error are essentially independent of the scaling. The reader may verify this for himself by a simple computation. Take a 3×2 least squares problem and solve it in four decimal digits on a hand calculator by forming the normal equations and solving them with Gaussian elimination. Note the rounding errors at each stage. Now multiply the second column by one hundred and repeat the calculations. Up to scaling factors that are powers of ten, exactly the same rounding errors will occur; the effect of the scaling is to scale the rounding errors, not to change them.

More precisely, what is actually shown by rounding error analysis is that if a numerically stable method is used to compute regression coefficients then the computed coefficients come from a matrix $X + E$, where the columns of E satisfy

$$\| e_j \| \leq c \cdot 10^{-t} \cdot \| x_j \| . \qquad (3.1)$$

Here t is the number of decimal digits carried in the computation and c is a constant, depending on n, p, and the details of the computer arithmetic. If we write this bound in the form

$$\frac{\| e_j \|}{\| x_j \|} \leq c \cdot 10^{-t} , \qquad (3.2)$$

then it says that the relative error in x_j introduced by rounding error depends only on the properties of the computer arithmetic and not on the initial scaling of the column.

Without the second observation above, the case for equal column scaling becomes less persuasive. It is true that the scaling approximately minimizes the condition number; but minimizing the condition number does not necessarily minimize a bound like (2.6), since a scaling that makes κ small may make $\| E \| / \| X \|$ large. It is only when we consider the error structure and the bound simultaneously that we can hope to make meaningful statements. We shall do just that in the next section.

4. Rounding Error and Collinearity

We begin this section by observing that the argument of §3 has a rather loose character. The first of the two observations is precise enough; the second is vague and false. But even if the second were truly and exactly worded, the connection between the two statements has not been made explicit. One feels that there ought to be a relation between condition number and rounding error and therefore what is good for either must be good for both. But in fact we have not been precise in stating what we are about.

To circumvent this problem let us focus on a specific question: *How does near collinearity enhance the effects of rounding error on computed regression coefficients?* In fact the material to answer this question is at hand. We have a measure $\kappa(X)$ of near collinearity in X. In (2.6) we have a relation between collinearity and accuracy. Finally, in (3.1) we have the structure of the error matrix E, when only errors due to rounding are considered. It will take only one more observation to bring these together in such a way as to suggest a natural scaling for computing the condition number.

The observation is that when E is due to rounding, $\| E \| / \| X \|$ will tend to be independent of the scaling of the columns of X. To see this first note that for any matrix E, $\| E \| \le \sqrt{p} \max\{ \| e_j \| \}$. It follows (3.1) that

$$\| E \| \le c \sqrt{p}\, 10^{-t} \max\{ \| x_j \| \} . \qquad (4.1)$$

Since $\| X \| \ge \max\{ \| x_j \| \}$, it follows that

$$\frac{\| E \|}{\| X \|} \le c \sqrt{p}\, 10^{-t} , \qquad (4.2)$$

a bound which is independent of scaling.

The argument is now short. If $\| E \| / \| X \|$ is independent of scaling, then we are free to use any scaling in (2.6). In particular, unit column scaling, which tends to minimize $\kappa(X)$ will tend to give the best bound. In other words, *if the condition number as a measure of collinearity is to be used to predict the effects of rounding error on regression coefficients, it should be computed with unit column scaling.*

The validity of this statement depends on the whether or not the bound (2.6) and (4.2) are realistic. We have already observed that although (2.6) gives away a lot, for the small errors encountered in rounding error analysis it is probably satisfactory. The fact that the scale independence suggested by (4.2) obtains in practice is supported by the details of the rounding error analysis that generated (3.1). Thus if the condition number, computed with unit column scaling, predicts that the regression coefficients are satisfactorily accurate, the result can be taken at face value.

5. Concluding Remarks

In the introduction of this paper we said that the problem of detecting collinearities can be resolved

> by deciding what deleterious effects of collinearity one wishes to avoid and then computing a measure of these effects for the problem at hand. If the effects are are acceptably small, one can continue with the analysis. If not, one must take special action.

This technique of focusing on specific problems is sound dogma, and the failure to observe it in §3 lead to confusion. Only when we posed a precise question in §4 were we able to obtain satisfactory answers to the problems introduced by the effect of scaling on condition numbers.

However, one pays a price for this success. Namely, one can pose many problems, and the answers may not all be compatible. Let us look at three ways in which our basic problem can change.

First let us change the problem of §4 by positing that the model has a constant term but we are *not* interested in the effects of rounding error on the regression coefficient coefficient corresponding to the constant term. How then should the condition number be computed to reflect the accuracy of the remaining coefficients? A careful analysis (which is beyond the scope of this paper) will suggest that unit column scaling should be applied to the original regression matrix, the matrix should then be centered, and finally the condition number should be computed from the centered matrix. Note that unit column scaling is not applied to the centered matrix before the condition number is computed. This means that, contrary to received opinion, correlation scaling is not appropriate for predicting the effects of rounding error on regression coefficients in models with a constant term.

A second way in which our problem can change is that we assess the effects of collinearity in a different way. For example, although the relative error

$$\frac{\| \bar{b} - b \|}{\| b \|} \qquad (5.1)$$

tells us a great deal about the largest components of $\bar{b}$, it tells us less about the smaller ones. If these are of concern, then a better measure will be the individual relative errors

$$\frac{|\bar{\beta}_j - \hat{\beta}_j|}{|\hat{\beta}_j|} . \qquad (5.2)$$

Here we end up with p separate problems, each having its separate answer.

A third way in which our problem can change is that we might forget about rounding error completely and ask how can (2.6) be used to predict the effects of errors from other sources on the regression coefficients. Again the problem of scaling must be reexamined. In (Stewart, 1983) I have given tentative reasons for believing that bound like (2.6) is most meaningful when the columns of X are scaled so that the columns of E are approximately equal — equal error scaling as opposed to unit column scaling. If this is true, then it must be concluded that near collinearity is not as basic a concept as might be wished, since a matrix may be deemed nearly collinear under one class of perturbations and may be well behaved under another.

The conclusion to be drawn from this is that we should not attempt to summarize something as complicated as collinearity in a single number. Instead we should look at all the techniques commonly used in regression analysis and analyze how collinearity effects them. If simplifying patterns emerge, well and good; but my belief is that several sets of numbers will be required to capture the effects of collinearity.

REFERENCES

Draper, N. R. and H. Smith (1981)
Applied Regression Analysis (2nd Ed.) John Wiley, New York.

Golub, G. H. and J. H. Wilkinson(1966)
"Note on the Iterative Refinement of Least Squares Solution," *Numerishce Mathematik* **9**, 139-148.

Golub, G. H. and C. Van Loan (1983)
Matrix Computations, Johns Hopkins, Baltimore.

Stewart, G. W. (1983)
"Rank Degeneracy," *SIAM Journal on Scientific and Statistical Computing* **5**, 403-413.

Turing, A. M. (1948)
"Rounding-off Errors in Matrix Processes," *The Quarterly Journal of Mechanics and Applied Mathematics* **1**, 287-308.

van der Sluis, A. (1969)
"Condition Numbers and Equilibration of Matrices," *Numerische Mathematik* **14**, 14-23.

COMPUTER SCIENCE AND STATISTICS:
The Interface, D.M. Allen (ed.)
Elsevier Science Publishers B.V. (North-Holland), 1986

BIVARIATE DENSITY ESTIMATION AND AUTOMATED STICK-PIN MAPS

Michael E. Tarter, William Freeman

Department of Biomedical and Environmental Health Sciences
University of California Berkeley and West Coast Cancer
Foundation, San Francisco

A variety of augmented scatter diagrams and stick-pin maps are described. These methods use a nonparametric bivariate density estimator to determine the color, representation and masking of data set elements. The identification of a single datum with a single "data point" is considered. It is shown that for some applications it may be both computationally and statistically useful to represent each datum with a spray consisting many individual symbols.

1. INTRODUCTION

Current microcomputers provide economical color as well as medium to high resolution graphical capability. The scatter diagram and stick-pin map can now be visualized as the pre-computer era ancestors of many new ways of displaying statistical information. This paper describes a series of experiments with a variety of augmented scatter diagrams and stick-pin maps which the new generation of computational hardware has made economically practical.

There seems to have been very little previous consideration of the intersection of the field of model-free or nonparametric curve estimation and the field of graphical methods in statistics. For example, none of the papers listed as references to the survey paper on graphical methods by Feinberg (1979), mentions the terms; curve estimation, p.d.f. or c.d.f. estimation or model-free methods. Even when a publication that concerns graphics contains material which is related to curve estimation, this relevance seems coincidental. For example, Trumbo (1981) carefully presents a theory for the coloring of bivariate statistical maps. Principle II of Trumbo's (1981) paper states that "Important differences in the levels of a statistical *variable should be represented by colors clearly perceived as different."*

In the present paper, color is discussed as a means of conveying information about an *estimated* bivariate density and not as a means of distinguishing values assumed by one or more random variates. Specifically a value $Z_i = \hat{f}(X_i, Y_i)$, where $\hat{f}$ is a bivariate density estimator and (X_i, Y_i) represents the $i-th$ of n members of a data set, is represented by color and other means. Note that Feinberg's (1979) classification of graphics lists histograms and scatterplots in the same Category 4. The one dimensional analog of the scatter plot is *not* a histogram but rather a line marked at univariate sample values. In this paper, $Z_i = \hat{f}(X_i, Y_i)$ can be envisaged as a generalization of a value obtained from a histogram as distinguished from a point of a scatterplot, e.g., (X_i, Y_i). In Feinberg's (1979) paper, contours are mentioned under the category of "graphs not involving data." In this paper a way of visualizing contours which are created from sprays of data is presented.

The reason the stick-pin map provides a particularly good framework for discussion, is that a stick-pin can be viewed as having a head which conveys graphic information to the viewer, attached to a point which associates this information with a location in a two dimensional space by a shaft of a length which could be made proportional to estimated probability density. It will be demonstrated that both the choice of head characteristics and the choice of pointer location can, in microcomputer applications, be made to suit a variety of applications.

The topics considered here can be viewed as extensions of two basic procedures. Kronmal and Tarter (1974, pp.377-381) present estimates of bivariate densities, one of which is shown in Figure 1. In essence, the production of this figure first involved the bivariate estimation of a density, e.g., $\hat{f}(x,y)$, as described in Tarter and Silvers, (1975) and secondly the graphing of $\hat{f}$. In the latter paper, several contour diagrams are presented which depict density estimators such as $\hat{f}$ shown in Figure 2. The routine used to produce Figure 2 traces each contour and only evaluates $\hat{f}$ at points near each contour. However, in Figure 1, $\hat{f}$ is evaluated at every x,y coordinate of a grid of points. This latter procedure uses simpler computer code, which is an important consideration for a subroutine designed to be moved easily to a variety of microcomputer systems. On the other hand, since the number of required evaluations of $\hat{f}$ increases quadratically with increase in graphical monitor resolution, the computer *time* demands of this simple code may be substantially greater than those of more complicated contouring routines, such as that utilized to obtain Figure 2.

The second procedure was developed by Tarter (1979) and depended on the observation that an estimated bivariate as well as univariate density could serve as a useful data transformation. Consider that the color or shape of each stick pin-head in a conventional stick-pin map can be chosen on the basis of density height estimated at the location of a data point. If one were only interested in the rare or unusual event, one could choose to insert a pin only at those points over which the estimated density is less than a constant. In an analogous way, Figure 3 was obtained after: 1) a bivariate density estimator $\hat{f}$ was computed. 2) A sequence $\{Z_i\}$ was obtained by using the bivariate density estimator $\hat{f}$ as a transformation. Specifically, the $i-th$ member of the sample $\{X_i, Y_i\}$ where $i=1,...n$ was associated with a value $Z_i = \hat{f}(X_i, Y_i)$. Each value of Z_i can be interpreted as an estimate of the sparseness or richness of density within a fixed size neighborhood of the point (X_i, Y_i). An exchange of variables option was exercised to plot the point pairs $\{(X_i, Z_i)\}\ i=1,...n$ alternately, $\{(Y_i, Z_i)\}\ i=1,...n$; could have been plotted as a preliminary to the next step. 4) An editing routine was used to select those $\{(X_i, Z_i)\}\ i=1,...n$ points where Z_i was greater than a constant. (This process masked all but the most sparsely distributed points. 5) The variable exchange option was again exercised to replace the display of the edited (X_i, Z_i) values with a display of corresponding (X_i, Y_i) values, i.e., the sparsely distributed subset of the original sample. A printer plot routine accompanies the program.

The basic feature which differentiates the method used to obtain Figure 2 from that used to obtain Figure 1 is that the end product of Figure 2 is a display of *data points* i.e., is associated with a sample; while Figure 1 illustrates an estimator of an underlying population. In essence the method used to generate Figure 2 goes full circle, i.e., starts with a sample and then uses a population density estimator to modify a display of sample elements. Displays such as Figure 1 and the contours of Figure 3, shed all reference to *individual* sample elements in order to convey information regarding the global or overall nature of an estimated density. As previously mentioned, the procedures to be described in this paper, with varying degrees of, success both display global distributional characteristics and the fine-structure of the sample. They also tend to resemble the routines used to produce Figure 1 and 2 and differ from contouring routines, in-so-far as they involve simple and transportable computer code.

2. GENERAL METHODOLOGY

The contours shown in Figure 3 were estimated from one thousand random variates generated from the three component mixture of bivariate densities, $(1/3)N(6,9,1.5,1.5,0.5)$ + $(1/3)N(10,10,1.5,1.5,-.7)$ + $(1/3)N(14,11,1.5,1.5,0.5)$, (the order of the parameter arguments of N is $\mu_x, \mu_y, \sigma_x, \sigma_y, \rho_{xy}$), using methods described in Tarter and Silvers (1975). The techniques to be described in this paper do not depend upon the computational tractability of the underlying density estimator. This is not the case with contouring methods which rely on gradient procedures and therefore the numerical or analytical tractibility of $\hat{f}$'s partial derivatives. Since any accurate bivariate density estimator can be used in conjunction with the methods to be described in this paper, for the sake of brevity, we will omit the specific steps used to obtain $\hat{f}$ and leave these steps to the tastes and needs of the reader.

Figure 4 was obtained from the same size sample and underlying distribution that was used to generate Figure 3. To obtain this graphical display all five techniques to be described in this paper were applied. These are: 1) Spraying, 2) Masking, 3) Banding, 4) Color and 5) Symbol differentiation. To implement all these techniques the fundamental idea which led to the generation of Figure 2 was utilized. Specifically the estimated value $Z_i = \hat{f}(X_i, Y_i)\ i=1,...,n$ was used to: 1) Pick the color of a display character 2) Determine whether a given point should or should not be displayed. 3) Select the number of display points to be associated with each datum, 4) Mask display points to better visualize the edges of an estimated terrace (this procedure is analogous to the trimming of the borders of a lawn.) and 5) Select the appropriate symbol for display purposes.

Note that unlike the display shown in Figure 1, $\hat{f}$ evaluations by these new procedures are required either at, or in the neighborhood of, n data points and not at all grid points. We have experimented with modifications of the methods used to obtain Figure 1 which were designed to use a series of evaluations of $\hat{f}$ over a widely spaced grid to determine the need for refinement. These routines not only required a cumbersome and lengthy code but failed to resolve detail for a variety of test patterns.

The methods to be described here have a tendency to emphasize data anomalies since, being elaborations of simple scatter diagrams and stick-pin maps, fine structure is clearly resolved. On the other hand, particularly when the spraying technique detailed in Section 3 is utilized, global population characteristics can usually be as clearly discerned by the new procedures as with contouring techniques (the latter tends to both smooth over sample fine structure and require a code highly dependent on the means used to obtain the bivariate estimator $\hat{f}(x,y)$).

Naturally in some systems a superposed scatter diagram and contour diagram may be a reasonable substitute for the new techniques described in this paper. Note however, that what appear to the eye as contours generated by the new methods are actually formed directly from the data points themselves. The previously mentioned Tarter, Silvers (1975) paper and considerable earlier work by Gregor (1969) and others deal with procedures for modifying and underlying density estimate to either increase or decrease the contrast between distribution components. A composite or overlay of the scatter diagram computed from one's original data and a contour diagram, in essence separates the head and point of each stick pin. On the other hand, since contours are actually formed from the scatter diagram or stick-pin head by the new method, it is easy to associate the effects of the contrast modification process upon individual or subgroups of points.

Before turning to the specific means of creating augmented scatter diagrams and stick-pin maps, it seems appropriate to summarize the following basic algorithm:

1) A bivariate estimate $\hat{f}(x,y)$ is obtained from the sample $\{X_i, Y_i\}\ i=1,...,n$.

2) The sequence $\{Z_i = f(X_i, Y_i)\}\ i=1,...,n$ i[illegible] obtained.

3) The $\{Z_i\}$ sequence is ranked.

4) The ranked values of the $\{Z_i\}$ sequence ar[illegible] used to determine the properties used to displa[illegible] each datum $\{X_i, Y_i\}\ i=1,...,n$.

The word "datum" rather than "data point" is use[illegible] above because is, as we shall see, a spray of point[illegible] can be usefully associated with a single datum.

The programs which generated the maps in thi[illegible] paper are written in Fortran 77 under the UNIX 4.[illegible] operating system and, with possibly minor altera[illegible] tions in the i/o portions the routines, can be com[illegible] piled with most alternative F77 compilers.

3. SPRAYING AND MASKING

We will now suggest that there may be considerabl[illegible] practical value to using several disconnected symbols to represent a single datum and even in some cases, representing some data with fewer "points" than other data.

Generally speaking the chief advantage of using a spray of points to represent a single datum is that at the edges of what we will later define as a contour, the spray can be *masked* in order to give the tracking eye useful information about the shape of the contour. As an analogy, consider that the user of a can of spraypaint delivers a cloud or scatter of droplets for each pull of the spraygun's trigger. Towards the center of a large area to be sprayed a single color, most of the droplets will usually reach the object to be painted. However, along the border separating two colors, masking tape or a masking tool is used to block off a significant portion of droplets. Now consider the important fact that when the painter knows that he is painting the interior of an object he or she need not be concerned about the use of a masking tool. This basic principal, when applied to statistical graphics, makes it computationally economical to program spraying techniques.

Specifically suppose five display points are used to represent a single datum where four of the five points are corners of a square centered at what for conventional procedures would be the fifth, i.e., the "data", point. Now define a contour of bivariate estimator $\hat{f}$ as the locus of (x,y) points where $\hat{f}(x,y)=C$ where C is some positive constant smaller than the largest value assumed by $\hat{f}$ over the entire x,y plane. In this discussion we will also assume that the contour is a single closed curve.

Consider two distinct contours associated with the loci $\hat{f}(x,y)=C_1$ and $\hat{f}(x,y)=C_2$ respectively, where $C_1 \neq C_2$. The region between these contours will be a bracelet shaped or banded subregion $R(C_1,C_2)$ of the x,y plane. Suppose, as in the previous section, the sequence $\{Z_i=\hat{f}(X_i,Y_i)\}\ i=1,\ldots,n$ is obtained and ranked to form the sequence $\{Z_{(i)}\}\ i=1,\ldots,n$. Under the previous assumption that each contour is a single closed curve, the indices $\{(i)\}$ of $\{Z_{(i)}\}$ associated with those data points (X_i,Y_i) which fall within $R(C_1,C_2)$ will form a consecutive sequence $S(C_1,C_2)$.

Now suppose a particular plotting character is to be placed at each of the five coordinates (X_i,Y_i), $(X_i+\varepsilon,Y_i)$, $(X_i,Y_i+\varepsilon)$, $(X_i-\varepsilon,Y_i)$, $(X_i,Y_i-\varepsilon)$, provided that each of these five points is within $R(C_1,C_2)$ (where ε is a small positive constant). Those values of these five coordinates which are most likely to protrude beyond the region $R(C_1,C_2)$ will be associated with $Z_{(i)}$ indices near the beginning or end of the sequence of indices $S(C_1,C_2)$. This fact allows one to construct a computer program which significantly reduces the computer time required to trim the edges of the five point datum spray.

For example, to obtain Figure 4, the following sequence of steps was utilized:

1) The one thousand random data elements from the distribution described in the previous section were used to obtain a bivariate estimator $\hat{f}$.

2) Each of the one thousand data elements, i.e., $(X_j,Y_j)\ j=1,\ldots,1000$, was transformed to $Z_j=\hat{f}(X_j,Y_j)$.

3) The set $\{Z_j\}\ j=1,\ldots,1000$ was ranked to form $\{Z_{(i)}\}$. We will henceforth define C_0 as the largest value assumed by $\hat{f}$ over the space within which $\hat{f}$ is to be displayed. The method which was used to produce Figures 4 through 8 was based on the specification of a set of regions in terms of $Z_{(i)}$, i.e., the kth region will contain those points for which $Z_{(i)}$ is in the half open interval $(u_k C_0, v_k C_0]$, where $u_k C_0$ is the left endpoint and $v_k C_0$ is the right endpoint of the kth interval (The sequence of u_k and v_k values is chosen by the user). The set of intervals used in Figure 4 through 10 of this paper was $(C_0,.95C_0]$, $(.95C_0,.90C_0]$, $(.90C_0,.80C_0]$, $(.70C_0,.55C_0]$, $(.55C_0,.35C_0]$, $(.35C_0,.10C_0]$, $(.10C_0,0.0]$. These were chosen in order to assure that standard uncorrelated normal data would generate bands one through six of approximately equal width. In Section 5 of this paper, practical reasons for using special procedures to choose the lowest band, here band seven, will be discussed. Only the 1st, 3rd, 5th, and 7th intervals are plotted using red, green, blue, black, respectively, for the color figures. As elaborated upon in the next section, the remaining intervals form the blank bands.

5) The inner and outer edges of the regions are masked as follows:

i) Let $Z_k^{(min)}$ and $Z_k^{(max)}$ be the smallest and largest values, respectively, of $Z_{(i)}$ in the kth region, i.e., $Z_k^{(min)}$ is the smallest value and $Z_k^{(max)}$ is the largest elements of the set $\{Z_{(i)} \mid v_k C_0 \leq Z_{(i)} \leq u_k C_0\}$. Thus, $v_k C_0$ and $u_k C_0$ represent the density estimate at the outer and inner bordering contours of the kth region, respectively.

ii) The area beyond the outer bordering contour of the kth region is masked by comparing the density at each of the four peripheral points of the spray to $v_k C_0$, i.e., only those points for which $\hat{f}$ is greater than or equal to $v_k C_0$, are plotted. Beginning with $Z_{(j)}$, where j is the smallest element of $S(C_k,C_{k+1})$, each datum is sequentially sprayed and masked until the masking process fails to reject any peripheral points m consecutive times.

iii) An analogous process to that described in ii is used to mask the area beyond the inner bordering contour of the kth region except that the starting value is Z_r, where r is the largest element of $S(C_k,C_{k+1})$, and the index r is sequentially decremented. The process terminates when none of the points in m consecutive five point sprays fail to be plotted.

One can conceive of examples where the plateau-like shape of $\hat{f}$ might cause the spraying process to terminate prematurely. Such problems are easily remedied by increasing the value chosen for m.

The ε used as part of the spraying process was specified as either a percent of the sample range or as a percent of the sample standard deviation. For all but a very few applications, ε will differ in the x and y directions (horizontal and vertical, respectively. Technically ε should be subscripted as either ε_x or ε_y. However, because in our program, both ε_x and ε_y are determined by a single user assigned multiplier of the estimated range or standard deviation, the use of subscripts was felt to be an unnecessary notation. In the examples, ε was chosen to be 2% of the sample range. In some applications involving natural rather than simulated data, it may be preferable to use the sample standard deviation or a scale parameter estimator which is robust with respect to outliers.

It should be mentioned that there are many alternative means that can be used to arrange the peripheral points, i.e. spray. However, the arrangement used in the examples is easy to program and gives useful regions. Notice the gaps in the 3rd and 5th contours of Figure 4 (green and blue contours, respectively). These gaps can be filled in by using a broader spray. For ninepoint spray (X_i,Y_i), $(X_i+\varepsilon_1,Y_i)$, $(X_i,Y_i+\varepsilon_1)$, $(X_i-\varepsilon_1,Y_i)$, $(X_i,Y_i-\varepsilon_1)$, $(X_i+\varepsilon_2,Y_i)$, $(X_i,Y_i+\varepsilon_2)$, $(X_i-\varepsilon_2,Y_i)$, $(X_i,Y_i-\varepsilon_2)$, where $\varepsilon_2=\alpha\varepsilon_1$, multiplier values α between two and four, were found to be effective. The rationale for the ninepoint spray is that the inner points, i.e., those offset by ε_1, give definition to the edge of the contour and the outer points, i.e., those offset by by ε_2, make the contours appear continuous. Of course, the broader the spray the greater the risk of masking features which provide clues about the fine-structure of the underlying density. Thus, the use of very broad sprays tends to negate the advantages of this new approach to contouring over the usual method of contouring illustrated in Figure 3. (This is true both in terms of processing time and resolution of detailed features of the data.)

We will now proceed to discuss and illustrate the fifth step in the above process; the choice of display symbol.

4. BLANK BANDING

An important and interesting aspect of these techniques is the decision about whether to plot a given data point. It may seem paradoxical that a datum can provide more information by not being represented than by being represented. Consider however, that possible data outliers will generally be represented by points in only one of the bands, i.e., the band associated with the smallest values of $\hat{f}$. Therefore, as long as the masking procedure described in the previous section does not blank out the lowest band, information about extraordinary data values, i.e., outliers, will be adequately conveyed even if data in other bands are not represented. As will be discussed in the last section of this paper it may be useful to mask some outliers.

There can however, be distributional details within a body of data which, although not what could be called global features, are of interest and importance. Consider, for example, what might be called a data *"dimple"*, i.e., a small set of values W_1 surrounded by a region W_2 such that underlying density $f(x,y) \ll f(u,v)$ where $(x,y)\in W_1$ and $(u,v)\in W_2$. Good and Gaskins 1980, use the word "dip" for the feature we call a "dimple". Because of the smoothing inherent in many nonparametric estimates $\hat{f}$ of f the presence of a dimple may not greatly influence $\hat{f}$, particularly if the kind of dimpling often referred to as digit preference pertains to one's data where a dimple is, in effect, in close proximity to a "raised area". However, since the methodology described here only uses display symbols at or near the coordinates of a datum, even if $\hat{f}$ is fairly constant at and around a dimple, the presence of a dimple will be apparent in many cases.

It has been our experience that data is either very dimpled, e.g., as would be the case for most situations where digit preference pertains, or has few if any dimples. Thus, if blank bands alternate with display bands, and if within diplay bands each datum is represented by one or more symbols, it is unlikely that data dimpling will go undetected since it is unlikely that all dimples will occur exclusively within blank bands.

Figures 6 through 8 illustrate the advantages of blank banding and masking. The first display of this sequence is a typical scatter diagram obtained from a one-thousand element random sample from the same three component mixed normal distribution used to obtain Figure 4. Figure 6 illustrates the use of blank banding where a single data point is used to represent a single datum. The next display illustrates the effect of spraying without masking. While the swarm of points shown in Figure 7 is more vivid than that shown in Figure 6, we have found that this same effect could be much more easily obtained by representing each point by a larger symbol, e.g., a circle or X. Finally, Figure 8 clearly illustrates the advantages of spraying and masking in terms of most user's ability to discern global distributional structure.

5. SYMBOLS, COLORS AND SPRAYING

When we first began the research described in this paper, our impression was that the most useful display characteristic which the $\{Z_{(i)}\}$ could determine would be found to be color. We now feel that banding and spraying can often yield so much information that the use of a color as opposed to monochrome display may be an unnecessary, albeit attractive, luxury.

The choice of which particular sequence of color or symbols to use to represent a particular sequence of $\hat{f}$ values is closely connected to representation of perspective as outlined in L. Gurry's book *Cezanne et L'expression L de L'espace*, 1950, page 6, which contains a brief history of the artists' use of color and other techniques to represent three dimensions on a two dimensional surface.

In general, warm colors such as *red* seem to be ideally suited to representation of points where $\hat{f}$ assumes large values, i.e., those which would be closest to the viewer if a three dimensional model rather than a two dimensional symbolic representation were used. Conversely, *blue and finally black* appear to be the best final colors to use, where as illustrated in Figure 4, black squares represent the outlying points where $\hat{f}$ values are smallest.

Use of black squares, dots or circles at low density levels, and larger sized "hats", "tildes", or "pulses" at higher levels, can simulate the visual clue that when viewed from above, if the height of a stick-pin whose head were the chosen colored symbol were proportionate to $\hat{f}$, then this stick-pin head would appear smaller.

It should be noted that many symbols have particular asymmetries which can be used to advantage. For example, *squares and "hats" tend to roughen* edges while "tildes" and "pluses" tend to "mask" well, i.e., lead the eye comfortably along a curved contour. Smoothness is usually desirable for moderate to high $\hat{f}$, i.e., stick-pin height, levels since it is for these values that global distributional features are often visualized. Therefore it seems ideal to use squares and other roughening symbols to represent outlying stick-pins, i.e., low values of f. Also, a symbol such as a "tilde" or "hat", which is longer than it is high, can enhance resolution if its longer dimension parallels that of the data display. For example, in Figure 4, since the overall display of points is longer than it is wide, it seemed advantageous to place tildes and hats as shown. Here again, artists using most engraving methods had learned to place the *burin* of their scratches parallel to the curves they wished the eyes of their viewers to scan. (Antraesian, Abrams 1971, p.376; Oxford 1971, p. 1188).

As one final point, it should be mentioned that there may often be reasons for using a blank band to represent points (X,Y) where $f(X,Y)<\delta$ for some small positive value of δ. We have often worked with sets of data where the presence of one or a few extreme outliers has led to an unsatisfactorily compressed view of the bulk of data points. It seems advisable in these situations to select the viewing area so that the bulk of the data are satisfactorily displayed, and to indirectly indicate outlying data that cannot be displayed. One method of doing this is to report the number of outliers outside the viewport. A more informative method is to place a special symbol just inside the border of the viewport at the intersection of the viewport border line and the line connecting the centroid of the lowest contour (or contours) and the outlier. If the graphical system available to the user is sufficiently flexiable, one can go a step further and let the size of the symbol or the choice of the symbol itself suggest the distance between the outlier and the symbol. For example, if the chosen symbol is a letter, then the greater this distance from the outlier to the viewport, the later the letter can be chosen from the alphabetic sequence.

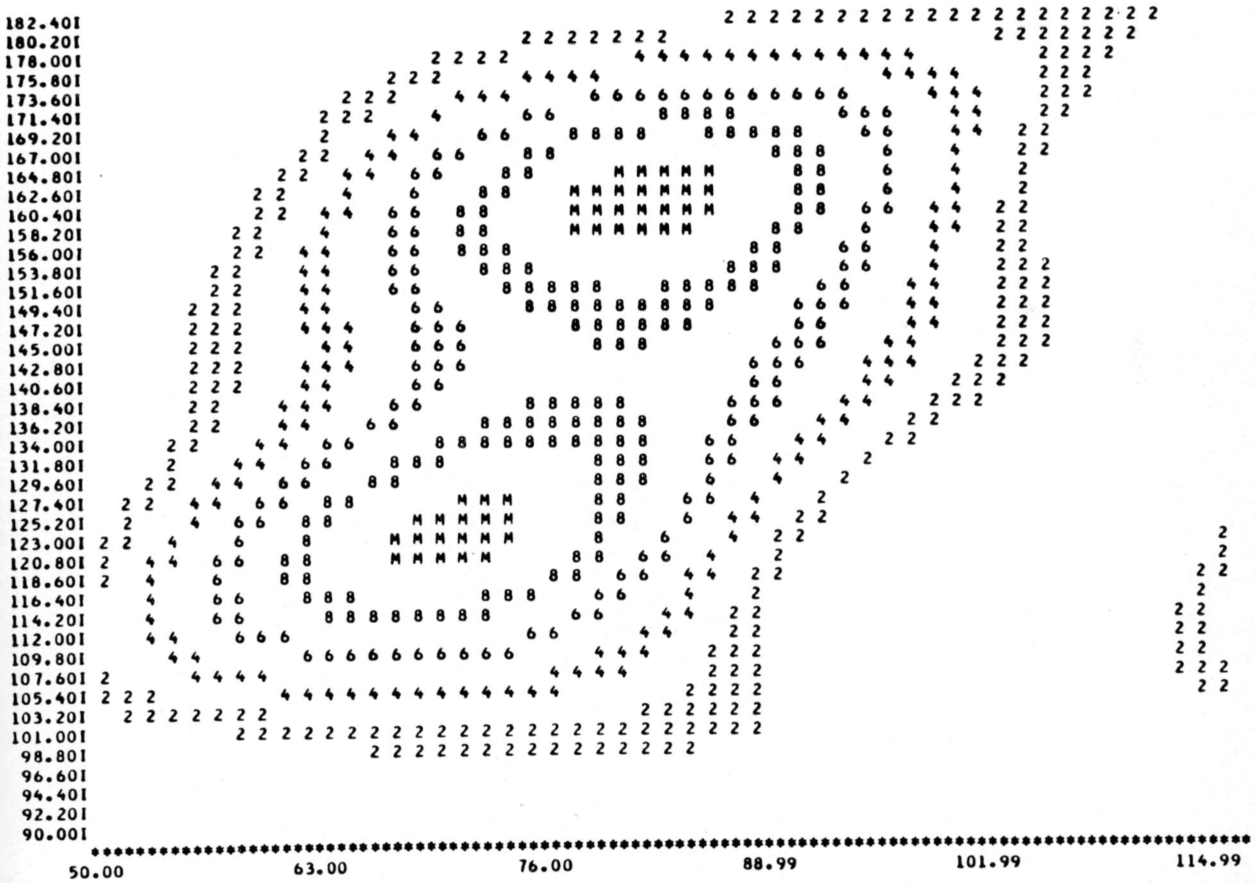

FIGURE 1. BIVARIATE PRINTER GRID PLOT SHOWING BANDED CONTOURS.

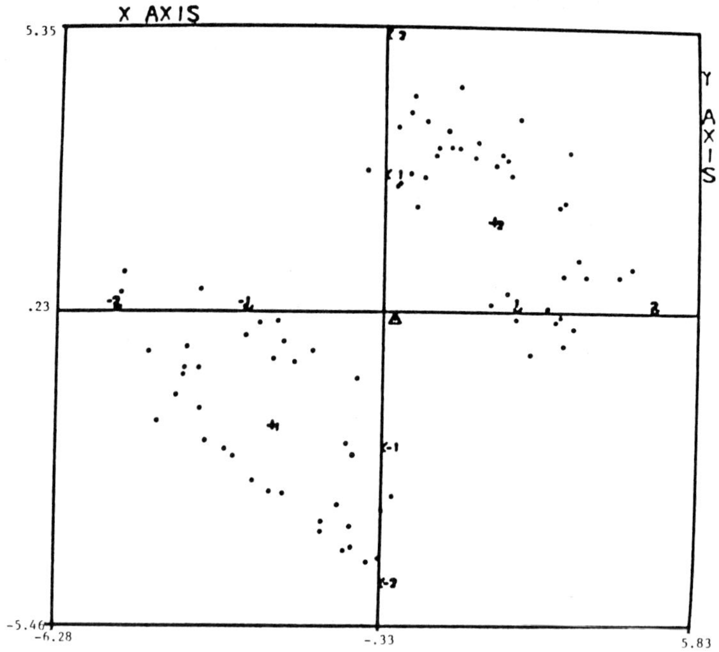

FIGURE 2. CONTOUR DIAGRAM OBTAINED FROM A NON-PARAMETRIC BIVARIATE DENSITY ESTIMATOR, n=1000.

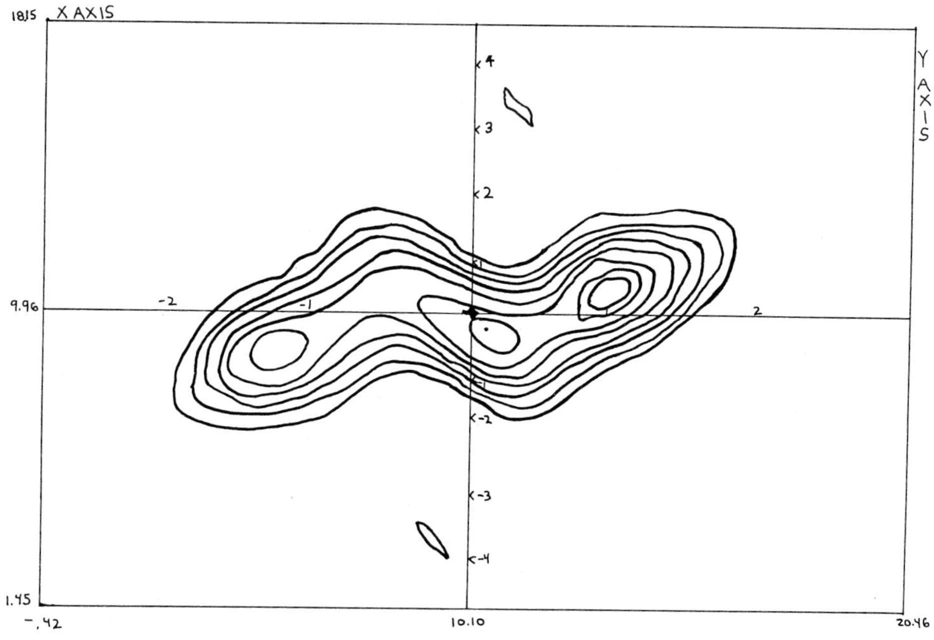

FIGURE 3. PENUMBRAL SCATTER DIAGRAM

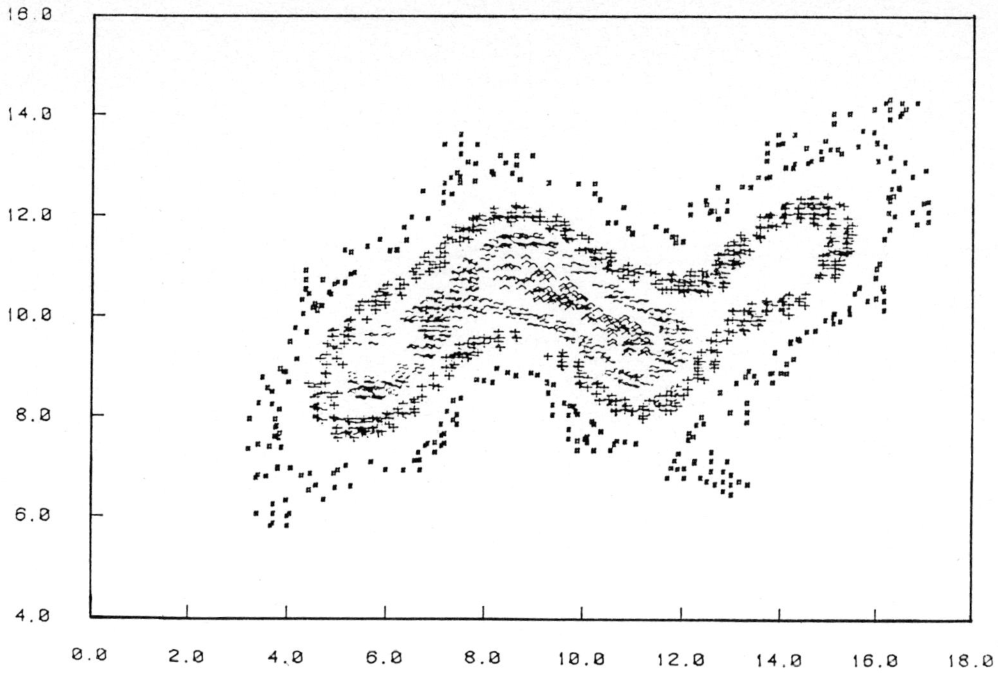

FIGURE 4. AUTOMATED STICK-PIN MAP - LEVELS DISTINGUISHED BY SYMBOL AND BANDING - SPRAYING AND MASKING USED.

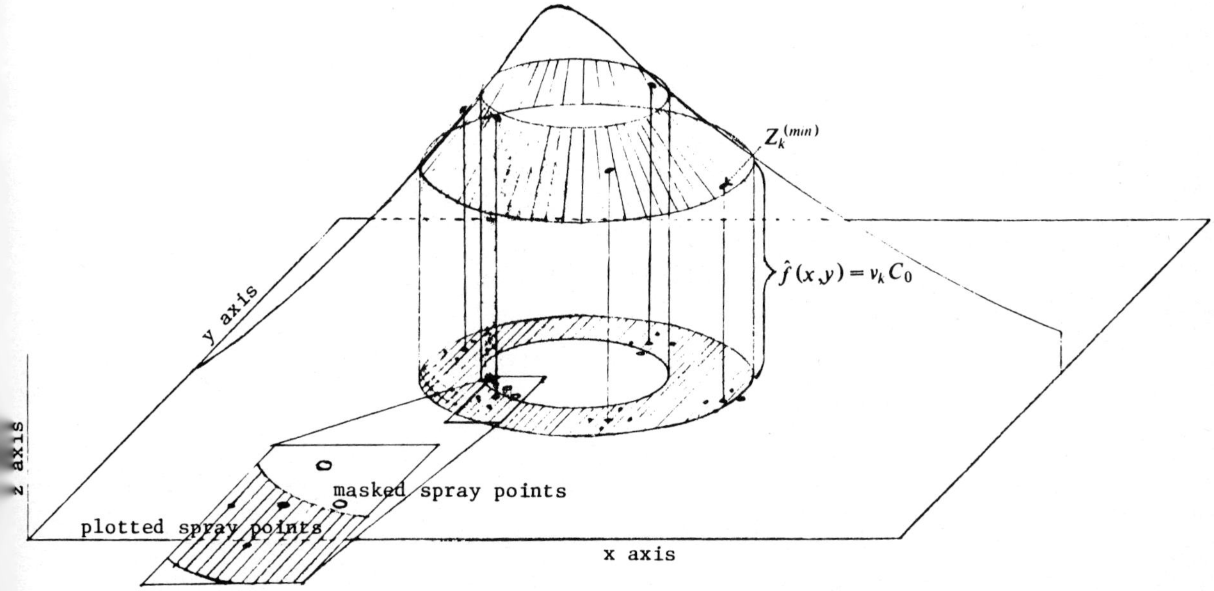

FIGURE 5. CONVENTIONAL SCATTER DIAGRAM.

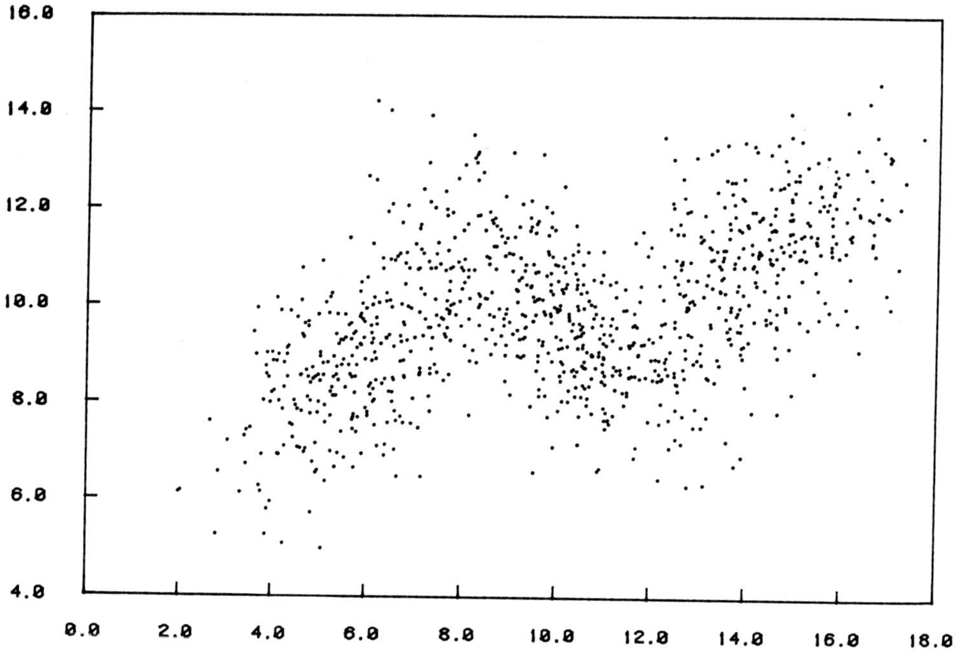

FIGURE 6. AUTOMATED STICK-PIN MAP - SPRAYING AND MASKING *NOT* USED.

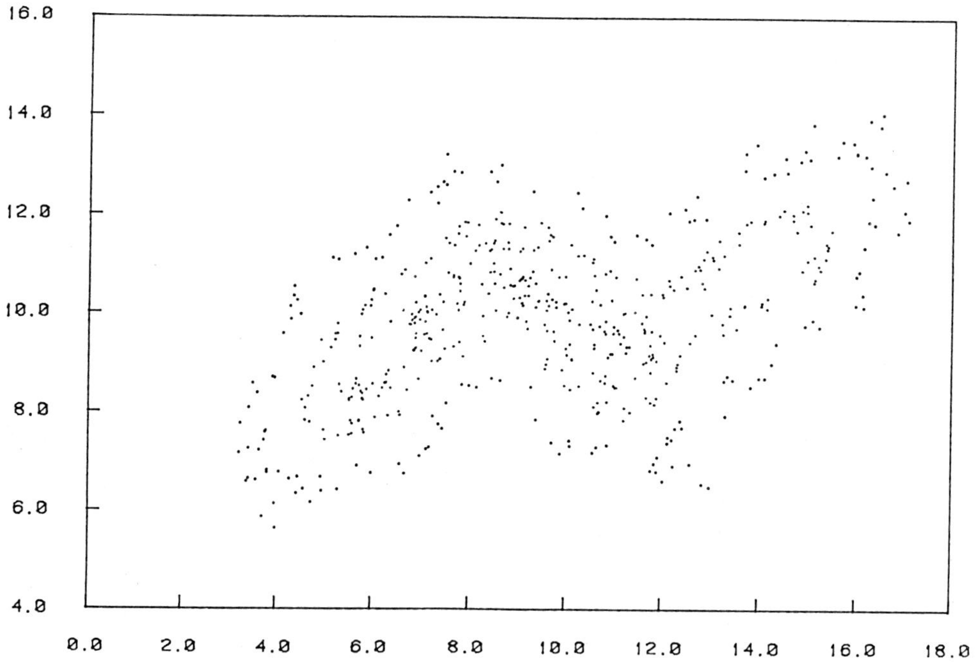

FIGURE 7. AUTOMATED STICK-PIN MAP - SPRAYING USED, MASKING NOT USED.

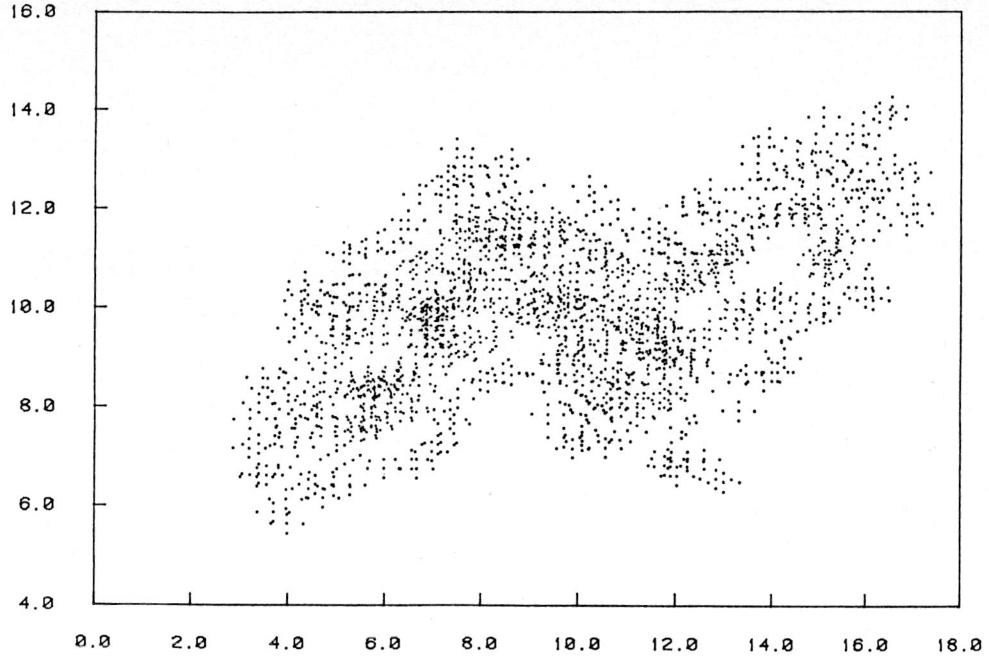

FIGURE 8. AUTOMATED STICK-PIN MAP - BOTH SPRAYING AND MASKING USED.

*Michael E. Tarter is Professor of Biostatistics, Department of Biomedical and Environmental Health Sciences University of California, Berkeley, California 94720. William Freeman is with the West Coast Cancer Foundation, San Francisco, California 94133. The authors would like to thank Ms. Yue-mei Ho and Paula Figdor for their technical assistance. Preparation of this paper is supported by National Cancer Institute Grant No.1 R01 CA35795-01A1.

REFERENCES

[1] ANTRAESIAN, GARO Z. and ADAMS, CLINTON(1971). *The Tamarind Book of Lithography: Art and Techniques*, Abrams, New York.

[2] FIENBERG, STEPHEN E.(1979). "Graphical Methods in Statistics", *The American Statistician*, 165, 33-4.

[3] GREGOR, J.(1969), "An Algorithm for the Decomposition of a Distribution into Gaussian Components", *Biometrics*, 25, 79-93

[4] GOOD, I.J. and GASKINS, R.A.(1980), "Density Estimation and Bump-Hunting by the Penalized Likelihood Method Exemplified by Scattering and Meteorite Data", *Journal of the American Statistical Association*, 75, 42-73.

[5] GUERRY, LILIANE(1950), *Cezanne Et L'expression De L'espace*, Paris: Flammarion.

[6] KRONMAL, RICHARD A. and TARTER, MICHAEL E.(1974), "The Use of Density Estimates Based on Orthogonal Expansions" in *Exploring Data Analysis*, ed. W.J. Dixon and W.L. Nicholson, Berkeley: University of California Press, 365-395.

Oxford English Dictionary (1971), Oxford University Press.

[7] SILVERS, A. and TARTER, MICHAEL E.(1975), "Implementation of Bivariate Gaussian Mixture Decompositions", *Journal of the American Statistical Association*, 70, 47-55.

[8] TARTER, MICHAEL E.(1979), "Density Estimation Applications for Outlier Detection", *Computer Programs in Biomedicine*, Elsevier: North-Holland Biomedical Press, 10, 55-60.

[9] TRUMBO, BRUCE E.(1981), " A Theory for Coloring Bivariate Statistical Maps", *The American Statistician*, 220, 35-4.

Statistical Computing Languages of the Future

Organizer: *Kenneth Berk*

Invited Presentations:

S as a Programming Environment for Data Analysis and Graphics
John M. Chambers

Integrated Programming Environments (Abstract Only)
John Alan McDonald and Jan Pedersen

The Monte Carlo Processor: Designing and Implementing a Language for Monte Carlo Work
David Alan Grier

COMPUTER SCIENCE AND STATISTICS:
The Interface, D.M. Allen (ed.)
© Elsevier Science Publishers B.V. (North-Holland), 1986

S as a Programming Environment for Data Analysis and Graphics.

John Chambers

AT&T Bell Laboratories
Murray Hill, New Jersey 07974

ABSTRACT

This paper discusses experience with the S system as an applications programming environment. It also considers, in the context of data analysis and graphics, the class of workstations called *integrated programming environments*. Current research on a merging of the needs of computing for data analysis with the attractive features of integrated programming environments is outlined.

1. Introduction.

This paper looks at interactive programming environments for applications using data analysis, graphics and related kinds of computing. The next two sections give a view of the history of S and of recent ideas in the general field of "integrated programming environments". In the context of the present conference we emphasize the experience gained by a substantial number of applications development groups from using S as a programming environment for their work. The last section outlines, necessarily briefly, current research aimed at combining the important features of integrated programming environments with facilities needed for quantitative (scientific) computing; for example, access to algorithms for numerical or graphical computations. Experience gained from the use of S as an applications programming environment for business research, data analysis, engineering projects and other applications is being used to guide the new design, particularly in terms of combining flexibility with run-time efficiency.

2. The S System.

S is a language and system for the interactive analysis of data, developed at Bell Laboratories and currently in use on the operating system. Two books [1; 2], describe respectively how to use the system for data analysis and graphics, and how to extend the system by incorporating new algorithms as S functions. The design of S and its relation to other work in computer science and in statistical computing are described in [3].

We designed S to enable and encourage good data analysis, by letting users look quickly and conveniently at many displays, summaries, and models for their data. In addition, we emphasized in our design the ability to *extend* S. Users could write S macros to encapsulate analyses that were to be repeated, possibly with differing arguments. They could develop new functions that interfaced to arbitrary algorithms (typically FORTRAN subroutines), not necessarily designed for use with S originally. Also, and unusually for such systems, S allowed easy creation of arbitrary new data structures to represent new analyses, plots, etc.

These facilities have made S into an *applications programming environment*, which a variety of groups, at Bell Labs, at AT&T and elsewhere (notably at universities), have used to create other, often more specialized, systems. We anticipated that this use would be made of S, and provided a number of features accordingly. (Besides those mentioned above, there are facilities for documenting user extensions, for writing menu-driven interfaces in S, and for incorporating S results in report-generation software.) In a typical scenario, a few of the more adventurous computer users in a local group find out about S, and begin to experiment with it for the needs of the group. After a while, these users decide to create some more-or-less canned facilities, built on S, that would then be a system to be used by other members of the group. In the two-tier user community resulting, the later users might have little direct contact with either S or the operating system.

The advantages of using S for such purposes are several. S is designed to be easy to use and highly interactive. It supports interactive graphics on a variety of devices. By using the macro facility, new analyses can be coded and tested easily. The ability to write compiled functions, interface to external algorithms, and define new data structures means that the extensions possible are unlimited. Feedback to us from about 20 applications projects has indicated that S has provided a substantial increase in the productivity of the developers compared either to programming in a language like C or FORTRAN or to the use of other, less flexible, systems.

This extensive experience on the part of applications developers has also contributed several new challenges to improve the system. Here, as often, there is a conflict between ease of implementation and efficiency of computation. Writing S macros is easy, particularly up to the point of trying to make the macros themselves "friendly" to the end users. But occasionally the computations involved are difficult to express in S. More frequently, serious inefficiencies can result when the macros are applied to sizable data or are themselves used in an iterative fashion. The usual cure attempted, to write the same calculations in a compiled function, helps in most cases but requires substantially greater programming activity on the developer's part.

The fundamental problem, to a large extent, is that the application developer is working not with one language and environment, but with three or four. Further, these languages inherit a degree of mutual inconsistency from the software tools used to create them. The current S environment depends heavily both on existing tools and on tools specially adapted for S. The macro processor is a version of the *m4* macro processor. The languages in which new functions and new algorithms for compilation with S are written are extensions of FORTRAN utilizing the Ratfor preprocessor and m4. Heavy use of tools was an important factor in making S work in the first place, and in the ease with which its design has adapted to rapidly evolving computing environments over the last five years or so. However, the price paid includes inconsistencies among the various levels of S as a programming environment.

The challenge for our current research is then to attack simultaneously:

- simplifying the application developer's view of the programming environment;
- making S more efficient for the kind of use described above.

Before outlining the implications of this challenge, let us look at another aspect of recent computing that points in an interestingly similar direction.

3. Integrated Programming Environments.

Recent evolution of high-powered and (relatively) high-priced personal workstations have produced examples, such as LISP machines and the Smalltalk-80 system, of *integrated programming environments*. Proponents of these systems assert, with considerable informal evidence in support, that the new environments allow users to be more productive in designing, implementing and testing new software. Specific features that distinguish integrated programming environments from earlier systems include:

- the user's processes operate in a single, persistent memory space (in contrast, for example, to communicating via files);
- the environment is based on a single language and corresponding set of programming facilities, for user-written and system facilities alike;
- system facilities (the "browser" in Smalltalk) allow users to examine, debug and change all the programs, user or system, in a highly interactive way.

The intent is to make the complete system easily visible, testable and open to user change, via a single *integrated* programming environment.

It is useful to compare this approach to the environment, which represents a popular current approach to interactive programming environments [e.g., 4].

- processes, in most cases, operate in separate address spaces and communicate via files and file-like connections;
- the environment emphasizes the use of multiple languages (e.g., S, the shell programs, C, FORTRAN, awk, ...), and especially the development and use of small, independent *software tools*.
- the most important virtue of the environment, for many uses, is that it does not get in the way, but provides a relatively clean and simple computing model in which users/programmers can do what they want;
- on a mundane level, is portable to a wider range of computers, including many that are an order of magnitude less expensive than current integrated workstations.

Parallel to the programming environment distinction is a dichotomy in programming *languages*. Languages like LISP, Smalltalk (regarded as a language) and Prolog are popular for the integrated programming environments. Conversely, languages like C, the Algol family, the FORTRAN family and Pascal are associated with "conventional" systems. If we label the two families of languages *interactive* and *algorithmic*, we can list characteristic contrasts:

- interactive languages tend to be used to build interactive systems, algorithmic languages to build algorithms or specific programs;
- algorithmic languages tend to use scientific notation, interactive languages some syntax related to logic, the lambda calculus or related forms;
- interactive languages tend to bind during execution (dynamically), algorithmic languages tend to some form of compile-and-load;
- most telling of all, probably, the families have different definitions of virtue: ease of use and adaptability for interactive languages versus correctness and efficiency for algorithmic languages.

Our interest is not to make a judgement of merit between the two approaches. Rather, we want to understand what features of each are most important to computing for data analysis, and how to obtain them.

Simply put, we would like the best of both. As the discussion in section 2 indicated, advantages of simple, highly-interactive program design and modification are very relevant to analytical computing. The learning barriers imposed by having to use several, partially inconsistent languages and a variety of (none too powerful) debugging tools seriously inhibit the development of applications systems. On the other hand, data analysis and graphics depend on a variety of algorithms and software tools for numerical calculations, graphics, and report generation. We estimated about 50,000 lines of support code for S [3]. A sizable fraction of that represents careful algorithmic design and implementation (usually not by us). Not having access to the languages like FORTRAN and C, in which such algorithms are written, would be crippling. Even if we had the energy to re-implement the algorithms, a prudent user would hesitate to trust the result, without a long sequence of testing.

In summary, both kinds of virtue are important in data analysis. We want ease of use, but we also need access to a variety of reliable algorithms and tools.

4. An Integrated Programming Environment for Data Analysis.

We believe that a consistent and achievable mixture of the virtues of both approaches outlined in section 3 will provide a substantially improved environment. Research is proceeding at Bell Laboratories on such an environment, using the experience with S as a starting point. This section briefly describes the new environment, a prototype of which has been written by the author. The essential characteristics of the environment are:

- a single analytical language, similar to the user language in S but allowing dynamic definition and modification of functions;
- a browsing and debugging environment in the same language;
- explicit inter-language interfaces that allow the use of existing or new algorithms written in languages like C and FORTRAN.
- similarly, an interface to the operating system tools (e.g., via the *pipe* mechanism [4, page 190]);

This environment can have, for the applications developer, much of the flavor of an integrated programming environment. Such developers will only rarely need to *design* algorithms or tools; rather, they will tend to *use* such software when it comes along. For their own design, the analytic environment will be much more effective.

As with the Interlisp or Smalltalk environments, programmers have access to the definition of the language interactively. In our system, we make use of the general hierarchical data structures to maintain both the definition of operators and functions in the language and also the tree of partially evaluated expressions during evaluation, all within the language itself. The fundamental operations of parsing, code generation (that is, optimization of the parsed expression), and evaluation are themselves accessible as functions in the language. In particular, there exists a definition of the semantics of evaluation written in the language. The prototype has a rudimentary version of a debugger, also written in the language. Important building blocks are datasets for the evaluation tree mentioned above and for the history of the user's interaction. Various functions use these datasets to examine and control evaluation; for example, a menu-oriented browser examines the evaluation tree (or any other hierarchical dataset), with facilities for editing any piece of the dataset.

The operators and functions in the language have definitions in the language. For efficiency, some functions are built-in (implemented by compiled C code), but equivalent definitions in the language exist (as in the case of evaluation itself), to permit verification or user modification. However, it is explicitly expected that algorithms for numerical, graphical and other calculations will be supplied as interfaces to C or FORTRAN code. Several approaches to implementing this interface are possible. The current prototype uses special functions in the language that map into suitable calls to subprograms in C or FORTRAN (or other low-level languages if needed). A table of currently used subprograms and a C-language routine that executes the actual subprogram call are generated by a function in the high-level language, from the parsed code for interpreted functions that interface to C or FORTRAN. New interpreted functions that do not invoke previously unseen subprograms do not require any special consideration. The best approach to invocations of *new* subprograms depends on the availability of dynamic loading in the local version of the loader.

Initial studies of the new system indicate substantial improvements in run-time efficiency, by comparison to similar computations in S, for many typical calculations found in application systems built on S. Future work will include studies of trade-offs between the ability to redefine everything dynamically and the desire to speed up a particular calculation. For example, while a function could be redefined within a loop and then reused in that same loop, this seems generally unlikely. (One can construct somewhat practical examples where it would make sense, however; for example, when a method is being modified based on previous iterations of the same method.) Given the assumption that function definitions remain constant, the code-generation phase can perform some optimizations of argument matching and other computations. Before deciding what options to pursue in these directions, we plan to study the performance of typical application computations to look for the important "hot spots".

In summary, experience so far has been encouraging that a programming environment can be designed to combine the ease of use and modification found in integrated programming environments with the access to algorithms needed for quantitative work and with sufficient run-time efficiency to support a variety of applications development.

References.

[1] Becker, Richard A., and Chambers, John M., *S: An Interactive Environment for Data Analysis and Graphics*, Wadsworth, Belmont CA, 1984.

[2] Becker, Richard A., and Chambers, John M., *Extending the S System*, Wadsworth, Belmont CA, 1985.

[3] Becker, Richard A., and Chambers, John M., "Design of the S System for Data Analysis", *Comm. ACM*, vol. 27 (May 1984), 486-495.

[4] Kernighan, Brian W. and Pike, Rob, *The UNIX programming Environment*, Prentice-Hall, 1984.

COMPUTER SCIENCE AND STATISTICS:
The Interface, D.M. Allen (ed.)

INTEGRATED PROGRAMMING ENVIRONMENTS

John Alan McDonald and Jan Pedersen

Department of Statistics
Stanford University

We argue that data analysis has much in common with *experimental programming* as described in the AI literature. It follows that integrated environments designed for experimental programming (such as Interlisp-D, Smalltalk, or Zetalisp) are more suited to data analysis than conventional operating systems (such as Unix).

COMPUTER SCIENCE AND STATISTICS:
The Interface, D.M. Allen (ed.)

THE MONTE CARLO PROCESSOR

DESIGNING AND IMPLEMENTING A LANGUAGE FOR MONTE CARLO WORK

David Alan Grier

Department of Statistics
University of Washington
Seattle, Washington 98195

In designing a new computer language for Monte Carlo Experimentation, one needs to include high level data structures, a large family of functions to generate random quantities and a wide range of control structures, but that is really the very least of it. Monte Carlo experiments should be designed, just like any other experiment, and hence a Monte Carlo Language should have a construct which can describe and perform a complicated experiment. In fact, it encourages researchers to design their experiments more carefully. Monte Carlo work tends to be computationally intensive, and hence a Monte Carlo Language cannot afford to be too inefficient. The Monte Carlo field is continuing to advance, and hence a new language must be able to adapt to changes.

The Monte Carlo Processor is a computer package designed to do Monte Carlo Experimentation. The heart of this package is a computer language called MCL, which is a descendent of the languages C and S. It is designed expressly for Monte Carlo Integration and Experimentation and more care has been spent on such issues as accuracy and compatibility with existing statistical software than is found in the existing discrete-value simulation languages, GPSS, Simula and Simscript. Unlike S, it is translated into FORTRAN and then compiled, and hence, considerably more efficient. The MCL language contains statements which describe experimental design, variance reduction techniques, random variable generation that are not found in more conventional higher level languages such as FORTRAN or Pascal.

1. INTRODUCTION

While there is a need to improve the computer systems we use to analyze data, there is an even greater need to improve the systems we use to do Monte Carlo Experiments. The use of Monte Carlo experiments in statistical research has increased in recent years and is now a fixed part of statistical research. A recent article surveying the use of Monte Carlo methods in recent years (Hauck and Anderson, 1984) claims that about 20% of the articles published in 1981 in several of the major journals (JASA, Applied Statistics, Biometrics, Biometrika and Technometrics) contained some form of Monte Carlo technique used to justify their methods or results. There are several reasons for the increased use of Monte Carlo techniques in statistical research. One is the increased accessibility of computers in the past decade. Another is greater prevalence of computationally intensive techniques such as the bootstrap. Certainly the most important reason, though, is the changing nature of statistics. Statisticians are now trying to find the properties of statistics in situations where the mathematical assumptions make the problem of determining the power of a test, or the variance or bias of an estimator very difficult if not completely intractable. The field of robust methods has contributed in this respect because researchers in that field are often interested in studying the properties of statistics when the nice, mathematically tractable assumptions break down.

The foundation work that has been done to produce statistical analysis packages such as SAS, BMDP, Minitab, S, as well as other packages, just hasn't been done for Monte Carlo work. This in part has something to do with the nature of Monte Carlo experimentation. The process of doing a Monte Carlo experiment has more steps in it than the process of analyzing data. One has to decide on a question, or set of questions to answer with a Monte Carlo experiment, design that experiment, write a program to perform that experiment and finally, analyze the results of that experiment. There are simply more parts to the process of doing Monte Carlo work than there often are to the process of analyzing data. There is more choice in how one puts the parts of an experiment together in a computer system.

In their article, Hauck and Anderson (1984) point out several problems in many Monte Carlo

The author wishes to thank Catherine Hurley for her help in preparing the talk and this paper, Andreas Buja and Richard Kronmal for listening to the author as he sorted out this project and Daijin Ko for his help, friendship and support.

Questions concerning this system may be addressed to the author at the University of Washington.

Studies. Researchers often use inferior algorithms to generate random numbers and do other tasks. Many times they are unaware of the properties of the algorithms they use. Hauck and Anderson point out an article published in 1981 containing a Monte Carlo Study which made use of the random number generator RANDU, an algorithm whose inferior properties have been known for 15 years. In addition, they don't always carefully design their experiments to explore the properties of the statistics in the situations in which they are interested. Parameters are chosen in haphazard ways which don't allow the researchers to draw the kinds of conclusions that they would like to draw. Finally, researchers don't analyze the results of their experiments with the kind of care that they should bring to a data set. Often results are published in tabulated form with little analysis, graphical display or summary. My system is designed to attempt to address these concerns. The core of the system works within the framework of the traditional statistical Design of Experiments setting. The experimenter is expected to design an experiment to answer some question about a statistic or family of statistics. My system will take a description of that experiment and perform it. It will also take the data produced by that experiment and load it into a statistical package, such as S or Minitab, for analysis.

2. BASIC PROBLEM

Before describing the system which I've been working on, it might be a good idea to present the type of problem which the system is designed to solve. There are at least three kinds of Monte Carlo simulations done in this world: Monte Carlo Experimentation to determine the properties of some kind of statistical procedure, Discrete Event Simulation in which a Queuing Network or Flow Chart Model is simulated on a computer and Monte Carlo Integration in which a complicated, multidimensional integral (such as ones found in Nuclear and Particle Physics) is estimated using Pseudo or Quasi Random Numbers. My system is designed to tackle the first kind of simulation, what I call Monte Carlo Experimentation, although many of the Monte Carlo problems done in Physics could be handled by it. There are several systems to do Discrete Event Simulation. While statistical Monte Carlo Experimentation could be done and while I want my system to have the capabilities to do it, it presently lacks certain attractive features. In order to do Discrete Event Simulation special data structures such as queues and calendars, coroutines and some kind of clock mechanism are often desirable. These capabilities to do Discrete Event Simulation will be installed at a later date.

It might be best to start with a simple example to illustrate the kinds of problems in Monte Carlo Experimentation. Suppose that we would like to compare to estimators of location, the sample median and the 10% trimmed mean. We are interested in deciding which is the better of these two estimators and in particular we would like to know which one is better to estimate the location of a small sample of data which comes from a symmetric long tailed distribution. In this case we will decide that the estimator which has the smallest variance is the best and we will use the Contaminated Normal family of distributions to study these estimators. (For our purposes here, a Contaminated Normal distribution is a distribution in which an observation comes from a Standard Normal $(1 - \gamma)$ 100% of the time and γ 100% of the time from a Normal distribution with a variance σ^2.

The general technique we will use will be to:

1) Generate a set of random data on the computer having a Contaminated Normal distribution.

2) Apply the median and the 10% trimmed mean to the random data set.

3) Replicate the process of generating data and applying the statistics, thereby collecting many estimates of the median and 10% trimmed mean.

4) Calculate the sample variance of our sample of medians and of our sample of 10% trimmed means.

5) Study the results.

The plan for our experiment is not quite complete. The Contaminated Normal family of distributions has two parameters, the fraction of contamination, γ, and the contamination variance, σ^2. There is the additional parameter of the size of our data sets, which we will call K. We must choose values for these parameters and organize these values into a Designed Experiment. We will choose six samples sizes, 10, 20, 30, 40, 50, 100. We are interested in small sample sizes, which the values 10 to 50 represent, but to make sure that we include all the sizes of interest we choose one larger sample size, 100. The design of our experiment will involve looking at every possible combination of the parameters, what is called a factorial design in the Design of Experiments literature. Finally, we have to choose the number of times we will replicate the experiment for each design point. A number of criteria are involved in choosing that value, most notably the amount of computing resources we have available, the amount of time we have open to us and the variance we'd like our final results to have (in this case, the variance of our estimates of the variance of the median and 10% trimmed mean). Often, the number of replications will have to be chosen from results of a small pilot study, a short, small version of the Monte Carlo Experiment. For our example, we will choose the number of replications to be 5000 for each point in our design. A brief summary of our experiment is found below in figure 1.

Statistics:
median
10% trimmed mean

Design:
Factorial

Distribution:
Contaminated Normal Distribution

Parameters:
K (Sample Size) 10,20,30,40,50,100
γ (Percent Contamination)
.01, .05, .1
σ^2 (Standard Deviation of
Contamination) 2,5,10,100

Replications:
5000

Figure 1

3. TYPICAL PROGRAM

The basic form of the program needed to perform our example experiment, or indeed any Monte Carlo Experiment is seen in Figure 2. It, at base, is a pair of nested loops. The innermost code is contained in a Replication Loop. That process of generating data and evaluating statistics is replicated a certain number of times (in the case of our example, 5000) for each point in the design. The results of interest to the experimenter (in our case the variances of the median and 10% trimmed mean) are then computed and stored. The outer loop, the design loop, performs the experiment for each point in the design (in our case, a factorial design).

Basic Program

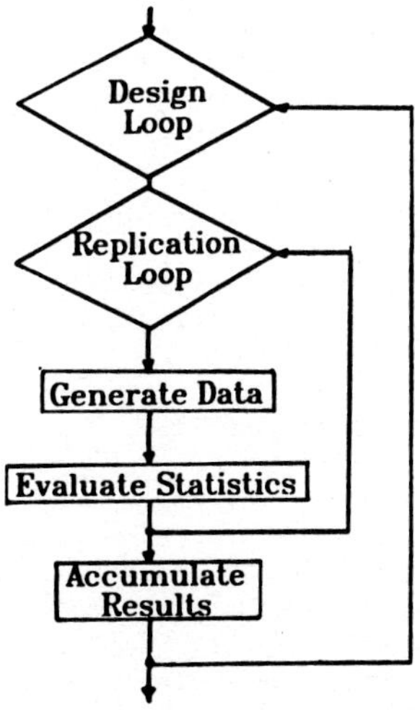

Figure 2

The system which I have been working on to do Monte Carlo Experimentation produces code which performs this program (or slightly more complicated versions of it).

3.1 PROBLEM TO BE SOLVED

The example described above points out many of the characteristics of Monte Carlo Experimentation, which my system tries to take into account. Perhaps the most important is that in Monte Carlo Experimentation we are studying a Mathematical Model for its own sake. We are not studying an approximation to a physical system. This has several implications for our system. We have strong mathematical assumptions. Our choice of distributions is done on a mathematical basis, not necessarily because they approximate a physical system, and hence the algorithms which must produce those random values must be provably correct. We have a careful experimental design because we are interested in the changes in the mathematical hypothesis. Finally, we have one thing which simplifies things somewhat for us, we are not doing Discrete Event Simulation and hence we don't have to worry about the problems of programming a system which needs to handle events occurring in time. We might be programming a problem which involves timeseries or time in a relatively simple way such as that but we need not be concerned with queues, servers or calendars or the problems associated with them. There is no need for a clock mechanism or coroutines.

4. THE MONTE CARLO PROCESSOR

The real purpose of this paper is to describe a new system which I have been building for the purpose of doing Monte Carlo Experimentation. That system takes as input, a description of a Monte Carlo Experiment, and produces as output a program to perform that experiment. After performing that experiment, the system then puts the results of the experiment into the form which a conventional statistical package can read. The researcher can then analyze the data produced by the experiment. The structure of the system is diagrammed in a flowchart in figure 3.

Monte Carlo Processor Structure

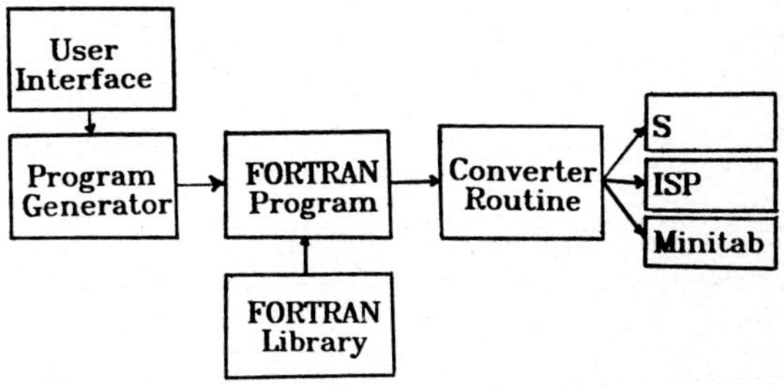

Figure 3

The core of the system is the Program Generator. It is a compiler which takes as input a program which describes a Monte Carlo Experiment. This program is written in a new language designed especially for this project. The syntax of this language is very similar to the language S (Becker and Chambers, 1984) and hence similar to the language C (Kernighan and Richie, 1977). There have been a few syntactic additions to enable a researcher to easily describe a Monte Carlo experiment. There are statements to describe the design of the experiment (the DESIGN statement), the parameters for the design, the number of replications for each design point and the quantities to be stored and accumulated for later analysis (the RETURN statement). The language is a functional language. New functions can be added with relative ease by anyone without recompiling the whole system.

An example of the code is shown in figure 4 below. It is a program to perform the experiment involving the median and the 10% trimmed mean which was described above.

Example Code

```
array(x,100)

design{factorial;
    k=(10,20,30,40,50,100),
    sig=(2,5,10,100),
    p=(.01,.05,.1);
    1000}

x <- rnorm(k)*
    (rber(k,p)*(sig-1)+1)

return{var; median(x) }

return{var; mean(x,.05) }
```

Figure 4

The first statement is the declaration of storage to hold the dataset. The structure x is a one dimensional array with maximum length 100. It can have any length between 1 and 100 and the functions in the system will use only the amount of data actually in the structure at any time. This language supports scalars, single dimensional arrays, multiple dimension arrays and compound structures made up of scalars and arrays.

The second statement describes the design of the experiment, defines the parameters and states the number of replications. In our example, we're doing a factorial experiment with parameters k (sample size), sig (standard deviation of contamination) and p (percent of contamination). The experiment is replicated 5000 times for each design point.

The third statement produces a sample from a contaminated normal distribution. The function rnorm produces a sample of length k from a standard normal distribution. The remainder of the statement calculates the standard deviation for each observation in the dataset. The rber function produces a set of k bernoulli (0,1) random variables which are 1 with probability p.

The third and fourth lines of code calculate the median and the 10% trimmed mean of each dataset and indicate that the system is to accumulate the variances of those means and medians.

4.1 THE GENERATED PROGRAM

The output from the Program Generator (the compiler) is a FORTRAN program which, along with a library of FORTRAN routines, actually performs the experiment. The compiler for this system does not compile directly into object code. The FORTRAN program, along with the FORTRAN library are compiled on the local machine's FORTRAN compilers. The prime reason for this scheme is portability. FORTRAN is one of the better defined and more portable languages. This system generates very conservative FORTRAN code, keeping close to the FORTRAN 77 standard and uses only well defined fixed format IO. This will keep the system from being tied down too closely to a specific machine. One can also send the FORTRAN output program to a high speed machine, such as a Cyber 205 or a Cray, for execution if the local computer proves too slow to do the desired experiment. FORTRAN also tends to be a fairly efficient language. Optimizing compilers can produce code which is quite good. Hence there is little to be gained by having this system produce object code.

4.2 THE FORTRAN LIBRARY

The FORTRAN Library is a collection of routines which perform much of the work to do the Monte Carlo Experiments. It contains routines to generate random numbers from various distributions, do matrix calculations, perform many of the tasks that one needs to do for computing statistics as well as calculate many of the conventional statistics. They have been carefully chosen and their methods and properties are well documented. This library is easily expandable. Using a simple table definition, a user can add a function to the library that will be recognized in the language.

4.3 THE CONVERTER ROUTINE

The Converter Routine manages the output data from the experiments. It can convert the data into a form which can be loaded into any one of the statistical packages like S, Isp, Minitab, BMDP, SAS or SPSS. It can also extract subsets of data or produce simple files of data which can then be fed into any application program.

4.4 THE USER INTERFACE

The User Interface controls the whole system. Using it, an experimenter can define new experiments, edit old ones, start experiments running, terminate experiments or temporarily suspend experiments to lighten the load on the computer and then restart them later. Within the User Interface, the experimenter controls the Converter Routine and can edit experimental output, direct output to a statistical package or edit output.

5. SUMMARY

The purpose of this system is really two-fold. It attempts to unify the body of information that a researcher needs to do statistical Monte Carlo Experimentation. Often an experimenter cannot do a good experiment without searching the literature of Statistics, Computer Science, Numerical Analysis and Operations Research. Its second goal is to improve the way Monte Carlo Experiments are performed and analyzed. It does this by casting the whole process of programming an experiment into the classic design of experiments framework and by giving the experimenter the support to help analyze the results. The result is to give a researcher greater freedom in preparing and performing Monte Carlo experiments. Rather than worrying about finding good random number generators or the details of coding a particular experimental design, the researcher is freed to work on questions more closely related to the study in question. There is more time to try different pilot studies to test ideas before doing a big Monte Carlo Experiment. It is easier to consider the use of variance reduction techniques, which may speed the experiment or improve its accuracy. Just as upper level computer languages free programmers from being concerned with many of the details of programming, this Monte Carlo System frees experimenters from the details of programming Monte Carlo Experiments.

REFERENCES

Andrews, D.F. et al., Robust Estimates of Location: Survey and Advances (Princeton University Press, Princeton, NJ, 1972).

Becker, Richard A. and Chambers, John M., S an Interactive Environment for Data Analysis and Graphics (Wadsworth, Belmont, CA, 1984).

Bratley, Paul, Fox, Bennett L. and Schrage, Linus E., A Guide to Simulation (Springer Verlag, New York, 1983).

Chambers, John, Computational Methods for Data Analysis (John Wiley, New York, 1977).

Grier, David, A Monte Carlo Processor for the S System, Proposal for General Exams, University of Washington, (1984).

Grier, David, A Monte Carlo Processor, Version 2.0, University of Washington Department of Statistics Technical Report No. 54, (October 1984).

Hauck, Walter W. and Anderson, Sharon, A Survey Regarding the Reporting of Simulation Studies, The American Statistician, Vol. 38, No. 3, (August 1984), 214 - 216.

Johnson, S.C., Yacc - yet another compiler compiler, CSTR 32 (Bell Laboratories, Murray Hill, NJ, 1975).

Kennedy, William J. and Gentle, James E., Statistical Computing (Marcel Dekker, New York, 1980).

Kernighan, Brian and Ritchie, Dennis, The C Programming Language (Prentice-Hall, Englewood Cliffs, NJ, 1978).

Knuth, Donald, The Art of Computer Programming, Volume 2, Second Edition (Addison-Wesley, Reading, MA, 1981).

Lesk, M.E., Lex - a lexical analyzer generator, CSTR 39 (Bell Laboratories, Murray Hill, NJ, 1975).

Rubenstein, Reuven Y., Simulation and the Monte Carlo Method (John Wiley, New York, 1981).

Multivariate Density Estimation and Regression

Organizer: *Daniel B. Carr*

Invited Presentations:

Choosing Smoothing Parameters for Density Estimators
David W. Scott

On a Class of Multivariate Density and Regression Estimators
Vassilios K. Klonias

COMPUTER SCIENCE AND STATISTICS:
The Interface, D.M. Allen (ed.)

Choosing Smoothing Parameters for Density Estimators

David W. Scott

Department of Mathematical Sciences
Rice University
Houston, Texas

For data analysis in one, two, and three dimensions, nonparametric density estimation has proven to be a powerful tool. A major practical problem in density estimation is the choice of smoothing parameters, to which the estimates are quite sensitive. There are three different approaches for choosing a smoothing parameter, assuming little *a priori* information: (i) interactive graphical evaluation of the smoothness of the density estimate or its derivatives; (ii) minimization of cross-validation criteria; and (iii) use of upper bounds as in oversmoothed density estimates. In this paper I describe these approaches, review theoretical results, and examine small-sample behavior.

1. Introduction

Automation of decisions required in statistical procedures is highly desirable. The resulting expert systems can be widely circulated and, contrary to popular belief, are likely to stimulate growth in the profession. More importantly, these systems encourage the user to understand the role of assumptions in statistical models and how to cope with situations where those assumptions fail. Statistical procedures currently recognized as capable of dealing with a broad range of models are often perceived as too difficult to use and subjective. The subjectivity is often embodied in the choice of a few parameters whose values reflect the expert's judgments about the data's peculiarities.

Multiple linear regression provides a typical example. This is a favored statistical procedure because it is fully automatic. But regression can only be viewed as automatic over a very limited range of probability models. Usually the model must be expanded to deal with outliers, influential points, and transformation of variables while simultaneously attempting to select an optimal subset of variables. Box and Cox (1964) introduced an additional parameter for each variable in their power-transformation family. Robust regression (Huber 1973) requires specification of an influence function, which in turn contains shape parameters. Handling influential points requires determining acceptable levels of leverage (Belsley, Kuh and Welsch 1980). Some of these ideas are addressed in an experimental expert system proposed by Gale and Pregibon (1983). Full automation of robust regression is clearly a large and difficult task, especially given current wisdom echoed by Carroll and Ruppert (1985) "that robust estimators should not be used blindly." But it is clear that robust regression is very important and even partial automation desirable.

In this paper, we focus on automatic parameter selection algorithms for nonparametric density estimators. Ideally, we desire procedures that take data and produce a nearly optimally smoothed density estimator for *finite sample sizes*. This problem is easier than the regression problem because nonparametric density estimators are robust (although some automatic selection procedures may not be). Thus we may hope to have a limited expert system for density estimation. In what follows we survey past attempts, describe current results, point to new results, and wonder whether in five years the consensus will be that "nonparametric density estimators should not be used blindly."

I will limit the discussion to histogram, series and kernel estimators, paying most attention to the histogram and to kernel estimators, of the usual form

$$\hat{f}(x) = \frac{1}{nh} \sum_{i=1}^{n} K\left(\frac{x-x_i}{h}\right) . \qquad (1.1)$$

I shall consider two kernels: the Gaussian kernel $(2\pi)^{-1/2}e^{-t^2/2}$ and the triweight kernel $\frac{35}{32}(1-t^2)^3 I_{[-1,1]}(t)$. The smoothing parameter in Equation (1.1) is the bandwidth h. For the histogram, the smoothing parameter is the bin width, which will also be denoted by h. Smoothing for series estimators may be controlled by the number of terms in the series expansion or by a bandwidth parameter similar to h or both.

The quality of the estimates will be estimated by the integrated mean squared error:

$$IMSE = \int E[\hat{f}(x)-f(x)]^2 dx .$$

Scott (1979) showed that use of a nonoptimal smoothing parameter, say a factor c times the optimal parameter, results in an *IMSE* increased by the factors

$$(c^3+2)/3c \quad \text{and} \quad (c^5+4)/5c \qquad (1.2)$$

for the histogram and kernel methods, respectively. In my experience, reasonable density estimates are within 10% of optimum. Hence, it is clear that only a fairly narrow range of values of the smoothing parameter is acceptable for any sample size, even $n=10^6$. The histogram is less sensitive than the kernel method to choice of smoothing parameter.

2. Survey of Pre-1980 Algorithms

2.1. Histogram Methods

The first automatic rule for choice of a smoothing parameter was given by Sturges (1926) for the histogram. His proposal was simple and elegant. Consider a histogram with k bins labeled $0,1,\ldots,k-1$. Then an "ideal" histogram would have $C(k-1,j)$ points in the jth bin; adding, the corresponding sample size is $n=\sum_j C(k-1,j)=2^{k-1}$. Hence the number of bins and bin width are given by

$$k = 1+\log_2 n \qquad (2.1a)$$

and

$$h = (sample\ range)/k , \qquad (2.1b)$$

respectively. This rule is given implicitly or explicitly in virtually every introductory textbook. Often the advice is given that a histogram should have between 5 and 20 bins (from which, I suppose, we infer that all samples contain between 2^4 and 2^{19} points).

Scott (1979) analyzed the *IMSE* of the histogram and found

$$IMSE = \frac{1}{nh} + \frac{1}{12} h^2 R(f') + 0(n^{-1}) , \quad (2.2)$$

where $R(\phi)$ denotes the squared L_2-norm of the function ϕ,

$$R(\phi) = \int_{-\infty}^{\infty} \phi(x)^2 dx ,$$

and is a measure of the "roughness" of ϕ. The first term in (2.2) is due to variance and the second to bias. From (2.2) it follows that optimally (asymptotically)

$$h^* = [6/R(f')]^{1/3} n^{-1/3} . \quad (2.3)$$

Comparing (2.3) and (2.1) we see that Sturges' rule asymptotically has far too few bins and that the *IMSE* (2.2) is dominated by errors due to bias.

It should be noted, however, that Sturges' rule is consistent, although not of optimal order. Hence, consistency results by themselves are not satisfactory.

Sturges based his arguments on the assumption that the data are nearly Gaussian. Tukey (1977) has advocated a similar role for the Gaussian density as a reference. Scott (1979) adopted this point of view and advocated using

$$h = 3.5 \hat{\sigma} n^{-1/3} . \quad (2.4)$$

Chen and Rubin (1984) have shown this rule is consistent if $EX^{6/5} < \infty$. Within the class of densities satisfying this and other technical constraints, rule (2.4) provides estimates of the optimal order. However, for very rough densities the rule can easily provide poor (usually oversmoothed) estimates. It is interesting to note that the textbook advice of between 5 and 20 bins when applied to Gaussian data corresponds roughly to $50 < n < 1500$, which is a more reasonable range of sample sizes.

Freedman and Diaconis (1981) proposed a more robust version of (2.4) based on the interquartile range (*IQR*):

$$h = 2\, IQR\, n^{-1/3} ,$$

which, generally, is at least 30% smaller than (2.4).

2.2. Kernel and Series Methods

Kernel estimators (1.1) were introduced by Rosenblatt (1956) and Parzen (1962). Several authors have proposed a rule that parallels (2.4) for Gaussian data with a Gaussian kernel:

$$h = 1.06 \hat{\sigma} n^{-1/5} . \quad (2.5)$$

This follows from the general result for nonnegative kernels:

$$IMSE = \frac{R(K)}{nh} + \frac{1}{4} \sigma_K^4 h^4 R(f'') + O(n^{-1}) \quad (2.6a)$$

and

$$h^* = [R(K)/\sigma_K^4 R(f'')]^{1/5} n^{-1/5} . \quad (2.6b)$$

Another informal procedure involves graphical inspection of estimates for a decreasing sequence of smoothing parameters. Generally, when the estimates begin to display high frequency noise, a good choice is a slightly larger smoothing parameter; see Tapia and Thompson (1978) for some examples.

2.2.1. Series

The first modern results for choosing nearly optimal data-based smoothing parameters came with the periodic series estimator:

$$\hat{f}(x) = \sum_{k=-m}^{m} w_k \hat{f}_k e^{2\pi i k x} , \quad (2.7)$$

where w_k are weights and $\hat{f}_k$ are estimates of the Fourier coefficients f_k. Kronmal and Tarter (1968) let $w_k \equiv 1$ and provided unbiased estimates of the *change* in the *IMSE* as m was increased. They also provided inclusion rules for the $\hat{f}_k$ terms. This anticipated the general unbiased estimates of the *IMSE* by Rudemo and Bowman, which are discussed in section 3.1. Unfortunately, as a smoothing parameter, m is a fairly crude choice. Hence the elegance of this result was somewhat obscured.

Wahba (1977, 1981) shifted the smoothing parameter away from m, which she took as $n/2$, to a continuously varying (smoothing) parameter λ in the weights w_k:

$$w_k = \frac{1}{1+\lambda(2\pi k)^4} .$$

Through unbiased estimates of f_k and $|f_k|^2$, Wahba provided asymptotically unbiased estimates of $IMSE(\lambda)$. Wahba proposed plotting $IMSE(\lambda)$ and choosing λ^* as the minimizer. This is essentially the thrust of modern kernel proposals, which differ by providing *exactly* unbiased estimates of the *IMSE* shifted by a constant. Wahba's algorithm has been illustrated in her papers and more extensively analyzed with Monte Carlo methods by Scott and Factor (1981). But the basic framework for automatic data-based density estimation was laid with series methods.

2.2.2. Kernel

It is well known that series estimators may be re-expressed as kernel estimates. For data in several dimensions, the kernel form is easier to deal with. In addition, very efficient algorithms for large samples such as the averaged shifted histogram (Scott 1985) approximate kernel estimators. Thus, cross-validation of general kernel estimators is required.

The first attempt at cross-validation of kernel estimators did not directly address *IMSE*, but used a modified maximum-likelihood criterion (Habbema, Hermans and van den Brock, 1974; Duin, 1976). Hermans and Habbema were particularly interested in constructing multivariate kernel estimates of medical data. Specially, the authors proposed a leave-one-out optimization problem:

$$\max_h \sum_{i=1}^{n} \log \hat{f}_i(x_i) , \quad (2.8)$$

where $\hat{f}_i(x_i)$ is the kernel estimator with x_i deleted and evaluated at $x = x_i$. Scott and Factor (1981) found the small-sample properties of (2.8) with Gaussian data were quite good, but that (2.8) was sensitive to outliers, as later proved by Schuster and Gregory (1981). Schuster has also proposed a related criterion based on random splits of the data, which seems promising empirically. The difficult proof of the consistency of (2.8) was provided by Chow et al. (1983), but Hall (1982) demonstrated the optimal order would not in general be realized.

In 1976, Jim Thompson suggested and I implemented an algorithm based on estimating *IMSE*. Notice in (2.6) that the only unknown quantity is $R(f'')$. We estimated this quantity by substituting the kernel estimate itself, which for a Gaussian kernel is given explicitly by

$$R(\hat{f}_h'') = \frac{3}{8\sqrt{\pi} n^2 h^5} \sum_{i=1}^{n} \sum_{j=1}^{n} (1 - \Delta_{ij}^2 + \Delta_{ij}^4/12) e^{-\Delta_{ij}^2/4} \quad (2.9)$$

where $\Delta_{ij} = (x_i - x_j)/h$. Then, following (2.6b), we formed the sequence:

$$h_+ = \left[\frac{R(K)}{h \sigma_K^4 R(\hat{f}_{h_c}'')} \right]^{1/5} , \quad (2.10)$$

where h_c and h_+ are the current and next iterates of h, respectively. We could have substituted (2.9) into the *IMSE* expression (2.6a) and proceeded as Wahba but choose instead this fixed point iteration. Not surprisingly, Scott and Factor (1981) found the small sample performance of (2.10) and Wahba's algorithm to be quite similar. Unfortunately, (2.9) does not provide a consistent estimator of $R(f'')$, but is positively biased (for small samples, this was unimportant). Removing this bias gives an algorithm in the spirit of Wahba (see Scott and Terrell 1985; also section 4.1). As an aside, $R(\hat{f})$ and $R(\hat{f}')$ are consistent, while $R(\hat{f}''')\rightarrow\infty$ when using h's given by (2.6b).

Silverman (1978) found a clever way to use the inconsistency of $\hat{f}_h''$ in his test graph procedure. He showed that the fluctuations in the second derivative should be of a certain fixed size for optimal h. By examining a series of plots of $\hat{f}_h''$ for a decreasing values of h, the size of the fluctuations may be guessed and an h chosen. This generalizes the visual inspection method described after Equation (2.6).

3. Algorithms since 1980

3.1. Unbiased cross-validation

A new twist in cross-validation came with the introduction of exactly (not asymptotically) unbiased estimates of the *IMSE* by Rudemo (1980) and Bowman (1984). Consider decomposing the $IMSE = E\int(\hat{f}-f)^2dx$ into three terms:

$$IMSE = E\int\hat{f}(x)^2dx - 2E\int\hat{f}(x)f(x)dx + \int f(x)^2dx\,. \quad (3.1)$$

Consider

$$\alpha(h) = \frac{1}{n}\sum_{i=1}^{n}\hat{f}_i(x_i) - \frac{2}{n}\sum_{i=1}^{n}\hat{f}(x_i)\,. \quad (3.2)$$

The authors show that (3.2) provides an unbiased estimate of the first two terms in (3.1) while the third term in (3.1) is constant. Plotting (3.2) provides an unbiased (pointwise) estimate of the true *IMSE* curve, shifted by the fixed (but unknown) constant $R(f)$. Again the cross-validation estimate is that h which minimizes the curve. Hall (1983) and Stone (1984) have shown the resulting estimates are not only consistent but asymptotically optimal. In practice, we should not expect very much difference between (3.2) and Wahba's proposal, since the bias in Wahba's *IMSE* estimator is quite small, of order n^{-4}.

This proposal has several remarkable features. First, it is applicable to any density estimator of the generalized kernel or delta type. Thus when applied to histograms, a sequence of smoothing parameters of order $n^{-1/3}$ results, while the sequence is of order $n^{-1/5}$ for nonnegative kernels, and of order $n^{-1/9}$ for appropriate negative kernels. Second, it avoids directly estimating terms such as $R(f')$ in (2.2) and includes the $O(n^{-1})$ terms as well. Third, it is easily extended to higher dimensions.

3.2. Example

For a histogram estimator, I examined the performance of (3.2) with very large samples of normal data. For equally spaced histograms with bin counts $\{n_k\}$, we must minimize

$$\alpha(h) = \frac{2}{nh} - \frac{1}{n^2h}\sum_k n_k^2\,. \quad (3.3)$$

In Figure 1, I have plotted $\alpha(h)$ for a $N(0,1)$ sample with $n=10{,}000$, for which $h^*=.162$. Exactly where to place the bins is a little problem, and I have chosen zero as a bin edge for all the histograms. Notice the minimum of the curve is close to $-R(f)=-1/2\sqrt{\pi}=-.2821$. But the amount of noise in the curve is (initially) surprising. We are actually looking among the obviously numerous local minima for the best h. Now it can be shown that the variation observed in Figure 1 approximates the standard deviation of the curve estimates about the true *IMSE* estimate - this variation is much less than the variance of the curve (3.3), which was shown by Rudemo (1980) to be of order $O(n^{-1})$; see Scott and Terrell (1985). Thus while the actual "best" local minimum is quite good in Figure 1, we may expect a large percentage of h's to be outside the interval $(.72h^*, 1.35h^*)$, even with such large samples; see Equation (1.2).

The corresponding curves in the kernel case do not exhibit the variation for individual samples because continuous kernels avoid problems due to the bin boundaries; however, the large variation exists and we cannot expect to obtain an h with desired accuracy for medium sample sizes with desired certainty. Thus the asymptotic optimality theorems do not translate into uniformly good small-sample properties; see also simulations by Bowman (1984).

4. Some Recent Work

4.1. Biased Cross-Validation

If we think of the procedures in Section 3.1 as "unbiased" cross-validation algorithms, then it is natural to think of Wahba's method as "biased" cross-validation. We have looked at some biased cross-validation algorithms in the spirit of the Scott-Tapia-Thompson procedure for histogram and kernel methods (Scott and Terrell, 1985). For histograms, we estimate

$$R(f') \leftarrow \frac{1}{n^2h^3}\sum_k(n_{k+1}-n_k)^2 - \frac{2}{nh^3} \quad (4.1)$$

and substitute in (2.2) to obtain

$$\hat{IMSE} = \frac{5}{6nh} + \frac{1}{12n^2h}\sum_{k=-\infty}^{\infty}(n_{k+1}-n_k)^2\,, \quad (4.2)$$

which may be compared to Equation (3.3). In Figure 2 (for the same sample as used in Figure 1) we plot the estimated *IMSE* (4.2). Notice the estimates are not only far less noisy, but also provide a good estimate of the true integrated squared error. The bias introduced is of lower order than the variance. Thus the roles of biased and unbiased cross-validation for finite sample sizes are not yet clear. Examples with certain mixture densities are more favorable to the unbiased procedures for samples $n<1000$.

For a fixed sample, both (3.3) and (4.2) converge to zero as $h\rightarrow\infty$. Now (3.3) is negative near h^* but (4.2) is clearly nonnegative. Hence (4.2) is actually minimized for $h=\infty$; we seek the local minimizer near h^*. We also expect $h=\infty$ to be a local minimum for (3.3). For small samples, the region in the neighborhood of the local minimum where (4.2) is convex may be very small or nonexistent when using the biased methods. This region is much larger for unbiased procedures. Recall the Scott-Factor simulation results where method (2.10) occasionally failed to have a solution. For these cases, the (oversmoothed) upper bounds given below are very useful.

4.2. Upper Bounds

Rules (2.4) and (2.5), which are based on Gaussian models, turn out to be close to upper bounds on smoothing parameters; see Terrell and Scott (1985). Under various constraints on scale measure, densities minimizing $R(f^{(k)})$ may be found. When substituted into expressions such as (2.3) and (2.6b), useful upper bounds may be obtained. For example, a histogram of a density with finite support (a,b) satisfies

$$h \leq (b-a)/(2n)^{1/3}\,. \quad (4.3)$$

Useful expressions exist for densities of infinite support, as well as for kernel estimators. Rules based on Gaussian models are only slightly narrower.

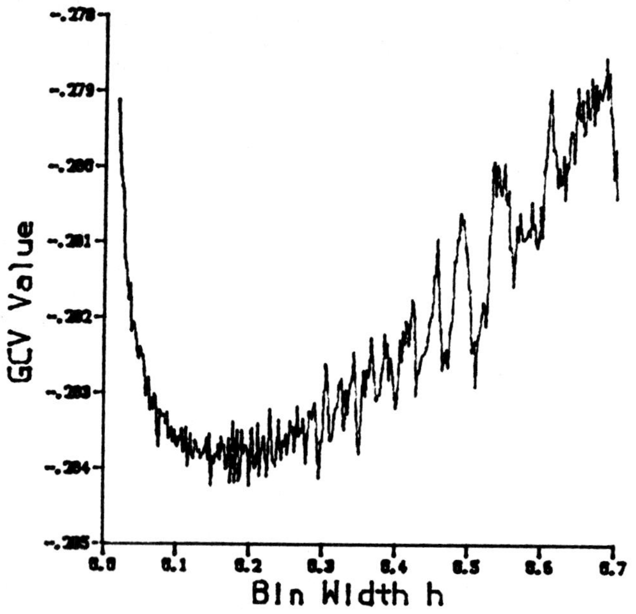

Figure 1.

Example of the unbiased cross-validation function for a histogram with Normal data and n = 10,000.

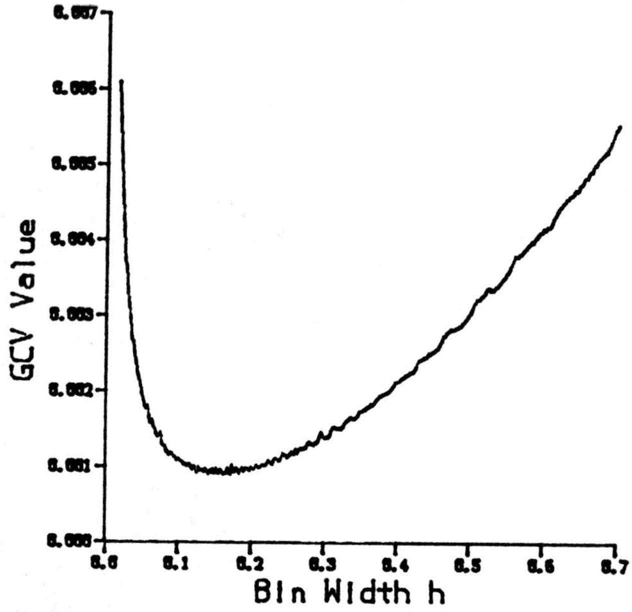

Figure 2.

Example of the biased cross-validation function with same data as in Figure 1.

For very small samples, these rules are probably as good as any. For very large samples, any inefficiency may not be important -- we may be willing to accept an oversmoothed *IMSE* of 10^{-6} even though the optimal *IMSE* could be 10^{-8}. This is because the oversmoothed density estimates will contain the important features of the true density, though somewhat flattened.

5. Discussion

Rice (1984) has investigated cross-validation results for the related problem of nonparametric kernel regression. But that is an easier problem to diagnose graphically, since the curve may be compared to the locations of the points. For kernel density estimates, some authors suggest comparison with a histogram but which histogram? It is possible to compare the integrated kernel estimate with the sample cdf, but the optimal smoothing parameters for the cdf and density are different. So cross-validation for the density is apparently not as easy a problem.

The univariate methods may be extended for choosin smoothing parameters for multivariate estimators (one for eac variable). In my experience where I choose smoothing parame ters by eye, I find the multivariate case is somewhat easier tha the univariate case because of interaction effects, which hel gauge changes in the density function for each paramete Perhaps cross-validation in this case will be no harder.

For large enough n, empirical evidence suggests the biased cross-validation algorithm works almost without fail. In other words, the smoothing parameter obtained is acceptably close to the optimal value for almost every sample. This should be conditioned by the obvious statement that the true density may contain very minute features not observable without more data. But such possibilities should not paralyze our willingness to use nonparametric methods. For smaller samples, the oversmoothed results are extremely useful, because if the proposed cross-validation value is greater than or much smaller than the upper bound, a clear signal for closer inspection has been given.

Can an expert system be built for density estimation? Yes, but as in the parametric regression case, it probably won't be blind.

6. Acknowledgments

I would like to thank Dan Carr, who organized this session. This research was supported by ONR, NASA/JSC, and ARO under grants N00014-85-K-0100, NCC 9-10, and DAAG-29-82-K0014, respectively.

7. References

(1) Belsley, D.A., Kuh, E. and Welsch, R.E. (1980), *Regression Diagnostics*, Wiley, New York.

(2) Bowman, A.W. (1984), An alternative method of cross-validation for the smoothing of density estimates, *Biometrika* **71**:353-360.

(3) Box, G.E.P. and Cox, D.R. (1964), An analysis of transformations, *JRSS-B* **26**:211-246.

(4) Carroll, R.J. and Ruppert, D. (1985), Transformations in regression: A robust analysis, *Technometrics* **27**:1-12.

(5) Chen, J. and Rubin, H. (1984), manuscript.

(6) Chow, Y.-S., Geman, S. and Wu, L.-D. (1983), Consistent cross-validated density estimation, *Ann. Statist.* **11**:25-38.

(7) Duin, R.P.W. (1976), On the choice of smoothing parameters for Parzen estimators of probability density functions, *IEEE Trans. Comp.* **C25**:1176-1179.

(8) Freedman and Diaconis (1981), On the histogram as a density estimator: L_2 theory, *Z. Wahrsch. verw. Gebiete* **57**:453-476.

(9) Gale, W.A. and Pregibon, D. (1983), An expert system for regression analysis, *Computer Science and Statistics: Proceedings of the 14th Symposium on the Interface* Heiner, et al., eds., Springer-Verlag, New York, 110-117.

(10) Habbema, J.D.F., Hermans, J. and van den Brock, K. (1974), A stepwise discriminant analysis program using density estimation, *Compstat, Proceedings in Computational Statistics* Weign, Physica Verlag, 101-110.

(11) Hall, P. (1982), Cross-validation in density estimation, *Biometrika* **69**:383-390.

(12) Hall, P. (1983), Large sample optimality of least squares cross-validation in density estimation, *Ann. Statist.* **11**:1156-1174.

(13) Huber, P.J. (1973), Robust regression: Asymptotics, conjectures and Monte Carlo, *Ann. Statist.* **1**:799-821.

(14) Kronmal, R.A. and Tarter, M.E. (1968), The estimation of probability densities and cumulatives by Fourier series methods, *J. Amer. Statist. Assoc.* **63**:925-952.

(15) Parzen, E. (1962), On estimation of a probability density function and mode, *Ann. Math. Statist.* **33**:1065-1076.

(16) Rice, J. (1984), Bandwidth choice for nonparametric regression, *Ann. Statist.* **12**:1215-1230.

(17) Rosenblatt, M. (1956), Remarks on some nonparametric estimates of a density function, *Ann. Math. Statist.* **27**:832-837.

(18) Rudemo, M. (1980), Empirical choice of histograms and kernel density estimators, *Scand. J. Statist.* **9**:65-78.

(19) Schuster, E.F. and Gregory, G.G. (1981), On the nonconsistency of maximum likelihood nonparametric density estimators, *Computer Science and Statistics: Proceedings of the 13th Symposium on the Interface*, W.F. Eddy, ed., Springer-Verlag, New York, 295-298.

(20) Scott, D.W. (1979), On optimal and data-based histograms, *Biometrika* **66**:605-610.

(21) Scott, D.W. (1985), Averaged shifted histograms: Effective nonparametric density estimators in several dimensions, *Ann. Statist.*, in press.

(22) Scott, D.W. and Factor, L.E. (1981), Monte Carlo study of three data-based nonparametric probability density estimators, *J. American Statistical Assoc.* **76**:9-15.

(23) Scott, D.W., Tapia, R.A. and Thompson, J.R. (1977), Kernel density estimation revisited, *J. Nonlinear Anal. Th. Meth. Appl.* **1**:339-372.

(24) Scott, D.W. and Terrell, G.R. (1985), Biased cross-validation, manuscript.

(25) Silverman, B.W. (1978), Choosing window width when estimating a density, *Biometrika* **65**:1-11.

(26) Stone, C.J. (1984), An asymptotically optimal window selection rule for kernel density estimates, *Ann. Statist.* **12**:1285-1297.

(27) Sturges, H.A. (1926), The choice of a class interval, *J. Amer. Statist. Assoc.* **21**:65-66.

(28) Tapia, R.A. and Thompson, J.R. (1978), *Nonparametric Probability Density Estimation*, Johns Hopkins, Baltimore.

(29) Terrell, G.R. and Scott, D.W. (1985), Oversmoothed nonparametric density estimates, *J. Amer. Statist. Assoc.* **80**:209-214.

(30) Tukey, J.W. (1977), *Exploratory Data Analysis*, Addison-Wesley, Reading, MA.

(31) Wahba, G. (1977), Optimal smoothing of density estimates, in *Classification and Clustering*, J. Van Ryzin, ed., Academic Press, New York, 423-258.

(32) Wahba,G (1981), Data-based optimal smoothing of orthogonal series density estimates, *Ann. Statist.* **9**:146-156.

COMPUTER SCIENCE AND STATISTICS:
The Interface, D.M. Allen (ed.)
© Elsevier Science Publishers B.V. (North-Holland), 1986

ON A CLASS OF MULTIVARIATE DENSITY AND REGRESSION ESTIMATORS

V. K. Klonias

Mathematical Sciences Department
The Johns Hopkins University
Baltimore, Maryland

We present a class of nonparametric multivariate maximum penalized likelihood estimators (MPLE) of a probability density functions. The estimates are multivariate splines with knots at the sample points. The numerical effort for the evaluation of the estimates is essentially independent of the dimension of the data. Under mild assumptions the MPLE's are seen to be consistent in a variety of metrics and with optimal rates of convergence. These density estimators lead naturally to a class of multivariate regression estimators. Some numerical examples are presented where the smoothing parameters are estimated from the data by an approach suited to these splines.

1. INTRODUCTION

Let $\underline{X}_1,\ldots,\underline{X}_n$ in $\mathbb{R}^p$, $p \in \mathbb{Z}_+$, be i.i.d. observations from a distribution function F with density f and let F_n denote the associated empirical distribution function. The nonparametric maximum penalized likelihood method of density estimation (MPLE), introduced by Good and Gaskins (1971 and 1980), suggests estimating f by the maximizer of the log-likelihood minus P(v), a roughness penalty functional which is usually applied on the square root of the density $v=f^{\frac{1}{2}}$. For example, $P(v) = \alpha \int (v')^2 + \beta \int (v'')^2$, where $\alpha,\beta \geq 0$ and at least one is strictly positive. In DeMontricher, Tapia and Thompson (1975), the existence and uniqueness of the MPLE's were rigorously established within the framework of the Sobolev spaces $W^{2,m} = \{u \in L_2(\mathbb{R})$ such that $\| u^{(m)} \|_2 < +\infty\}, m \in \mathbb{Z}_+$, where $L_2(\mathbb{R})$ denotes the space of square integrable functions and $\|u\|_2 = (\int u^2)^{\frac{1}{2}}$. For discretized MPLE's see Tapia and Thompson (1978) and Scott, Tapia and Thompson (1980). For penalties on log f see Silverman (1982). We follow here the setting in Klonias (1984), and discuss the construction of the multivariate MPLE's, their consistency, numerical evaluation and data-based choice of the smoothing parameters.

2. THE ESTIMATORS

For the estimation of the probability density in a nonparametric setting, the likelihood can be considered as a functional with argument ranging over a suitable space of density functions f. If no smoothness conditions are imposed on f, the likelihood is unbounded and the unconstraint maximum likelihood "solution" can be represented as an average of Dirac deltas centered at the observations, i.e., it is the distributional derivative δ_n of F_n. In fact, the classical kernel estimates of the density can be viewed as approximations to S_n.

The MPLE's u of $v = f^{\frac{1}{2}}$ considered here are solutions to the following optimization problem

(2.1) $\max\{\Sigma_{i=1}^n \log u(\underline{X}_i)^2 - \lambda \int_{\mathbb{R}^p} |\tilde{u}|d\mu, u \in H\}$
subject to $u(\underline{X}_i) \geq 0,\ldots,n,$

where $\tilde{u}$ denotes the Fourier transform of $u \in L_2(\mathbb{R}^p)$, $H = \{u \in L_2(\mathbb{R}^p) : \int_{\mathbb{R}^p} |\tilde{u}|d\mu < +\infty\}$, μ is a positive measure on $\mathbb{R}^p$ dominated by the Lebesgue measure, and $\lambda > 0$ is such that $\int_{\mathbb{R}^p} u^2 = 1$. The optimization problem (2.1) has a unique solution given implicitly by

$$u(\underline{x}) = \lambda^{-1} \Sigma_{i=1}^n u(\underline{X}_i)^{-1} \kappa(\underline{x}-\underline{X}_i),\ \underline{x} \in \mathbb{R}^p,$$

where the function κ is determined by $\tilde{\kappa}\, m = 1$, $m = \mu'$, i.e., the MPLE u is a spline function with knots at the sample points. The smoothing parameters enter through the penalty functional by letting μ depend on $h_1,\ldots,h_p > 0$, i.e., we consider $m(h_1 t_1,\ldots,h_p t_p)$. Then, the MPLE u is of the form

(2.2) $u(\underline{x}) = \lambda^{-1} \Sigma_{i=1}^n u(\underline{X}_i)^{-1} (h_1 \ldots h_p)^{-1}$
$k((x_1 - X_{i1})/h_1,\ldots,(x_p - X_{ip})/h_p),$

where k can be any real function on $\mathbb{R}^p$, which integrates to one and $\tilde{k} > 0$. The MPLE of the density function is $f_n = u^2$.

The flexibility in the choice of the penalty functional in (2.1), allows a variety of kernels k in (2.2), which can be chosen in ways that allow for clearer definition of the "peaks" and "valleys" of the density estimates. In particular we can choose

$$k(\underline{x}) = [1-c(\underline{x}^T\underline{x}-1)]\phi(\underline{x}),\ \underline{x} \in \mathbb{R}^p,$$

where ϕ denotes the p-variate standard normal density, a kernel corresponding to c=0. For c=1, essentially we subtract u" from a u based on ϕ, resulting in a spline with improved performance at the concave and convex parts of the density surface. The value of c=½, results in a kernel with zero second moment and in an estimate with enhanced rates of convergence. Note that in the last two cases the MPLE u may assume negative values over areas that the data is very sparse. The density estimate f_n however, remains a proper probability density.

Under mild moment and smoothness assumptions on the underlying density f, the MPLE's are consistent, with optimal rates of convergence, in a variety of senses, e.g., in the Hellinger distance, L_1, L_2, uniform and Sobolev (corresponding to H) norms. Analogous results can be derived for the derivatives of f_n.

Note that once $u(\underline{x}_i, \underline{y}_i)$, i=1,...,n have been determined, it is straightforward to compute the corresponding nonparametric regression estimator $\hat{m}(x) = \int y f_n(x,y)dy / \int f_n(x,y)dy$, for details see Klonias (1984). When the kernel k in (2.2) is a product of univariate kernels, these regression estimates have the appealing property of reducing to the classical nonparametric kernel regression estimators, when the smoothing parameters corresponding to the Y's are let to go to zero. For example, if $k(x,y) = k_1(x)k_2(y)$ the MPLE of m(x) is given by

$$\hat{m}(x) = \{\Sigma_{i=1}^n \Sigma_{j=1}^n w_{ij}(x)Y_i\} / \Sigma_{i=1}^n \Sigma_{j=1}^n w_{ij}(x),$$

where,

$$w_{ij}(x) = [u(X_i,Y_i)u(X_j,Y_j)]^{-1} k_1\left(\frac{x-X_i}{h_1}\right) k_1\left(\frac{x-X_j}{h_1}\right) (k_2 * k_2)\left(\frac{Y_i-Y_j}{h_2}\right),$$

where * denotes convolution. Then, letting $h_2 \to 0$ we obtain the kernel regression estimate

$$\hat{m}_0(x) = \{\Sigma_{i=1}^n k_1^2\left(\frac{x-X_i}{h_1}\right) Y_i\} / \{\Sigma_{i=1}^n k_1^2\left(\frac{x-X_i}{h_1}\right)\}.$$

3. NUMERICAL EVALUATION OF THE MPLE'S

Note that (2.2) defines the spline u implicitly and we need to evaluate u at the sample points. To this end we set $\underline{x} = \underline{X}_i$, i=1,...,n in (2.2) and obtain the following system of nonlinear equation:

$$(3.1)\quad q_i^{-1} = \Sigma_{j=1}^n q_j k((X_{i1}-X_{j1})/h_1, \ldots, (X_{ip}-X_{jp})/h_p),$$

where $q_i = (\lambda h_1 \cdots h_p)^{-\frac{1}{2}} u(\underline{X}_i)^{-1}$, i=1,...,n.

Note that the q_i's do not depend on λ which is then determined by the equation $\int u^2 = 1$, i.e.,

$$\lambda = \Sigma_{i=1}^n \Sigma_{j=1}^n q_i q_j (k*k)((X_{i1}-X_{j1})/h_1, \ldots, (X_{ip}-X_{jp})/h_p).$$

Note that the $\underline{q} = (q_1,\ldots,q_n)^T$ which solves (3.1) is the unique solution to the following optimization problem:

$$(3.2)\quad \min \{\underline{q}^T \Sigma\, \underline{q} - \Sigma_{i=1}^n \log q_i^2,\ \underline{q} \in \mathbb{R}^n\}$$

$$\text{subject to } q_i \geq 0,\ i=1,\ldots,n,$$

where the $(i,j)^{th}$ entry of the positive definite matrix Σ is given by:

$$k((X_{i1}-X_{j1})/h_1, \ldots, (X_{ip}-X_{jp})/h_p).$$

The algorithm we use to solve (3.2) is based on a truncated-Newton method, described in Nash (1982), for details see Klonias and Nash (1983). Note that the dimension of the data influences only the computation of Σ, so that the numerical effort of solving (3.2) does not increase significantly with the dimension of the random variable. For n=200 and p=2, the solution of (3.2) requires CPU time on a VAX 11/780 of the order of 40 seconds.

For the data based choice of the smoothing parameters we propose to max $\{\lambda(h_1,\ldots,h_p),\ h_1,\ldots,h_p > 0\}$, for details see Klonias (1984). In the graphs that follow, when h_1, h_2 are estimated from the data, h_1, h_2 were chosen as the minimizers of

$$\|u_\lambda - u_n\|_2^2 = [(\lambda/n)^{\frac{1}{2}} - 1]^2,$$

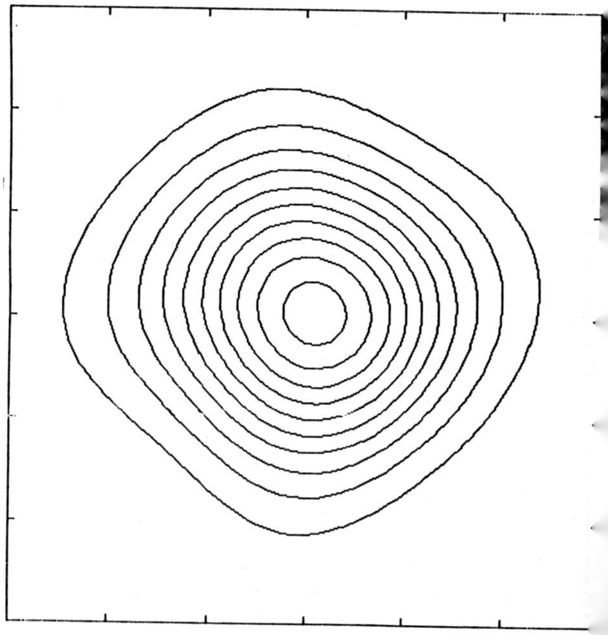

FIGURE 1. n=100; based on kernel (3.3).

where u_λ denotes the MPLE and u_n denotes the solution to problem (2.1) with $\lambda=n$, the asymptotic value of λ. The CPU time required for the numerical evaluation of the MPLE with data based h_1,h_2, for n=200, is of the order 80 seconds.

The data for the graphs that follow was generated using the IMSL routine GGNML with DSEED's 255866175 and 1949292845.

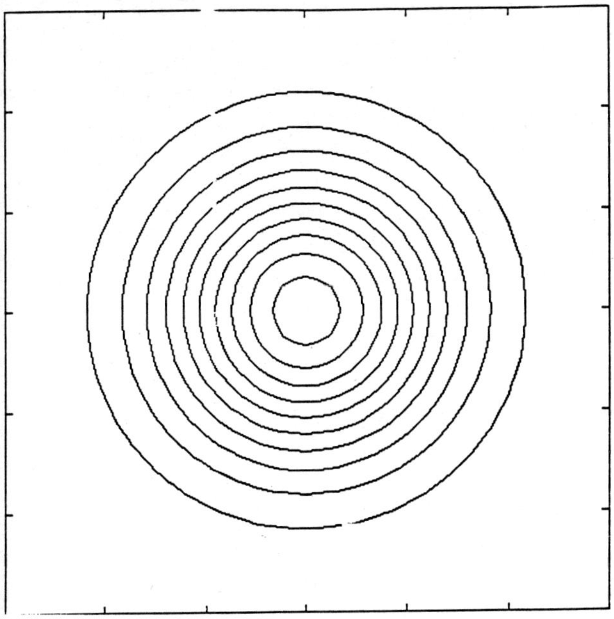

FIGURE 2. The N(0,0;1,1;0) surface.

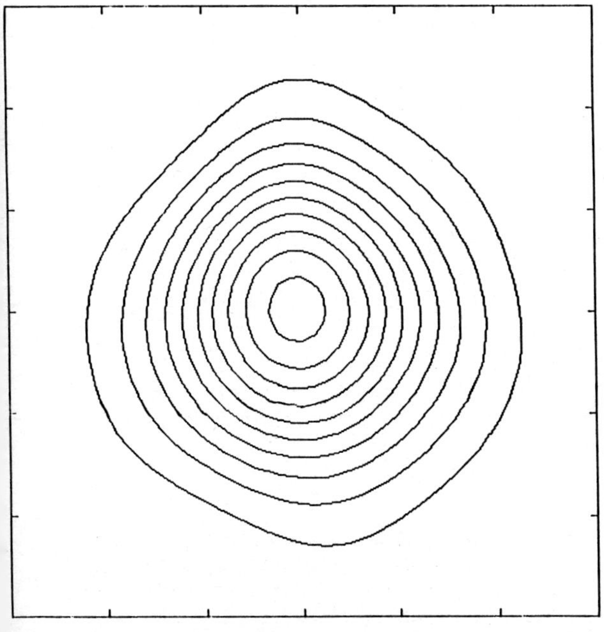

FIGURE 3. n=200; data based choice of h_1,h_2; based on kernel (3.3).

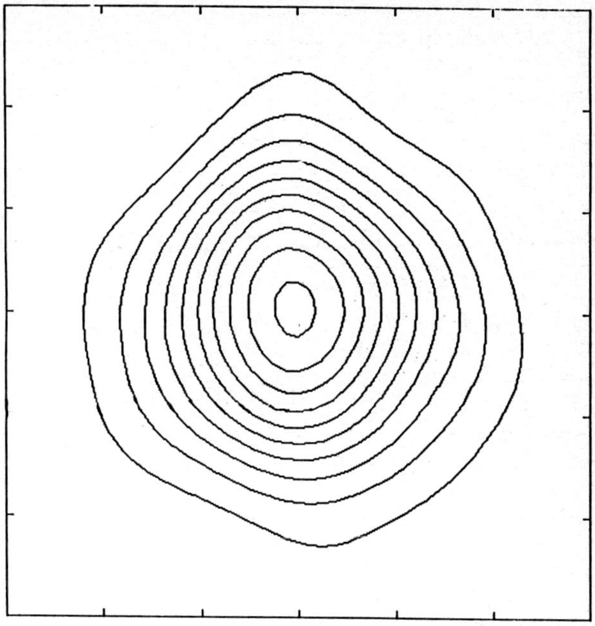

FIGURE 4. n=200; based on the standard normal kernel.

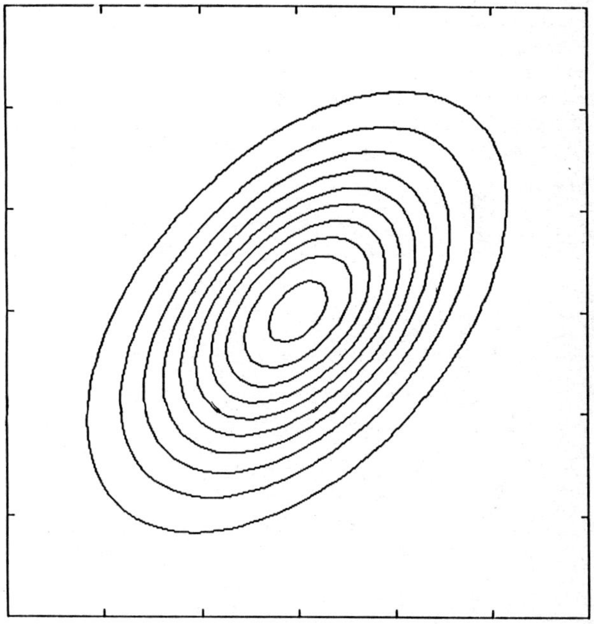

FIGURE 5. The N (0,0;1,1;½) surface.

Figures 2, 5 and 8 are the underlying surfaces which we estimate by the surfaces in Figures 1, 3, 4 and in Figures 6, 7 and in Figure 9 respectively. In Figures 3, 6, 9 the smoothing parameters have been estimated from the data, as described earlier. The estimates in Figures 1, 3, 6, 9 are based on the following kernel:

$$(3.3)\quad k(\underline{x}) = (4-\underline{x}^T\underline{x})\exp\{-\underline{x}^T\underline{x}/2\}4\pi,\ \underline{x}\in \mathbb{R}^2.$$

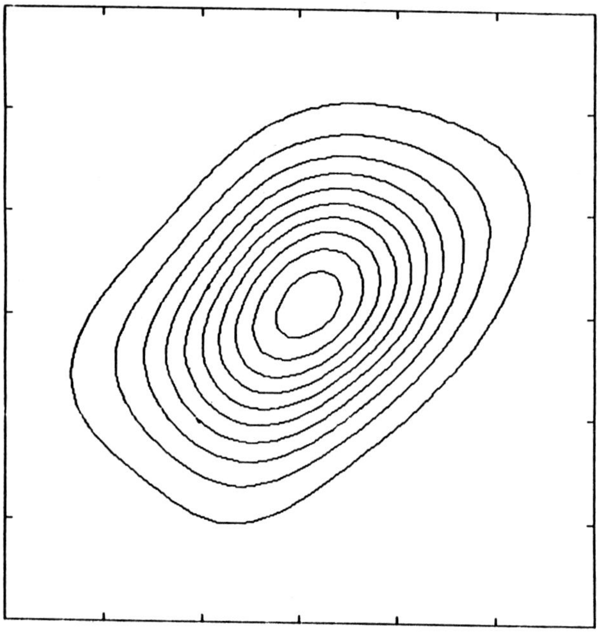

FIGURE 6. n=100; data based choice of h_1, h_2; based on kernel (3.3).

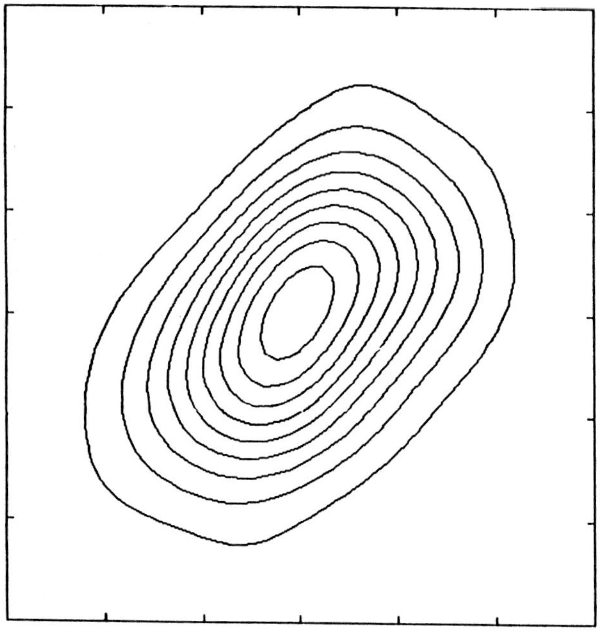

FIGURE 7. n=200; based on the standard normal kernel.

4. REFERENCES

[1] DeMontricher, G.F., Tapia, R.A. and

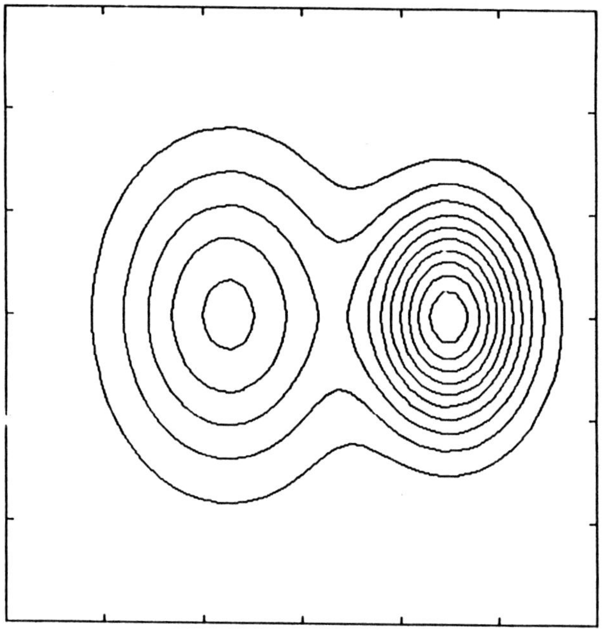

FIGURE 8. The (½)N(0,0;1,1;0) + (½)N(3,0;1,1;0) surface.

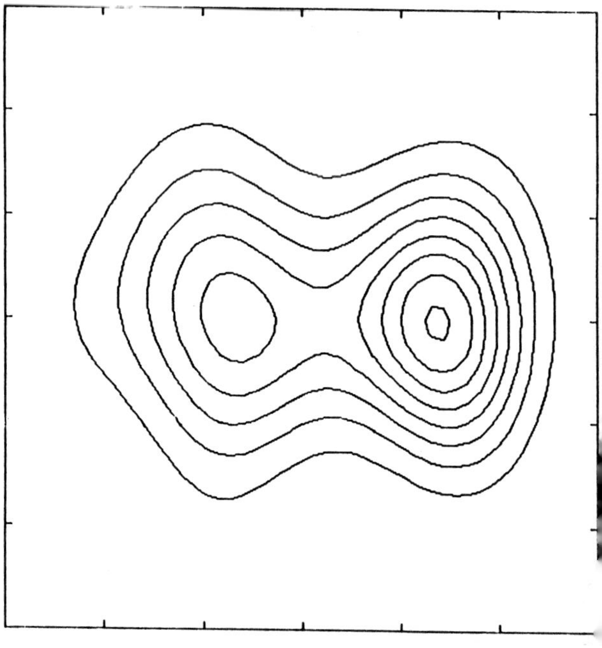

FIGURE 9. n=200; data based choice of h_1, h_2; based on kernel (3.3).

Thompson, J.R., Nonparametric maximum likelihood estimation of probability densities by penalty function methods, Ann. Statistics 3 (1975) 1329-1348.

[2] Good, I.J. and Gaskins, R.A., Nonparametric roughness penalties for probability densities, Biometrika 58 (1971) 255-277.

[3] Good, I.J. and Gaskins, R.A., Density estimization and bumphunting by the penalized likelihood method exemplified by scattering and meteorite data, (invited paper) J. Amer. Statist. Assoc. 75 (1980) 42-73.

[4] Klonias, V.K., On a class of nonparametric density and regression estimators, Ann. Stast. 12 (1984) 1263-1284.

[5] Klonias, V.K. and Nash, S.G., On the numerical evaluation of a class of nonparametric density and regression estimators. Tech. Report No. 376, Department of Mathematical Sciences, The Johns Hopkins University.

[6] Leonard, T., A Bayesian method for histograms, Biometrika 60 (1978) 297-308.

[7] Nash, S.G., Preconditioning of truncated-Newton methods, Tech. Report 371, Dept. of Mathematical Sciences, The Johns Hopkins Univ. (1982).

[8] Scott, D.W., Tapia, R.A., and Thompson, J. R., Nonparametric probability density estimation by discrete maximum penalized-likelihood criteria, Ann. Statist. 8 (1980) 820-832.

[9] Silverman, B.W., On the estimation of a probability density function by the maximum penalized likelihood method, Ann. Statist. 10 (1982) 795-810.

[10] Tapia, R.A. and Thompson, J.R., Nonparametric Probability Density Estimation (The Johns Hopkins University Press, Baltimore and London, 1978).

Categorical Data

Organizer: *J. Richard Landis*

Invited Presentations:

Categorical Data Analysis Strategies Using SAS Software
William M. Stanish

Categorical Data Analysis in BMDP: Present and Future
Morton B. Brown

Log-Linear Modeling with SPSS
Clifford C. Clogg and Mark P. Becker

Fitting Multinomial Regression Models to Categorical Data
Christopher Cox

COMPUTER SCIENCE AND STATISTICS:
The Interface, D.M. Allen (ed.)

CATEGORICAL DATA ANALYSIS STRATEGIES USING SAS SOFTWARE

William M. Stanish

SAS Institute, Inc.
Cary, NC, USA

This paper reviews current methods of categorical data analysis, and illustrates how SAS® software can be used to perform the analyses. Topics include: randomization methods for testing hypotheses under a minimum of assumptions, linear and log-linear modeling of categorical responses, weighted-least-squares estimation methods for investigating the variation among functions of proportions, maximum-likelihood estimation using Newton-Raphson and iterative proportional fitting, repeated measures analysis, stratified analysis, logistic regression, and the analysis of data from complex sample surveys. Examples of each type of analysis are given.

1. INTRODUCTION

The capabilities of SAS software for categorical data analysis have increased dramatically over the past few years. The capabilities discussed in this paper are available in Version 5 of the software, scheduled for release in the middle of 1985. The primary procedures for categorical data analysis are

- CATMOD procedure (replaces FUNCAT)
- FREQ procedure (replaces TFREQ)
- IML procedure (replaces MATRIX).

The CATMOD procedure does general linear modeling of categorical data, including linear models, log-linear models, logistic regression, and repeated measures analysis. The FREQ procedure does analysis of association and stratified analysis. The IML procedure encompasses an interactive matrix language that makes it relatively easy to program any customized analysis that is desired.

The remaining sections of this paper are divided as follows.

2. Two-Way Contingency Tables
3. Stratified Analysis
   3.1 Partial Association Testing
   3.2 Estimation of Relative Risk
4. General Linear Model Analysis
5. Log-Linear Models, Maximum Likelihood
6. Models for Ordinal Data
7. Repeated Measures Analysis
8. Complex Sample Survey Data Analysis

For conservation of space, the printed output displayed in this paper for any given problem is generally only a small portion of the amount produced by the procedure.

2. TWO-WAY CONTINGENCY TABLES

For two-way contingency tables, PROC FREQ does an analysis of association that is divided into two or more parts. The first part contains test statistics and p-values for testing the null hypothesis of no association between the two variables. The second part contains measures of association for estimating the strength of any association that might be present.

In choosing measures of association to use in analyzing a two-way table, one should consider the study design, the measurement scale of the variables, the type of association that each measure is designed to detect, and any assumptions required for valid interpretation of a measure. For further information on choosing measures of association for a specific set of data, see Goodman and Kruskal (1979), or Bishop, Fienberg, and Holland (1975, Chapter 11).

Similar comments apply to the choice and interpretation of the test statistics. For example, the Mantel-Haenszel chi-square statistic requires an ordinal scale for both variables, and is designed to detect a linear association. The Pearson chi-square, on the other hand, is appropriate for all variables, and can detect any kind of association, but it is less powerful for detecting a linear association because its power is dispersed over a greater number of degrees of freedom (except for 2 by 2 tables).

For 2 by 2 tables, PROC FREQ also computes estimates of relative risk and their corresponding confidence intervals. For two dichotomous variables, D and E, the relative risk of D is defined as

$$RR = Prob(D=yes|E=yes)/Prob(D=yes|E=no).$$

Because the definition of relative risk involves conditional probabilities, the estimation procedure depends on which variable, if either, was fixed by the study design. The FREQ procedure therefore gives different estimates for different study designs.

- For case-control studies (D fixed, E random), the estimator of the common relative risk is the common odds ratio.

- For cohort studies (E fixed, D random) and for cross-sectional studies (D and E both random), there is a direct estimator of the common relative risk.

See the SAS User's Guide: Statistics(1985) for computational formulas and references for all of the test statistics and measures of association.

Example

The following control statements read some hypothetical data and request an analysis of association from PROC FREQ.

```
DATA;
   INPUT FACTOR $ DISEASE $ COUNT;
   CARDS;
YES YES 19
YES NO  53
NO  YES 13
NO  NO  65
PROC FREQ ORDER=DATA;
   WEIGHT COUNT;
   TABLE FACTOR*DISEASE / ALL;
```

Figure 1 displays the contingency table printed by PROC FREQ, and Figure 2 shows the corresponding statistics. The statistics indicate a nonsignificant(α=.10) association, with a relatively small correlation coefficient (.12). The relative-risk estimates suggest that those who are exposed to the factor of interest are at least one and a half times more likely to get the disease than those who are not exposed to the factor.

Figure 1
TABLE OF FACTOR BY DISEASE

```
FACTOR     DISEASE

FREQUENCY|
 PERCENT |
 ROW PCT |
 COL PCT |YES     |NO      |  TOTAL
---------+--------+--------+
YES      |     19 |     53 |     72
         |  12.67 |  35.33 |  48.00
         |  26.39 |  73.61 |
         |  59.38 |  44.92 |
---------+--------+--------+
NO       |     13 |     65 |     78
         |   8.67 |  43.33 |  52.00
         |  16.67 |  83.33 |
         |  40.63 |  55.08 |
---------+--------+--------+
TOTAL          32      118      150
            21.33    78.67   100.00
```

Figure 2
STATISTICS FOR TABLE OF FACTOR BY DISEASE

STATISTIC	DF	VALUE	PROB
CHI-SQUARE	1	2.109	0.146
LIKELIHOOD RATIO CHI-SQUARE	1	2.114	0.146
CONTINUITY ADJ. CHI-SQUARE	1	1.569	0.210
MANTEL-HAENSZEL CHI-SQUARE	1	2.095	0.148
FISHER'S EXACT TEST (1-TAIL)			0.105
(2-TAIL)			0.166
PHI		0.119	
CONTINGENCY COEFFICIENT		0.118	
CRAMER'S V		0.119	

STATISTIC	VALUE	ASE
GAMMA	0.284	0.186
KENDALL'S TAU-B	0.119	0.081
STUART'S TAU-C	0.097	0.067
SOMERS' D C|R	0.097	0.067
SOMERS' D R|C	0.145	0.098
PEARSON CORRELATION	0.119	0.081
SPEARMAN CORRELATION	0.119	0.081
LAMBDA ASYMMETRIC C|R	0.000	0.000
LAMBDA ASYMMETRIC R|C	0.083	0.075
LAMBDA SYMMETRIC	0.058	0.052
UNCERTAINTY COEFFICIENT C|R	0.014	0.019
UNCERTAINTY COEFFICIENT R|C	0.010	0.014
UNCERTAINTY COEFFICIENT SYM	0.012	0.016

ESTIMATES OF THE RELATIVE RISK (ROW1/ROW2)

TYPE OF STUDY	VALUE	95% CONFIDENCE BOUNDS	
CASE-CONTROL	1.792	0.811	3.962
COHORT (COL1 RISK)	1.583	0.844	2.969
COHORT (COL2 RISK)	0.883	0.745	1.047

3. STRATIFIED ANALYSIS

The FREQ procedure provides an analysis of the relationship between two variables, after adjusting for the effect of potential confounding variables. Stratified analysis is similar to the process of fitting a regression model that relates some function of the dependent variable to a linear combination of the independent variable and the confounding variables. The advantage of stratified analysis over regression is twofold: (1) you can adjust for the effect of the confounding variables without being forced to estimate parameters for them, and (2) you can get a much clearer picture of the patterns of interaction and the sources of variation since you can look at statistics from the individual strata.

For specifying a stratified analysis of the relationship between variables C and D, after adjusting for variables A and B, the required statements are

```
PROC FREQ;
   TABLES A*B*C*D / ALL;
```

On the basis of these statements, one stratum is formed for each combination of the levels of variables A and B. For each stratum, a contingency table of C by D is printed, together with test statistics and measures of association. Lastly, the FREQ procedure prints the statistics that summarize the information across the strata in an efficient way. The following sections pertain to these summary statistics.

3.1 Partial Association Testing

The class of generalized Cochran-Mantel-Haenszel (CMH) statistics (Landis, Heyman, and Koch 1978) is an important class of statistics for testing no partial association in a stratified analysis. They have several major advantages.

- The assumptions required for their validity are minimal. They do not require a linear model, nor do they assume any parametric form for the observed data. They require only fixed row and column margins for the contingency table in each stratum, and these fixed margins can be obtained by design or by conditional distribution arguments.

- They do not require a large sample size within each stratum. They have a chi-square distribution when the null hypothesis of no partial association is true and when the effective overall sample size is large.

- The statistics depend on scores for the row and column variables. The scores give flexibility with respect to the alternative hypothesis being tested, and they allow the choice of parametric or nonparametric analyses.

CMH statistics have low power for detecting an association in which the patterns of association for some of the strata are in the opposite direction of the patterns displayed by other strata. Thus, a nonsignificant CMH statistic suggests either that there is no association, or that no pattern of association had enough strength or consistency to dominate any other pattern.

The formulas for the CMH statistics are given in the SAS User's Guide: Statistics (1985). For additional information on the development of CMH statistics, see Cochran (1954), Mantel and Haenszel (1959), Mantel (1963), Birch (1965), Landis, Heyman, and Koch (1978).

The FREQ procedure computes the following types of CMH statistics, reflecting different alternative hypotheses.

The correlation statistic (df=1)

The correlation statistic, with one degree of freedom, is also known as the Mantel-Haenszel statistic. This statistic requires that both the row and column variables be ordinally scaled, and the alternative hypothesis is that there is a linear association in at least one stratum. When there is only one stratum, the Mantel-Haenszel statistic reduces to $(N-1)r^2$, where r is a correlation coefficient (either Pearson or Spearman, depending on whether the scores are parametric or nonparametric).

The ANOVA statistic (df=R-1)

This statistic requires that the column variable lie on an ordinal (or interval) scale. The mean column score is computed for each row of the table, and the alternative hypothesis is that, for at least one stratum, the mean scores of the R rows are unequal. In other words, the statistic is sensitive to location differences among the R distributions of the column variable.

When there is only one stratum, this CMH statistic is essentially an analysis-of-variance (ANOVA) statistic in the sense that it is a function of the variance ratio F statistic. If nonparametric scores are specified in this case, then the ANOVA statistic is identical to a Kruskal-Wallis test.

If there is more than one stratum, then the CMH statistic corresponds to a stratum-adjusted ANOVA or Kruskal-Wallis test. In the special case where there is one subject per row and one subject per column in the contingency table of each stratum, this CMH statistic is identical to Friedman's chi-square.

The general association statistic (df=(R-1)(C-1))

This statistic is always interpretable because it does not require an ordinal scale for either variable. The alternative hypothesis is that, for at least one stratum, there is some kind of association. When there is only one stratum, then the general association CMH statistic reduces to $\{(N-1)/N\}Q_P$, where Q_P is the Pearson chi-square statistic.

Example

As an example of partial association testing, we consider data from a study of the treatment of duodenal ulcer (Grizzle, Starmer, and Koch 1969). Specifically, interest lies in the question of whether there is an association between treatment and the severity of an undesirable complication of treatment called dumping syndrome. As indicated in Figure 3, severity is ordinally scaled (none, slight, moderate), and treatment is also ordinally scaled since the treatments correspond to the percentage of the stomach removed during a surgical operation. The hospital at which surgery was done represents a potential confounding variable which needs to be controlled in the analysis.

Figure 3 shows the control statements required to do both a parametric and a nonparametric stratified analysis. As shown in Figure 4, the general-association and the analysis-of-variance CMH statistics are nonsignificant (α=.05), but the correlation statistics are significant (p<.02). This indicates that there is a linear association in at least one of the strata, and it illustrates the value of having statistics that have their power concentrated on narrowly defined alternative hypotheses.

Figure 5 displays the correlation results from the individual strata. The source of correlation and the pattern of interaction is very clear: the linear association between treatment and severity arises only from hospital 2.

Figure 3

```
*-------------------------------------------*
|                                           |
|          DUMPING SYNDROME DATA            |
|          ---------------------            |
|                                           |
| INDEPENDENT VARIABLES                     |
| ---------------------                     |
|   1. TREATMENT(OPERATION)                 |
|      A. DRAINAGE AND VAGOTOMY             |
|      B. 25% RESECTION AND VAGOTOMY        |
|      C. 50% RESECTION AND VAGOTOMY        |
|      D. 75% RESECTION                     |
|   2. HOSPITAL(1, 2, 3, 4)                 |
|                                           |
|                                           |
| DEPENDENT VARIABLE                        |
| ------------------                        |
|   SEVERITY OF DUMPING SYNDROME            |
|      (NONE, SLIGHT, MODERATE)             |
|                                           |
|  REFERENCE: GRIZZLE, ET AL.(1969),        |
|             BIOMETRICS 25, 489-504.       |
*-------------------------------------------*;

PROC FREQ  ORDER=DATA;
 WEIGHT WT;
 TABLES HOSPITAL*TRT*SEVERITY / ALL;
 TABLES HOSPITAL*TRT*SEVERITY / ALL SCORES=RANK;
 TITLE 'ANALYSIS OF DUMPING SYNDROME DATA';
```

Figure 4
ANALYSIS OF DUMPING SYNDROME DATA

SUMMARY STATISTICS FOR TRT BY SEVERITY
CONTROLLING FOR HOSPITAL

COCHRAN-MANTEL-HAENSZEL STATISTICS (BASED ON TABLE SCORES)

STATISTIC	ALTERNATIVE HYPOTHESIS	DF	VALUE	PROB
1	NONZERO CORRELATION	1	6.340	0.012
2	ROW MEAN SCORES DIFFER	3	6.590	0.086
3	GENERAL ASSOCIATION	6	10.598	0.102

COCHRAN-MANTEL-HAENSZEL STATISTICS (BASED ON RANK SCORES)

STATISTIC	ALTERNATIVE HYPOTHESIS	DF	VALUE	PROB
1	NONZERO CORRELATION	1	5.583	0.018
2	ROW MEAN SCORES DIFFER	3	7.234	0.065
3	GENERAL ASSOCIATION	6	10.598	0.102

TOTAL SAMPLE SIZE = 417

Figure 5
Correlation Analysis by Stratum

Hospital	Sample Size	Pearson Correlation	Mantel-Haenszel Chi-Square	DF	Prob
1	148	0.10	1.57	1	.21
2	105	0.26	7.06	1	.01
3	74	0.05	0.16	1	.69
4	90	0.09	0.66	1	.42

3.2 Estimation of Relative Risk

As in the case of a single two-way contingency table, the estimate of relative risk depends on the study design, and thus PROC FREQ gives separate estimates for the different designs. Also, it uses two different methods to obtain the estimate and its corresponding confidence interval.

- Mantel-Haenszel estimate, with a test-based confidence interval
- Logit estimate, with a precision-based confidence interval

A major advantage of the Mantel-Haenszel (MH) estimator over the logit estimator (Woolf 1955, Haldane 1955) is that cell frequencies of zero pose no computational problem for the MH estimator. Thus, there is no need to add 1/2 to certain cell frequencies, as is sometimes necessary with the logit estimator and its corresponding confidence interval.

The test-based confidence interval has some theoretical problems because it is based on the assumption that the Cochran-Mantel-Haenszel test statistic has a chi-square distribution, which is true only when the null hypothesis of no partial association is true. However, from a practical point of view, the bias seems to be very small when the parameter of interest does not differ greatly from 1 (say, for 1/4 < RR < 4).

The formulas for the estimators are given in the SAS User's Guide: Statistics(1985). For additional information on stratified analysis, relative risk estimation, and confidence interval estimation, see Kleinbaum, Kupper, and Morgenstern (1982).

Example

These data are from a detergent preference study (Cox 1970). See Figure 6 for a description of the dependent and independent variables, and Figure 7 for a listing of the data and the control statements required to do a stratified analysis with PROC FREQ. The question of interest for this example is the following. Is there an association between preferred brand of laundry detergent and previous usage of Brand M, after controlling for the softness and the temperature of the laundry water, and if so, what is the magnitude of the relationship?

Figure 8 displays the contingency table for stratum 1, and Figure 9 shows the page of summary statistics from the printed output. The CMH statistic is highly significant, indicating very strong evidence of a partial association between preferred brand and previous usage of Brand M. This study was a cross-sectional study, and the contingency tables are set up with PREF=M in the first column of each table. Thus, we refer to the COL1 RISK section of the output for estimation of relative risk. The results indicate that, on the average, previous users of Brand M laundry detergent are about 1.3 (=1/.75) times more likely to prefer Brand M than those who are not previous users.

Figure 10 shows a relative–risk analysis by stratum. The results indicate a fair amount of interaction, with strata 1, 2, and 3 having similar estimates (.65, .65, .61), with strata 4 and 5 displaying a weaker association (.80, .80), and with stratum 6 showing no association (.99). Given the large sample sizes within each stratum, one could use the CATMOD procedure to do modeling of the relative risk estimates.

Figure 6

```
*-------------------------------------------------------------*
|                 DETERGENT PREFERENCE STUDY                   |
|                                                              |
|                                                              |
| DEPENDENT VARIABLE                       VALUES              |
| ------------------                       --------            |
| BRAND = BRAND PREFERRED                  M, X                |
|                                                              |
|                                                              |
| INDEPENDENT VARIABLES                    VALUES              |
| ------------------                       --------            |
| SOFTNESS = SOFTNESS OF WATER             SOFT, MED, HARD     |
| PREV = PREVIOUS USER OF BRAND M?         YES, NO             |
| TEMP = TEMP OF LAUNDRY WATER             HIGH, LOW           |
|                                                              |
|                                                              |
| FROM: RIES AND SMITH, CHEMICAL ENGINEERING                   |
|       PROGRESS 59(1963), PP. 39-43.                          |
|                                                              |
|       COX(1970) THE ANALYSIS OF BINARY DATA. P.38            |
*-------------------------------------------------------------*;
```

Figure 7

```
TITLE 'DETERGENT PREFERENCE STUDY';
DATA DETERG;
  INPUT SOFTNESS$ BRAND$ PREV$ TEMP$ COUNT @@;
  CARDS;
SOFT  X  YES  HIGH  19    SOFT  X  YES  LOW  57
SOFT  X   NO  HIGH  29    SOFT  X   NO  LOW  63
SOFT  M  YES  HIGH  29    SOFT  M  YES  LOW  49
SOFT  M   NO  HIGH  27    SOFT  M   NO  LOW  53
MED   X  YES  HIGH  23    MED   X  YES  LOW  47
MED   X   NO  HIGH  33    MED   X   NO  LOW  66
MED   M  YES  HIGH  47    MED   M  YES  LOW  55
MED   M   NO  HIGH  23    MED   M   NO  LOW  50
HARD  X  YES  HIGH  24    HARD  X  YES  LOW  37
HARD  X   NO  HIGH  42    HARD  X   NO  LOW  68
HARD  M  YES  HIGH  43    HARD  M  YES  LOW  52
HARD  M   NO  HIGH  30    HARD  M   NO  LOW  42

PROC FREQ;
  WEIGHT COUNT;
  TABLE SOFTNESS*TEMP*PREV*BRAND / ALL;
```

Figure 8

```
              DETERGENT PREFERENCE STUDY

               TABLE 1 OF PREV BY BRAND
        CONTROLLING FOR SOFTNESS=HARD TEMP=HIGH

PREV       BRAND

FREQUENCY|
 PERCENT |
 ROW PCT |
 COL PCT |M       |X       |  TOTAL
---------+--------+--------+
NO       |     30 |     42 |     72
         |  21.58 |  30.22 |  51.80
         |  41.67 |  58.33 |
         |  41.10 |  63.64 |
---------+--------+--------+
YES      |     43 |     24 |     67
         |  30.94 |  17.27 |  48.20
         |  64.18 |  35.82 |
         |  58.90 |  36.36 |
---------+--------+--------+
TOTAL          73       66      139
            52.52    47.48   100.00
```

Figure 9

DETERGENT PREFERENCE STUDY

SUMMARY STATISTICS FOR PREV BY BRAND
CONTROLLING FOR SOFTNESS AND TEMP

COCHRAN-MANTEL-HAENSZEL STATISTICS (BASED ON TABLE SCORES)

STATISTIC	ALTERNATIVE HYPOTHESIS	DF	VALUE	PROB
1	NONZERO CORRELATION	1	19.769	0.000
2	ROW MEAN SCORES DIFFER	1	19.769	0.000
3	GENERAL ASSOCIATION	1	19.769	0.000

ESTIMATES OF THE COMMON RELATIVE RISK (ROW1/ROW2)

TYPE OF STUDY	METHOD	VALUE	95% CONFIDENCE BOUNDS	
CASE-CONTROL	MANTEL-HAENSZEL	0.569	0.444	0.730
(ODDS RATIO)	LOGIT	0.569	0.442	0.731
COHORT	MANTEL-HAENSZEL	0.753	0.664	0.853
(COL1 RISK)	LOGIT	0.752	0.662	0.854
COHORT	MANTEL-HAENSZEL	1.326	1.171	1.501
(COL2 RISK)	LOGIT	1.305	1.150	1.482

THE CONFIDENCE BOUNDS FOR THE M-H ESTIMATES ARE TEST-BASED.

BRESLOW-DAY TEST FOR HOMOGENEITY OF THE ODDS RATIOS

CHI-SQUARE = 8.053 DF = 5 PROB = 0.153

TOTAL SAMPLE SIZE = 1008

Figure 10

Detergent Preference Study
Relative Risk Analysis by Stratum

Stratum	SOFTNESS	TEMP	Sample Size	Estimated Relative Risk	Pearson Chi-square
1	HARD	HIGH	139	.649	7.05
2	HARD	LOW	199	.653	8.09
3	MED	HIGH	126	.612	8.56
4	MED	LOW	218	.799	2.54
5	SOFT	HIGH	104	.798	1.55
6	SOFT	LOW	222	.988	0.01

4. GENERAL LINEAR MODEL ANALYSIS

The CATMOD procedure fits linear models to general functions of categorical data. It does so by facilitating transformations of an initial proportion vector(p) to a function vector(F), and by estimating the parameters of the linear model $F(\pi) = X\beta$, where π is the vector of underlying probabilities. CATMOD uses one of two estimation methods:

- weighted-least-squares estimation, available for all types of response functions
- maximum-likelihood estimation, available for logistic regression and log-linear models.

Both methods of estimation are BAN (best asymptotic normal), and therefore they are asymptotically equivalent. After the parameters are estimated, CATMOD computes a goodness-of-fit test, as well as Wald statistics for testing model effects (such as main effects and interactions) and other null hypotheses of interest.

The theory for the weighted-least-squares estimation and the general linear modeling may be found in Grizzle, Starmer, and Koch(1969). The theory for the maximum-likelihood estimation and the log-linear modeling is in Fienberg(1980) and Bishop, Fienberg, and Holland(1975). The computational formulas used by CATMOD can be found in the SAS User's Guide: Statistics(1985).

One can analyze almost any functions of the original proportions, including logits, marginal probabilities, marginal logits, means, cumulative probabilities, cumulative logits, survival probabilities, kappa statistics, odds ratios, risk ratios, etc. Some of the most common analyses use linear response functions (for linear models) or logit response functions (for logistic regression and log-linear models). The two examples in this section illustrate a linear model and a logistic regression. Log-linear models and repeated measures analysis are dealt with in separate sections.

Example

The first example is a linear model analysis of the detergent preference data used in Section 3. The control statements required to fit a main-effects model are

```
PROC CATMOD;
  RESPONSE 1 0;
  WEIGHT COUNT;
  MODEL BRAND = SOFTNESS PREV TEMP;
  TITLE2 'LINEAR MAIN-EFFECTS MODEL';
```

Figure 11 shows part of the output printed by CATMOD. The design matrix X contains columns corresponding to the main effects in the model statement. The analysis-of-variance table shows that the model fits the data adequately (Q=8.26, df=7), and that the PREV and TEMP main effects are statistically significant (α=.05). The analysis of individual parameters gives the parameter estimates and their standard errors. The estimated covariance matrix and the correlation matrix of the estimated parameters are also computed upon request.

Figure 11
DETERGENT PREFERENCE STUDY
LINEAR MAIN-EFFECTS MODEL

SAMPLE	RESPONSE FUNCTIONS	DESIGN MATRIX 1	2	3	4	5
1	0.416667	1	1	0	1	1
2	0.381818	1	1	0	1	-1
3	0.641791	1	1	0	-1	1
4	0.58427	1	1	0	-1	-1
5	0.410714	1	0	1	1	1
6	0.431034	1	0	1	1	-1
7	0.671429	1	0	1	-1	1
8	0.539216	1	0	1	-1	-1
9	0.482143	1	-1	-1	1	1
10	0.456897	1	-1	-1	1	-1
11	0.604167	1	-1	-1	-1	1
12	0.462264	1	-1	-1	-1	-1

ANALYSIS OF VARIANCE TABLE

SOURCE	DF	CHI-SQUARE	PROB
INTERCEPT	1	1004.93	0.0001
SOFTNESS	2	0.24	0.8859
PREV	1	20.96	0.0001
TEMP	1	3.95	0.0468
RESIDUAL	7	8.26	0.3100

ANALYSIS OF INDIVIDUAL PARAMETERS

EFFECT	PARAMETER	ESTIMATE	STANDARD ERROR	CHI-SQUARE	PROB
INTERCEPT	1	0.50797	0.016024	1004.93	0.0001
SOFTNESS	2	-.002562	.0218391	0.01	0.9066
	3	.0103542	.0217608	0.23	0.6342
PREV	4	-.071088	.0155288	20.96	0.0001
TEMP	5	.0319446	0.016071	3.95	0.0468

Example

The second example is a logistic regression analysis of the same data. The response functions to be analyzed are the logits, but the required control statements do not include a response statement since logits are the default response functions:

```
PROC CATMOD;
  WEIGHT COUNT;
  MODEL BRAND = SOFTNESS PREV TEMP
        / NOPROFILE  NODESIGN  NOPARM  ML;
  TITLE2 'LOGIT MAIN-EFFECTS MODEL';
```

The ML specification in the MODEL statement requests maximum–likelihood estimation of the parameters.

Figure 12 shows the maximum–likelihood analysis of the data. The initial estimates (iteration 0) of the parameters are the weighted-least-squares estimates, and subsequent estimates are printed for each Newton-Raphson iteration until convergence is achieved. The goodness-of-fit test in the analysis-of-variance table is the likelihood-ratio test, and it shows that the model fits the data (Q=8.23, df=7). With respect to the significance of the main effects in the model, the Wald statistics based on the maximum–likelihood estimates for the logit model are very similar to those based on weighted-least-squares estimates for the linear model. Predicted cell frequencies are also computed by CATMOD, if requested.

Figure 12
DETERGENT PREFERENCE STUDY
LOGIT MAIN-EFFECTS MODEL

CATMOD PROCEDURE

MAXIMUM LIKELIHOOD ANALYSIS

ITERATION	SUB ITERATION	-2 LOG LIKELIHOOD	CONVERGENCE CRITERION
0	0	1372.72	1
1	0	1372.72	8.55E-07
2	0	1372.72	6.97E-14

PARAMETER ESTIMATES

ITERATION	1	2	3	4	5
0	0.0301634	-.0099095	0.0391404	-0.281692	0.1277
1	0.0301777	-.0094684	0.0400479	-0.283508	0.128326
2	0.0301777	-.0094683	0.0400483	-0.283508	0.128326

ANALYSIS OF VARIANCE TABLE

SOURCE	DF	CHI-SQUARE	PROB
INTERCEPT	1	0.21	0.6491
SOFTNESS	2	0.22	0.8976
PREV	1	19.70	0.0001
TEMP	1	3.73	0.0534
LIKELIHOOD RATIO	7	8.23	0.3129

5. LOG-LINEAR MODELS, MAXIMUM LIKELIHOOD

General log-linear modeling, with hierarchal or nonhierarchal models, can be done by the CATMOD procedure. Both weighted-least-squares and maximum-likelihood (ML) estimation are available. CATMOD uses Newton-Raphson iteration to obtain its maximum-likelihood estimates. If one has a large hierarchal model, then iterative proportional fitting (IPF) is a more efficient method of ML estimation, and the IPF function in the IML procedure can be used for this purpose.

The basic log-linear model for one population may be expressed as

$$\pi = \exp(X\beta) \; / \; 1'\exp(X\beta)$$

where π is the vector of multinomial probabilities for the population. Because of the restriction that the probabilities add to one, an equivalent way of expressing the model is

$$F(\pi) = C \log(\pi) = CX\beta = X^{*}\beta$$

where

$$C = (I_{r-1}, -1_{r-1}) .$$

But $F(\pi)$ is simply the vector of generalized (or multiple) logits for the population probabilities. Thus, the latter equations show that fitting a log-linear model on the probabilities is equivalent to fitting a linear model on the generalized logits. Such a transformation brings log-linear modeling into a general linear modeling framework, so that the power and flexibility of a program such as CATMOD can be brought to bear on log-linear models.

In particular, the generalization of log-linear models to multiple populations is totally straightforward with CATMOD. Multiple populations are formed on the basis of independent (or design) variables, and a separate multinomial distribution is assumed for each population. The model equations for such multiple population log-linear models can be found in Imrey(1985) and Imrey, Koch, and Stokes(1981).

Imrey(1985) illustrates the use of the CATMOD procedure for numerous logit and log-linear model applications, including multiple logistic models, quasi-independence, proportional odds models, and a repeated measures (split-plot) analysis of marginal logits. Imrey also discusses some of the technicalities of CATMOD, including the role of the REPEATED statement in log-linear model analysis, the treatment of structural vs. random zeros, and alternative formulations of logistic models in terms of log-linear models.

Example

The example is a simple one-population study in which each subject was given three different drugs, and their response (F=Favorable, U=Unfavorable) to each was recorded (Koch et al. 1977). The following control statements set up the data set and specify a maximum-likelihood analysis of a log-linear model:

```
DATA DRUGS;
   INPUT DRUGA $ DRUGB $ DRUGC $ COUNT @@;
   CARDS;
F F F 6    F F U 16    F U F 2    F U U 4
U F F 2    U F U  4    U U F 6    U U U 6

PROC CATMOD;
  WEIGHT COUNT;
  RESPONSE OUT=PRED;
  MODEL DRUGA*DRUGB*DRUGC = _RESPONSE_
   / ML COVB PRED=FREQ;
  REPEATED
   / _RESPONSE_ = DRUGA DRUGB DRUGC DRUGA*DRUGB;
  TITLE  'ONE-POPULATION DRUG STUDY';
  TITLE2 'MLE ANALYSIS OF THE JOINT FREQUENCIES';
```

The RESPONSE statement specifies the analysis of generalized logits and the creation of an output data set containing predicted values. The responses to the three drugs are designated as dependent variables by their appearance on the left-hand side of the MODEL statement.

The _RESPONSE_ keyword in the MODEL statement indicates that the model is to include sources of variation based on the levels of the dependent variables. The REPEATED statement is used only to define the _RESPONSE_ effect in terms of the usual log-linear model main effects and interactions. (When there is no repeated measurement involved in the study, then the term REPEATED is a misnomer, but the definition of _RESPONSE_ is nonetheless placed on the REPEATED statement.) The specified model contains main effects for each of the three drugs, together with the DRUGA*DRUGB interaction. The MODEL statement also requests maximum - likelihood analysis, predicted cell frequencies, and the estimated covariance matrix of the parameter estimates.

Figure 13 shows the results of the maximum-likelihood analysis, with the final parameter estimates appearing in the row corresponding to the last iteration. The analysis-of-variance table gives the likelihood-ratio goodness-of-fit test, together with Wald statistics for testing the individual effects in the model. Figure 14 contains the estimated covariance matrix of the parameter estimates, along with the table of predicted cell frequencies and their standard errors.

Figure 13
ONE-POPULATION DRUG STUDY
MLE ANALYSIS OF THE JOINT FREQUENCIES

MAXIMUM LIKELIHOOD ANALYSIS

ITERATION	SUB ITERATION	-2 LOG LIKELIHOOD	CONVERGENCE CRITERION
0	0	173.046	1
1	0	173.029	9.53E-05
2	0	173.029	5.50E-09

PARAMETER ESTIMATES

ITERATION	1	2	3	4
0	0.143617	0.143617	-0.305493	0.5127[illegible]
1	0.151448	0.151448	-0.314281	0.49811[illegible]
2	0.151534	0.151534	-0.314304	0.49810[illegible]

ANALYSIS OF VARIANCE TABLE

SOURCE	DF	CHI-SQUARE	PROB
DRUGA	1	0.80	0.3726
DRUGB	1	0.80	0.3726
DRUGC	1	4.12	0.0423
DRUGA*DRUGB	1	8.59	0.0034
LIKELIHOOD RATIO	3	1.75	0.6266

Figure 14
ONE-POPULATION DRUG STUDY
MLE ANALYSIS OF THE JOINT FREQUENCIES

COVARIANCE OF ESTIMATES

	1	2	3	4
1	0.0288813	-0.0127843	1.506E-18	-.00236594
2	-0.0127843	0.0288813	1.063E-18	-.00236594
3	1.506E-18	1.063E-18	0.023958	-3.776E-19
4	-.00236594	-.00236594	-3.776E-19	0.0288813

PREDICTED VALUES FOR RESPONSE FUNCTIONS AND FREQUENCIES

		OBSERVED		PREDICTED		
AMPLE	FUNCTION NUMBER	FUNCTION	STANDARD ERROR	FUNCTION	STANDARD ERROR	RESIDUAL
1	1	0	0.57735	-.022473	0.473928	.0224729
	2	0.980829	0.478714	0.606136	0.358854	0.374693
	3	-1.09861	0.816497	-1.32176	0.588056	0.223144
	4	-.405465	0.645497	-.693147	0.499978	0.287682
	5	-1.09861	0.816497	-1.32176	0.588056	0.223144
	6	-.405465	0.645497	-.693147	0.499978	0.287682
	7	0	0.57735	-.628609	0.309567	0.628609
	F1	6	2.28416	7.65217	1.94305	-1.65217
	F2	16	3.23029	14.3478	2.69608	1.65217
	F3	2	1.38313	2.08696	0.899293	-.086957
	F4	4	1.91107	3.91304	1.54809	.0869566
	F5	2	1.38313	2.08696	0.899293	-.086957
	F6	4	1.91107	3.91304	1.54809	.0869566
	F7	6	2.28416	4.17391	1.3353	1.82609
	F8	6	2.28416	7.82609	2.11709	-1.82609

Iterative proportional fitting of the model is also available, and may be desirable for those situations in which the contingency table is very large and the hierarchal model contains a great many parameters. For this example, the required control statements for IPF estimation of the parameters are

```
PROC IML;
   TITLE2 'IPF ESTIMATION OF THE FREQUENCIES';
   DIM = {2 2 2};
   TABLE = { 6 16 2 4 ,
             2  4 6 6 };
   CONFIG = { 1 2 ,
              0 3 };
   CALL IPF(FIT,STATUS,DIM,TABLE,CONFIG);
   PRINT "OBSERVED FREQUENCIES ";   PRINT TABLE;
   PRINT "ESTIMATED FREQUENCIES";
   PRINT FIT{FORMAT=7.5};
```

Figure 15 shows that the cell frequencies predicted from the IPF algorithm are identical to those obtained from the Newton-Raphson algorithm in Figure 14.

Figure 15
ONE-POPULATION DRUG STUDY
IPF ESTIMATION OF THE FREQUENCIES

OBSERVED FREQUENCIES

TABLE	COL1	COL2	COL3	COL4
ROW1	6.0000	16.0000	2.0000	4.0000
ROW2	2.0000	4.0000	6.0000	6.0000

ESTIMATED FREQUENCIES

FIT	COL1	COL2	COL3	COL4
ROW1	7.65217	14.3478	2.08696	3.91304
ROW2	2.08696	3.91304	4.17391	7.82609

Since the IPF analysis yields only the estimated cell frequencies, one might be interested in running a general linear model analysis of the predicted cell frequencies in order to obtain other useful information such as (1) Wald statistics for the individual effects in the model, and (2) the maximum-likelihood estimate of the covariance matrix of the estimated parameters. The required control statements are the same as those used previously, except that the observed frequencies are replaced by the predicted frequencies in the WEIGHT statement:

```
*-------------------------------------------*;
DATA PREDICT; SET PRED; IF _TYPE_='FREQ';
DATA DRUG2; MERGE DRUGS PREDICT;
*-------------------------------------------*;
PROC CATMOD;
  WEIGHT _PRED_;
  MODEL DRUGA*DRUGB*DRUGC = _RESPONSE_
   / COV ML COVB PRED=FREQ;
  REPEATED
   / _RESPONSE_ = DRUGA DRUGB DRUGC DRUGA*DRUGB;
  TITLE2 'ANALYSIS OF IPF-ESTIMATED FREQUENCIES';
```

The results, shown in Figures 16 and 17, are essentially the same as the previous results, except that only one Newton-Raphson iteration is required for convergence, and the goodness-of-fit statistic is zero, as are the residuals.

Figure 16
ONE-POPULATION DRUG STUDY
ANALYSIS OF IPF-ESTIMATED FREQUENCIES

MAXIMUM LIKELIHOOD ANALYSIS

ITERATION	SUB ITERATION	-2 LOG LIKELIHOOD	CONVERGENCE CRITERION
0	0	173.029	1
1	0	173.029	0

PARAMETER ESTIMATES

ITERATION	1	2	3	4
0	0.151534	0.151534	-0.314304	0.498108
1	0.151534	0.151534	-0.314304	0.498108

ANALYSIS OF VARIANCE TABLE

SOURCE	DF	CHI-SQUARE	PROB
DRUGA	1	0.80	0.3726
DRUGB	1	0.80	0.3726
DRUGC	1	4.12	0.0423
DRUGA*DRUGB	1	8.59	0.0034
LIKELIHOOD RATIO	3	0.00	1.0000

Figure 17
ONE POPULATION DRUG STUDY
ANALYSIS OF IPF-ESTIMATED FREQUENCIES

COVARIANCE OF ESTIMATES

	1	2	3	4
1	0.0288826	-0.0127841	0	-.00236742
2	-0.0127841	0.0288826	0	-.00236742
3	0	0	0.0239583	0
4	-.00236742	-.00236742	0	0.0288826

PREDICTED VALUES FOR RESPONSE FUNCTIONS AND FREQUENCIES

		OBSERVED		PREDICTED		
SAMPLE	FUNCTION NUMBER	FUNCTION	STANDARD ERROR	FUNCTION	STANDARD ERROR	RESIDUAL
1	1	-.022473	0.508389	-.022473	0.473942	0
	2	0.606136	0.444381	0.606136	0.35887	0
	3	-1.32176	0.779066	-1.32176	0.588076	0
	4	-.693147	0.619139	-.693147	0.5	0
	5	-1.32176	0.779066	-1.32176	0.588076	0
	6	-.693147	0.619139	-.693147	0.5	0
	7	-.628609	0.606103	-.628609	0.30957	0
	F1	7.65217	2.52571	7.65217	1.94305	0
	F2	14.3478	3.14207	14.3478	2.69608	0
	F3	2.08696	1.41148	2.08696	0.899302	0
	F4	3.91304	1.89214	3.91304	1.54811	0
	F5	2.08696	1.41148	2.08696	0.899302	0
	F6	3.91304	1.89214	3.91304	1.54811	0
	F7	4.17391	1.94812	4.17391	1.33536	0
	F8	7.82609	2.54845	7.82609	2.11722	0

6. MODELS FOR ORDINAL DATA

A recent book by Agresti(1984) focuses on analysis methods that can be used whenever there are ordinally-scaled variables to be analyzed. Two of the primary methods of analysis recommended by Agresti can be done with the CATMOD procedure:

- log-linear models, using generalized logits
- logit models, using cumulative logits.

Cumulative logits (logits of cumulative probabilities) are monotonically increasing (or decreasing), so that the ordinal nature of the dependent variable is automatically incorporated into those functions. Generalized logits, on the other hand, do not inherently reflect the ordinal nature of the dependent variable, and therefore the ordinality must be built into the design matrix in a general linear model.

Regardless of whether the model is logit or log-linear, structural models can be built that reflect certain hypotheses and take into account the scaling of other variables in the analysis. The following discussion assumes a two-way table, with the dependent (column) variable always presumed to be ordinally scaled. Three of the most important structural models are as follows.

- INDEPENDENCE MODEL --- For the log-linear model, this structural model implies an odds ratio of 1 for every choice of two rows and two columns. For the logit model, it implies an odds ratio of 1 for every possible dichotomy of the column variable and every pair of rows.

- ROW-EFFECTS MODEL --- This model is used when the row variable lies on a nominal scale. Compared to the independence model, it contains one additional parameter for each row. For a log-linear model with integer column scores, it implies that the odds ratio for 2 adjacent columns and for any 2 rows is a function of the difference between the row parameters. For a logit model, it implies that the odds ratio for any 2 rows is a function of the difference between the row parameters, regardless of which collapsing is used to form a dichotomy of the column variable.

- UNIFORM-ASSOCIATION MODEL ---This model is used when the row variable lies on an ordinal scale. Compared to the independence model, it contains one additional parameter, β, that measures the association between the two variables. For a log-linear model with integer row and column scores, it implies that the odds ratio for any 2 adjacent columns and any 2 adjacent rows is exp(β). Such a model is also called an equal adjacent odds ratio model. For a logit model with integer row scores, it implies that the odds ratio for any 2 adjacent rows is exp(β), regardless of which collapsing is used to form a dichotomy of the column variable. Such a model is also called a proportional odds model.

All of these ordinal models can be generalized to the case of multiple variables.

Example

The methods are illustrated with the dumping syndrome data introduced in Section 3. The dependent variable, severity, is ordinally scaled (with values NONE, SLIGHT, and MODERATE), and the independent variable, treatment, is also ordinally scaled since the treatments correspond to the percentage of the stomach removed during a surgical operation (0, 25, 50, 75). Thus, a uniform-association model is most appropriate for these data, and that is the type of structural model fitted here. The variable HOSPITAL is ignored in order to illustrate the two-variable models.

Figure 18 shows the control statements required to fit the log-linear uniform-association model. The third column of the design matrix reflects, in a multiplicative way, the ordinal scales of the variables treatment and severity { (2 1) kronecker (1 2 3 4) }. Figure 19 displays the results of the maximum-likelihood analysis, showing that the model fits well. The operation effect is now significant (p=.01) due to the facts that the ordinal nature of treatment has been exploited and there is some linear association between treatment and severity. The maximum-likelihood estimate of the uniform-association parameter β (-.162) converts to a uniform odds ratio estimate of exp(-.162)=0.85 .

Figure 18

```
PROC CATMOD ORDER=DATA;
   TITLE2 'LOGLINEAR UNIFORM ASSOCIATION';
   WEIGHT WT;
   POPULATION TRT;
   MODEL  SEVERITY  =  ( 1  0  2 ,
                         0  1  1 ,
                         1  0  4 ,
                         0  1  2 ,
                         1  0  6 ,
                         0  1  3 ,
                         1  0  8 ,
                         0  1  4 )
    ( 1 = 'INTERCEPT1',
      2 = 'INTERCEPT2',
      3 = 'OPERATION' )
    / FREQ ONEWAY ML PREDICT=FREQ;
```

Figure 19
ANALYSIS OF DUMPING SYNDROME DATA
LOGLINEAR UNIFORM ASSOCIATION

ANALYSIS OF VARIANCE TABLE

SOURCE	DF	CHI-SQUARE	PROB
INTERCEPT1	1	39.74	0.0001
INTERCEPT2	1	31.47	0.0001
OPERATION	1	6.15	0.0132
LIKELIHOOD RATIO	5	4.59	0.4680

ANALYSIS OF INDIVIDUAL PARAMETERS

EFFECT	PARAMETER	ESTIMATE	STANDARD ERROR	CHI-SQUARE	PROB
MODEL	1	2.4672	0.391377	39.74	0.0001
	2	1.43336	0.255521	31.47	0.0001
	3	-.162621	.0655858	6.15	0.0132

Figure 20 shows the control statements required to fit the logit uniform-association model. Figure 21 displays the results of the weighted-least-squares analysis, showing that the estimate and the test of β are very similar to those obtained from the log-linear model. Although the first two columns of the design matrix are parameterized differently than those in the log-linear model, they span the same space.

Figure 20

```
PROC CATMOD ORDER=DATA;
   TITLE2 'LOGIT UNIFORM ASSOCIATION';
   WEIGHT WT;
   RESPONSE    1 -1  0  0 ,
               0  0  1 -1    LOG     1 0 0 ,
                                     0 1 1 ,
                                     1 1 0 ,
                                     0 0 1 ;
   DIRECT TRTMNT;
   MODEL SEVERITY =  _RESPONSE_  TRTMNT;
```

Figure 21
ANALYSIS OF DUMPING SYNDROME DATA
LOGIT UNIFORM ASSOCIATION

CATMOD PROCEDURE

SAMPLE	FUNCTION NUMBER	RESPONSE FUNCTION	DESIGN MATRIX 1	2	3
1	1	-0.555526	1	1	1
	2	-2.54273	1	-1	1
2	1	-0.635989	1	1	2
	2	-1.94591	1	-1	2
3	1	-0.109199	1	1	3
	2	-2.10006	1	-1	3
4	1	0.0186921	1	1	4
	2	-1.73827	1	-1	4

ANALYSIS OF VARIANCE TABLE

SOURCE	DF	CHI-SQUARE	PROB
INTERCEPT	1	45.73	0.0001
RESPONSE	1	142.29	0.0001
TRTMNT	1	6.37	0.0116
RESIDUAL	5	4.57	0.4712

(Fig. 21 continued on next page)

(Fig. 21 continued from previous page)

ANALYSIS OF INDIVIDUAL PARAMETERS

EFFECT	PARAMETER	ESTIMATE	STANDARD ERROR	CHI-SQUARE	PROB
INTERCEPT	1	1.73389	0.256394	45.73	0.0001
RESPONSE	2	-.855483	.0717182	142.29	0.0001
TRTMNT	3	-0.22157	.0877552	6.37	0.0116

Figure 22 shows the summary of tests of the uniform-association parameter β, all obtained from CATMOD. Regardless of whether one uses a logit or a log-linear model, maximum-likelihood or weighted-least-squares estimation, Wald or likelihood-ratio tests, the results are essentially the same. A similar conclusion can be drawn from Figure 23, which displays the results of the estimation of β and the uniform odds ratio, exp(β).

Figure 22
Analysis of Dumping Syndrome Data
Results of Testing the Uniform Association Parameter β

Type of Analysis	Type of Estimation	Type of Test Statistic	Test Statistic	Prob
Loglinear	WLS	Wald	5.96	0.01
Loglinear	MLE	Wald	6.15	0.01
Loglinear	MLE	LRT*	6.29	0.01
Logit	WLS	Wald	6.37	0.01

*LRT = G^2(Independence) - G^2(Uniform Association Model)
= 10.88 - 4.59
= 6.29

Figure 23
Analysis of Dumping Syndrome Data
Results of Estimating the Uniform Association Parameter

Type of Analysis	Type of Estimation	Estimate of β	Standard Error	exp(β)
Loglinear	WLS	-0.160	0.066	0.85
Loglinear	MLE	-0.163	0.066	0.85
Logit	WLS	-0.222	0.088	0.80

Another powerful method of dealing with an ordinal dependent variable is to analyze the mean score of that variable for each population, rather than analyzing a set of logits (Grizzle, Starmer, and Koch 1969). If an independent variable is nominally scaled in such an analysis, then it is treated as a main effect, and the analysis is sensitive to differences among the levels of that variable with respect to the mean scores. If it is ordinally scaled, then it is treated in a quantitative way by a single column in the design matrix, and the extent of its linear association with the dependent variable is measured by the corresponding parameter.

In the following analysis of the dumping syndrome data, the scores for SEVERITY are 0 for none, 0.5 for slight, and 1 for moderate. The variables hospital and treatment are both regarded as nominally scaled:

```
PROC CATMOD ORDER=DATA;
   WEIGHT WT;
   RESPONSE  0  0.5  1;
   MODEL SEVERITY = TRT HOSPITAL;
   TITLE2 'MAIN-EFFECTS MODEL';
```

The results, shown in Figure 24, indicate a significant treatment effect. However, if treatment is regarded as ordinally scaled (by its appearance in a DIRECT statement):

```
PROC CATMOD ORDER=DATA;
   WEIGHT WT;
   DIRECT TRTMNT;
   RESPONSE  0  0.5  1;
   MODEL SEVERITY = TRTMNT HOSPITAL;
   TITLE2 'LINEAR OPERATION EFFECT';
```

then the results (Figure 25) show even stronger evidence of association.

Figure 24
ANALYSIS OF DUMPING SYNDROME DATA
MAIN-EFFECTS MODEL

ANALYSIS OF VARIANCE TABLE

SOURCE	DF	CHI-SQUARE	PROB
INTERCEPT	1	248.77	0.0001
TRT	3	8.90	0.0307
HOSPITAL	3	2.33	0.5065
RESIDUAL	9	6.33	0.7069

Figure 25
ANALYSIS OF DUMPING SYNDROME DATA
LINEAR OPERATION EFFECT

ANALYSIS OF VARIANCE TABLE

SOURCE	DF	CHI-SQUARE	PROB
INTERCEPT	1	18.28	0.0001
TRTMNT	1	8.60	0.0034
HOSPITAL	3	2.31	0.5098
RESIDUAL	11	6.63	0.8284

7. REPEATED MEASURES ANALYSIS

The CATMOD procedure has a number of features that facilitate repeated measures analysis. They include

- a REPEATED statement that allows one to

- identify and name repeated measurement factors
- a very general modeling specification that allows repeated measures to be modeled in any fashion
- shorthand specification of commonly used response functions in repeated measures analysis, such as marginal probabilities, marginal logits, and means.

Repeated measures methodology and the corresponding CATMOD capabilities are reviewed in Stanish and Koch (1984). Numerous examples are given there and in the SAS User's Guide: Statistics(1985). The following example is a simple illustration of an analysis of marginal probabilities.

Example

These data are from a study of the effect of advertising on sales (Bishop, Fienberg, and Holland 1975, p. 274). At each of two time points, subjects were asked if they had seen an advertisement for a specific product and if they had bought that product. The question of interest is: what is the effect on sales of the time between the two interviews, seeing the first advertisement, and seeing the second advertisement.

The first model is simply a saturated model to assess the significance of the main effects and interactions of the independent variables and the repeated measurement factor. The required control statements to read the data and fit the saturated model are as follows:

```
DATA A;
   INPUT SEE1 $ SEE2 $ BUY1 $ BUY2 $ COUNT @@;
   CARDS;
NO  NO  YES YES  95       NO  NO  YES NO   15
NO  NO  NO  YES   6       NO  NO  NO  NO  493
YES YES YES YES  83       YES YES YES NO    8
YES YES NO  YES  22       YES YES NO  NO   68
YES NO  YES YES  35       YES NO  YES NO    7
YES NO  NO  YES  11       YES NO  NO  NO   28
NO  YES YES YES  25       NO  YES YES NO   10
NO  YES NO  YES   8       NO  YES NO  NO   32

PROC CATMOD ORDER=DATA;
   WEIGHT COUNT;
   RESPONSE MARGINALS;
   MODEL BUY1*BUY2 = SEE1|SEE2|_RESPONSE_;
   REPEATED TIME 2;
   TITLE 'ADVERTISING DATA---SATURATED MODEL';
```

The results, shown in Figure 26, indicate that some of the interactions are nonsignificant($p>.10$). That fact, together with an examination of the marginal probabilities of buying at the two time points, leads one to a reduced model that contains two primary effects. One is an effect due to seeing at least one ad, which may reflect, in part, exposure to the medium (or media) in which the ads appear. The other is an incremental effect of the first ad on the probability of buying the second product, which may reflect, in part, exposure to the company selling the products. The control statements required to fit this reduced model are

```
PROC CATMOD ORDER=DATA;
   WEIGHT COUNT;
   POPULATION SEE1 SEE2;
   RESPONSE MARGINALS;
   MODEL BUY1*BUY2 = ( 1 0 0 ,
                       1 0 0 ,
                       1 1 0 ,
                       1 1 0 ,
                       1 1 0 ,
                       1 1 1 ,
                       1 1 0 ,
                       1 1 1 )
        ( 1 = 'P(BUY1 | NO ADS SEEN)',
          2 = 'SEEING AT LEAST ONE AD',
          3 = 'EFFECT OF AD #1 ON BUY2')
        / FREQ PRED;
   TITLE 'ADVERTISING DATA---REDUCED MODEL';
```

The results of fitting the reduced model are shown in Figures 27 and 28. The four populations are based on whether or not the subjects saw the two advertisements. The printed response functions are the marginal probabilities of buying the two products. The analysis-of-variance table indicates that the model fits ($p=.40$) and that all of the effects are statistically significant ($p<.05$). The parameter estimates and the predicted marginal probabilities are given in Figure 28.

Figure 26
ADVERTISING DATA---SATURATED MODEL

ANALYSIS OF VARIANCE TABLE

SOURCE	DF	CHI-SQUARE	PROB
INTERCEPT	1	468.85	0.0001
SEE1	1	33.60	0.0001
SEE2	1	12.40	0.0004
SEE1*SEE2	1	12.72	0.0004
TIME	1	1.06	0.3025
SEE1*_RESPONSE_	1	4.13	0.0420
SEE2*_RESPONSE_	1	0.04	0.8459
SEE1*SEE2*_RESPONSE_	1	0.23	0.6300
RESIDUAL	0	0.00	1.0000

NOTE: _RESPONSE_ = TIME

Figure 27
ADVERTISING DATA---REDUCED MODEL

POPULATION PROFILES

SAMPLE	SEE1	SEE2	SAMPLE SIZE
1	NO	NO	609
2	NO	YES	75
3	YES	NO	81
4	YES	YES	181

(Fig. 27 continued on next page)

(Fig. 27 continued from previous page)

RESPONSE FREQUENCIES

SAMPLE	RESPONSE NUMBER 1	2	3	4
1	95	15	6	493
2	25	10	8	32
3	35	7	11	28
4	83	8	22	68

SAMPLE	FUNCTION NUMBER	RESPONSE FUNCTION	DESIGN MATRIX 1	2	3
1	1	0.180624	1	0	0
	2	0.165846	1	0	0
2	1	0.466667	1	1	0
	2	0.44	1	1	0
3	1	0.518519	1	1	0
	2	0.567901	1	1	1
4	1	0.502762	1	1	0
	2	0.58011	1	1	1

ANALYSIS OF VARIANCE TABLE

SOURCE	DF	CHI-SQUARE	PROB
P(BUY1 | NO ADS SEEN)	1	133.21	0.0001
SEEING AT LEAST ONE AD	1	113.38	0.0001
EFFECT OF AD #1 ON BUY2	1	9.10	0.0026
RESIDUAL	5	5.15	0.3973

Figure 28
ADVERTISING DATA---REDUCED MODEL

ANALYSIS OF INDIVIDUAL PARAMETERS

EFFECT	PARAMETER	ESTIMATE	STANDARD ERROR	CHI-SQUARE	PROB
MODEL	1	0.171152	.0148293	133.21	0.0001
	2	0.32091	.0301385	113.38	0.0001
	3	.0758221	.0251339	9.10	0.0026

PREDICTED VALUES FOR RESPONSE FUNCTIONS

SAMPLE	FUNCTION NUMBER	OBSERVED FUNCTION	OBSERVED STANDARD ERROR	PREDICTED FUNCTION	PREDICTED STANDARD ERROR	RESIDUAL
1	1	0.180624	.0155891	0.171152	.0148293	.0094719
	2	0.165846	.0150719	0.171152	.0148293	-.005306
2	1	0.466667	.0576066	0.492062	.0262379	-.025395
	2	0.44	.0573178	0.492062	.0262379	-.052062
3	1	0.518519	.0555174	0.492062	.0262379	.0264564
	2	0.567901	.0550409	0.567884	.0286547	1.7E-05
4	1	0.502762	.0371641	0.492062	.0262379	.0107003
	2	0.58011	.0366846	0.567884	.0286547	.0122262

8. COMPLEX SAMPLE SURVEY DATA ANALYSIS

Currently, the CATMOD procedure is not suited for the analysis of complex sample survey data because CATMOD computes covariance estimates under the assumption that the frequencies were obtained by stratified simple random sampling. However, the IML procedure can be used for such an analysis because it contains a very powerful programming language. This makes it straightforward to program a general linear modeling algorithm with any desired capabilities. Since there are already SAS procedures available to compute weighted probability and covariance estimates for complex sample survey applications (PROC SURREGR© and PROC SESUDAAN©), the IML program could be used to read a function vector and its estimated covariance matrix, and then do general linear modeling of the function vector. Such an IML program has been written, and it is listed in the appendix.

Example

These data are from the blood lead subsample of Second National Health and Nutrition Examination Survey (denoted NHANES II, Reference: McDowell, et al. 1981). Only the data for persons under age 18 in one stratum (out of 32) are considered here (Landis and Lepkowski 1984). The levels of the dependent and independent variables are given in Figure 29. The question of interest is: to what extent are the variables age, race, and income related to the presence of elevated levels of lead in the blood?

The weighted probability and covariance estimates were computed with PROC SURREGR (Landis and Lepkowski 1984). The IML program WLS was then run with three input data sets in order to fit a saturated model via weighted least squares.

- One data set, called INPUT, contains the proportion vector and its estimated covariance matrix. For this example, the estimates were typed in directly (Figure 30), but ordinarily, the estimates would be contained in an output data set created by PROC SURREGR.
- Another data set, called DESIGN, contains the design matrix (Figure 31).
- A third data set, called TEST, contains C matrices for testing the hypotheses $C\beta=0$, together with labels for the hypotheses (Figure 32).

Figure 33 shows that the analysis is invoked simply by calling the IML program WLS. Figure 33 also displays the control statements required to fit a reduced model.

Figures 34 and 35 give the results of the saturated model analysis. Included in the output are the estimated parameters and their standard errors, the predicted functions and their standard errors, the goodness-of-fit test, and the analysis-of-variance table that contains tests for all the $C\beta=0$ hypotheses specified in the TEST data set.

The results of the reduced model analysis, shown in Figure 36, indicate that the fit of the model is barely adequate (p=.09). The age, race, and income effects are all statistically significant (α=.05), with the race and income effects being significantly more important for the younger age group.

Figure 29
NHANES II BLOOD LEAD SAMPLE

Dependent Variable	Levels	
Blood Lead Level	<20 µg/dl	20+

Independent Variables	Levels	
Age	<=6	6--17
Race	Black	White
Income	<$10,000	$10,000+

Figure 30

```
*---CREATE DATASET FOR THE PROPORTIONS AND THE COVARIANCE MATRIX---;
DATA INPUT;
   INPUT P1-P8;
   CARDS;

 0.590463217E+00   0.384720414E+00   0.273269035E+00   0.136457611E+00
 0.175965918E+00   0.154594680E+00   0.109570229E+00   0.426890868E-01

 0.168974410E-02   0.910836183E-03   0.473521399E-03   0.432496330E-03
 0.253181532E-03   0.590977558E-03   0.288562763E-04   0.108497325E-03

 0.910836183E-03   0.182321079E-02   0.306286708E-04   0.101834710E-03
-0.269055648E-03   0.884567192E-03  -0.143051974E-03  -0.432938662E-04

 0.473521399E-03   0.306286708E-04   0.105623295E-02   0.476541924E-03
 0.194276655E-03  -0.775871606E-04   0.378714792E-03   0.170211691E-03

 0.432496330E-03   0.101834710E-03   0.476541924E-03   0.396282470E-03
 0.229308563E-03   0.118240173E-03   0.254232226E-03   0.143883020E-03

 0.253181532E-03  -0.269055648E-03   0.194276655E-03   0.229308563E-03
 0.137699912E-02   0.525303046E-03   0.227391392E-03   0.174629455E-03

 0.590977558E-03   0.884567192E-03  -0.775871606E-04   0.118240173E-03
 0.525303046E-03   0.196178735E-02   0.314218078E-04   0.837499251E-04

 0.288562763E-04  -0.143051974E-03   0.378714792E-03   0.254232226E-03
 0.227391392E-03   0.314218078E-04   0.423029598E-03   0.987408359E-04

 0.108497325E-03  -0.432938662E-04   0.170211691E-03   0.143883020E-03
 0.174629455E-03   0.837499251E-04   0.987408359E-04   0.117652321E-03
;
```

Figure 31

```
*---CREATE DATASET FOR THE DESIGN MATRIX---;

DATA DESIGN;
 INPUT X1-X8;
 CARDS;
      1  1  1  1  1  1  1  1
      1  1  1 -1  1 -1 -1 -1
      1  1 -1  1 -1  1 -1 -1
      1  1 -1 -1 -1 -1  1  1
      1 -1  1  1 -1 -1  1 -1
      1 -1  1 -1 -1  1 -1  1
      1 -1 -1  1  1 -1 -1  1
      1 -1 -1 -1  1  1  1 -1
```

Figure 32

```
*-----CREATE DATASET FOR THE HYPOTHESIS TESTS-----;

DATA TEST;
   INPUT TITLE $ 1-24 N C1-C8;
   CARDS;
AGE                       1   0  1  0  0  0  0  0  0
RACE                      1   0  0  1  0  0  0  0  0
INCOME                    1   0  0  0  1  0  0  0  0
AGE * RACE                1   0  0  0  0  1  0  0  0
AGE * INCOME              1   0  0  0  0  0  1  0  0
RACE * INCOME             1   0  0  0  0  0  0  1  0
AGE * RACE * INCOME       1   0  0  0  0  0  0  0  1
ALL INTERACTIONS ZERO     4   0  0  0  0  1  0  0  0
                          .   0  0  0  0  0  1  0  0
                          .   0  0  0  0  0  0  1  0
                          .   0  0  0  0  0  0  0  1
;
```

Figure 33

```
*------CALL THE MACRO TO DO THE ANALYSIS------;
TITLE 'ANALYSIS OF COMPLEX SAMPLE SURVEY DATA';
TITLE2 'SATURATED MODEL';
WLS


*----FIT A NEW MODEL---CHANGE DESIGN MATRIX----;
TITLE2 'NESTED MAIN EFFECTS MODEL';
DATA DESIGN;
   INPUT X1-X6;
   CARDS;
         1  1  1  0  1  0
         1  1  1  0 -1  0
         1  1 -1  0  1  0
         1  1 -1  0 -1  0
         1 -1  0  1  0  1
         1 -1  0  1  0 -1
         1 -1  0 -1  0  1
         1 -1  0 -1  0 -1
;

*--CREATE DATASET FOR THE NEW HYPOTHESIS TESTS--;
DATA TEST;
   INPUT TITLE $ 1-28 N C1-C6;
   CARDS;
MODEL|INTERCEPT              5   0  1  0  0  0  0
                             .   0  0  1  0  0  0
                             .   0  0  0  1  0  0
                             .   0  0  0  0  1  0
                             .   0  0  0  0  0  1
AGE                          1   0  1  0  0  0  0
RACE(AGE)                    2   0  0  1  0  0  0
                             .   0  0  0  1  0  0
RACE(AGE GROUP 1)            1   0  0  1  0  0  0
RACE(AGE GROUP 2)            1   0  0  0  1  0  0
RACE(AGE1) = RACE(AGE2)      1   0  0  1 -1  0  0
INCOME(AGE)                  2   0  0  0  0  1  0
                             .   0  0  0  0  0  1
INCOME(AGE GROUP 1)          1   0  0  0  0  1  0
INCOME(AGE GROUP 2)          1   0  0  0  0  0  1
INCOME(AGE1) = INCOME(AGE2)  1   0  0  0  0  1 -1
;


*----CALL THE MACRO TO DO THE ANALYSIS----;
WLS
```

Figure 34
Analysis of Complex Sample Survey Data
Saturated Model

Estimated Parameters	Std Errors	Predicted Functions	Std Errors
0.2335	0.0184	0.5905	0.0411
0.1128	0.0127	0.3847	0.0427
0.0930	0.0147	0.2733	0.0325
0.0539	0.0108	0.1365	0.0199
0.0484	.0094979	0.1760	0.0371
0.0318	.0069225	0.1546	0.0443
.0029277	.0095636	0.1096	0.0206
0.0143	.0067561	0.0427	0.0108

Figure 35
ANALYSIS OF COMPLEX SAMPLE SURVEY DATA
SATURATED MODEL

ANALYSIS OF VARIANCE TABLE

SOURCE	DF	CHI-SQR	PROB
AGE	1	78.7373	0.0001
RACE	1	40.2621	0.0001
INCOME	1	24.9236	0.0001
AGE * RACE	1	25.9616	0.0001
AGE * INCOME	1	21.0859	0.0001
RACE * INCOME	1	0.0937	0.7595
AGE * RACE * INCOME	1	4.4832	0.0342
ALL INTERACTIONS ZERO	4	51.0695	0.0001

Figure 36
ANALYSIS OF COMPLEX SAMPLE SURVEY DATA
NESTED MAIN EFFECTS MODEL

ESTIMATED PARAMETERS (B)

0.2220
0.1138
0.1325
0.0365
0.0786
0.0313

GOODNESS-OF-FIT TEST

SOURCE	DF	CHI-SQR	PROB
RESIDUAL	2	4.7773	0.0918

(Fig. 36 continued)

ANALYSIS OF VARIANCE TABLE

SOURCE	DF	CHI-SQR	PROB
MODEL|INTERCEPT	5	269.8816	0.0001
AGE	1	82.3638	0.0001
RACE(AGE)	2	58.1268	0.0001
RACE(AGE GROUP 1)	1	57.8449	0.0001
RACE(AGE GROUP 2)	1	5.5449	0.0185
RACE(AGE1) = RACE(AGE2)	1	27.0114	0.0001
INCOME(AGE)	2	63.0541	0.0001
INCOME(AGE GROUP 1)	1	61.3218	0.0001
INCOME(AGE GROUP 2)	1	12.2894	0.0005
INCOME(AGE1) = INCOME(AGE2)	1	17.2909	0.0001

9. ACKNOWLEDGEMENTS

The author is indebted to Dr. J. Richard Landis, who organized the categorical data analysis session, and who provided a copy of his tutorial (with Lepkowski) on complex sample survey data analysis. The Graphic Arts Department of SAS Institute was instrumental in preparing the camera-ready copy of this manuscript.

SAS is the registered trademark of SAS Institute Inc., Cary, NC, USA

PROC SURREGR and PROC SESUDANN are copyrighted by Research Triangle Institute.

10. REFERENCES

1. Agresti, Alan (1984), Analysis of Ordinal Categorical Data, New York: John Wiley & Sons, Inc.

2. Birch, M. W. (1965), "The Detection of Partial Association, II: the General Case," Journal of the Royal Statistical Society, B, 27, 111-124.

3. Bishop, Y., Fienberg, S.E., and Holland, P.W. (1975), Discrete Multivariate Analysis: Theory and Practice, Cambridge: The MIT Press.

4. Breslow, N. E. and Day, N. E. (1980), Statistical Methods in Cancer Research, Volume 1: The Analysis of Case-Control Studies, Lyon, International Agency for Research on Cancer.

5. Cochran, William G. (1954), "Some Methods for Strengthening the Common χ^2 Tests," Biometrics, 10, 417-451.

6. Cox, D.R. (1970), The Analysis of Binary Data, New York: Halsted Press.

7. Fienberg, S.E. (1980), The Analysis of Cross-Classified Categorical Data, Second Edition, Cambridge, Massachusetts: The MIT Press.

8. Goodman, L.A. and Kruskal, W.H. (1979), Measures of Association for Cross Classification, New York: Springer-Verlag. (Reprints of JASA articles above).

9. Grizzle, J.E., Starmer, C.F., and Koch, G.G. (1969), "Analysis of Categorical Data by Linear Models," Biometrics 25, 489-504.

10. Haldane, J.B.S. (1955), "The Estimation and Significance of the the Logarithm of a Ratio of Frequencies," Annals of Human Genetics, 20, 309-314.

11. Imrey, P.B. (1985), "SAS Software for Log-Linear Models," Proceedings of the Tenth Annual SAS Users Group International Conference, Cary, NC: SAS Institute Inc.

12. Imrey, P.B., Koch, G.G., and Stokes, M.E. (1981), "Categorical Data Analysis: Some Reflections on the Log Linear Model and Logistic Regression. Part I: Historical and Methodological Overview," International Statistical Review 49, 265-283.

13. Imrey, P.B., Koch, G.G., and Stokes, M.E. (1982), "Categorical Data Analysis: Some Reflections on the Log Linear Model and Logistic Regression. Part II: Data Analysis," International Statistical Review 50, 35-63.

14. Kleinbaum, David G., Kupper, Lawrence L., and Morgenstern, Hal (1982) Epidemiologic Research: Principles and Quantitative Methods, Belmont, California: Wadsworth, Inc.

15. Koch, G.G., Landis, J.R., Freeman, J.L., Freeman, D.H., and Lehnen, R.G. (1977), "A General Methodology for the Analysis of Experiments with Repeated Measurement of Categorical Data," Biometrics 33, 133-158.

16. Landis, J.R., Heyman, E.R., and Koch, G.G. (1978), "Average Partial Association in Three-way Contingency Tables: a Review and Discussion of Alternative Tests," International Statistical Review, 46,237-254.

17. Landis, J.R. and Koch, G.G. (1977), "The Measurement of Observer Agreement for Categorical Data," Biometrics 33, 159-174.

18. Landis, J.R. and Lepkowski, J.M. (1984). Tutorial on The Analysis of Categorical Data from Complex Sample Surveys. Unpublished.

19. Landis, J.R., Stanish, W.M., Freeman, J.L. and Koch, G.G. (1976), "A Computer Program for the Generalized Chi-Square Analysis of Categorical Data Using Weighted Least Squares, (GENCAT)," Computer Programs in Biomedicine 6, 196-231.

20. Mantel, N. and Haenszel,W. (1959), "Statistical Aspects of the Analysis of Data from Retrospective Studies of Disease," Journal of the National Cancer Institute, 22, 719-748.

21. Mantel, N. (1963), "Chi-square Tests with One Degree of Freedom; Extensions of the Mantel-Haenszel Procedure," Journal of the American Statistical Association, 58, 690-700.

22. McDowell, A., Engel, A., Massey, J.T., and Maurer, K. (1981). "Plan and Operation of the Second National Health and Nutrition Examination Survey, 1976-80." Vital and Health Statistics, Series 1, No. 15, DHHS Publication No.(PHS) 81-1317, Public Health Service, Washington: U.S. Government Printing Office.

23. SAS User's Guide: Statistics (1985), Version 5 Edition, Cary, NC: SAS Institute Inc.

24. Stanish, W.M. and Koch, G.G. (1984), "The Use of CATMOD for Repeated Measurement Analysis of Categorical Data," Proceedings of the Ninth Annual SAS Users Group International Conference, Cary, NC: SAS Institute Inc.

25. Wald, A. (1943), "Tests of Statistical Hypotheses Concerning General Parameters when the Number of Observations is Large," Transactions of the American Mathematical Society 54, 426-482.

26. Woolf, B. (1955), "On Estimating the Relationship between Blood Group and Disease," Annals of Human Genetics, 19, 251-253.

APPENDIX

```
*-----------------------------------------------------------------------;
* DEFINE A MACRO FOR WEIGHTED LEAST SQUARES ANALYSIS THAT CAN ACCEPT   ;
*   DIRECT INPUT OF A FUNCTION VECTOR AND ITS COVARIANCE MATRIX         ;
*-----------------------------------------------------------------------;
MACRO WLS

  *---------------------------------------------------------------------;
  * READ IN THE VECTOR OF PROPORTIONS AND THE COVARIANCE MATRIX         ;
  *---------------------------------------------------------------------;
  PROC IML;                          * INVOKE THE MATRIX PROCEDURE      ;
  USE INPUT; READ ALL INTO IN;       * READ FROM INPUT DATA SET         ;
  Q=NCOL(IN);                        * NUMBER OF PROPORTIONS            ;
  F = (IN(|1,|))';                   * WEIGHTED PROPORTION VECTOR       ;
  VF = IN(|2:Q+1,|);                 * COVARIANCE OF PROPORTIONS        ;
  VF_INV = INV(VF);                  * INVERSE COVARIANCE MATRIX        ;

  *---------------------------------------------------------------------;
  * READ IN DESIGN MATRIX, AND SET UP FOR WEIGHTED LEAST SQUARES        ;
  *---------------------------------------------------------------------;
  USE DESIGN; READ ALL INTO X;       * READ DESIGN MATRIX               ;
  T=NCOL(X);                         * NUMBER OF COLUMNS IN X           ;
  H = X' * VF_INV * X;               * CROSSPRODUCT OF X WITH X         ;
  G = X' * VF_INV * F;               * CROSSPRODUCT OF X WITH F         ;
  TOT = F' * VF_INV * F;             * TOTAL SUM OF SQUARES             ;
  W = (H||G) // (G'||TOT);           * COMPUTE SWEEP MATRIX             ;

  *---------------------------------------------------------------------;
  * GET WEIGHTED LEAST SQUARES SOLUTION AND GOODNESS-OF-FIT TEST        ;
  *---------------------------------------------------------------------;
  B = SWEEP(W,1:T);                  * SWEEP TO SOLVE THE EQUATIONS     ;
  BETA = B(|1:T,T+1|);               * VECTOR OF ESTIMATED PARAMETERS;
  VBETA = B(|1:T,1:T|);              * COVARIANCE MATRIX OF BETA        ;
  SEBETA = SQRT(VECDIAG(VBETA));     * STANDARD ERRORS OF BETA          ;
  RES = B(|T+1,T+1|);                * RESIDUAL SUM OF SQUARES          ;
  RANK = SUM(VECDIAG(VBETA)¬=0);     * COMPUTE DF FOR CHI-SQUARE        ;
  DFRES = Q - RANK;                  * RESIDUAL DEGREES OF FREEDOM      ;
  FHAT = X * BETA;                   * PREDICTED PROPORTIONS            ;
  VFHAT = X * VBETA * X';            * COVARIANCE MATRIX OF FHAT        ;
  SEFHAT = SQRT(VECDIAG(VFHAT));     * STANDARD ERRORS OF FHAT          ;
  START;                             * START IF-THEN-ELSE MODULE        ;
  IF DFRES=0 THEN PVAL=0;            * SPECIAL CASE--SATURATED MODEL    ;
  ELSE PVAL = 1-PROBCHI(RES,DFRES);  * COMPUTE CORRESPONDING P-VALUE    ;
  FINISH; RUN;                       * FINISH IF-THEN-ELSE MODULE       ;

*-----------------------------------------------------------------------;
* PRINT INPUT DATA, DESIGN MATRIX, AND ESTIMATED PARAMETERS             ;
*-----------------------------------------------------------------------;
RESET NONAME;                         * NO NAMES ON MATRICES            ;
PRINT "OBSERVED FUNCTIONS";             PRINT F;                        ;
PRINT " "; PRINT " "                                                    ;
PRINT "COVARIANCE OF FUNCTIONS";        PRINT VF{FORMAT=E12.};          ;
PRINT " "; PRINT " "                                                    ;
PRINT "DESIGN MATRIX";                  PRINT X{FORMAT=BEST8.};         ;
PRINT " "; PRINT " "                                                    ;
PRINT "ESTIMATED PARAMETERS (B)";       PRINT BETA;                     ;
PRINT " "; PRINT " "                                                    ;
PRINT "ESTIMATED STANDARD ERRORS OF B"; PRINT SEBETA;                   ;
PRINT " "; PRINT " ";                                                   ;
PRINT "PREDICTED FUNCTIONS";            PRINT FHAT;                     ;
PRINT " "; PRINT " "                                                    ;
PRINT "ESTIMATED STANDARD ERRORS OF PREDICTED FUNCTIONS";               ;
PRINT SEFHAT; PRINT " ";                PRINT " ";                      ;

*-----------------------------------------------------------------------;
* PRINT INPUT DATA AND RESULTS OF MODEL FITTING                         ;
*-----------------------------------------------------------------------;
SRT = "SOURCE  ";                     * SOURCE OF VARIATION             ;
DFT = "      DF";                     * DEGREES OF FREEDOM              ;
CHT = " CHI-SQR";                     * CHI-SQUARE STATISTIC            ;
PRT = "   PROB       ";               * P-VALUE                         ;
R='RESIDUAL'                                                            ;
START                                                                   ;
IF DFRES>0 THEN DO                                                      ;
  PRINT "GOODNESS-OF-FIT TEST"                                          ;
  PRINT  R{COLNAME=SRT} DFRES{COLNAME=DFT FORMAT=BEST8.}
         RES{COLNAME=CHT FORMAT=8.4} PVAL{COLNAME=PRT FORMAT=8.4};
  END
ELSE PRINT "GOODNESS-OF-FIT TEST PERFECT -- SATURATED MODEL"            ;
FINISH; RUN                                                             ;
```

```
*-----------------------------------------------------------------------;
* DEFINE A HYPOTHESIS TESTING MODULE                                    ;
*-----------------------------------------------------------------------;
START TESTS;                         * START DEFINING THE TEST MODULE;
DO I = 1 TO NROW(CM);                * LOOP THROUGH THE MATRICES        ;
  N = CM(|I,1|);                     * NUMBER OF ROWS IN C MATRIX       ;
  C = CM(|I:I+N-1,2:T+1|);           * SET UP THE C MATRIX              ;
  C = BLOCK(C,1);                    * AUGMENT THE C MATRIX             ;
  WC = C * B * C';                   * COMPUTE THE SWEEP MATRIX         ;
  WC(|N+1,N+1|) = 0;                 * ZERO LOWER RT-HAND ELEMENT       ;
  EC = WC(|1:N,N+1|);                * COMPUTE ESTIMATED CONTRAST       ;
  VEC = WC(|1:N,1:N|);               * COVARIANCE MATRIX OF EC          ;

  *---------------------------------------------------------------------;
  * COMPUTE THE CHI-SQUARE TEST STATISTIC FOR THIS HYPOTHESIS           ;
  *---------------------------------------------------------------------;
  B2 = SWEEP(WC,1:N);                * SWEEP THE MATRIX                 ;
  QC = B2(|N+1,N+1|);                * COMPUTE TEST STATISTIC           ;
  RANK = SUM(VECDIAG(B2)¬=0)-1;      * COMPUTE DF FOR CHI-SQUARE        ;
  PVAL = 1 - PROBCHI(QC,RANK);       * COMPUTE CORRESPONDING P-VALUE    ;
  IF PVAL<.0001 THEN PVAL=.0001;     * ADJUST FOR ROUNDOFF ERROR        ;

  *---------------------------------------------------------------------
  * ACCUMULATE THE RESULTS IN THE ANOVA TABLE
  *---------------------------------------------------------------------
  R = R || TITLE(|I|);               * STORE LABEL IN ROWNAME MATRIX
  SOURCE=SOURCE//(RANK||QC||PVAL);   * STORE RESULTS IN ANOVA TABLE
  I = I + N - 1;                     * INCREMENT TO GET NEXT CONTRAST
  END;                               * END PROCESSING OF CONTRAST
  FINISH;                            * FINISH DEFINING TEST MODULE

*-----------------------------------------------------------------------
* COMPUTE HYPOTHESIS TESTS AND SET UP ANOVA TABLE
*-----------------------------------------------------------------------
USE TEST;                            * MAKE TEST THE CURRENT DATASET
READ ALL INTO CM{ROWNAME=TITLE};     * READ IN ALL C MATRICES
SOURCE = 1:3;                        * INITIALIZE TABLE
RUN TESTS;                           * RUN THE HYPOTHESIS TESTS
R = (R(|,2:NCOL(R)|))';              * DELETE FIRST ELEMENT OF R
SOURCE=SOURCE(|2:NROW(SOURCE),|);    * DELETE FIRST ROW OF TABLE
DF2 = SOURCE(|,1|);                  * COLUMN FOR DEGREES OF FREEDOM
CH2 = SOURCE(|,2|);                  * COLUMN FOR CHI-SQUARE
PR2 = SOURCE(|,3|);                  * COLUMN FOR P-VALUE

*-----------------------------------------------------------------------
* PRINT ANALYSIS OF VARIANCE AND OTHER HYPOTHESIS TESTS
*-----------------------------------------------------------------------
PRINT " "; PRINT " ";
PRINT "ANALYSIS OF VARIANCE TABLE";
PRINT   R{COLNAME=SRT}  DF2{COLNAME=DFT FORMAT=BEST8.}
          CH2{COLNAME=CHT FORMAT=8.4} PR2{COLNAME=PRT FORMAT=8.4};
```

COMPUTER SCIENCE AND STATISTICS:
The Interface, D.M. Allen (ed.)
© Elsevier Science Publishers B.V. (North-Holland), 1986

CATEGORICAL DATA ANALYSIS IN BMDP: PRESENT AND FUTURE

Morton B. Brown

Department of Biostatistics
University of Michigan
Ann Arbor, Michigan 48109

The BMDP series of statistical computer programs currently contains two programs for categorical data analysis. One (P4F) enables the user to analyze two-way frequency tables by various statistics, including measures of association and of prediction, or multiway tables by fitting hierarchical log-linear models. The other (PLR) can be used to fit logistic models to data using arbitrary design matrices, provided the response variable is dichotomous. Both programs have features to build models in a sequential fashion, such as in a stepwise manner.

The development of P4F and its precursors is described in relation to the evolving methodology of analyzing two-way and multiway frequency tables. Issues of computational accuracy are contrasted with those of statistical validity.

A new program for categorical data analysis is being developed. Its features include an ability to fit linear, log-linear and logistic models. The specification of the models will be either by macro-level keywords or by design matrices. Both ordinal and nominal variables can be used in the models. The models will be fitted by either weighted least squares, iteratively reweighted least squares or iterative proportional fitting. Methods for semi-automatic model-building will be included.

1. INTRODUCTION

The availability of computer software for the analysis of data summarized as frequency tables has changed dramatically within the last decade. Prior to 1975 the major software packages only computed statistics for two-way tables, and these were limited to tests for independence (the chi-squared test and Fisher's exact test) and related statistics.

The first major package to provide more general methods to analyze contingency tables was BMDP [9]. Its initial program for frequency table analysis, P1F, was a conversion of a program BMD02S from the earlier Biomedical Computer Programs [8]. In the next six years programs were added and several (including P1F) were made obsolete by the development of P4F (see Table 1).

P1F incorporated measures of association and of optimal prediction for two-way tables, but otherwise remained unchanged from BMD02S. P2F was added to allow models of quasi-independence in the two-way table. Included in P2F were stepwise algorithms for the identification of extreme cells [3]. The third program P3F was developed to fit log-linear models to data in multiway contingency tables using an iterative proportional fitting algorithm [14]. Since BMDP was not an interactive package, the user needed an easy way to identify the subset of models that should be fitted to the data. This led to tests of marginal and partial association [2,5].

Table 1: The development of computer programs for the analysis of frequency tables.

1964		
	BMD02S:	CONTINGENCY TABLE ANALYSIS
1975		
	P1F:	TWO-WAY CONTINGENCY TABLES
1976		
	P1F:	TWO-WAY CONTINGENCY TABLES -- MEASURES OF ASSOCIATION
	P2F:	TWO-WAY CONTINGENCY TABLES -- EMPTY CELLS AND DEPARTURES FROM INDEPENDENCE
	P3F:	MULTIWAY FREQUENCY TABLES -- THE LOG-LINEAR MODEL
1979		
	PLR:	STEPWISE LOGISTIC REGRESSION
1981		
	P4F:	TWO-WAY AND MULTIWAY FREQUENCY TABLES -- MEASURES OF ASSOCIATION AND THE LOG-LINEAR MODEL (COMPLETE AND INCOMPLETE TABLES)

Support of P1F, P2F and P3F was discontinued when P4F was released.

In 1981 P4F was released [10]. P4F combined the strengths of the previous programs (P1F, P2F and P3F) into a single program. In addition to the features described above, it included a more flexible manner of identifying structural zeros, a stepwise algorithm for model selection, methods to identify extreme cells or strata and the Mantel-Haenszel statistic when a set of 2x2 tables are analyzed. Since its release we have made corrections that affect the computations for data in sparse tables [7] and in tables with structural zeros.

2. THE ANALYSIS OF TWO-WAY TABLES

The first version of P1F included many measures of association (or correlation) and prediction. In retrospect, these measures and their standard errors were computed without considering the implications of the sampling framework. For example, the estimate of the standard error of the correlation coefficient used a formula that assumed that the data were normally distributed instead of summarized in a contingency table.

Brown and Benedetti [6] studied various approximations for the standard errors of measures of correlation and association for data summarized as contingency tables. Using the delta method [12,13], they derived asymptotic standard error formulas for the product-moment correlation and Spearman rank correlation. In addition, they found a modification that appeared to be less optimistic when used to test the null hypothesis that the correlation or association is zero.

Brown and Benedetti (unpublished) used the same type of expansion to derive formulas for the asymptotic standard errors of measures of prediction under the null hypothesis and added these formulas to the program in 1977, but unfortunately the small-sample behaviors of these statistics were not checked by simulation at that time. After simulations showed that the test statistics did not have reasonable empirical sizes under the null distribution, these asymptotic standard error formulas for predictive measures were eliminated in 1981.

Table 2: Example of a two-way frequency table from Dixon ([9], page 293)

CELL FREQUENCY COUNTS

	SECTION			
	DR. A	DR. B	DR. C	TOTAL
ATTITUDE				
WORSE	1	1	11	13
WORSE-NC	1	0	10	11
NOCHANGE	8	4	16	28
NC-BETTR	11	7	5	23
BETTER	1	8	3	12
TOTAL	22	20	45	87

Table 2 presents an example of a two-way frequency table from the first version of the BMDP manual. The data in this table are reanalyzed by the current program P4F. Statistics printed by P1F and/or P4F are listed in Table 3. As can be seen from the table, some statistics, primarily those involving standard errors, have changed since the inital release of P1F. The date of the change is indicated.

The only statistics modified were the uncertainty coefficients. In deriving standard errors for these coefficients, Brown [4] noted that the coefficients were not normalized to lie in the range from zero to one. The asymmetric coefficient was unbounded, whereas the symmetric coefficient could not exceed one-half. Modifications in these coefficients were made to normalize them to lie in the range from zero to

Table 3: A comparison of statistics produced by P1F and P4F.

Unchanged:

CHI-SQUARE	MAXIMUM LIKELIHOOD CHI-SQUARE
PHI	CONTINGENCY COEFFICIENT C
CRAMER'S V	YULE'S Q AND Y
CROSS-PRODUCT RATIO	
FISHER'S EXACT TEST (1-TAIL and 2-TAIL)	
MCNEMAR'S TEST OF SYMMETRY	

Added:

TETRACHORIC CORR (added 1977)
RELATIVE RISK -- MANTEL-HAENSZEL (added 1981)
KAPPA (added 1982)

Changed:

A) ASSOCIATION AND CORRELATION

	VALUE	ASE		VAL/ASE0	
Date of release:		1975	1977	1975	1977
CORRELATION	-.374	.101	.082	-3.72	-4.35
SPEARMAN CORR	-.422	.098	.087	-4.29	-4.94
GAMMA	-.478	.100	*	-4.80	-4.81
KENDALL TAU-B	-.344	.098	.073	-4.71	-4.81
STUART TAU-C	-.355	.074	*	-4.81	*
SOMERS D	-.384	.085	*	-4.54	-4.81
"	-.307	.063	*	-4.85	-4.81

B) OPTIMAL PREDICTION AND UNCERTAINTY

	VALUE		ASE		VAL/ASE0	
Date of release:	1975	1977	1975	1977	1975	1981
LAMBDA-SYM	.178	*	.089	*	2.00	N/A
LAMBDA-ASYM	.119	*	.088	*	1.35	N/A
LAMBDA-*-ASYM	.144	.179	.099	.075	1.46	N/A
TAU-ASYM	.094	*	.030	*	3.13	N/A
UNCERTAIN-SYM	.082	.164	.0	.046	0.	N/A
UNCERTAIN-ASYM	1.912	.137	.830	.039	2.30	N/A

* unchanged N/A no longer printed

one. (The change in lambda-star was due to an error in programming.)

Once Brown and Benedetti [6] derived improved estimators for the standard errors under the null hypothesis of the measures of association and correlation, we included two different standard errors (ASE and ASE0) for each statistic. Under the heading ASE is the asymptotic standard error to be used in building confidence intervals for the expected value of the statistic. A test of the hypothesis that the expected value of the statistic is zero is given by the ratio of the statistic to its asymptotic standard error under the null hypothesis (ASE0); this ratio is printed under the heading VAL/ASE0.

The above history raises several issues. The changes in the formulas occurred as a result of work by Benedetti and myself. Some packages avoid the problem by not including standard errors while others use formulas that are inappropriate for the sampling framework. The casual user of a statistical program does not have the ability to evaluate the quality or source of approximations used within a program, especially when asymptotic expansions are involved. Also, it is difficult to check whether formulas are correctly implemented. Although now there are more journals that will accept articles that evaluate the quality of approximations or compare programs, these articles are not read widely by the community that uses these programs for analysis. What are the program developers' responsibilities to the research community that uses and trusts the software developed?

Table 4: Some capabilities of P4F.

FORMS OF INPUT:
- CASEWISE
- AS CELL FREQUENCIES
- AS A MULTIWAY TABLE

TWO-WAY COMPLETE TABLE:
- ALL STATISTICS DESCRIBED ABOVE

TWO-WAY INCOMPLETE TABLE:
- MODELS OF QUASI-INDEPENDENCE
- IDENTIFICATION OF EXTREME CELLS

MULTIWAY TABLES:
- LOG-LINEAR MODELS
- MODEL SCREENING AND BUILDING

- IDENTIFICATION OF EXTREME CELLS
- IDENTIFICATION OF EXTREME STRATA

- SPECIFICATION OF STRUCTURAL ZEROS
- SPECIFICATION OF INITIAL FIT MATRIX

- PARAMETER ESTIMATION OF LOG-LINEAR MODELS
- STD ERRORS FOR THE PARAMETER ESTIMATES
- COVARIANCE MATRIX OF PARAMETER ESTIMATES

- CELL DEVIATES

3. THE CAPABILITIES OF P4F

The program P4F was planned to replace all the categorical programs previously developed (P1F, P2F and P3F). Many of the capabilities of P4F are listed in Table 4.

Since P4F can be used to fit log-linear models to multiway frequency tables, it is often used to analyze or reanalyze data that are already summarized in a (multiway) frequency table. Therefore, three methods of input are acceptable: raw data in a case-by-variable format, processed data as cell indices and frequencies and final data summarized as cell counts in a frequency table.

All the statistics for the two-way table were carried over from P1F to P4F. The Mantel-Haenszel and kappa statistics were added.

A major goal for the development of P4F was to make available to a wide audience the ability to describe the relationships among the factors of a multiway frequency table by log-linear models. There was a need to provide an easy manner to specify models and to identify possible models.

Log-linear models are specified by listing the factors or interactions in the minimal configuration. If a redundant list is provided, the extra terms will be ignored. All models are assumed to be hierarchical. That is, if a higher-order interaction is specified, all lower-order interactions and main effect that are specified by subsets of the interaction are automatically included in the model.

Since there are many possible log-linear model when a table is multidimensional, it was necessary to include some methods that aid in the identification of models. When the table is two- or three-way it is possible to enumerate and evaluate all the possible hierarchical models at a reasonable cost and time. However, for four-way and higher tables it is necessary to screen the interactions for those likely to contribute to the final model. Brown [5] (see also [2]) proposed using tests of marginal and partial association to screen the interactions. These tests are computed by P4F when the appropriate keyword is specified.

In addition, the user can request that effects and/or interactions be added or deleted from a base model in a stepwise manner. This option is very useful when used in conjuction with the tests of marginal and partial association. The tests are used to screen for a starting (base) model and then the stepwise procedure is used to evaluate the effect of adding or deleting terms from the model.

The user can identify cells that are to be treated as structural zeros; these cells are excluded from all analyses. Brown [3] presented an algorithm to identify extreme cells (outliers) such that at each step the most extreme cell was eliminated and treated thereafter as a structural zero. To evaluate the influence of these extreme cells, the expected values of these cells were estimated from the log-linear model fitted to all cells as yet not eliminated and not defined as structural zeros. In P4F each cell defined as a structural zero will have its expected value estimated in the manner described for eliminated cells. This is similar to the calculation of deleted residuals in regression.

The usual manner in which the parameters of the log-linear model are estimated within P4F is by applying the ANOVA formulas to the logarithms of the estimated expected values. This solution is not possible when either structural zeros are specified or at least one of the marginal cells in a configuration of the model is zero; i.e., there are zero expected values. In either of these situations P4F forms a variance-covariance matrix and estimates the parameters by sweeping (or partially sweeping) this matrix. This procedure will give correct estimates, although the solution may no longer be unique; i.e., the problem may be overparameterized [7].

Some of the limitations of P4F are described in Table 5.

Sparse data in contingency tables can cause problems of numerical accuracy and of statistical interpretation. A sparse table is one in which there are many cells with small expected values and one or more observed zeros. When the pattern of observed zeros creates zeros in a marginal subtable corresponding to one of the configurations in the model, there can be numerical problems in the estimation of parameters, of expected values and of degrees of freedom [7]. Care in implementations of the algorithms can alleviate some of the numerical problems, but cannot guarantee their absence. Overparameterized models with nonestimable parameters can occur.

Table 5: Known problems and limitations of P4F

SPARSE TABLES:
WHEN MARGINAL ZEROS OCCUR, TWO MODELS BEING COMPARED MAY DIFFER IN THEIR SETS OF CELLS WITH FITTED VALUES EQUAL TO ZERO

STD ERRORS MUST BE OBTAINED BY INVERTING INFORMATION MATRIX -- MAY REQUIRE TOO MUCH MEMORY

NONHIERARCHICAL MODELS:
CANNOT BE FITTED

ORDINAL CATEGORICAL VARIABLES:
CANNOT BE TAKEN INTO ACCOUNT (EXCEPT FOR MEASURES OF ASSOCIATION IN TWO-WAY TABLE)

The small expected values affect the distribution theory of the statistics. The distribution theory underlying the chi-square statistics is large-sample asymptotic theory which is inappropriate for statistics based on sparse tables. Also, when the model is overparametrized, the computer program will print out a solution, but there are many other equally good alternate solutions with differing parameter estimates. One approach often used is to augment each cell by a constant. Although this approach eliminates the numerical problems, it leaves the problems of inference untouched.

P4F uses an iterative proportional fitting algorithm to estimate the expected values of a log-linear model which restricts the models that can be specified and fitted. For example, all models must be hierarchical. In addition, models that incorporate the ordering of indices, such as those described by Agresti [1], are not available.

4. DESIGNING A NEW PROGRAM

Given the rapid strides in developing new models for categorical data, it is necessary to develop more flexible computer programs that will allow the fitting of such models.

Some general goals for a program are:

1) To make available new statistical methodology. For example, Goodman and Kruskal [11,12,13] proposed statistics, such as the gamma, lambda and tau, to estimate relationships among the indices in the two-way frequency table. Other have proposed alternate measures. As long as these measures did not appear in computer programs, it was difficult to evaluate their usefulness. To interpret the meaningfulness of the statistics, it is necessary to compute their standard errors and z-scores.

2) To provide aids for the unsophisticated user. For example, special purpose programs to fit log-linear models (ECTA, GLIM, etc) assume that the user knows which model is to be fitted to the data based on an a priori knowledge of the variables. Identification of the appropriate model was made by testing effects in the model or by a stepwise procedure. The rationale behind tests of marginal and partial association in P4F [2,5] is to enable the investigator to screen all the possible interactions for their 'maximal' effect and thus order them in importance.

3) To be easy for a novice to use. This last consideration is critical when planning a new program. For example, how should models be specified in the general case where the model may be nonhierarchical or when the factors are ordinal or when the dependent variable is ordinal.

When the only programs available analyzed data in two-way tables and the only statistic computed was the chi-square, it was reasonable to assume that, if the user can run the program, s/he can understand the output. When there is a program such as P4F with a relatively simple means to specify options, users can request options that produce results which they are not trained to interpret correctly. When planning a new program that starts where P4F stops, which audience should be addressed:
--the unsophisticated user in an applied area,
--the sophisticated user in the applied area,
--the statistician with a masters degree, or
--the advanced practitioner of statistics.

A requirement to specify design matrices explicitly would indicate that the last group is the target audience. The presence of a totally automatic model search routine would allow all groups to use the program and possibly not understand the results. Therefore, there is a need to allow different levels of sophistication of usage, where users at the lowest level would not need access to all the options (and probably would not desire the excluded options).

Models that are not hierarchical, such as those of marginal symmetry, cannot be fitted within P4F. In addition, the internal structure between cells cannot be specified to P4F. Therefore, when repeated observations are taken on a variable and each repetition is not treated as a separate index, P4F is unable to analyze the data.

Several forms of models have been proposed for categorical data. The two most commonly used at this time are the log-linear model where

$$\ln p = \text{linear model}$$

and the logistic regression model where

$$\ln [p/(1-p)] = \text{linear model}.$$

Alternative models include writing on the left-hand side either p or the odds-ratio or some other function of one or more p's.

When the independent variables, or factors, are not ordered, the usual representation of the linear model is the same as that of an analysis of variance model. The only difference is that in the log-linear model the logarithm of the expected value, and not the expected value itself, has a linear form. When one or more factors are ordered, it may be possible to write the linear model using a reduced set of variables (such as the lower-order terms of an orthogonal decomposition) for that factor, or the models of Agresti [1].

Classically, statistics and biostatistics have been concerned with fitting models to data such that the deviations of the observations from the model are mutually independent. More recently, models have been developed to allow for repeated observations from individuals. In these models it is recognized that the repeated observations from an individual have less variation than a similar set of observations, each obtained from a different individual. Repeated measures models for categorical data have primarily treated the situation when there is a single response variable, such as voting preference, observed over time for a group of individuals. The models that are fitted to the data, and hypotheses tested, describe change over time. General models for repeated measures will be able to be fitted to the data in the new program.

Several methods of fitting the log-linear model to categorical data will be available:

i) Maximum likelihood (ML) using the iterative proportional fitting algorithm (IPF). This method is limited to fitting hierarchical models.

ii) ML using a Newton-Raphson algorithm (NR). This method may require computing a large covariance matrix at each iteration.

iii) Weighted least squares (WLS). These estimates are not maximum likelihood. The method does not require iteration but the same covariance matrix is needed as for the NR algorithm.

Table 6 summarizes many of the attributes of the program that is being developed.

Table 6: Attributes of the new program

MODELS THAT CAN BE FITTED:
LINEAR
LOG-LINEAR
LOGISTIC

MODEL SPECIFICATION BY:
MACRO-LEVEL KEYWORDS
DESIGN MATRICES

VARIABLES CAN BE:
NOMINAL
ORDINAL

ALGORITHMS:
ML USING IPF
ML USING IRWLS
WEIGHTED LEAST SQUARES

MODEL-BUILDING:
SEMI-AUTOMATIC
INTERACTIVE

TYPES OF MODELS:
POISSON
MULTINOMIAL
REPEATED MEASURES

REFERENCES

[1] Agresti, A., Analysis of Ordinal Categorical Data. (Wiley, New York, 1984).

[2] Benedetti, J.K. and Brown, M.B., Strategies for the selection of log-linear models. Biometrics 34 (1978) 680-686.

[3] Brown, M.B., Identification of the sources of significance in two-way contingency tables. Applied Statistics 23 (1974) 405-413.

[4] Brown, M.B., The asymptotic standard errors of some estimates of uncertainty in the two-way contingency table. Psychometrika 40 (1975) 291-296.

[5] Brown, M.B., Screening effects in multidimensional contingency tables. Appl. Statist. 25 (1976) 37-46.

[6] Brown, M.B. and Benedetti, J.K., Sampling behavior of tests for correlation in two-way contingency tables. J. Amer. Statist. Assoc. 72 (1977) 309-315.

[7] Brown, M.B. and Fuchs, C., On maximum likelihood estimation in sparse contingency tables. Computational Statistics and Data Analysis 1 (1983) 3-15.

[8] Dixon, W.J., Editor, BMD Biomedical Computer Programs. (University of California Press, Berkeley, 1964).

[9] Dixon, W.J., Editor, BMDP Biomedical Computer Programs, P-series. (University of California Press, Berkeley, 1975).

[10] Dixon, W.J. et al, Editors, BMDP Statistical Software. (University of California Press, Berkeley, 1981).

[11] Goodman, L.A. and Kruskal, W.H., Measures of associations for cross-classification. J. Amer. Statist. Assoc. 49 (1954) 732-764.

[12] Goodman, L.A. and Kruskal, W.H. Measures of associations for cross-classification. III. Approximate sampling theory. J. Amer. Statist. Assoc. 58 (1963) 310-364.

[13] Goodman, L.A. and Kruskal, W.H., Measures of associations for cross-classification. IV. Simplification of asymtotic variances. J. Amer. Statist. Assoc. 67 (1972) 415-421.

[14] Haberman, S.J., Log-linear fit for contingency tables, algorithm AS51. Appl. Statist. 21 (1972) 218-224.

COMPUTER SCIENCE AND STATISTICS:
The Interface, D.M. Allen (ed.)

LOG-LINEAR MODELING WITH SPSS^x

Clifford C. Clogg and Mark P. Becker

The Pennsylvania State University
University Park, Pennsylvania 16802[1]

The recently released software package SPSS^x contains two procedures for log-linear analysis of contingency tables, LOGLINEAR and HILOGLINEAR. LOGLINEAR is based on Haberman's (1979) program FREQ, and it uses a Newton-Raphson algorithm for calculating maximum likelihood estimates. LOGLINEAR is probably the most general computer program for log-linear analysis now included in major software packages. HILOGLINEAR is based on the iterative-proportional-fitting (IPF) algorithm and is restricted to hierarchical models that can be expressed in terms of fitted marginals. We evaluate these two procedures according to the following criteria: (1) What can be done with the procedures? (2) Does the available documentation give a suitable description of those capabilities? (3) What should SPSS^x have done? (Or, what should they do with these procedures in the future?) (4) What diagnostics and/or warnings are available or could be made available given current knowledge?

1. INTRODUCTION

In 1979 Haberman introduced a computer program called FREQ that "can be used to compute maximum likelihood estimates for any log-linear model" (Haberman, 1979, p. 571). What he meant was that his program could be used to obtain ML fits for any model for contingency tables that is additive in the logarithms of cell frequencies, when the cell frequencies arise from Poisson, multinomial, or product-multinomial sampling schemes. There were three main advantages of FREQ in relation to others that existed in the 1970s:
1. It calculated adjusted (truly standardized) residuals (cell by cell) and generalized adjusted residuals for contrasts among cells.
2. It allowed for adjustment of Poisson frequencies for differential cell-by-cell exposures, thus permitting log-linear analysis of rates of rare events.
3. The Cholesky factorization of the estimated information matrix at successive steps in the Newton-Raphson algorithm was done with great care, and analysts were thereby alerted to non-existence problems and related problems that arise from sparse data and/or from specifications of quasi-log-linear models.

The main disadvantage of FREQ was that users had to supply the model matrix (or design matrix) in complete detail, a difficulty that prevented its widespread use.

In 1983 the FREQ program was incorporated in the LOGLINEAR procedure of SPSS^x. The most obvious difference between LOGLINEAR and FREQ is that in the former the model matrix can be created with only a small number of commands using symbolic representations for the types of contrasts that are to be employed. The Kronecker product operations that build the model matrix from the variable contrasts are performed automatically. Many options are available for specifying contrasts, quantitative covariates may be added to a model quite easily, logit-type models (or multinomial-response models) can be readily distinguished from the wider class of log-linear models for the cell frequencies, normal probability plots for residuals can be obtained, and an analysis of dispersion including asymmetric measures of association for logit-type models is available. LOGLINEAR is not designed to be a stand-alone exploratory analysis procedure. But once the contingency table -- including both the variables and the categories used for each -- and a relatively small number of models for this table are specified, LOGLINEAR is probably the best ("most general") program for log-linear models currently in existence.

Below we describe briefly what LOGLINEAR can do, whether the documentation provides a satisfactory description of its capabilities, and what could be done to improve the program in the light of current knowledge. It is not our purpose to compare LOGLINEAR with other programs. In our experience, analysis of contingency tables in practical research settings usually requires the use of more than one procedure from more than one software package. And it should be acknowledged that computing for contingency table models is very primitive compared to computing for linear models. We are a long way from having computational equipment that is as flexible -- and as believable -- as the procedures REG and GLM in the SAS package. And we are even further from the development of intelligent software like the REX program of Bell Laboratories for regression analysis (Hahn, 1985; Gale and Pregibon, 1982). Our goal is not to make invidious comparisons but rather to assess strengths and weaknesses of the particular program under review. More borrowing of ideas among software developers is called for, and we hope that the present review points to areas where such borrowing is most likely to be beneficial.

2. The General Log-linear Model

LOGLINEAR, like its predecessor FREQ, works with the following general formulation of the log-linear model for frequency data. Suppose that there are J "groups" with the number n_j of observations per group, $j = 1,\ldots,J$, fixed either by the sampling scheme or by conditioning. Suppose further that there are I levels of response, which may represent crossed or nested combinations of response variables. Let $n_{ij} \geq 0$ denote observed frequency in a given response-group combination, $m_{ij} = E(n_{ij})$, w_{ij} a fixed "weight", z_{ij} a dummy variable taking on the value 0 if the i-th response in the j-th group is a structural zero ($m_{ij} = 0$) or is to be fitted perfectly ($\hat{m}_{ij} = n_{ij}$), otherwise taking on the value 1. Finally, let x_{ijk}, $1 \leq k \leq K$, denote the k-th column of the relevant model matrix, where K is the number of parameters to be estimated. The general model is

$$\log (z_{ij} m_{ij}/w_{ij}) = \alpha_j + \sum_{k=1}^{K} \beta_x x_{ijk}.$$

Special cases of this model include the following:

I. Log-linear models for complete contingency tables: $z_{ij} = w_{ij} = 1$, all i and j, $J = 1$. (All variables are responses or "dependent" variables.)

II. Log-linear models for incomplete tables ("quasi-log-linear models"): $z_{ij} = 0$ for $(i,j) \varepsilon S$ where S denotes structurally empty response-group levels, $z_{ij} = 1$ for $(i,j) \varepsilon S^c$, $J = 1$.

III. Multinomial-response models: $J > 1$ (the dichotomous response logit model is obtained when $I = 2$).

IV. Poisson (or rate) models: w_{ij} = exposure (e.g., time in months) for rare event count n_{ij}. (Here, m_{ij}/w_{ij} is the rate of the rare event for the (i,j) combination, and we will usually want to take $J = 1$.)

Cell-by-cell residuals, $n_{ij} - \hat{m}_{ij}$, are examined by comparing them to the estimated asymptotic standard deviation $s(n_{ij} - \hat{m}_{ij})$, and generalized residuals compare $\Sigma_{i,j}\, c_{ij}(n_{ij} - \hat{m}_{ij})$ to $s(\Sigma_{i,j}\, c_{ij}(n_{ij} - \hat{m}_{ij}))$, where $\Sigma_{i,j}\, c_{ij} = 0$. See Haberman (1973, 1978). Dispersion in multinomial responses (marginal and conditional) is analyzed using the entropy and concentration measures (Haberman, 1982). The program gives estimated parameter values, chi-squared statistics (Pearson and likelihood-ratio), the variance-covariance matrix of parameter estimates (from the information matrix), correlations obtained from them, and a variety of output options.

Estimation is by the Newton-Raphson method, which as programmed is essentially based on iteratively re-weighted least squares (with weights that take account of the fixed weights w_{ij} and the approximations for $\hat{m}_{ij}$ obtained from a previous cycle). If $z_{ij} = 0$ for some response-group combination, or if $n_j = \Sigma_i\, n_{ij} = 0$ (no observations in the j-th group), the procedure actually eliminates (gives zero weight to) the given response-group combination, or the responses in the j-th group, respectively.

All analyses of contingency tables based on frequentist perspectives are plagued by the problem of sparse data, regardless of the estimation method used (weighted least squares and ML being the two most popular methods). It is useful to distinguish two extreme types of sparse data:

Type I. One or more of the $n_{ij} = 0$ but $n_j > 0$ for all j, $j = 1, \ldots, J$.
Type II. Some $n_j = 0$, but $n_{ij} > 0$ for all response-group combinations where $n_j > 0$.

These conditions are specified so that they pertain to the multinomial-response model, but similar conditions apply to log-linear models for the set of cell frequencies, in which case the condition $n_j = 0$ should be replaced by the condition that observed values of some sufficient statistics take on the value zero. A third case would have some $n_j = 0$ (no responses for some groups) and some $n_{ij} = 0$ for response-group combinations that are actually observed (where $n_j > 0$). To our knowledge, all programs now in existence give zero weight to responses in a void group ($n_j = 0$), and estimability may or may not be affected by this. ML procedures will check estimates of m_{ij} at each cycle t (say $m_{ij}(t)$) when sparse data of Type I occur. If $m_{ij}(t) = 0$ then most programs will give zero weight to that response-group combination <u>in all successive</u> iterations. (Curiously, a re-examination of the offending (i,j) estimated count in cycles after the first one where the problem occurs does not seem to be carried out.) This effective deletion (fitting a zero expected count) might lead to a rank problem for the matrix of the x_{ijk}, and when this occurs smart programs will delete -- rather arbitrarily it turns out -- one or more "columns" of the model matrix. Most computational problems in ML fitting arise in sparse data situations: when $n_{ij} > 0$ for all i and j, there are no problems at all theoretically (Haberman, 1974), and computation is straightforward. In our opinion, the chief computational problem in contingency table analysis based on ML methods is diagnosing when sparse data (Type I or Type II) creates an estimability or rank problem. As we shall see below, LOGLINEAR can be improved on, although what it currently does is probably better than what similar programs do. Diagnostic warnings concerning such problems, at least intelligible ones, are virtually nonexistent, not just in LOGLINEAR but in all other procedures or programs we have used.

3. Specifying Models in LOGLINEAR

To illustrate the flexibility of LOGLINEAR, consider the case with three categorical varibles A, B, and C. Examples of models in each general case (I - IV) described above will be given. These examples can of course be done in a variety of ways; we only intend to convey the flavor of modeling with LOGLINEAR here.

Case I. Log-linear models for contingency tables. The model of no 3-factor interaction (no "second-order interaction") can be estimated by the following two commands:

```
LOGLINEAR A(1,3) B(1,3) C(1,3)/        (3.1)
DESIGN = A, B, C, A BY B, A BY C, B BY C/
```

Each variable is assumed to be trichotomous. The first statement says that there is "one group" (J = 1) or equivalently that each variable is a response. The model matrix is filled with two columns for the main effects of A, by including "A" in the DESIGN statement. Four columns are used for each interaction. The default coding of variable contrasts leads to parameter estimates that correspond to deviations from means. In Goodman's (1970) notation, λ_1^A and λ_2^A will be estimated, for example, and $\hat{\lambda}_3^A$ (which is not estimated) is given by $-(\hat{\lambda}_1^A + \hat{\lambda}_2^A)$. (An easy modification of the program would be to include as an option a feature that would calculate the redundant parameter estimates as well as their standard errors.)

Now suppose that the levels of all three variables are equally spaced, and we wish to examine the model that has linear-by-linear interaction structure. The simplest way to do this is to use orthogonal polynomials to code each variable; this is done by specifying CONTRAST(A) = POLYNOMIAL, etc. Then the DESIGN statement is replaced by

```
DESIGN = A, B, C, A(1) BY B(1), A(1) BY C(1),
         B(1) BY C(1),                  (3.2)
         A(1) BY B(1) BY C(1)/
```

The term "A(1)" denotes the linear orthogonal contrast for A, for example. This model has linear-by-linear 2-factor interactions and linear-by-linear-by-linear 3-factor interaction. It is related to models considered in Haberman (1974), Goodman (1984), and Clogg (1982).

Case II. Quasi-log-linear models. Suppose that cells (1,1,1) and (2,2,2) are structural zeroes, or are to be fitted perfectly because they are "outliers". Either of the above models can be examined recognizing the set S of structural zeroes; this is done by specifying the z_{ij} of the previous section. The CWEIGHT command in LOGLINEAR can be used to convey this information to the program. If Z is the vector with entries z_{ij} (= 0 for structural zeroes, 1 for others), then specifying

```
CWEIGHT = Z/
```

prior to the DESIGN statement will cause the program to analyze a quasi-log-linear model. The quasi-independence model (in three dimensions) would be specified by

```
DESIGN = A, B, C/,                       (3.3)
```

for example, and quasi-log-linear models analogous to those in (3.1) or (3.2) can be analyzed as well. LOGLINEAR calculates parameter estimates for quasi-log-linear models, unlike some programs based on the iterative-proportional-fitting algorithm, and if the pattern of blanked out cells creates rank problems in the model matrix, the program will recognize the difficulty and delete one or more parameters from the model. This should alert the user to potential problems in interpreting parameter values (contrasts of log-estimated counts). It essentially solves the problems in calculating degrees of freedom for chi-squared statistics when such problems arise. (The special problem of dealing with separable subtables created by particular patterns of structural zeroes--see Goodman (1968) --is solved without difficulty.)

Case III. Multinomial-response (logit-type) models. Responses are distinguished from "factors" or independent variables with the BY specification in the LOGLINEAR command. Suppose A is the response variable and that B and C are factors with joint BC levels fixed by sampling design or conditionally fixed by the researcher's wish to examine only the "effects" of B and C on A. Suppose first that we are only interested in the first two levels of A; perhaps level 3 of A represents a "don't know" response or censored observations. The additive dichotomous logit model is specified by:

```
LOGLINEAR A(1,2) BY B(1,3) C(1,3)/
DESIGN = A, A BY B, A BY C/              (3.4)
```

The "BY" fixes the n_j, j = 1, ..., 9, where n_1 = sample total with B = 1 and C = 1, ..., n_9 = the sample total with B = 3 and C = 3. This command essentially determines the α_j values in (2.1). A model with A trichotomous (perhaps now including the observations censored in the previous model) is obtained by replacing "A (1,2)" with "A(1,3)".

Now suppose that level 3 of A represents a "don't know" response. The researcher wants to examine contrasts of A=1 versus A=2 taking account of the censoring that takes place in the model of (3.4). A natural way to do this exploits the "special" contrast specification:

```
CONTRAST(A) = SPECIAL (3*1, 1 -1 0, 1 1 -2)/ (3.5)
```

The contrast (1, -1, 0) is of special interest, and the contrast (1, 1, -2) can be used to examine the difference between non-censored and censored observations. Now suppose that we wish to examine linear effects of B and C as in (3.2). The appropriate model will be estimated by the following commands:

```
LOGLINEAR A(1,3) BY B(1,3) C(1,3)/
DESIGN = A, A BY B(1), A BY C(1)/        (3.6)
```

Case IV. Poisson models. Now suppose that A, B, and C denote risk factors, and the frequencies in the cross-classification of these risk factors denote event counts (e.g., deaths). Suppose further that the cell-by-cell exposures (e.g., person months) are collected in a vector W. The command "CWEIGHT = W" adjusts the cell counts for the exposures. If each factor is quantitative with equal spacing, a model of interest could be:

```
LOGLINEAR A(1,3) B(1,3) C(1,3)/
CWEIGHT = W/
CONTRAST(A) = POLYNOMIAL/
CONTRAST(B) = POLYNOMIAL/                (3.7)
CONTRAST(C) = POLYNOMIAL/
DESIGN = A(1), B(1), C(1)/
```

If m_{stu} is the expected count in cell (s,t,u) and w_{stu} is the corresponding exposure in the A x B x C table, the model estimated above is equivalent to:

$$\log(m_{stu}/w_{stu}) = \log(r_{stu}) = \beta + \beta_1 s + \beta_2 t + \beta_3 u,$$

an additive log-rate model with linear effects of each risk factor. It is very difficult to estimate such a rate model using the IPF algorithm advocated in Laird and Olivier (1981). But as Laird and Olivier note, Poisson log-linear models are closely related to the familiar proportional-hazards model.

Covariates. An attractive feature of LOGLINEAR is the covariate option. If X is a quantitative covariate or dummy variable, it may be added to the model by using a WITH specification. For example, suppose we wish to examine the linear effect of X on the log-odds that A = 1 instead of A = 2. A modification of the model given in (3.4) might be as follows:

```
LOGLINEAR A(1,2) BY B(1,3) C(1,3) WITH X/
DESIGN = A, A BY B, A BY C, A BY X/   (3. )
```

4. Some Simple Diagnostic Tests

Maximum likelihood or other estimation methods derived from frequentist theory can be difficult to apply to sparse data. Table 1 gives three simple examples of sparse data in 2x2x2 contingency tables. These data can be studied either in terms of logit models (C the response and A and B the factors) or in terms of the equivalent log-linear models. MLE's do not exist for the additive logit model (model of no 3-factor interaction) applied to Table 1a. MLE's do not exist for the saturated logit (or log-linear) model applied to Table 1c. For Table 1b the theory is less clearcut; the zero counts for responses on C when A=B=1 amount to giving zero weight to that response pattern in a logit model. Because of this the main effects of A and B on the logits of C are not simultaneously estimable. We treat all three cases with the corresponding models discussed above as nonexistence problems, however, recognizing that nonexistence might not be the preferred term for Table 1b.

Clogg, Rubin, and Weidman (1985) use these three contingency tables to compare eight popular logit regression or log-linear analysis programs. The LOGLINEAR procedure in SPSS^X was one of the programs considered. The following discussion indicates that there are some problems with LOGLINEAR at least in the area of providing diagnostic information.

For Table 1a and using the additive logit model (model of no 3-factor interaction), LOGLINEAR prints chi-squared values of 0.00, 2 degrees of freedom, and two zero fitted frequencies corresponding to the sampling zeroes. From Haberman (1974a) these are the correct answers. This model would have 1 df if no more than one sampling zero occurs (or if all counts are positive) and most researchers would like to know why the correct answer is df = 2. Neither the program output nor the documentation provide any help on this matter. The two main effects are not simultaneously identifiable: the LOGLINEAR fixup deletes the B-C interaction term (for B's effect on C), but of course the A-C interaction term could have been deleted with equal justification. It is only because the B-C interaction information was stored in the "last" entry in the relevant arrays or matrices that this parameter value was deleted. (Incidentally, LOGLINEAR prints "." for both parameter values and standard errors for deleted parameter values.) The only diagnostic message given by the program is "ML did not converge," but this diagnostic is misleading. The program did give the correct--and exact--ML solution for the expected frequencies, which in this case are merely the observed frequencies. Researchers might conclude that the A-C interaction was estimated appropriately and that the B-C interaction is zero, but of course such an inference would be incorrect. The estimated value of the A-C interaction does not refer to the contrast of log-frequencies that is used to define the original model. The point is that the user is left in the dark concerning what the program did, what the results mean, and what could be done to remedy the problem.

For Table 1c using the saturated model, the output is again somewhat misleading. The MLE's do not exist for the saturated model when there are sampling zeroes, so some indication of this would be expected. Here is what LOGLINEAR gives. The program gives the correct chi-squared value (0.00) and the correct df (df = 0). But even though the MLE's of the parameters do not exist, LOGLINEAR print outs estimates for them along with standard errors. The standard errors are large and the parameter values are nonsensical, so some researchers would recognize that there is a kind of identifiability problem. But no warning messages or diagnostics are printed.

The additive logit model (model of no 3-factor interaction) was applied to Table 1b. LOGLINEAR gives chi-squared values of 0.00, which is correct. But most ML advocates would say that the model applied to Table 1b is equivalent to blanking out the two sampling zeroes because the ML solution will estimate these frequencies as zeroes. The model would be redefined and reparameterized for the remaining six cells. When this is done, the additive logit model is saturated relative to these six cells, so df = 0 should be reported. Nevertheless, LOGLINEAR gives df = 1. It is curious that a chi-squared value that has to be zero for such a sparse table would be said to have one degree of freedom. And once the two sampling zeroes are removed, the parameter values that would be calculated no longer refer to standard contrasts of the logits. LOGLINEAR nonetheless prints parameter values and standard errors with no warning that they do not refer to the contrasts originally specified in formulating the model.

To summarize, LOGLINEAR does not do a good job in reporting results obtained from elementary examples with nonexistent MLE's. Diagnostics are virtually nonexistent. Users who suspect problems in their output (suspicious parameter values and/or standard errors, or unanticipated degrees of freedom) will have to turn to an experienced consultant to answer their questions. To put this evaluation in proper perspective, however, it should be noted that LOGLINEAR performed at least as well as the seven other programs examined in Clogg, Rubin, and Weidman (1985). More internal checks for consistency and more intelligible diagnostic messages are required in all of these programs.

5. Suggestions for Improvement

Another procedure in $SPSS^X$ can be used for analysis of categorical data too: HILOGLINEAR, a program based on the IPF (iterative-proportional-fitting) algorithm. The "HI" is not a salutation, but stands for hierarchical models; this procedure can calculate ML fits for hierarchical models having observed marginals as sufficient statistics. HILOGLINEAR was evidently prepared to serve as an exploratory screening procedure that could be used to select models for further study in LOGLINEAR. At present, however, HILOGLINEAR appears to be quite preliminary and we cannot recommend it. The procedure does not calculate parameter estimates for unsaturated models; because of this, the procedure can never stand alone even if the researcher is interested in the kinds of models that can be considered with the procedure. The program does not calculate degrees of freedom correctly for incomplete tables: the example in the $SPSS^X$ documentation (one of the classic examples--see Goodman (1968) and Clogg (1985)) reports incorrect df because it does not recognize separable subtables. There are both forward selection and backward elimination model search options. A general recommendation is that HILOGLINEAR should be greatly improved and expanded; the P4F program in BMD provides a good example of what should be incorporated.

We have the following recommendations for improving LOGLINEAR, most of which can be implemented easily:

1. Improve the documentation. How covariates may or may not be used is unclear from the published report. There are no examples with continuous covariates. There are few references to the literature. There is little indication that the CWEIGHT command can be used to adjust Poisson counts for exposures, no indication that the program provides a flexible procedure for analysis of rates.
2. Output: multinomial-response models are alternatives to discriminant analysis. Since multinomial-response models (logit-type models) are convincing alternatives to linear discriminant analysis (Press and Wilson, 1978), it would be helpful if output from such models could be arranged to facilitate practical discriminant analysis. This would involve obtaining the predicted proportions in the I response levels for each of the J groups and assessing their variability (prediction intervals) under the model. This is easy to do. Output from programs dealing exclusively with dichotomous logistic regression models (SAS: LOGIST or PREDICT, BMD: PLR) already facilitates such analysis.
3. Input-Output: linear contrasts of parameters and the associated variance-covariance matrix. If β is the vector of parameter estimates, linear contrasts of the form L β can be used to advantage. Such linear contrasts can be tested using Wald statistics. Since the variance of β is already calculated, this creates no special problem. Various specifications of L could be used to examine how a given model might be simplified (exploratory use), to examine collapsibility of categories (Suman, 1985), and to perform simultaneous tests on sets of parameters

without resorting to the comparison of nested models and likelihood-ratio tests.

4. Output: variances of measures of association. Haberman (1982) derived the approximate distributions for both entropy and concentration measures of association. This information should be added to LOGLINEAR.
5. Input: adding fractional counts to the data. It is easy to add the same constant to all cell counts (e.g., ½), and there is some justification for doing so when saturated models are considered (Goodman, 1970). Adding constants to the frequencies can be interpreted from a Bayesian perspective; the prior is either beta or Dirichlet. Adding the same constant to all counts shrinks the data toward equiprobability. In a logit model this shrinks all parameter values, including the constant, toward zero. More flexible priors that are model based are discussed in Clogg, Rubin, and Weidman (1985). Simple changes in LOGLINEAR would allow implementation of these. (The most obvious choice in a logit model is to add constants to "successes" and "failures" in proportion to the marginal distribution of the response.)
6. Programming: internal checks. As the examples in the previous section show, there are problems when even simple tables with sparse data are analyzed. The program does not seem to "correct" for zero observed group totals in multinomial-response models, or at least does not do so all of the time. For tables of high dimension there should be additional checks on a cycle-to-cycle basis for estimability. We believe, but cannot prove, that it is not sufficient in general to let conclusions reached in one cycle about estimability dictate model redefinition (parameter deletion) in all subsequent cycles.
7. Diagnostics: warning messages, cautionary remarks. The only warning we have seen in using LOGLINEAR is "ML did not converge." This is not informative enough. There are many other messages that should be given, particularly when sparse data problems arise. Some information about possible rank problems in the information matrix would be helpful as well. (Perhaps such diagnostics could be borrowed from those in wide use for the X'X matrix in regression.) These problems are ignored in the technical documentation.

In spite of the criticisms noted above, LOGLINEAR is a good program for the analysis of contingency tables. In our opinion, researchers who have access to both LOGLINEAR and BMD's program P4F will be able to deal with most contingency table problems that are likely to arise in practice.

Table 1. Three 2x2x2 Contingency Tables with Sampling Zeroes

AB	1a. C=1	1a. C=2	1b. C=1	1b. C=2	1c. C=1	1c. C=2
11	0	3	0	0	0	3
21	9	4	9	4	9	4
12	6	3	6	3	6	3
22	5	0	5	3	4	1
	n = 30		n = 30		n = 30	

REFERENCES

[1] Clogg, C.C., Some models for the analysis of association in multiway cross-classifications having ordered categories, Journal of the American Statistical Association, 77, (1982) 803-815.

[2] Clogg, C.C., Quasi-independence, Encyclopedia of Statistical Science, Vol. 7 (forthcoming) (1985).

[3] Clogg, C.C., Rubin, D.B. and Weidman, L., Simple Bayesian Methods for the Analysis of Logistic Regression Models, unpublished manuscript (1985).

[4] Gale, W.A. and Pregibon, D., An expert system for regression analysis, in Proceedings of the 14th Symposium on the Interface of Computer Science and Statistics, New York Springer--Verlag (1982) 110-117.

[5] Goodman, L.A., The analysis of cross-classified data: independence, quasi-independence, and interactions on contingency tables with or without missing entries, Journal of the American Statistical Association, 63 (1968) 1091-1131.

[6] Goodman, L.A., The multivariate analysis of qualitative data: interactions among multiple classifications, Journal of the American Statistical Association, 65 (1970) 226-256.

[7] Goodman, L.A., The Analysis of Cross-Classified Data Having Ordered Categories (Harvard University Press, Cambridge, Mass. 1984)

[8] Haberman, S., The analysis of residuals in cross-classified tables, Biometrics, 29 (1973) 205-220.

[9] Haberman, S., The Analysis of Frequency Data (University of Chicago Press, Chicago, 1974a).

[10] Haberman, S., Log-linear models for frequency tables with ordered classifications, Biometrics, 30 (1974b) 589-600.

[11] Haberman, S., Analysis of Qualitative Data:

Vol. 1. Introductory Topics (Academic Press, New York 1978).

[12] Haberman, S., Analysis of Qualitative Data: Vol. 2. New Developments (Academic Press, New York 1979).

[13] Haberman, S., Analysis of dispersion of multinomial responses, Journal of the American Statistical Association, 77 (1982) 568-580.

[14] Hahn, G.J., More intelligent statistical software and statistical expert systems: future directions, The American Statistician, 39 (1985) 1-8.

[15] Laird, N. and Olivier, D., Covariance analysis of censored survival data using log-linear analysis techniques, Journal of the American Statistical Association, 76, (1981) 231-240.

[16] Press, S.J. and Wilson, S., Choosing between logistic regression and discriminant analysis, Journal of the American Statistical Association, 73 (1978) 699-705.

[17] Suman, V.J., Using Wald Statistics to Assess Collapsibility in Contingency Tables, Masters Paper, Dept. of Statistics, The Pennsylvania State Univ. (1985).

[1] The senior author was supported in part by NSF grants SES-7823759 and SES-8303838 from the Division of Social and Economic Sciences, National Science Foundation. The authors are not affiliated with SPSS, Inc.

COMPUTER SCIENCE AND STATISTICS:
The Interface, D.M. Allen (ed.)

FITTING MULTINOMIAL REGRESSION MODELS TO CATEGORICAL DATA

Christopher Cox

Division of Biostatistics
and Division of Toxicology
in the Department of Radiation Biology and Biophysics
University of Rochester

Parametric models for the multinomial distribution are considered within the larger family of regular exponential family models. This allows a unified approach to fitting multinomial regression models using algorithms based on iteratively reweighted least squares. A useful example is provided by the family of continuation ratio models. Generalized linear models are considered as an important special case of the exponential family which provides an approach to categorical data based on log-linear models.

1. INTRODUCTION

We consider the theory and practice of maximum likelihood estimation for multinomial regression models. These are parametric models for data obtained by measuring a categorical response in the presence of possibly multiple explanatory variables. The appropriate sampling scheme is what is known as product multinomial sampling. We discuss the exponential family formulation of such models and review the fitting of general exponential family models using iteratively reweighted least squares. The discussion is based on the approach of Jennrich and Moore, (1975) which deserves to be more widely recognized. We include a slightly simpler derivation of their results, which basically consist of a formal identification of the maximum likelihood problem with a weighted nonlinear least squares problem. This yields an iteratively reweighted Gauss-Newton algorithm for the computation of maximum likelihood estimates and asymptotic standard errors. We illustrate the theory with an example using continuation ratio models for ordinal data (Fienberg, 1980.) We also discuss the generalized linear models of Nelder and Wedderburn (1972) and the resulting fitting algorithm, which is also based on iteratively reweighted least squares. This subclass has the advantage of being much more analogous to ordinary linear models and is the basis for the GLIM statistical computing system. Finally, we consider the analysis of categorical data using GLIM, which rests on the assumption of Poisson sampling.

2. MULTINOMIAL REGRESSION MODELS

We consider a random n-dimensional vector Y having a multinomial distribution as a member of a regular exponential family. To this end we write the density of Y as

$$p(y,\eta) = pr[Y_1=y_1,\ldots,Y_n=y_n] =$$
$$\exp[\sum_i \eta_i y_i - y.\ln(\sum_j e^{\eta_j}) + \ln\binom{y.}{y}],$$

where $y. = \Sigma y_j$ and the multinomial probabilities can be computed from the natural parameters as

$$\pi_i = e^{\eta_i}/\sum_j e^{\eta_j}.$$

We assume that the n-dimensional natural parameters η depends on $p \leq n$ parameters θ and write $E_\theta(Y) = \mu(\theta)$ and $Var_\theta(Y) = \Sigma(\theta)$. For the multinomial distribution these are $\mu(\theta) = y.\pi(\theta)$ and $\Sigma(\theta) = y.[D(\pi)-\pi\pi']$. Differentiating under the integral sign we have the standard results

$$\mu = -\partial d/\partial\eta \text{ and } \Sigma = -\partial^2 d/\partial\eta^2.$$

The likelihood equations for the regular exponential family likelihood are

$$s(\theta) = \frac{\partial\eta'}{\partial\theta} y + \frac{\partial d'}{\partial\theta} = 0 .$$

To transform these equations we first apply the chain rule to the previous expressions for the mean and variance functions to obtain

$$\frac{\partial d'}{\partial \theta} = -\frac{\partial \eta'}{\partial \theta}\, \mu$$

$$\frac{\partial \mu}{\partial \theta} = \frac{\partial}{\partial \theta}\left(-\frac{\partial d'}{\partial \eta}\right) = \Sigma \frac{\partial \eta}{\partial \theta} .$$

Now let Σ^- be a symmetric generalized inverse of Σ, satisfying $\Sigma\Sigma^-\Sigma = \Sigma$. In the multinomial case we have $\Sigma^- = (y.)^{-1}D^{-1}(\pi)$. Then since $a'\Sigma=0$ implies $\mathrm{Var}(a'(Y-\mu))=0$, we have $\Sigma\Sigma^-(Y-\mu)= Y-\mu$, ($Y-\mu$ is in the range of Σ) with probability one. Combining these results we may write the likelihood equations as

$$0 = s(\theta) = \frac{\partial \eta'}{\partial \theta}(y-\mu)$$

$$= \frac{\partial \eta'}{\partial \theta}\, \Sigma\Sigma^-(y-\mu)$$

$$= \frac{\partial \mu'}{\partial \theta}\, \Sigma^-(y-\mu) \quad .$$

These are the normal equations for the nonlinear least squares problem: minimize $(y-\mu)'\Sigma^-(y-\mu)$. It also follows from these results that

$$\Sigma\Sigma^- \frac{\partial \mu}{\partial \theta} = \frac{\partial \mu}{\partial \theta},$$

allowing us to write the information matrix as

$$I(\theta) = \mathrm{Var}\ s(\theta) = \frac{\partial \mu'}{\partial \theta}\, \Sigma^- \frac{\partial \mu}{\partial \theta} = \frac{\partial \eta'}{\partial \theta}\, \Sigma\, \frac{\partial \eta}{\partial \theta} .$$

Therefore the Fisher scoring algorithm becomes

$$\Delta(\theta) = I^{-1}(\theta)\ s(\theta) = \left(\frac{\partial \mu'}{\partial \theta}\, \Sigma^- \frac{\partial \mu}{\partial \theta}\right)^{-1} \frac{\partial \mu'}{\partial \theta}\, \Sigma^-(y-\mu),$$

which is an iteratively reweighted Gauss-Newton algorithm for the nonlinear least squares problem . Asymptotic standard errors, obtained by inverting the information matrix, may be computed from the usual standard errors given by the Gauss-Newton algorithm if we omit the residual mean square

$$\hat{\sigma}^2 = (y-\mu)'\Sigma^-(y-\mu)/(n-p).$$

For the multinomial distribution the numerator is just the Pearson chi-square statistic for goodness-of-fit. Recent work on quasi-likelihood models (McCullagh and Nelder, 1983) suggests that if $\hat{\sigma}^2$ is not reasonably close to one, e.g., is significantly larger than one, then the asymptotic standard errors should be corrected by multyiplying by $\hat{\sigma}$.

In practice one can fit general exponential family models using any weighted regression program which can be iterated after recomputation of the weights. This can be done for example in MINITAB. Nonlinear regression programs which impliment the Gauss-Newton algorithm are easier to use provided they allow iterative computation of the weights. Such programs are available in BMDP, SAS and GENSTAT. To use such a program one must specify the quantities μ, $\partial\mu/\partial\theta$ and Σ^- (means, derivatives and weights). We used the program BMDP3R (Dixon et al., 1981) which also allows the use of a loss function as a termination criterion. The natural loss function is the deviance,

$$G^2 = -2[\sum_i y_i \log(\hat{\pi}_i) - y_i \log(y_i/y.)] \quad ,$$

where $\hat{\pi}_i$ are the estimated probabilities. This is the likelihood ratio statistic, with n-1-p d.f., of the current to the saturated model which estimates the multinomial probabilities by the observed proportions.

3. AN EXAMPLE: CONTINUATION RATIO MODELS

If the ordering of the categories 1,...,n is not arbitrary (or even if it is), the n-1 conditional probabilities known as continuation ratios (Fienberg, 1980) are defined as

$$p_i = \sum_{j>i} \pi_j \Big/ \sum_{j\geq i} \pi_j \ .$$

Continuation ratio models are just logit models for these conditional probabilities. In the framework of product multinomial sampling we have multinomial probabilities $\{\pi_{ij} | 1 \leq i \leq R,\ 1 \leq j \leq C\}$, satisfying $\Sigma_j \pi_{ij}=1$ for $1 \leq i \leq R$, and continuation ratios p_{ij}. The model is specified by writing the logits, ℓ_{ij}, as functions of the explanatory variables and parameters θ. For example Fienberg (1980) considers data on 3 levels of educational attainment. The explanatory variables are age (2 levels), race (2 levels) and father's education (4 levels). The data consist of counts of the three levels of the response variable for each combination of the three explanatory variables, for a total of 16 trinomials (32 d.f.). Fienberg (1980) considers, among others, an 18 parameter model having different parameters for each of the two continuation ratios. The model includes main effects for each of the three factors, as well as an interaction between father's education and race. In an obvious notation the model is given by

$$\ell_{arfc} = \mu^c + \alpha_a^c + \beta_r^c + \gamma_f^c + (\beta\gamma)_{rf}^c ,$$

where c(=1,2) denotes continuation ratio and the appropriate constraints (e.g., $\alpha_1^c = 0$) are imposed for identifiability.

As Fienberg (1980) points out, this model can be fitted as a separate pair of logit models for the two conditional probabilities. We fitted the entire model using the nonlinear regression program BMDP3R. The means (probabilities), derivatives, weights and loss function are supplied to the program in a FORTRAN subroutine (Figure 1). The multinomial probabilities π_i may be computed for $i \geq 2$ ($\pi_1=1-p_1$) from the relation

$$\pi_i = (1-p_i) \prod_j^{i-1} p_j \qquad (p_n = 0) ,$$

or more easily from the recursion

$$\pi_i = (1-p_i) \frac{p_{i-1}}{1-p_{i-1}} \pi_{i-1} ,$$

which can also be used for the computation of derivatives. The program was run with initial values of 0.1 for all parameters and converged in 8 iterations to a G^2 of 18.6, which agrees with the value of 18.5 given by Fienberg (1980) in Table 6-11. This example is also discussed in Cox (1985).

There is of course no reason why one should have different parameters for different continuation ratios, nor need the model for the logits be linear. Consider, for example, the following multiplicative interaction model for an RxC table under product multinomial sampling,

$$\ell_{ij} = \beta_j + \gamma_i + \theta_i \delta_j$$

$$i=1,\ldots,R; \qquad j=1,\ldots,C-1,$$

subject to the identifiability constraints $\beta_1=0$, $\delta_1=0$, for a total of 2R+2C-4 parameters. This model cannot be fitted as a series of logit models for the continuation ratios. As an illustration we consider a 4x4 table discussed by Cox and Chuang (1984). The data consist of ratings (poor, fair, good, very good to excellent) of four analgesic drugs. The data are given in Figure 2. The γ and θ parameters now model differences between drugs, while the β and δ parameters model differences between continuation ratios. Figures 1-2 display the FORTRAN and BMDP programs for fitting a model with eight constraints, (6 d.f.) which essentially identify the first two and the last two drugs. Here again convergence was fairly rapid (Figure 3) with initial values taken from a previous, unconstrained fit. The deviance G^2 = 9.58 with 6 d.f., as well as parameter estimates and asymptotic standard errors are given in Figure 3. Observed and predicted proportions (Figure 4) can be extracted for the computation of standardized residuals.

4. GENERALIZED LINEAR MODELS - A SPECIAL CASE

This class of models is important because of its useful similarities to ordinary linear models and because it forms the basis of the GLIM statistical system (Baker and Nelder, 1978). Two additional assumptions are required for a generalized linear model. The first is that the components of the random vector Y are independent. This means that we can factor the likelihood so that the function

$$d(\eta) = \sum_i d_i(\eta_i),$$

and $\mu_i = -\partial d_i/\partial\eta_i$,

$$\mathrm{Var}(Y_i) = -\partial^2 d_i/\partial\eta_i^2 = \partial\mu_i/\partial\eta_i,$$

and Σ is a nonsingular, diagonal matrix. The second assumption is that $\eta_i=f(\psi_i)$, where f is a monotone link function and ψ is the linear predictor, $\psi=X\theta$, where X is a matrix of predictor variables. Thus on the appropriate scale we are dealing with a linear regression problem although not with the usual error structure.

Nelder and Wedderburn (1972) develop an iteratively reweighted least squares algorithm for fitting generalized linear models by defining a working dependent variable

$$z = \psi+\left(\frac{\partial\mu}{\partial\psi}\right)^{-1}(y-\mu) ,$$

and a diagonal matrix of weights

$$W = \frac{\partial\mu'}{\partial\psi}\,\Sigma^{-1}\frac{\partial\mu}{\partial\psi} .$$

With these definitions the likelihood equations can be written as $X'Wz = X'WX\theta$, which are the normal equations for the linear least squares problem: minimize $(z-X\theta)'W(z-X\theta)$.

The Fisher scoring algorithm can be rewritten as

$$\Delta\theta = (X'WX)^{-1}X'W(z-\psi) ,$$

or since $\psi=X\theta$, as

$$\theta+\Delta\theta = (X'WX)^{-1}X'Wz.$$

Thus each iteration yields the next approximation, rather than the increment. Again omitting the residual mean square the variances of the least square estimates are also correct since $X'WX = (\partial\mu'/\partial\theta)\Sigma^{-1}(\partial\mu/\partial\theta)$.

Because of the assumption of statistical independence the natural error structure in GLIM for categorical data is the Poisson distribution. The connection between Poisson and multinomial models is well known (McCullagh and Nelder, 1983). An approach using the multinomial distribution is possible by using composite link functions (Thompson and Baker, 1981) although this is much more involved and, we believe, more awkward than the method of fitting expected values discussed previously. Examples of log-linear models with Poisson errors may be found in Nelder and Wedderburn (1972) and McCullagh and Nelder (1983).

5. ACKNOWLEDGEMENTS

Debra Jacobson exercised considerable skill in preparing the manuscript. Work was supported by NIEHS grant ES-01248.

REFERENCES:

[1] Baker, R.J. and Nelder, J.A. General Linear Interactive Modelling (GLIM). Release 3. (Numerical Algorithms Group, Oxford, 1978).

[2] Cox, C., Computation of maximum likelihood estimates by iteratively reweighted least squares: a spectrum of examples, BMDP Statistical Software, Technical Report No. 82 (1985).

[3] Cox, C. and Chuang, C., Comparison of analytical approaches for ordinal data from pharmaceutical studies, Statistics in Medicine 3 (1984) 273-285.

[4] Dixon, W.J., Brown, M.B., Engelman, L., Frane, J.W., Hill, M.A., Jennrich, R.I. and Toporek, J.D., BMDP Statistical Software. (University of California Press, Los Angeles, 1981).

[5] Fienberg, S.E., The Analysis of Cross-Classified Categorical Data, 2nd ed. (M.I.T. Press, Cambridge, Massachusetts, 1980).

[6] Jennrich, R.I. and Moore, R.H., Maximum likelihood estimation by means of nonlinear least squares. Proceedings of the Statistical Computing Section. Amer. Stat. Ass. (1975) 57-65.

[7] McCullagh, P. and Nelder, J.A. Generalized Linear Models (Chapman and Hall, London, 1983).

[8] Nelder, J.A. and Wedderburn, R.W.M., Generalized linear models. J. R. Statist. Soc. A. 135 (1972) 370-384.

[9] Thompson, R. and Baker, R.J., Composite link functions in generalized linear models. Appl. Statist. 30 (1981) 125-131.

Figure 1 FORTRAN program for the computation of means (probabilities), derivatives and weights (variances) for a 14 parameter continuation ratio model. The program is used with BMDP3R.

```
      SUBROUTINE P3RFUN(F,DF,P,X,N,KASE,NVAR,NPAR,IPASS,XLOSS,IDEP)
      IMPLICIT REAL*8 (A-H,O-Z)
      DIMENSION DF(NPAR),P(NPAR),X(NVAR),DCP(20),PDP(20),PDF(20)
      DO 30 I=1,NPAR
      DCP(I) = 0.0
   30 CONTINUE
      IF(X(1) .EQ. 4.)GO TO 35
      ICP = IDINT(X(1))
      TH = P(1)*X(2) + P(2)*X(3) + P(3)*X(4) + P(4)*X(5) + P(ICP+8)
     x  + P(5)*P(ICP+11)*X(2) + P(6)*P(ICP+11)*X(3)
     x  + P(7)*P(ICP+11)*X(4) + P(8)*P(ICP+11)*X(5)
      CP = 1./(1. + DEXP(-TH))
      DCP(1) = CP*(1. - CP)*X(2)
      DCP(2) = CP*(1. - CP)*X(3)
      DCP(3) = CP*(1. - CP)*X(4)
      DCP(4) = CP*(1. - CP)*X(5)
      DCP(5) = CP*(1. - CP)*P(ICP+11)*X(2)
      DCP(6) = CP*(1. - CP)*P(ICP+11)*X(3)
      DCP(7) = CP*(1. - CP)*P(ICP+11)*X(4)
      DCP(8) = CP*(1. - CP)*P(ICP+11)*X(5)
      DCP(ICP+8) = CP*(1. - CP)
      DCP(ICP+11) = CP*(1. - CP)*(P(5)*X(2) + P(6)*X(3)
     x + P(7)*X(4) + P(8)*X(5))
   35 IF(X(1) .EQ. 4.)CP = 0.0
      IF(X(1) .EQ. 1.)PF = 1.
      IF(X(1) .EQ. 1.)PP = 0.5
      F = (1. - CP)*PP/(1. - PP)*PF
      DO 40 I=1,NPAR
      DF(I) = -DCP(I)*PP/(1. - PP)*PF
      IF(X(1) .EQ. 1.)GO TO 40
      DF(I) = DF(I) + (1. - CP)/(1. - PP)**2*PDP(I)*PF
      DF(I) = DF(I) + (1. - CP)*PP/(1. - PP)*PDF(I)
   40 CONTINUE
      PF = F
      PP = CP
      DO 50 I=1,NPAR
      PDP(I) = DCP(I)
      PDF(I) = DF(I)
   50 CONTINUE
      X(9) = X(7)/F
      XLOSS = -2.*X(6)*DLOG(F/X(8))
      RETURN
      END
```

Figure 2 BMDP3R program for fitting a 14 parameter continuation ratio model. Redundancy is avoided by the use of constraints. This model has 8 constraints and 6 independent parameters.

```
 /PROBLEM          TITLE IS 'CONTINUATION RATIO MODELS - ANALGESIC DATA'.
/INPUT             VARIABLES ARE 7.
                   CASES ARE 16.
                   FORMAT IS FREE.
 /VARIABLE         ADD IS 2.
                   NAMES ARE CPID,X1,X2,X3,X4,NUM,DEN,RATIO,CASEWT.
 /TRANSFORM        RATIO = NUM/DEN.
                   CASEWT = 0.5.
 /REGRESS          DEPENDENT IS RATIO.
                   WEIGHT IS CASEWT.
                   PARAMETERS ARE 14.
                   MEANSQUARE IS 1.0.
                   LOSS.
 /PARAMETER        NAMES ARE GAMMA1,GAMMA2,GAMMA3,GAMMA4,THETA1,THETA2,
                        THETA3,THETA4,BETA1,BETA2,BETA3,
                        DELTA1,DELTA2,DELTA3.
              INITIALS ARE 1.63,1.63,1.0,1.00,.00,.00,-.00,-0.00,0,
                    0,-1.66,0,1,1.00.
                    CONST = (10)1          K=0.
                    CONST = (9)1           K=0.
                    CONST = (12)1          K=0.
                    CONST = (13)1          K=1.
                    CONST = (1)1    (2)-1  K=0.
                    CONST = (3)1    (4)-1  K=0.
                    CONST = (5)1    (6)-1  K=0.
                    CONST = (7)1    (8)-1  K=0.
 /END
1 1 0 0 0  5 30
2 1 0 0 0  1 30
3 1 0 0 0 10 30
4 1 0 0 0 14 30
1 0 1 0 0  5 31
2 0 1 0 0  3 31
3 0 1 0 0  3 31
4 0 1 0 0 20 31
1 0 0 1 0 10 31
2 0 0 1 0  6 31
3 0 0 1 0 12 31
4 0 0 1 0  3 31
1 0 0 0 1  7 29
2 0 0 0 1 12 29
3 0 0 0 1  8 29
4 0 0 0 1  2 29
/END
/FINISH
```

Figure 3 Edited output from BMDP3R showing convergence steps, parameter estimates, and asymptotic standard errors.

ITERATION NUMBER	INCREMENT HALVINGS	LOSS FUNCTION	GAMMA1	GAMMA2	GAMMA3	GAMMA4	THETA1	THETA2
0	0	30.1428	1.630000	1.630000	1.000000	1.000000	0.000000	0.000000
1	0	10.1381	1.629240	1.629240	0.926800	0.926800	0.663940	0.663940
2	0	9.59077	1.629241	1.629241	0.927986	0.927986	0.823133	0.823133
3	0	9.58209	1.629241	1.629241	0.927987	0.927987	0.834557	0.834557
4	0	9.58209	1.629241	1.629241	0.927987	0.927987	0.834613	0.834613
5	0	9.58209	1.629241	1.629241	0.927987	0.927987	0.834613	0.834613
6	2	9.58209	1.629241	1.629241	0.927987	0.927987	0.834613	0.834613
7	2	9.58209	1.629241	1.629241	0.927987	0.927987	0.834613	0.834613
7	5	9.58209	1.629241	1.629241	0.927987	0.927987	0.834613	0.834613

ITERATION NUMBER	INCREMENT HALVINGS	LOSS FUNCTION	THETA3	THETA4	BETA1	BETA2	BETA3	DELTA1
0	0	30.1428	0.000000	0.000000	0.000000	0.000000	-1.660000	0.000000
1	0	10.1381	-0.705323	-0.705323	0.000000	0.000000	-1.331262	0.000000
2	0	9.59077	-0.600175	-0.600175	0.000000	0.000000	-1.595623	0.000000
3	0	9.58209	-0.599483	-0.599483	0.000000	0.000000	-1.626028	0.000000
4	0	9.58209	-0.599483	-0.599483	0.000000	0.000000	-1.626029	0.000000
5	0	9.58209	-0.599483	-0.599483	0.000000	0.000000	-1.626029	0.000000
6	2	9.58209	-0.599483	-0.599483	0.000000	0.000000	-1.626029	0.000000
7	2	9.58209	-0.599483	-0.599483	0.000000	0.000000	-1.626029	0.000000
7	5	9.58209	-0.599483	-0.599483	0.000000	0.000000	-1.626029	0.000000

ITERATION NUMBER	INCREMENT HALVINGS	LOSS FUNCTION	DELTA2	DELTA3
0	0	30.1428	1.000000	1.000000
1	0	10.1381	1.000000	1.000000
2	0	9.59077	1.000000	1.157625
3	0	9.58209	1.000000	1.147819
4	0	9.58209	1.000000	1.148077
5	0	9.58209	1.000000	1.148077
6	2	9.58209	1.000000	1.148077
7	2	9.58209	1.000000	1.148077
7	5	9.58209	1.000000	1.148077

ITERATION 7 HAS THE SMALLEST LOSS FUNCTION(SUBJECT TO CONSTRAINTS, IF ANY).

PARAMETER	ESTIMATE	ASYMPTOTIC STANDARD DEVIATION	TOLERANCE
GAMMA1	1.629241	0.345844	0.2168177975
GAMMA2	1.629241	0.345844	0.2135888060
GAMMA3	0.927987	0.286495	0.2349297025
GAMMA4	0.927987	0.286495	0.2430840954
THETA1	0.834613	0.625208	0.1032794369
THETA2	0.834613	0.625208	0.1019756658
THETA3	-0.599483	0.421468	0.1871953269
THETA4	-0.599483	0.421468	0.1920284580
BETA1	0.000000	0.000000	0.3822374771
BETA2	0.000000	0.000000	0.4731990749
BETA3	-1.626029	0.457673	0.2382297510
DELTA1	0.000000	0.000000	0.4452469641
DELTA2	1.000000	0.000000	0.3323873391
DELTA3	1.148077	0.640712	0.2434495500

Figure 4 Observed and predicted proportions from a 6 parameter fit. Standardized residuals can be computed and examined for lack of fit.

CASE NO. LABEL	PREDICTED RATIO	STD DEV OF PRED VALUE	OBSERVED RATIO	RESIDUAL	COOK DISTANCE	CASEWT	CPID	X1
1	0.163934	0.047401	0.166667	0.002732	0.000116	183.000000	1.000000	1.000000
2	0.065574	0.031694	0.033333	-0.032240	0.053444	457.500000	2.000000	1.000000
3	0.213115	0.052432	0.333333	0.120219	0.149656	140.769232	3.000000	1.000000
4	0.557377	0.063596	0.466667	-0.090710	0.011252	53.823529	4.000000	1.000000
5	0.163934	0.047401	0.161290	-0.002644	0.000121	189.100000	1.000000	0.000000
6	0.065574	0.031694	0.096774	0.031200	0.056607	472.750000	2.000000	0.000000
7	0.213115	0.052432	0.096774	-0.116341	0.156159	145.461538	3.000000	0.000000
8	0.557377	0.063596	0.645161	0.087784	0.011463	55.617647	4.000000	0.000000
9	0.283333	0.058174	0.322581	0.039247	0.011240	109.411764	1.000000	0.000000
10	0.300000	0.059161	0.193548	-0.106452	0.074239	103.333333	2.000000	0.000000
11	0.333333	0.060858	0.387097	0.053763	0.015390	93.000000	3.000000	0.000000
12	0.083333	0.035681	0.096774	0.013441	0.008205	372.000000	4.000000	0.000000
13	0.283333	0.058174	0.241379	-0.041954	0.010434	102.352941	1.000000	0.000000
14	0.300000	0.059161	0.413793	0.113793	0.069095	96.666667	2.000000	0.000000
15	0.333333	0.060858	0.275862	-0.057471	0.014397	87.000000	3.000000	0.000000
16	0.083333	0.035681	0.068966	-0.014368	0.007329	348.000000	4.000000	0.000000

Software and Interactive Systems

Organizer: *Sally E. Howe*

Invited Presentations:

Graphical Analysis of Proportional Poisson Rates
Brian S. Yandell

C-LAB, An Interactive System for Cluster Analysis
Marvin B. Shapiro and G. D. Knott

Scatterplot Matrix Techniques for Large N
Daniel B. Carr, Richard J. Littlefield, and Wesley L. Nicholson

Status of the NBS Guide to Available Mathematical Software
Sally E. Howe

COMPUTER SCIENCE AND STATISTICS:
The Interface, D.M. Allen (ed.)

GRAPHICAL ANALYSIS OF PROPORTIONAL POISSON RATES

Brian S. Yandell

University of Wisconsin - Madison

We present graphical tools for examining proportionality of a Poisson process rate to a baseline from a group of similar processes. We examine smooth deviations from this baseline using smoothing splines for general linear models. An example of egg-laying rates for leafhoppers is examined in some detail.

1. Introduction

This paper concerns inference for nonstationary Poisson rates which are "almost" proportional to a common baseline. It provides a means for "pre-smoothing" rate estimates to avoid some of the common problems of estimating functions with large curvature at certain places.

One may believe that a group of female potato leafhoppers in the same fluctuating temperature regime (Hogg, 1984) would oviposit at rates which rose and fell at roughly the same time. That is, one would suppose that the oviposition rates would be proportional to a common baseline rate. One could estimate this baseline rate, and then estimate the individual curves by simply determining the constant of proportionality, as was done by Bartoszyński et al. (1981). However, one might want to examine the proportionality as a function of time to determine whether or not it is constant.

We propose a method to estimate this proportionality over time. Although many approaches are possible (Clevenson and Zidek, 1977; Hastie and Tibshirani, 1984), we develop our estimators in the framework of penalized maximum likelihood (Good and Gaskins, 1971; O'Sullivan, Yandell, and Raynor, Jr., 1984).

Section 2 formulates the problem of proportional rates. Penalized maximum likelihood estimators for the baseline rate and proportionality terms are developed in Section 3. Section 4 briefly presents diagnostic tools. The methods are applied to leafhopper oviposition data in Section 5.

2. Proportional Poisson Rates

An individual leafhopper i, $i=1, \cdots, r$, may lay Y_{ij} eggs at time t_j, $t_1 < \cdots < t_n$. The count Y_{ij} is assumed Poisson with mean $h_i(t_j)$, which may be nonstationary. We focus on the model

$$h_i(t) = h^o(t)\, a_i(t),\ t \geq 0,\ i=1, \cdots, r. \qquad (2.1)$$

Proportional rates would correspond to constant a_i, with $h^o(.)$ being the baseline rate. Taking logarithms yields

$$\log(h_i(t)) = \log(h^o(t)) + \log(a_i(t)), \qquad (2.1)$$

or, reparameterizing one has

$$\theta_i(t)) = \theta^o(t) + \alpha_i(t),\ t \geq 0,\ i=1, \cdots, r. \qquad (2.2)$$

The degree to which the a_i, or α_i, are not constant corresponds to how much the proportional rates assumption is violated. This suggests that one could evaluate the degree of nonproportionality by estimating a_i, or equivalently α_i, and plotting these against time.

2.1. Log Likelihood

The likelihood can be written down and decomposed into pieces so that, subject to constraints, we can have a separate likelihood for each individual proportionality term. The overall log likelihood

$$\frac{1}{nr}\sum_i\sum_j Y_{ij}[\log(Y_{ij}) - \theta^o(t_j) - \alpha_i(t_j)]$$

can be reexpressed as the sum of

$$L(\theta^o) = \frac{1}{n}\sum_j Y_{+j}[\log(Y_{+j}/r) - \theta^o(t_j)] \qquad (2.3)$$

and

$$\frac{1}{nr}\sum_i\sum_j Y_{ij}[\log(rY_{ij}/Y_{+j}) - \alpha_i(t_j)] \qquad (2.4)$$

Throughout this paper, "+" indicates sum over the intended index. Note that (2.3) is a Poisson penalized likelihood, and (2.4) is a multinomial penalized likelihood conditional on Y_{+j}. In other words, Y_{ij} is binomial $(Y_{+j}, a_i(t_j)/r)$. This suggests splitting (2.4) into r terms of the form

$$L(\alpha_i) = \frac{1}{n}\sum_j Y_{ij}\log\left[\frac{Y_{ij}}{\mu_{ij}}\right] + (Y_{+j} - Y_{ij})\log\left[\frac{Y_{+j} - Y_{ij}}{Y_{+j} - \mu_{ij}}\right]$$

in which $\mu_{ij} = a_i(t_j)Y_{+j}/r$. Thus the log likelihood can be split into $r+1$ terms, for θ^o and for α_i, $i=1, \cdots, r$, with the the restriction that $\sum a_i(t)/r = 1$.

3. Penalized Maximum Likelihood Estimates

We now impose a penalty on the estimators to insure a certain smoothness not guaranteed by the likelihood as written. The penalized maximum likelihood estimate (MPLE) for the baseline rate (Bartoszyński et al., 1981; O'Sullivan, Yandell, and Raynor, Jr., 1984) can be found by minimizing, for fixed λ,

$$L(\theta^o, \lambda) = L(\theta^o) + \lambda J(\theta^o) \qquad \mathbf{(3.1)}$$

in which J() is an appropriate penalty function, typically

$$J(f) = \int (f^{(m)}(t))^2\, dt \qquad (3.2)$$

with m = 1 or 2 for penalty on the slope or curvature, respectively. A large value of the penalty, or smoothing, parameter λ forces θ^o to be nearly linear, while a small λ allows θ^o to interpolate the data.

The smoothing splines incorporate a prior belief that the true curve is smooth in a certain sense. The smoothing parameters λ are chosen by means of generalized cross validation (Craven and Wahba, 1979), which tries to minimize the mean square error, forcing a tradeoff between bias and variance.

Similar expressions can be written down for determining the MPLE of α_i for each i,

$$L(\alpha_i,\lambda_i) = L(\alpha_i) + \lambda_i J(\alpha_i) \tag{3.3}$$

When $m=1$ and $\lambda_i=\infty$, the constant MPLEs are

$$\bar{\alpha}_i = \log(rY_{i+}/Y_{++}),\ i=1,\cdots,r.$$

The estimation problem can be split into $r+1$ minimization problems, for θ^o and for α_i, $i=1,\cdots,r$, provided we are willing to ignore the restriction that $\sum a_i/r=1$. Of course, such a restriction could be imposed, but it would place awkward constraints on the smoothing penalties.

3.1. Data over Time Intervals

The data considered in Section 5 is grouped by 2 or 3 day intervals. With this design imbalance, the estimates of θ^o and of α_i may be biased, depending on the pattern of grouping. However, the unconditional expectation of the estimates is unbiased provided that the pattern of grouping is independent of the state of an individual. We can adjust the penalized likelihood expressions in a natural way to account for the reduced data, namely,

$$L(\theta^o) = \frac{1}{n}\sum_j Y_{+j}[\log(Y_{+j}/d_{+j})-\theta^o(t_j)] \tag{3.4}$$

$$L(\alpha_i) = \frac{1}{n_i}\sum_j Y_{ij}\log\left[\frac{Y_{ij}}{\mu_{ij}}\right]+(Y_{+j}-Y_{ij})\log\left[\frac{Y_{+j}-Y_{ij}}{Y_{+j}-\mu_{ij}}\right]$$

in which $\mu_{ij} = a_i(t_j)d_{ij}Y_{+j}/d_{+j}$. That is, for each i, there were n_i distinct times t_j at which counts Y_{ij} were made. These counts encompass d_{ij} days each, and the proportion of days for i out of the total count Y_{+j} is d_{ij}/d_{+j}. These technical adjustments were used for computing, but are not pursued further in this paper.

3.2. Survival and Oviposition

Throughout the leafhopper study, individuals died. Thus group size declined over time. These deaths can affect the estimate of the "baseline" rate h^o, as well as the proportionality terms a_i, even if all the rates are constant. This problem is most profound for small groups, such as in the latter portion of the leafhopper experiment.

A simple solution shown in the data analysis section is to factor out a step function from the baseline rate, with steps at times of death. This can be easily accomplished with partial splines (Shiau, 1985; Wahba, 1983a). Appropriate modifications can then be made to (2.4) based on the estimated step sizes. A serious danger arises in overparameterizing the model with steps for each individual.

4. Diagnostics for Poisson Rates

We propose an ad hoc "confidence interval" and log likelihood residuals for graphical inspection of proportionality. At present we have no concrete results, but support these tools by analogy to other work.

Several diagnostics have been proposed for penalized maximum likelihood in the linear (least squares) model with i.i.d. errors. Wahba (1983b) proposed pointwise confidence intervals based on a Bayesian model with normal errors. Carmody, Eubank, and Thombs (1984) proposed jackknife confidence intervals which performed poorly in comparison to the intervals of Wahba (1983b). Other diagnostics based on residuals (Eubank, 1984; Gunst and Eubank, 1983) naturally extend diagnostics for unpenalized problems. Recent work of Cox (1984) offers strong approximation of the penalized least squares estimator in the i.i.d. case, under certain conditions on the design points and smoothing parameter, which lead to simultaneous confidence bands if one ignores bias. Another direction based on a supremum penalty for the regression function (Knafl, Sacks, and Ylvisaker, 1983ab) yields bias-corrected simultaneous confidence bands: here, bias is accounted for by a bias correction.

We adapt Wahba (1983b) to the non-i.i.d. case and argue in an ad hoc fashion that this might have reasonable properties for our problem. We consider the model

$$X = g+\epsilon,\ g\,\tilde{}\,N(0,(n\lambda)^{-1}\Sigma_{gg}),\ \epsilon\,\tilde{}\,N(0,\Sigma),$$

with Σ diagonal. The posterior estimator of g is

$$\hat{g} = E(g\mid X) = \Sigma_{gg}(\Sigma_{gg}+n\lambda\Sigma)^{-1}X = H_\lambda X. \tag{4.1}$$

The covariance is derived in an analogous fashion as

$$COV(g\mid X) = (I+H_\lambda)\Sigma_{gg}/(n\lambda) = H_\lambda\Sigma. \tag{4.2}$$

This suggests an approximate 95% confidence interval for g_i

$$\hat{g}_j \pm 1.96\,\sigma_j\sqrt{h_{jj}(\lambda)} \tag{4.3}$$

Now suppose, for fixed i, we let $X_j=\log(Y_{ij}/(Y_{+j}-Y_{ij}))$ and approximate the covariance to first order,

$$\sigma_j^2 = 2r\exp(-\alpha_i(t_j))/Y_{+j},\ j=1,\cdots,n.$$

The estimated confidence interval for $\alpha_i(t_j)$ becomes

$$\hat{\alpha}_{i\lambda}(t_j)\pm 1.96\sqrt{2h_{jj}(\lambda_i)r\exp(-\hat{\alpha}_{i\lambda}(t_j))/Y_{+j}} \tag{4.4}$$

This approach has some problems, as the solution to the penalized log likelihood is not the same as the solution to a logit regression with normal errors. We will pursue this in later work using ideas of Leonard (1982).

We propose an ad-hoc test of the hypothesis of constant proportionality by computing the difference in deviances between the smooth and constant estimates,

$$D(i,\lambda) = 2[L(\bar{\alpha}_i)-L(\hat{\alpha}_{i\lambda})],\ i=1,\cdots,r, \tag{4.5}$$

with $\hat{\alpha}_{i\lambda}()$ being the spline estimate of $\alpha_i()$ for fixed smoothing parameter λ and $\bar{\alpha}_i$ the estimate for constant α_i. In other words, $D(i,\lambda)$ is simply the deviance between the constant and the smoothed logit models. We suppose that this statistic may have approximately a chi-square distribution with degrees of freedom $(n-1)-trace(I-H_\lambda)$. We will compare this with the usual likelihood ratio statistic, $D(i)=2L(\bar{\alpha}_i)$ with $n-1$ degrees of freedom, in the data analysis section.

Expression (4.5) suggests examining the deviance contributions at t_j (Green, 1984; Pregibon, 1981)

$$\pm[2Y_{ij}(\log(Y_{ij})-\hat{\alpha}_{i\lambda}(t_j))]^{1/2}, \tag{4.6}$$

with the sign the same as that of $Y_{ij}-\exp(\hat{\alpha}_{i\lambda}(t_j))Y_{+j}/r$. For given t_j, this is approximately $N(0,1)$: thus large positive or negative values suggest significant deviations. However, the graphical "tests" at different t_j are highly correlated, and a graphical plot of t versus logit residuals cannot be viewed as a global test.

5. Data Analysis

We consider data from a laboratory experiment conducted by Hogg (1984) in which female potato leafhoppers were kept in controlled laboratory conditions at one of three fluctuating temperature regimes. We focus here only on the cold regime. We examine the baseline for the 23 females in this group along with the proportional term for two of these females. A more complete analysis is in progress jointly with David Hogg, Entomology Department, UW-Madison, who kindly offered the data he collected.

All individuals have grouped records, that is counts of eggs for 1-3 day intervals. Also, individuals were removed from the study by death, either natural or accidental (due to handling). We assume that the grouping does not introduce any bias in the estimation of the baseline rate, and that we are interested in the baseline rate and proportionality terms at any time only for those leafhoppers which were alive. We initially proceed as if survival did not affect bias, and later correct for survival as indicated in Section 3.3.

Figure 5.1 shows the baseline rate and the rates for individuals 22 and 23. Note the rise to a fairly constant rate, with gradual decay. The raw proportionality for individuals 22 and 23 are plotted alongside curve estimates with penalties for slope and for curvature in Figures 5.2-3. The curve estimate based on a penalty for non-zero slope appear much rougher than the curves based on curvature penalty. Approximate 95% pointwise confidence intervals for the proportionality estimates, based on the curvature penalty, are shown in Figures 5.4-5.

The likelihood ratio statistics with degrees of freedom and p-value are shown in Table 5.1. Note the great reduction in degrees of freedom for the penalized curves, while the deviances stay fairly high. Figure 5.6-7 show the logit deviances over time.

Table 5.1 Smooth Deviances

	Deviance	d.f.	log(λ)
#22:			
constant	188.21	68.	∞
m = 1 (slope)	117.66	14.88	-6
m = 2 (curvature)	99.49	8.83	-12
#23:			
constant	113.99	64.	∞
m = 1 (slope)	63.56	5.87	-4
m = 2 (curvature)	62.36	3.35	-8

We conclude with curve estimates for the baseline once one adjusts for the survival process. Figure 5.8 shows the naive and adjusted baseline rate estimates for the cold regime. One sees that survival has little effect on the baseline rate for most of the experiment, though estimates at the later times can be affected.

Acknowledgements

This work was supported in part by USDA-CSRS grant 511-100. Computing was performed on the Statistics VAX 11/750 Research Computer at the University of Wisconsin. Discussions with David Hogg were most helpful.

References

Bartoszyński, R., Brown, B. W., McBride, C., and Thompson, J. R. (1981), "Some Nonparametric Techniques for Estimating the Intensity Function of a Cancer Related Nonstationary Poisson Process," *Ann. Statist.*, 9, 1050-1060.

Carmody, T. J., Eubank, R. L., and Thombs, L. A. (1984) "Jackknife Confidence Intervals for Smoothing Splines." Technical Report, Dept. of Statistics, So. Methodist U..

Clevenson, M. L., and Zidek, J. V. (1977), "Bayes Linear Estimators of the Intensity Function of the Nonstationary Poisson Process," *Journal of the American Statistical Association*, 72, 112-120.

Cox, D. D. (1984) "Gaussian Approximation of Smoothing Splines." Technical Report#743, Dept. of Statistics, U. of Wisconsin.

Craven, P., and Wahba, G. (1979), "Smoothing Noisy Data with Spline Functions: Estimating the Correct Degree of Smoothing by the Method of Generalized Cross-Validation," *Numer. Math.*, 31, 377-403.

Eubank, R. L. (1984), "The Hat Matrix for Smoothing Splines," *Statist. & Probab. Letters*, 2, 9-14.

Good, I. J., and Gaskins, R. A. (1971), "Non-Parametric Roughness Penalties for Probality Densities," *Biometrika*, 58, 255-277.

Green, P. J. (1984), "Iteratively Reweighted Least Squares for Maximum Likelihood Estimation, and Some Robust and Resistant Alternatives (with Discussion)," *J. Roy. Statist. Soc. B*, 46, 149-192.

Gunst, R. F., and Eubank, R. L. (1983) "Regression Diagnostics and Approximate Inference Procedures for Penalized Least Squares Estimators." Technical Report#181, Dept. of Statistics, So. Methodist U..

Hastie, T. J., and Tibshirani, R. J. (1984) "Generalized Additive Models." Technical Report#98, Div. of Biostatistics, Stanford U..

Hogg, D. B. (1984) "Potato Leafhopper (Homoptera: Cicadellidae) Immature Development, Life Tables, and Population Dynamics Under Fluctuating Temperature Regimes." *Environmental Entomology*, . (submitted)

Knafl, G., Sacks, J., and Ylvisaker, D. (1983a) "Uniform Confidence Intervals for Regression Estimates." Technical Report, Center for Statistics and Probability, Northwestern U..

Knafl, G., Sacks, J., and Ylvisaker, D. (1983b) "Model Robust Confidence Intervals II." Technical Report#55, Center for Statistics and Probability, Northwestern U..

Leonard, T. (1982) "An Empirical Bayesian Approach to the Smooth Estimation of Unknown Functions." Technical Report#2339, Math. Research Center, U. Wisconsin.

O'Sullivan, F., Yandell, B. S., and Raynor, Jr., W. J. (1984) "Automatic Smoothing of Regression Functions in Generalized Linear Models." Technical Report#734, Dept. of Statistics, U. of Wisconsin.

Pregibon, D. (1981), "Logistic Regression Diagnostics," *Annals of Statistics*, 9, 705-724.

Shiau, J-J H. (1985), "Smoothing Spline Estimation of Functions with Discontinuous D-th Derivatives". (Ph.D. thesis in preparation, U. Wisconsin, Madison)

Wahba, Grace (1983a) "Cross Validated Spline Methods for the Estimation of Multivariate Functions from Data on Functionals." Technical Report#722, Dept. of Statistics, U. of Wisconsin.

Wahba, G. (1983b), "Bayesian "Confidence Intervals" for the Cross-Validated Smoothing Spline," *J. Roy. Statist. Soc. B*, 45, 133-150.

Figure 5.1

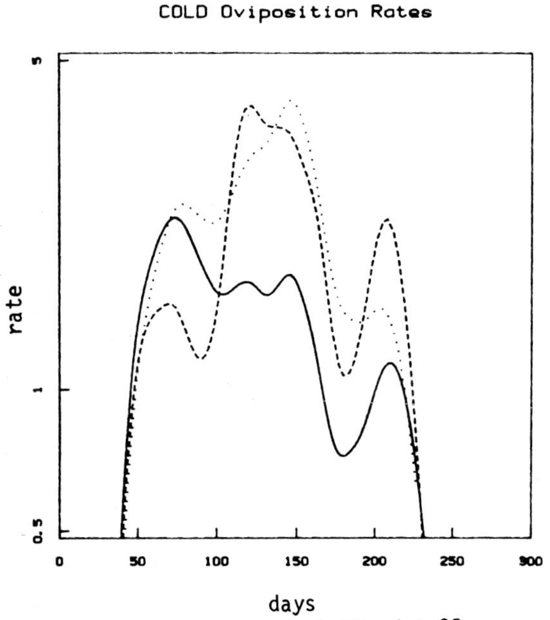

solid=baseline; dash=22; dot=23

Figure 5.2

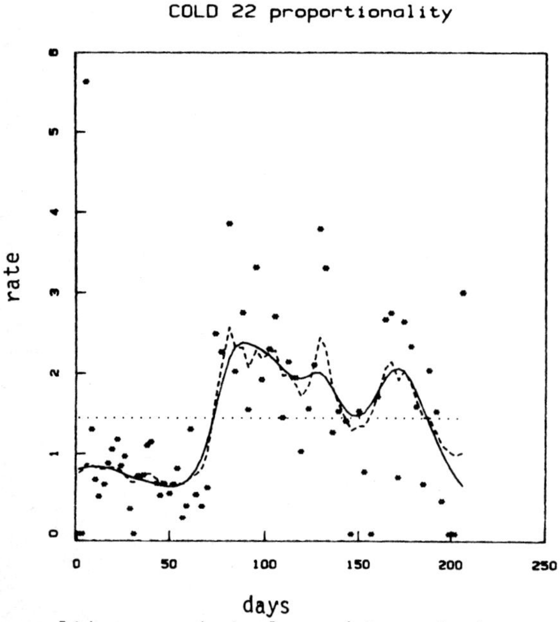

solid=curve; dash=slope; dot=constant

Figure 5.3

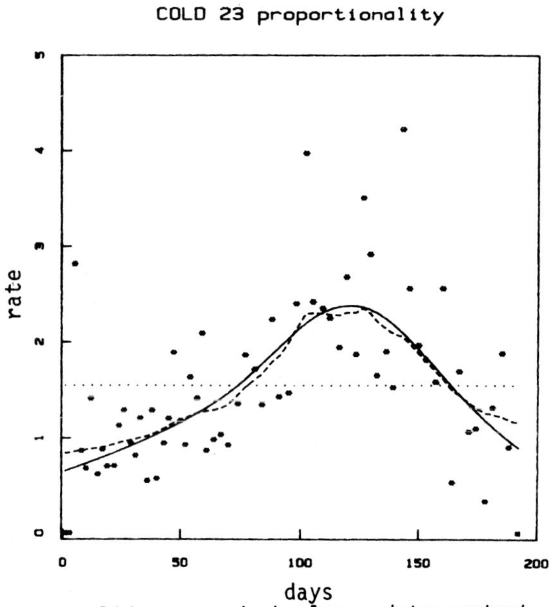

solid=curve; dash=slope; dot=constant

Figure 5.4

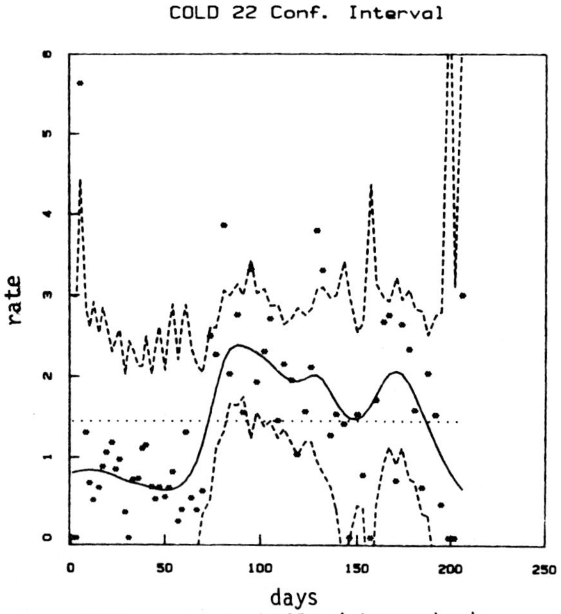

solid=curve; dash=CI; dot=constant

Figure 5.5

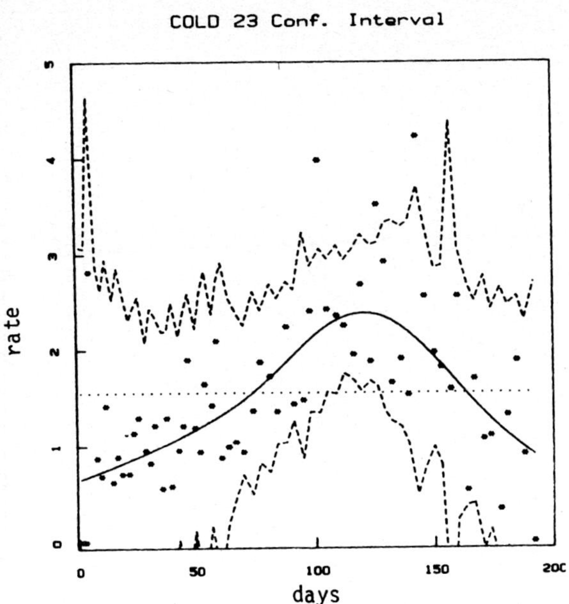

solid=curve; dash=CI; dot=constant

Figure 5.7

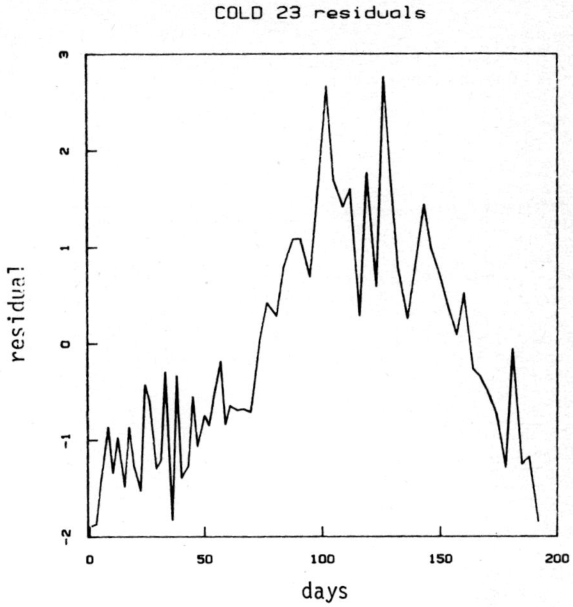

Figure 5.6

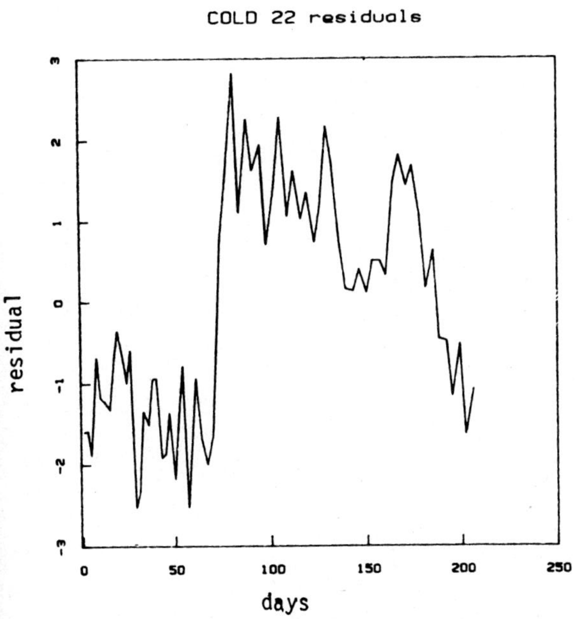

Figure 5.8

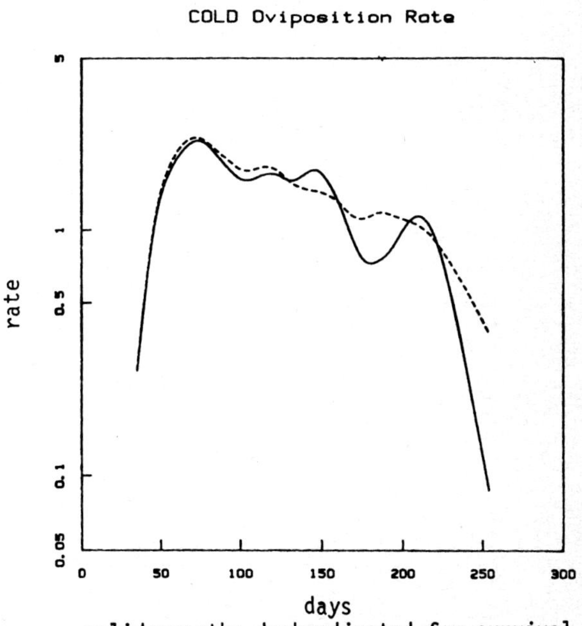

solid=smooth; dash=adjusted for survival

COMPUTER SCIENCE AND STATISTICS:
The Interface, D.M. Allen (ed.)
Elsevier Science Publishers B.V. (North-Holland), 1986

C-LAB, AN INTERACTIVE SYSTEM for CLUSTER ANALYSIS

M.B. SHAPIRO and G.D. KNOTT

Division of Computer Research and Technology, NIH, Bethesda, MD 20205

The C-LAB system, for doing cluster analysis and related work, is described. It differs from other cluster analysis systems in that (1) it is interactive, (2) it has its own built-in language in which user algorithms can be programmed, (3) it contains many built-in functions for matrix manipulation, numerical analysis and statistics, and (4) it is display oriented, having commands for producing publication quality clustering diagrams.

1. INTRODUCTION

C-LAB is an on-line Clustering LABoratory facility that runs on DECSYSTEM-10 and -20 computers. It consists of a collection of subroutines which implement many of the most commonly used techniques in cluster analysis, plus some miscellaneous related methods. C-LAB is a subset of operators in the MLAB (Modeling LABoratory) system (Knott 1979), which has its own high-level language for writing programs. C-LAB differs from other cluster analysis packages in three main ways: (1) it is interactive, (2) it has a built-in language (the MLAB language), and (3) it is display oriented. MLAB provides matrix manipulation and display facilities and has many built-in functions useful in statistics and numerical analysis. C-LAB is run on display terminals and since it is interactive, results and drawings are seen as they are computed. MLAB has its own commands for drawing pictures, and these are supplemented by C-LAB operators for preparing the output of cluster analysis algorithms for drawing. A user can program his own algorithms not available in C-LAB. As pointed out by Anderberg (1973) most cluster analysis methods are relatively easy to program. Such special algorithms can be programmed as subroutines (called DO files in MLAB) and invoked to process specific data.

2. THE C-LAB OPERATORS

There are many aspects to cluster analysis, including the choice of data units, variables, clustering criteria, and of what to cluster, the method of homogenizing variables, the computation of similarity measures and clustering algorithms, and, finally, the presentation and interpretation of the results. These aspects are dealt with by C-LAB as described in the following.

Most C-LAB operators work on a data matrix, where each row represents a data point (also called a sample or an object) in n dimensions. The columns are called variables (or features or attributes). Data to be clustered must have similar scales of values; C-LAB has two operators for scaling.

There are many measures of similarity (or dissimilarity) between pairs of objects, the most common being the euclidean distance. C-LAB has an L^p distance metric built in as a dissimilarity operator. Other dissimilarity measures are usually easy to program.

The basic idea of cluster analysis is to partition a set of n-dimensional points representing measurements or descriptive values of an object (e.g. measurements of different parts of a plant, or symptoms of a disease) into groups called clusters. The number of and nature of the clusters may or may not be specified, and the clusters are to be discovered. Also of interest are the properties of the points which determine to which cluster they belong.

The usual paradigm for cluster analysis is to define a similarity measure or metric, d(x,y), which produces a numerical measure of how similar the two points x and y are. The choice of such a metric can be crucial and is, of course, left to the user. Once the metric is chosen clusters can be defined in various ways, based on grouping similar points together.

The main part of C-LAB consists of the operators for doing clustering. There are operators for each of the three broad categories of clustering algorithms: hierarchical clustering, non-hierarchical clustering, and approaches using graph theory. The hierarchical clustering operators are those for computing and drawing dendrograms. Clusters are determined by visually examining the drawing; there is no algorithm in C-LAB for selecting clusters from dendrograms. Non-hierarchichal clustering is done in C-LAB using a variant of the K-means algorithm: objects are put into separate clusters, using a minimum variance optimizing criterion, and information about each cluster is then printed out, rather than drawn as it is for dendrograms. A graph theory approach to clustering is implemented in C-LAB through the minimal spanning tree operator and related operators for "breaking" certain "inconsistent" tree edges. Clusters are then defined as the resulting subtrees.

Graphical output is a specialty of MLAB and there are a number of C-LAB operators used for displaying results as drawings. In addition to the standard MLAB facilities for drawing graphs there are C-LAB operators which compute matrices from which dendrograms, minimal spanning trees, and Chernoff faces (Chernoff 1973) can be drawn. There are two operators for reducing the dimensionality in a set of data and they can be used to obtain a plot of the data in 2 dimensions, and in 3 dimensions also, since there are commands for drawing pictures in 3D.

The C-LAB operators are organized into six categories: scaling, feature reduction, cluster analysis, output, triangulation, and miscellaneous. At present the following operators are available:

<u>Scaling</u>	<u>Feature reduction</u>
AUTOSCALE	FISHERRANK
RANGESCALE	PRCOMP
	NLM
<u>Cluster analysis</u>	
MST	<u>Output</u>
INCONSISTENT	CLUSTERINFO
TREECLUSTERS	DENCURVE
ALINKAGE	FACESCURVE
CLINKAGE	TREECURVE
CENTROID	
WARD	<u>Triangulation</u>
KMEANS	DELAUN
	DELCURVE
<u>Miscellaneous</u>	VORCURVE
CLUSTERERROR	VORSTAT
COPHEN	
DISTANCES	

Auto-scaling and range-scaling are used for scaling the variables of a data matrix by normalizing them or by putting them in a 0-1 range. The FISHERRANK operator is the standard Fisher discriminant ratio, and is used for ranking the variables of a data set according to their ability to discriminate between known categories for the data. PRCOMP and NLM perform principal components and non-linear mapping algorithms, and are used for reducing the dimensionality of a set of data. The non-linear mapping algorithm is from Chang and Lee (1973).

For the cluster analysis operators, MST, INCONSISTENT, and TREECLUSTERS are used for computing a minimal spanning tree, then for finding "inconsistent" edges in that tree, "breaking" them and determining the resulting clusters. This approach is based on the work of Zahn (1971). MST, ALINKAGE, CLINKAGE, CENTROID, and WARD are the operators for computing dendrograms based on single linkage, average linkage, complete linkage, centroid linkage, and Ward's method. KMEANS is one of the many variants of the K-means algorithm, this one taken from Hartigan (1975).

The CLUSTERERROR operator computes the cluster error from a given clustering solution, such as computed by the K-means operator. COPHEN is used to compute the correlation between the dissimilarity matrix for a data set and a dendrogram computed for the data. (There are operators in MLAB for computing correlation and covariance matrices.) DISTANCES is an algorithm for computing Minkowski's distance metric, and is used for creating a dissimilarity matrix. The euclidean distance metric is most commonly used.

The output operators compute matrices that contain a summary of information about clusters (CLUSTERINFO), or matrices used to draw dendrograms (DENCURVE), Chernoff faces (FACESCURVE), or minimal spanning trees (TREECURVE). Examples of the use of DENCURVE and TREECURVE are given below.

In addition to the operators directly related to cluster analysis there are four used for the triangulation of a set of points in the plane. The triangulation is done by the DELAUN operator (for Delaunay triangulation, defined below), the triangulation drawing by DELCURVE, the computing and drawing of nearest neighbor (Voronoi or Dirichlet) regions by VORCURVE, and statistics related to the Voronoi regions are computed by VORSTAT. The triangulations algorithms are from Lee and Schachter (1980) and Shapiro (1981).

3. THE MLAB LANGUAGE

The MLAB language is extensive. Only a brief introduction to the statements and operators that would likely be used by C-LAB users is given here, however there are also operators for matrix manipulation, curve fitting, differential equation solving, and integration of functions, plus commands for input and output, including drawing pictures.

Assignment statements are similar to those in other computationally-oriented languages, having the form

    variable-name = expression

where the variable is a scalar or matrix depending on whether the expression is a scalar or matrix one. MLAB is a higher level language than FORTRAN or BASIC, and no declaration of variables is needed. Expressions have the same form as in other high level languages, for example a root of a quadratic equation would be expressed as

```
(-B+SQRT(B^2-4*A*C))/(2*A)
```

Scalars and matrices are created and manipulated through assignment statements and through the use of built-in and user-defined functions. Operators for matrices include the following:

```
A & B        Concatenate matrix B below A.
A &´ B       Concatenate matrix B to the
                  right of A.
A * B        Ordinary matrix multiplication.
A *´ B       Multiply corresponding elements.
A´           ´ indicates the transpose.
```

Some of the commonly used built-in functions are the following:

```
A:B:C                The values A,A+C,A,A+2C,
                     ...,B. If C is omitted
                     then C=1 is assumed.
NROWS(X)             The number of rows in
                     matrix X.
NCOLS(X)             The number of columns in
                     matrix X.
READ(DATA,M,N)       Input data from file
                     DATA into an MxN matrix.
SORT(X,C)            Sort matrix X, using
                     column C as the key.
SUM(I,A,B,E)         The sum of expression E
                     for index I running from
                     A to B. Usually E
                     contains index I.
CORR(X)              The correlation matrix
                     for matrix X.
LIST(E1,E2,...,En)   A one-column matrix with
                     n elements. E1,E2,...,En
                     are expressions.
CROSS(1:M,1:N)       An MxN matrix containing
                     (1,1) (1,2) ... (M,N).
```

Specific rows and/or columns of matrices can be referenced, as in

```
X(I,J)               The I,J th element of
                     matrix X.
Y ROW 1:5            Rows 1 to 5 of Y.
                     (":" indicates through.)

Z ROW A:B COL C:D    Columns C to D of rows
                     A to B of Z.
```

The essential ingredients in most MLAB programs are the function statements. Functions are defined as in the following:

```
FCT F(X)=A*X^2+B*X+C         A quadratic
                             function.
FCT G(T)=A*EXP(-B*H(T))      H is a previously
                             defined function.
FUNCTION MAX(A,B)=IF A<B THEN B ELSE A
                                   Max of A and B.
```

Functions are computed using the ON and POINTS operators, as in

```
U = F ON 2:9           U is a column vector of
                       (F(2),F(3),...,F(9)).
V = POINTS(F,A:B)      A:B in column 1,
                       (F(A),...,F(B)) in
                       column 2.
```

Pictures are drawn with the DRAW statement, which specifies a matrix of coordinates to be drawn in a window. The window specifies the position of an imaginary box around the data on the display screen, e.g.

```
WINDOW W, 10 BY 20, AT 0,0
```

indicates that window W is 10 data units by 20 data units, with the lower left of the screen having coordinates 0,0. Thus the point (5,10) would be plotted in the middle of the screen. The STRING statement is used to display characters, as in the following:

```
STRING "ABC" IN W, AT 5,10      ABC is drawn
                                starting at
                                5,10.
```

The DRAW statement has a number of options. Some of those that are used with C-LAB are illustrated in the following, where Z is a 2 column matrix of (x,y) coordinates and W is a window as described above.

```
DRAW Z IN W, LINE 1 The points are connected
                        by a solid line.
DRAW Z IN W, LINE 0, LABEL WITH 1:NROWS(Z)
                    The points in Z are labeled
                    with consecutive integers
                    and the points are not
                    connected (LINE 0).
DRAW Z IN W, LINE 6   Line type 6 specifies
                      lifting the pen between
                      curve segments.
```

4. EXAMPLES

Five examples are given here to illustrate the type of programming and picture drawing associated with C-LAB. As can be seen, quite a bit is accomplished in a few statements.

4.1 Jaccard´s Coefficient

For presence-absence data, association coefficients are used for similarity measures. Jaccard´s coefficient is illustrated here. For two m-vectors X and Y containing 0 and 1 values, J is computed as a/(a+b+c), where

a=the number of places where both X and Y are 1
b=the number of places where X=1 and Y=0
c=the number of places where X=0 and Y=1.

```
FCT A(X,Y)=SUM(I,1,M,X[I] AND Y[I])
FCT B(X,Y)=SUM(I,1,M,X[I] AND NOT Y[I])
FCT C(X,Y)=SUM(I,1,M,NOT X[I] AND Y[I])
X=READ(DATA1,M); Y=READ(DATA2,M)
AA=A(X,Y)                 "compute a"
J = AA/(AA+B(X,Y)+C(X,Y))
```

4.2 Finding Nearest Neighbors

The euclidean distances (squared) of point Q (nx1) to each of the points in mxn matrix X are computed and sorted and put into matrix IX, which also contains the corresponding indices in column 2. Thus, after the code below is executed, the index of the point closest to Q is in IX[1,2] and the distance of it to Q is in IX[1,1].

```
FUNCTION DIST(I)=SUM(J,1,N,(X[I,J]-Q[J])^2)
D = DIST ON 1:M
IX = SORT(D &´ 1:M,1)
```

4.3 Drawing a Dendrogram

In this example an average linkage dendrogram is drawn. First the steps are explained, then the C-LAB statements for executing the steps are given. (The other linkages would be done similarly, changing only step 4.) The dendrogram is shown in Figure 1.

(1) Create a 1x1 window in which the dendrogram is to be drawn.

(2) Input the mxn data into matrix X, in this case 20x5.

(3) Compute a dissimilarity matrix D for the data.

(4) Compute (m-1)x3 matrix A defining the dendrogram.

(5) Use A to create matrix Q with coordinates (in columns 1 and 2)and labels (in column 3) for drawing the dendrogram.

(6) Draw the dendrogram, using columns 1 and 2 of Q to draw the lines and column 3 for the labels at the top.

```
WINDOW W, 1 BY 1, AT 0,0
X = READ(DATA,20,5)
D = DISTANCES(X)
A = ALINKAGE(D)
Q = DENCURVE(A)
DRAW Q COL 1:2 IN W, LINE 6,
       LABEL WITH Q COL 3
```

The distances shown at the left of the dendrogram are drawn separately. They range from 0 to the maximum value found in (column 3 of) matrix A. 900 is used here. The X coordinate for these numbers in the window is .02, and the Y coordinates go from .05 to .95. The numbers are drawn as follows:

```
L = .02 &´ (.05:.95:.3)
DRAW L IN W,LINE 0,LABEL WITH 900:300:-300
```

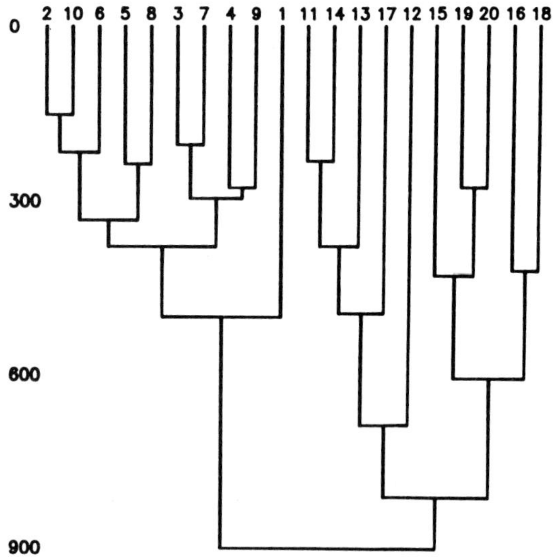

Figure 1: Average linkage dendrogram for 20x5 data

4.4 Drawing a Minimal Spanning Tree on a Non-linear Map

The NLM (non-linear mapping) operator is used to project the points in some higher dimension to 2 dimensions, preserving the interpoint distance relationships as much as possible. There is inevitably some distortion, and one way of assessing it is to superimpose a minimal spanning tree, with edge length labels, on the non-linear map, since the edges in the tree represent nearest neighbors connections. This type of combination was suggested by Kruskal (1977). It is easily done in C-LAB, as is described in the following steps. The C-LAB statements are shown at the end. Figure 2 shows a tight cluster containing points 1 to 10 and a loose cluster of points 11 to 20, and indicates that the non-linear map represents the data well.

(1) The algorithm starts with the mapped points in the 0-1 range, then on each iteration the points can move out of that range. Therefore a window is set up

covering the region -2 to +2 in the X and Y directions.

(2) Input the data (here 20 points in 5 dimensions) into matrix X.

(3) Compute the dissimilarity matrix D for the data.

(4) Map X into 2 dimensions (matrix N).

(5) Draw the 2 dimensional representation, labeling the points with their point indices.

(6) From D compute a 29x3 matrix M defining the minimal spanning tree for X. Columns 1 and 2 of M contain the point indices of each edge, and column 3 is the edge length.

(7) Use matrix M to compute a 3 column matrix Q of coordinates (in columns 1 and 2) and edge labels (column 3) for drawing the tree.

(8) Draw the minimal spanning tree edges, using matrix N as the coordinates of the points.

```
WINDOW W, 4 BY 4, AT -2,-2
X = READ(DATA,20,5)
D = DISTANCES(X)
N = NLM(D)
DRAW N IN W,LINE 0,LABEL WITH 1:NROWS(X)
M = MST(D)
Q = TREECURVE(M COL 1:2,0,N)
DRAW Q COL 1:2 IN W, LINE 6,
      LABEL WITH Q COL 3
```

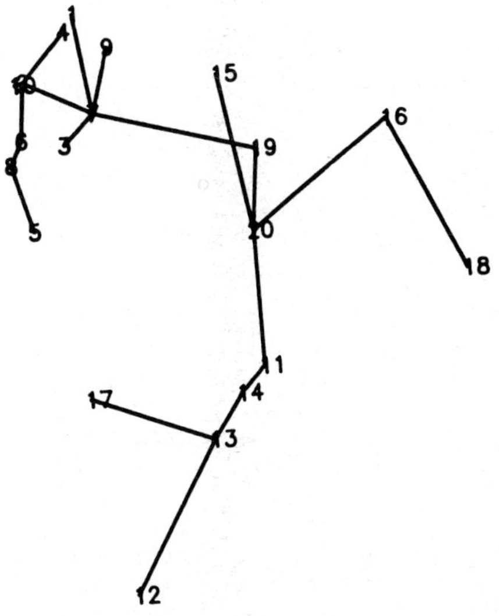

igure 2: A minimal spanning tree superimposed on a non-linear mapping of a 20x5 data set

4.5 Voronoi Regions

The computation and drawing of Voronoi diagrams is an area of much recent interest. They define nearest neighbor regions of planar data, and can be used to develop fast algorithms for finding nearest neighbors of a point. See Sibson (1980) for a discussion of their use.

For a set of vertices in two dimensions the nearest neighbor region for a given vertex i is defined as the locus of points closer to i than to any other coordinate. Such an area is called the Voronoi region for vertex i and, if bounded, is a convex polygon with on the average 6 sides. The edges of the Voronoi polygon for a given vertex are the perpendicular bisectors of the edges from that vertex to adjacent vertices. The Voronoi tesselation for a set of points, i.e. its collection of Voronoi polygons, can be found by first computing the Delaunay triangulation of the points, which is a maximal connection of the points with the restriction that no two edges intersect. For a set of m points there are 2(m-1)-h triangles, where h is the number of points on the convex hull of the m points.

There are many ways to do a triangulation. The Delaunay triangulation, which can be used to find Voronoi regions, has the special property that the circumcircle of each of its triangles does not contain any other point than its three vertices in its interior. The Voronoi region for a given point can be found by connecting the centers of adjacent Delaunay triangles for which that point is a vertex. Thus a Voronoi diagram can be constructed by first doing a Delaunay triangulation of the points, then connecting the centers of adjacent triangles.

A description of the steps used in C-LAB to create a Voronoi diagram (Figure 3) is given here. The C-LAB statements follow. A set of 16 coordinates is used as the data.

(1) The data is in the 0-1 range. Set up a window for drawing it.

(2) Input the 16x2 data into matrix X.

(3) Compute a 6 column matrix D defining the Delaunay triangulation of X. Each row in D defines a triangle in the triangulation; columns 1 to 3 contain the indices of the three adjacent triangles (in a counterclockwise direction), and columns 4 to 6 the indices of the points at the vertices of the triangle.

(4) From X and D compute a 2 column matrix V of coordinates for drawing the Voronoi diagram.

(5) Draw the Voronoi diagram. Line type 6 is used to lift the pen between segments of the diagram.

(6) Label the points.

```
WINDOW W, 1 BY 1, AT 0,0
X = READ(DATA,16,2)
D = DELAUN(X)
V = VORCURVE(X,D)
DRAW V IN W, LINE 6
DRAW X IN W,LINE 0,LABEL WITH 1:16
```

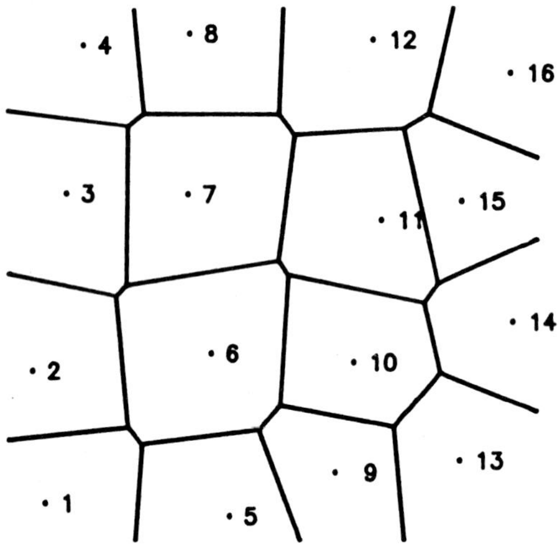

Figure 3: Voronoi diagram for 16 points in the plane

5. Discussion

Blashfield et al (1982) present a good summary of the current state of cluster analysis software. They divide such software into 5 categories and include C-LAB in with cluster analysis packages, the main ones being CLUSTAN (Wishart 1978), NT-SYS (Rohlf, Kishpaugh, and Kirk 1974), and CLUS (Rubin and Friedman 1967). (There is one mistake made in describing C-LAB: it has 5, not 3, linkage methods.) C-LAB does fit in that category more than in any other, but it has some unique features that separate it from all the other cluster analysis software discussed. These are that it is used interactively, has a built-in, high level language, and is oriented around graphics terminals, having built-in capabilities for drawing high quality pictures. However, C-LAB does not have the extensive set of clustering commands available in some other packages, notably CLUSTAN and NT-SYS. This is somewhat overcome by the fact that the C-LAB user can in many cases program his own special algorithms. Thus, whereas C-LAB has only one operator (DISTANCES) for computing dissimilarity values, it is usually easy to program others, as illustrated above for the Jaccard coefficient. This language feature can be considered a plus, but it also means that a beginner would have more trouble than with say CLUSTAN, which has 38 different similarity measures available.

Being interactive, C-LAB is designed to be used differently than other cluster analysis systems. Rather than the user having to know beforehand the exact series of computations to be done, succeeding steps are based on current results. The value of this feature depends on the particular work being done.

The usefulness of the graphical capabilities of the C-LAB language can be attested by the fact that most of the techniques found in Everitt (1978) for displaying multivariate data are either already available as C-LAB operators or are easily programmed. The former include the operators PRCOMP (principal components analysis), NLM (non-linear mapping), MST, CLINKAGE, ALINKAGE, CENTROID, and WARD (hierarchical clustering), and FACESCURVE (Chernoff faces). The latter include probability plots (Gerson 1975), Andrews plots (1972), and biplots (Gabriel 1971).

Copies of the system documentation and the MLAB/C-LAB program are available by writing the authors.

REFERENCES

ANDERBERG, M.R. (1973), Cluster Analysis for Applications, Academic Press, New York.

ANDREWS, D.F. (1972), "Plots of High Dimensional Data", Biometrics, 28, 125-136.

BLASHFIELD, R.K., ALDENDERFER, M.S. and MOREY, L.C. (1982), Cluster Analysis Software, in Handbook of Statistics 2, Krishnaiah,P.R. and Kanal,L.N. Eds., North-Holland, New York, 245-266.

CHERNOFF, H. (1973), "The Use of Faces to Represent Points in K-dimensional Space Graphically", J. Am. Stat. Assoc., 68, 361-368.

EVERITT, B.D. (1978), Graphical Techniques for Multivariate Data, North-Holland, New York.

GABRIEL, K.R. (1971), "The Biplot Graphic Display of Matrices with Applications to Principal Components Analysis", Biometrika, 58, 453-467.

GERSON, M. (1975), "The Techniques and Uses of Probability Plotting", The Statistician, 4, 235-257.

HARTIGAN, J.A. (1975), Clustering Algorithms, John Wiley & Sons, New York, Chapter 4.

KNOTT, G.D. (1979), "MLAB - A Mathematical Modeling Tool", Comp. Programs Biomed., 10, 271-280.

KRUSKAL, J. (1977), "The Relationship between Multidimensional Scaling and Clustering", in Classification and CLustering, Van Ryzin J., Ed.,Academic Press, New York, 17-44.

LEE, D.T. and SCHACHTER, B.J. (1980), "Two Algorithms for Constructing a Delaunay Triangulation", Intl. J. Comp. and Inf. Sci., 9, 219-242.

ROHLF, F.J., KISHPAUGH, J. and KIRK, D. (1974), NT-SYS User´s Manual, State University of New York at Stonybrook, Stonybrook.

RUBIN, J. and FRIEDMAN, H. (1967), "CLUS: A CLuster Analysis and Taxonomy System, Grouping, and Classifying Data", IBM Corporation, New York.

SHAPIRO, M. (1981), "A Note on Lee and Schachter´s Algorithm for Delaunay Triangulation", Intl. J. Comp. and Inf. Sci., 10, 413-418.

SIBSON, R. (1980), "The Dirichlet Tessalation as an Aid in Data Analysis", Scand. J. Statist. 7, 14-20.

WISHART, D. (1978), CLUSTAN IC User´s Manual, Program Library Unit of Edinburgh University, Edinburgh.

ZAHN, C.T. (1971), "Graph-Theoretical Methods for Detecting and Describing Gestalt Clusters", IEEE Trans. C-20, 68-86.

COMPUTER SCIENCE AND STATISTICS:
The Interface, D.M. Allen (ed.)

SCATTERPLOT MATRIX TECHNIQUES FOR LARGE N

D. B. Carr, R. J. Littlefield, and W. L. Nicholson

Pacific Northwest Laboratory
Richland, WA 99352

High-performance interaction with scatterplot matrices is a powerful approach to exploratory multivariate data analysis. For a small number of data points, real-time interaction is possible and overplotting is usually not a major problem. However, when the number of plotted points is large, display techniques that deal with overplotting and slow production are important. This paper addresses these two problems in the context of display devices that have a color look-up table. Topics include compromised brushing, film loops, and density representation by gray-scale or by symbol area. The paper also discusses techniques that are generally applicable, including interactive graphical subset selection from any collection of scatterplots, and comparison of scatterplot matrices.

1. INTRODUCTION

A scatterplot matrix for p variate data is the ordered display of p*(p-1) scatterplots as shown below in Exhibit 1. Since 1980, many descriptions of scatterplot matrices have appeared in statistical graphics literature.[1, 2,3,4,5,6]. With different names and modest variations, the important themes prevail. The two themes are 1) scatterplot matrices provide an effective approach to exploratory multivariate data analysis and 2) scatterplot matrices can be enhanced to provide more information. Undoubtedly, scatterplot matrices and a variety of enhancement procedures, including transformations, smoothings, missing

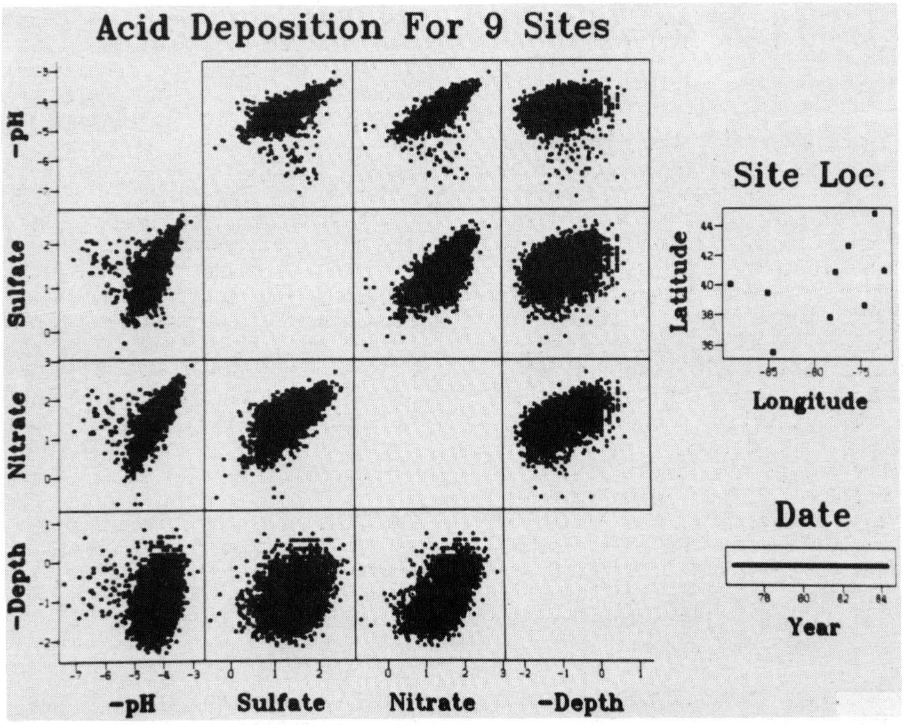

Exhibit 1: Scatterplot collection. Data in the scatterplot matrix are multiple measurements on individual rain samples collected at nine sites in the ADS (Acid Deposition System) network.[7] Nitrate and Sulfate measurements are ion concentrations expressed in logarithms of micro-moles per liter. Depth is rain gage depth in logarithms of millimeters. Two additional plots show site location in degrees and collection dates in decimal year. With 4109 points (minus some missing data) the overplotting is substantial.

data representation, and interactive subset selection and representation will find their way into an increasing number of statistical packages and into common use. The purpose of this paper, which is a sequal to [5], is to elaborate on interaction, density representation and display techniques that are helpful in representing a large number of points, and to exploit the color look-up table on color raster display devices.

An objective of "new" graphical techniques is to make the discovery of significant patterns in data easier and more likely. Once a pattern is found, analyst ingenuity can typically produce an alternative display that shows the same pattern and meets with publication regulations. Thus, readers of publications often have little exposure to techniques that are particularly effective in the interactive exploratory setting. Until electronic journals become available, the gap between what is useful in exploratory data analysis and what can be portrayed in static monochrome journals will remain. In this paper, the importance of the scatterplot matrix and graphical interaction is assumed. Little space is devoted to description and interpretation of data to prove that patterns were found that could not have been found any other way. Thus, what is shown here does not convey the speed or power of the scatterplot matrix for finding significant patterns.

2. LARGE N

What is large depends on the frame of reference. If available plotting space for a scatterplot is a one inch square, 500 points can seem large. For our purposes, N is large if plotting time is much greater than real time, if straight forward plots can have an extensive amount of overplotting, or if computation times are long. Exhibit 1 provides an example. If there were no missing data, each scatterplot would contain 4109 points. With fourteen plots, the total number of points in the display exceeds 50000. Currently, few if any display devices can display this number of points in real time (in a fraction of a second). Thus, with commonly available display software/hardware, 50000 is large. The exhibit also fits the other definitions of large. Substantial overplotting can be inferred from plotted area, the dot size and the number of points, 4109. Computation times are also long for some operations. For instance obtaining graphically specified subsets from 4109 points does not take a lot of time, but obtaining lower, middle, and upper smooths [8], say using LOWES [9], does. Thus, the data set for Exhibit 1 qualifies as large under all three counts. Because display speed and computing capabilities can be expected to improve dramatically, the key definition of large concerns overplotting.

Large data sets are common. Many monitoring studies generate large quantities of data. In a substantial subclass of such studies, the same measurements on variables are obtained at different times and/or at different spatial locations. Large data sets then arise from pooling data across temporal and spatial indicies. Exhibit 1 shows data from 9 sites in just one of several acid rain deposition monitoring networks. Other monitoring examples include seismic networks, multispectral satellite images, and so on.

From visual appearances, Exhibit 1 might more appropriately be called a scatterplot collection. That is, Exhibit 1 illustrates plots in addition to a scatterplot matrix. The indexing parameters of site location and time are shown in separate plots. An addition would be an underlayed map in the site locations plot. Whatever the layout, there are two key concepts: (1) the background data structure is an N x P matrix of data; and (2) interactive graphical subset selection can be driven from any plot.

3. INTERACTIVE GRAPHICAL SUBSET SELECTION AND REPRESENTATION

One of the most powerful enhancement procedures for scatterplot matrices is to distinguish subsets of data for comparison against each other or against the whole set. Interactive graphical subset selection is particularly convenient. Four approaches to graphical subset selection have been described in the literature. The first [10] involves picking a point in a plot, and having the subset include the k nearest neighbors. The second approach defines the subset by specifying a rectangular region which contains the subset. The third approach [5,11] involves drawing a polygon around desired points. The fourth approach [6], called "brushing", uses an interactively specified rectangular region that can be swept through the plot to define arbitrarily shaped regions. Points falling in the region are in the subset.

Omitting the nearest neighbor approach which is largely algorithmic, brushing is the most convenient form of subset selection and encompasses the other approaches. Brushing is particularly advantageous when selected points are distinguished in real time. However when the number of points gets large, significant time is required to find and redisplay all points falling within the sweep of the brush and the display lags behind the brush. With polygon selection, defining lines can be drawn in real time. For finding points, computations are reduced since a simple boundary is involved, as opposed to a long sequence of rectangles generated by brushing. For storage purposes the polygon definition is more compact than a long sequence of rectangles or a vector indicating which points were chosen. In addition, polygon definition can be readily modified when applied to revised data bases. Consequently, for large data sets, polygon selection is the method of greater practicality.

Color is generally accepted as a good method for distinguishing a small number of subsets. When display speed is a problem, as in brushing, approaches can be taken that lead to compromised but rapidly produced displays. The underlying trick is to redisplay only selected points and leave remaining points alone. In a monochrome setting, two types of dots can be used as shown in the top row of Exhibit 2.[6]

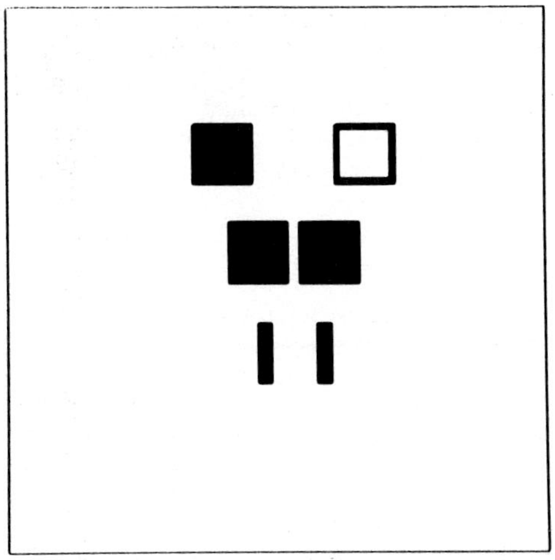

Exhibit 2: "Exclusive Or" Dots. The top row shows two dot types, filled and open. The second row shows two disjoint filled dots. The third row shows partially overplotted filled dots. The invisible fourth row corresponds to any even number of identically positioned dots of the same state.

Filled dots represent points in the selected set and open dots represent points in the complement set. Writing on the bits in the central portion of each dot using the "exclusive or" operator causes the dot to switch its filled/open state. The speed is obtained at a price. Consider the two filled dots in each of rows two, three and four in Exhibit 2. The second row is fine. In the third row dots are partially overwritten, so what is visible has a different shape. The invisible fourth row shows the blank created by perfect overwriting of any two or any even number of dots of the same state. The "exclusive or" approach is not desirable for large N problems since the approach requires large dots with visible interiors and misrepresents overplotted points. Color displays provide more alternatives.

Color raster display devices with a color look-up table allow the definition of multiple plotting surfaces with control over their priority (color overplotting/mixing control) and visibility. To rapidly display a chosen subset, the selected subset can be written in a higher priority (color overwrite) surface. This is also a compromise. A better display would also distinguish regions by a third color when selected points overplot points in the complement set. This three-color plot unfortunately requires replotting all data. One approach is to work initially with the fast display. The color mixture version can be written in hidden surfaces as a background process. When the color mixture version is complete, it can then be substituted for the approximate version. Erasure or removal of a small chosen subset can also be handled by plotting with the background color in a higher priority surface. Unfortunately, when overplotting between the chosen set and it's complement is substantial, such plots become unacceptable. The only recourse seems to be direct plotting of the complement set. Thus, even color devices do not solve all the speed bottlenecks of brushing.

For study and comparison purposes the simultaneous representation of more than one set becomes desirable. Figures 3a and 3b show an example. After noting the similarity of the X1 versus X6 and X1 versus X7 plots in Exhibit 3a, a pencil shaped region was selected in each of the two plots. The two selected sets are shown by using open circles and filled squares in Exhibit 3b. Disjoint sets and apparent symmetry in the X6 versus X7 plot came as a surprise. The example suggests that the subset selection tool should be in the hands of those who understand the particle physics experiment. Then symmetry would probably be taken as given and more subtle patterns would be of interest.

Color can be used to handle more types of overplotting. Suppose two sets A and B have been graphically specified as in Exhibit 3b. Depending on the specification, the intersection, denoted A•B, may not be null. This creates 4 sets of points, A•~B, B•~A, A•B and ~A•~B, and potentially 11 types (6 pairs, 4 triples and 1 quadruple) of overplotting. Since eleven colors is too many colors to distinguish rapidly, we chose colors in Exhibit 4 to represent subsets and overplotting.

The separation of the two sets is then shown as the absence of yellow. In contrast to Exhibit 3b, the relationship of two sets to the rest of the data is conveyed in a single picture.

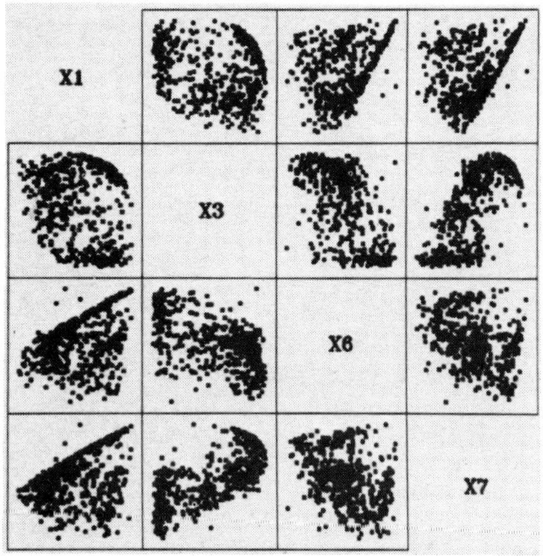

Exhibit 3a: Particle physics data scatterplot matrix. The four variables (out of seven) partially describe individual replications of a high energy particle physics scattering experiment.[12] Units have been altered by taking logarithms of absolute values. Note the similarity of the two right most plots in the top row.

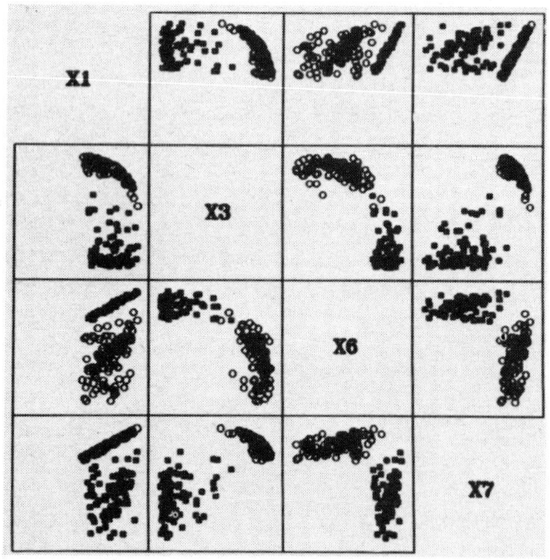

Exhibit 3b: Two subsets. Pencil shaped sets were selected in each of the two right most plots in the top row. Open circles and filled squares show the two sets. The two sets have no elements in common and the X6 versus X7 plot shows symmetry.

Subset	Colors
A·~B	Green
B·~A	Red
~A·~B	Blue
A·B	Yellow
A·~B with ~A·~B	Cyan(roughly)
B·~A with ~A·~B	Magenta(roughly)
everything else	Yellow

Exhibit 4: Colors for Subsets and Overplotting

When brushing is impractical and several subsets are to be compared, another approach is available. Each subset can be written in a different plotting surface. Then surfaces are cycled by changing the color look-up table. This film loop approach is described in more detail in [13,14]. In general, sequences of views are easier to follow if there is continuity between views. In the subset context, putting a composite view between each of the subset views facilitates comparison.

4. DENSITY REPRESENTATIONS

Section 4 discussed overplotting for different subsets. No distinction was made if a displayed point came from one data point or 10000 data points. This section addresses overplotting of multiple points from the same set.

The three basic strategies in dealing with overplotting are 1) to plot open circular symbols 2) to alter the data to reduce overplotting and 3) to represent the point density. Plotting open circular symbols [8] and jittering the data [3] are helpful techniques for small data sets, but are inadequate for large N plots. To represent a large number of points, some form of density representation is required.

In representing bivariate densities a common approach is to bin the data and to indicate the bin counts. For printer plots symbols such as those in Exhibit 5 are often used with rectangular binning regions that correspond to space allocated for line printer characters.

Counts	Symbol
0	blank
1	*
2-9	2-9
>9	+

Exhibit 5: Plotter Symbols Representing Counts

Since the amount of ink used is only remotely related to the density, a "simple" visual process cannot be used in assessing and comparing local densities. When the goal is visual assessment of local density, two approaches are available: either symbol intensity represents the count, or symbol area represents the count.

4.1 Interactive Gray Scale Density Representations

Using gray-scale intensity to show counts is a common practice in the field of image processing [15,16], which offers the opportunity for real time exploration based on data density. The process starts by binning data (see Exhibit 6a) into say a 256 X 256 matrix. Then the density estimate is often smoothed using fast algorithms such as the shifted histogram [16, 17,18] or the Fast Fourier Transform (FFT). Next, the image is written into display device memory. This involves assigning elements of the matrix to specific pixels and the transformation of the estimated density to discrete pixel values. The correspondence from pixel values (density) to gray scale can be manipulated in real time via the color look-up table. Exhibit 6b shows an image corresponding to a different transfer function between the pixel values (horizontal axis) and the grey scale (vertical axis). The menu at right provides options for altering the transfer function and for contouring based on interactively chosen density levels. A fast way to find the density levels is to create a spike transfer function (Exhibit 6c) that can be moved left and right with a mouse. The corresponding rough contours are shown in real time.

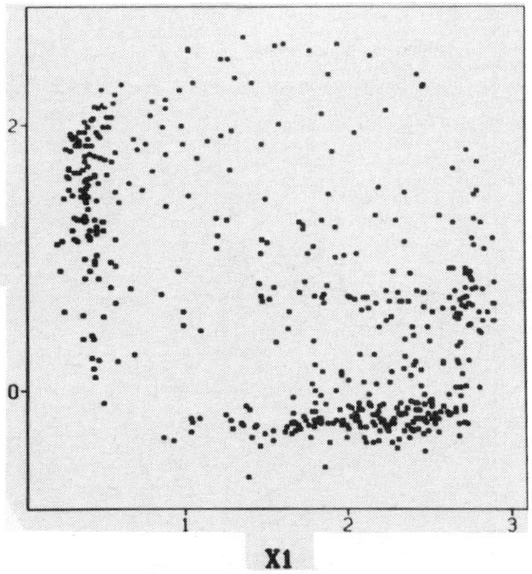

Exhibit 6a: Binned Particle Physics Data. The data is described in Exhibit 3a. Here, the first two variables have been binned into a 256 X 256 matrix.

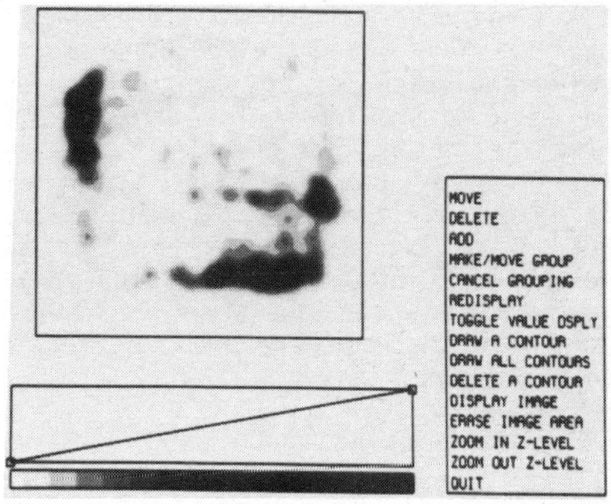

Exhibit 6b. A Gray-Scale Density Representation. The transfer function at the bottom defines the correspondence between density and the gray level for each pixel. The transfer function can be interactively manipulated to call attention to different density regions.

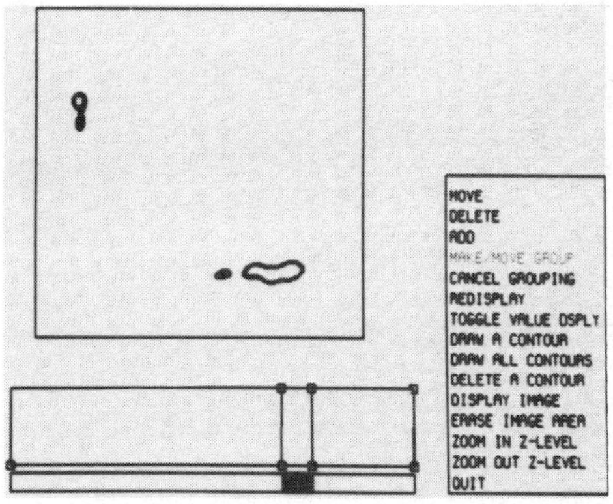

Exhibit 6c. "Empirical" Contours. Movement of a spike transfer function allows real-time investigation of different "empirical" contours.

In our implementation, the density estimation and interactive transfer function manipulation routines were added as functions to S, a statistical package from AT&T Bell Laboratories.[19] Chosen contour levels are easily passed on to a contouring routine in S. Other variations of the transfer function can be used to call attention to high- or low-density regions. Gray scale density images can also be examined rapidly for scatterplot collections.

When the above process is applied to large N problems, little changes except that binning takes longer and density estimates often have a few extreme values. When the range of pixel values is limited to say 256, a linear mapping from the density estimates results in low gray-scale resolution for low density regions. S can be used to provide square root or other transformation before the density is transferred to display device memory. A useful approach in density representation is to treat densities above (or below) a specified value the same. We call this "blunting", and provide it thru the menu under the item rescaling Z. Thus gray-scale methods are well suited for handling large sample sizes.

4.2 DENSITY REPRESENTATION BY SYMBOL AREA

The use of area is another choice for direct visual representation of the local density. Area is not perceived as accurately as several other visual variables [20] but in this context area provides a reasonable choice. The technique of representing density with area can be found in various guises in the literature. One variant [3,8] is shown in Exhibit 7a. In this variant, the binning region is a square and the symbol is a sunflower. The number of petals of the sunflower indicates the number of points in the region. Except for regions with a single point which are represented with a dot, and except for overplotting of line segments, the amount of ink used (plotting area) by the symbol is proportional to the counts. When such a plot is compared to Exhibit 7b which uses hexagonal bins or to original data in Exhibit 7c, point locations in the sunflower plot with square bins appears stretched out in the vertical and horizontal directions. The hexagon bins seem to represent the data more faithfully. Bin shape is a type of two dimensional smoothing parameter. The density estimate bias reduction for using hexagons instead of squares is approximately 4 percent.[21] This would not account for the large visual discrepancy in Exhibits 7a and 7b. The exact placement of the binning lattices can make a difference. However, the discrepancy most likely results from emphasis of human-preferred visual directions, horizonal and vertical, by square bins and round symbols within the bins. For this reason, we prefer hexagon bins. The additional cost for hexagon binning is small as can be seen by the algorithm in the appendix. Given that hexagon bins are to be used, the next question concerns the symbol. We prefer a filled hexagon whose area is proportional to the count, as shown in Exhibit 7d. This provides a general impression of density. Some may complain that the exact count has been lost. In an interactive environment, if one really wants to know the exact count, the best procedure is graphical selection of the desired area and a query to the computer, "how many?". Exhibit 7d differs from other area representations in that points are shown exactly when 3 or fewer points fall in a region. A more difficult variant is to plot each hexagon symbol as close as possible to the center of mass within each hexagonal bin

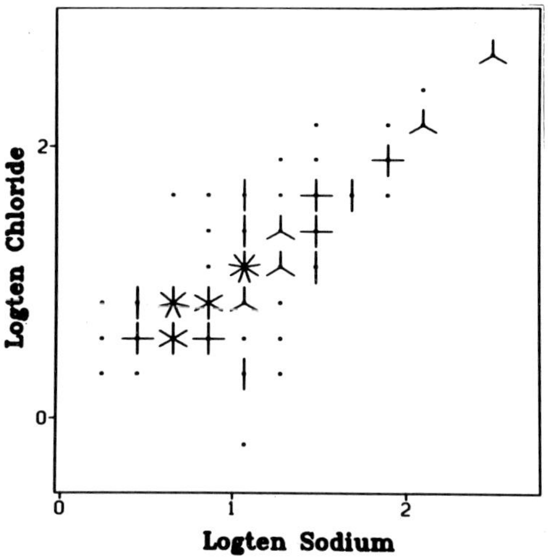

Exhibit 7a: Sunflowers in square bins. Data are paired sodium and chloride ion concentration measurements in logarithms of micro-moles per liter from individual rain samples collected at Acid Deposition Site 152A -- Indian River, Delaware.[7] Note the visual impact in horizontal and vertical directions.

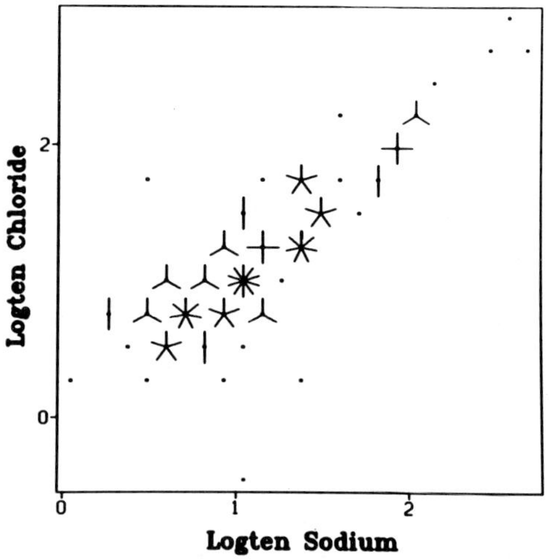

Exhibit 7b: Sunflowers in hexagon bins. The binning lattice still detracts, but the plot looks closer to that in Exhibit 7c.

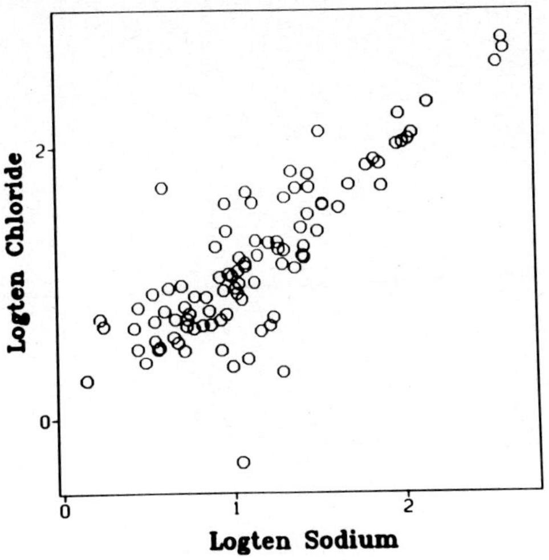

Exhibit 7c: Original data. Actual overplotting is not substantial.

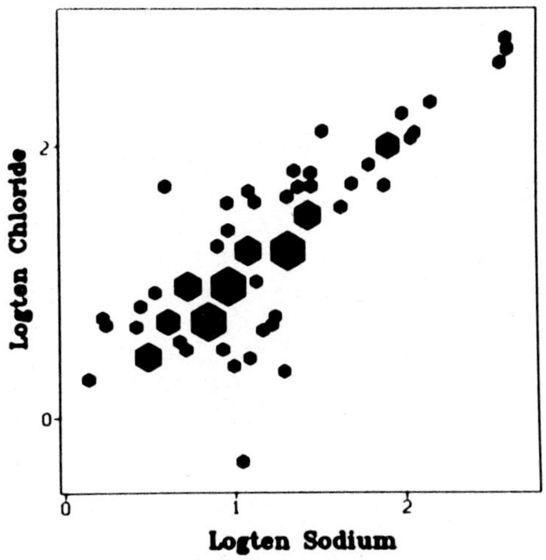

Exhibit 7d: Hexagon symbols in hexagon bins. The hexagon symbol area is portional to the count in each hexagonal bin. For bins with three or fewer counts, individual points are represented by single-count-sized hexagons, and are plotted at data coordinates. Thus, modest overplotting is tolerated.

while keeping the symbol completely inside the bin. Since even the hexagon lattice structure can be distracting, approaches that break it up are worth considering. Exhibit 8 shows a hexagon density representation of the original data plotted in Exhibit 1. Note that the number of symbols actually plotted is much less than that in the original. Thus, this display can be produced on a pen plotter. Of course the most important aspect is that regions of high density are now evident. When hexagons are written into display device memory, pixel values can be assigned that correspond to count intervals. The color look-up table can be used to alter the displayed intensity in real time, just as in the gray-scale discussion. Alternatively, a few easily distinguished colors can be used to call out selected density intervals.[15] When the number of points is large density scaling becomes an issue. The maximum hexagonal area displayed corresponds to the largest count and just fills its bin. With this fixed point and the area being proportional to the count, low density symbols become smaller than device resolution. In such cases a single pixel can be shown as in Exhibit 8. As in Section 4.2, this identical treatment for a range of densities is called blunting. The blunting of high densities is also useful as are other transformations that emphasize selected portions of densities, for example low count regions. For binned data, transforming counts and redisplaying them is a rapid operation. Other procedures such as smoothing can be adapted to binned data with great computational savings and little loss of accuracy.

4.3 COMPARISON OF SCATTERPLOT MATRICES USING HEXAGON SYMBOLS

Two scatterplot matrices can be compared by juxtaposition. Lower left and upper right triangles can contain two distinct but commonly scaled subsets.[22] Variable order is reversed for one data set so that no mental rotation about forty-five degree lines is required for comparison. However, it is still desirable to place corresponding plots closer together. The hexagon representation above allows this to be done by overplotting. Suppose two data sets are to be compared. Hexagon lattice points for the two data sets can be made identical by selecting a common scale and using the same bin size. If one data set is considered the reference data set, counts of the other can be scaled so that the total counts are the same for the two sets. Then the two displays can be overplotted, one set in red, one in green and overlap in grey. This maintains the scatterplot matrix context and makes scanning for differences easy. Other displays can be considered such as direct display of functions of non-zero and non-infinite count ratios.

4.4 LOOSE ENDS

A thorough treatment of density representations should include results concerning smoothing parameters for density estimation, human perception of density representations, and should balance these in view of plot production speed and device resolution. In this presentation only a few pointers to the literature are given. Scott [23] provides

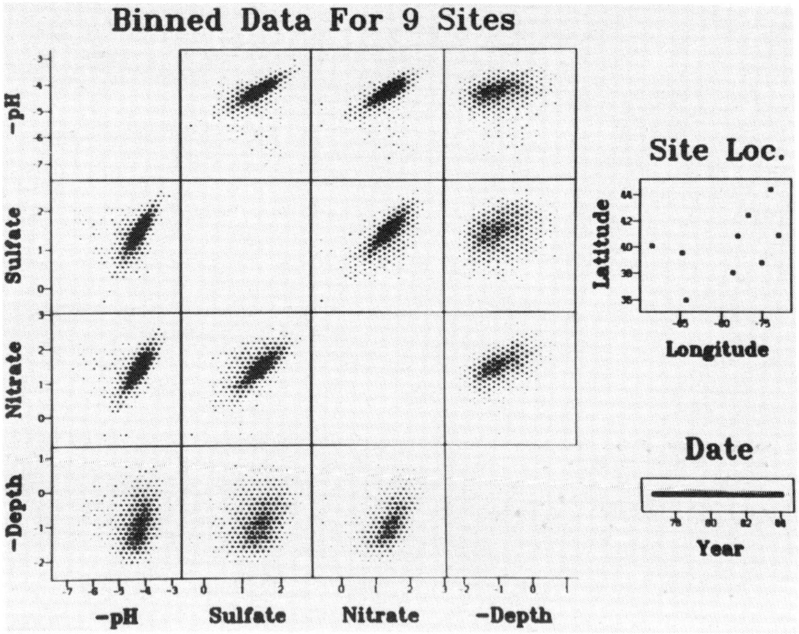

Exhibit 8: Hexagon Area Density Representation. The area of each hexagon is proportional to the count. The largest hexagon fills hexagonal binning regions. Small counts are represented by degenerate hexagons. Below a certain count, all counts are represented by a single square pixel. This is an example of blunting. Regions of high density can be readily identified.

results concerning the choice of smoothing parameters for bivariate densities and is a portal to the general literature. In terms of gray-scale perception, numerous references are available.[15,16] In assessing the response to circle sizes, some studies show that humans respond to area raised to the 0.7 power, but when the comparison areas are in view, there is little reason to use other than a linear correspondence between the variable represented and the area of the circle.[24] Presumably this applies to hexagons also. With small hexagons, more is involved than the comparison of two areas. As dots approach 70 per inch as when viewed from 12 inches away it becomes possible to respond to gray level even though individual dots are visible.[25] Our hexagon centers are further apart than this, but in regions where displayed hexagons are separated by roughly 0.01 inches (our raster display device resolution), some gray-scale impression is induced. Higher resolution devices such as laser printers provide opportunity to capitalize on the human impression of gray scale as conveyed through area symbols.

With the focus on large N, the question arises, "Why not sample?" The answer is that there are trade-offs. Certainly sampling reduces display problems and provides cross validation opportunities. However, sampling can hide aspects of fine structure that do not necessarily get hidden by binning. An advantage of large N is that patterns begin to emerge in low-density regions of data. It is precisely these low-density patterns that are destroyed by sampling. Another argument against sampling is that obtaining representative samples when pooling over temporal and spatial strata can require substantial work. Thus, both small N sample plots and large N plots have merits.

5. SUMMARY

The display of a large number of points in a scatterplot matrix has been a problem. The problem manifests itself in terms of hidden point density, long computation times for selected enhancement operations, and slow displays. The difficulties are ameliorated by computing and displaying densities. One density representation codes density as gray-scale. With real-time graphical manipulation of a color look-up table, attention can be focused on different density regions and real-time "empirical" contours can be obtained. A second density representation codes density as the size of hexagon symbols as shown within hexagonal binning regions.

Both approaches are useful in the context of scatterplot matrices and scatterplot collections where density exploration, subset selection and subset representation are of interest. The representation of a million or more points in each plot is feasible.

For subset selection in the large N context, real-time brushing is not feasible and polygon selection is the method of choice. Since display times are less than real time, the color look-up table can be used to store different subset displays for real-time review or to display subsets simultaneously, with careful control of color mixing. Thus, the color look-up table is a useful tool in the context of large N.

6. ACKNOWLEDGEMENTS

The authors would like to thank Janis Littlefield and Vern Crow for substantial software contributions and Paul Tukey for suggesting the hexagonal binning algorithm in the appendix. Work supported by Applied Mathematical Sciences of the U.S. Department of Energy under Contract DE-AC06-76RLO 1830.

7. APPENDIX - ALGORITHM SKETCH FOR HEXAGONAL BINNING

Binning algorithms are available for various lattices in dimensions two through eight.[26] The following sketches a fast implementation of the hexagonal binning motivated in [26]. Details like reading data, scaling data, handling missing data, zeroing count arrays, and checks for pathological conditions are left to the implementor. Since the hexagons are typically envisioned in a (square) plotted version of data, raw data coordinates are presumed scaled into [0,1] and resulting vectors of length N are denote SX and SY. The original minimums and ranges are given by XMIN, YMIN, XR, and YR. SIZE is a user-specified scaling parameter that indicates roughly the number of bins along the X axis.

```
C     SIZEMAX=100
C     IMAX=SIZEMAX/SQRT(3.)+1
C     JMAX=SIZEMAX+1

      PARAMETER IMAX=58,JMAX=101,NMAX=10000
      DIMENSION SX(NMAX),SY(NMAX)
      INTEGER LAT1(0:IMAX,0:JMAX),LAT2(0:IMAX,0
     * :JMAX)

      C1=SIZE/SQRT(3.)

      DO K=1,N
      X=SIZE*SX(K)
      Y=C1*SY(K)

C     Compute Two Candidate Lattice Points

      J=X+.5
      I=Y+.5
      J2=X
      I2=Y

C     Select the Nearest

      IF( (X-J)**2 + 3.*(Y-I)**2 .LT.
     * (X-J-.5)**2 + 3.*(Y-I-.5)**2) THEN
      LAT1(I,J)=LAT1(I,J)+1
      ELSE
      LAT2(I2,J2)=LAT2(I2,J2)+1
      ENDIF

      ENDDO

C     The Lattice points in original data
C      coordinates for non-zero counts in LAT1
C     and LAT2 are obtained as follows:

C     CONSTANTS
C     C2=SQRT(3.)*YR/SIZE, C3=XR/SIZE

C     LAT1(I,J)
C     Y=C2*I+YMIN, X=C3*J+XMIN

C     LAT2(I,J)
C     Y=C2*(I+.5)+YMIN, X=C3*(J+.5)+XMIN
```

8. REFERENCES

[1] Tukey, P.A. and J.W. Tukey, Graphical Display of Data Sets in Three or More Dimensions, Chapters 10, 11, and 12 in Interpreting Multivariate Data, ed. V. Barnett, (Wiley, Chichester, United Kingdom, 1981).

[2] Tukey, J.W. and P.A. Tukey, Some Graphics for Studying Four- Dimensional Data, Computer Science and Statistics: Proceedings of the 14th Symposium on the Interface, (Troy, NY, Springer-Verlag, New York, 60-66, 1983).

[3] Chambers, J.M., W.S. Cleveland, B. Kleiner, and P.A. Tukey, Graphical Methods For Data Analysis, (Duxbury Press, Boston, MA 1983).

[4] Tufte, E., The Visual Display of Quantitative Information. (Graphics Press, Cheshire, CT, 1983).

[5] Carr, D.B., and W.L. Nicholson, Graphical Interaction Tools for Multiple 2- and 3-Dimensional Scatterplots, Computer Graphics ' 84: Proceedings of the 5th Annual Conference and Exposition of the National Computer Graphics Association, Inc. ISBN 0-941514-05-6, Vol. 2, 748-752, (May 1984, Anaheim, CA).

[6] Becker, R.A., and W.S. Cleveland, Brushing a Scatterplot Matrix: High-Interaction Methods for Analyzing Multidimensional Data, Statistics Technical Report No. 7, AT&T Bell Laboratories, Murray Hill, NJ (1985).

[7] Watson, C.R., and A.R. Olsen, Acid Deposition System (ADS) for Statistical Reporting, System Design and Coding Manual, Pacific Northwest Laboratory, PNL-4286, Richland, WA, (1984).

[8] Cleveland, W.S., and R. McGill, The Many Faces of a Scatterplot, JASA 79:388 (1984) 807-822.

[9] Cleveland, W.S., Robust Locally Weighted Regression and Smoothing Scatterplots, JASA 74:368 (1979) 829-836.

[10] Friedman, J.H., J.A. McDonald and W. Stuetzle, An Introduction to Real-Time Graphical Techniques for Analyzing Multivariate Data, in Proceedings of the 3rd Annual Conference and Exposition of the National Computer Graphics Association, Inc., Vol. 1, (June 1982), Anaheim, CA, 421-427.

[11] Littlefield, R.J., Basic Geometric Algorithms for Graphic Input, In Proceedings of the 5th Annual Conference and Exposition of the National Computer Graphics Association, Inc. ISBN 0-941514-05-6, Vol. 2, (Anaheim, CA, 767-776, May 1984).

[12] Friedman, J.H. and J.W. Tukey, A Projections Pursuit Algorithm for Exploratory Data Analysis, IEEE Transaction on Computers C-23 (1974) 811-890.

[13] Littlefield, R.J., Stereo and Motion in the Display of Three Dimensional Scattergrams, Proceedings of the Eighth Annual Computer Graphics Conference of the Engineering Society of Detroit, (Detroit, MI. 13-17, May 4-6, 1982).

[14] Nicholson, W.L., and R.J. Littlefield, The Use of Color and Motion to Display Higher-Dimensional Data, In Proceedings of the 3rd Annual Conference and Exposition of the National Computer Graphics Assoc., Inc. Vol. 1, (Anaheim, CA, 476-485, June 1982).

[15] Gonzalez, R. C. and P. Wintz, Digital Image Processing: (Addison-Wesley Publishing Company, Reading, MA, 1977).

[16] Pratt, W.K., Digital Image Processing, (John Wiley & Sons. New York, NY, 1978).

[17] Scott, D.W., and J.R. Thompson, Probability Density Estimation in Higher Dimensions, Computer Science and Statistics: Proceedings 15th Symposium on the Interface, Houston, TX, North Holland Publishing Co., New York, 173-179, 1983).

[18] Scott, D.W., Average Shifted Histograms: Effective Nonparametric Density Estimation in Several Dimensions, Technical Report No. 83-101, Rice University, Houston, TX, (1984).

[19] Becker, R.A. and J.M. Chambers, S - A Language and System for Data Analysis and Graphics: Wadsworth Advanced Book Program (Belmont, CA, 1984).

[20] Cleveland, W.S., and R. McGill, Graphical Perception: Theory, Experimentation, and Application to the Development of Graphical Methods, JASA 79:387 (1984) 531-554.

[21] Scott, D.W., A Note on Choice of Bivariate Histogram Bin Shape, Technical Report No. 85-311-3, Rice University, Houston, TX, (1985).

[22] Carr, D.B., W.L. Nicholson, R.J. Littlefield, and D. L. Hall, Interactive Color Display Methods for Multivariate Data, Pacific Northwest Laboratory, PNL-SA-1311, Richland, WA, (1985).

[23] Scott, D.W., Frequency Polygons: Theory and Applications, Technical Report No. 82-476. Rice University, Houston, TX, (1982). To appear in JASA.

[24] Cleveland, W.S., C.S. Harris and R. McGill, Judgments of Circle Sizes on Statistical Maps, JASA 77 (1982) 541-547.

[25] Castner, H.W. and A.H. Robinson, Dot Area Symbols in Cartography: The Influence of Pattern on Their Perception. American Congress on Surveying and Mapping, Cartography Division. Technical Monograph CA-4, 1969.

[26] Conway, J.H. and N.J.A. Sloane, Fast Quantizing and Decoding Algorithms for Lattice Quantizer and Codes, IEEE Transactions on Information Theory, Vol. IT-28, No. 2, (March 1982), 227-231.

COMPUTER SCIENCE AND STATISTICS:
The Interface, D.M. Allen (ed.)
Elsevier Science Publishers B.V. (North-Holland), 1986

STATUS OF THE NBS GUIDE TO AVAILABLE MATHEMATICAL SOFTWARE

Sally E. Howe

National Bureau of Standards

The Guide to Available Mathematical Software (GAMS) is a classification scheme, a data base system, and a printed catalog. GAMS provides a framework for both the end-user scientist and the software maintainer to handle large quantities of mathematical and statistical software.

The extensive problem-oriented GAMS classification scheme provides a structure for organizing software for general purpose mathematical and statistical computations. The software currently cataloged in GAMS consists of approximately 2400 programs, subprograms, and interactive systems in some two dozen libraries. These libraries are available on a variety of computers. Data about the software and about library availability are stored in a relational data base and are maintained using a variety of software tools. Users access the data via an on-line query system based on the classification scheme. The printed GAMS catalog organizes information about the software according to the classification scheme and in several other useful ways.

1. INTRODUCTION

A vast body of reliable and well-designed computer software for solving many standard mathematical and statistical problems now exists. This software is a crucial resource for scientific computing through saving time and money, expanding the scope of problems which can be routinely solved by applied scientists, and insuring that the most up-to-date and reliable numerical algorithms are used.

Collections of mathematical and statistical software are now available in many scientific computing centers. These collections are often large and diverse, and thereby create several software management problems, including acquisition, maintenance, and documentation.

The NBS solution to these problems is the result of the Guide to Available Mathematical Software (GAMS) project. GAMS is joint work with Ronald F. Boisvert and David K. Kahaner and consists of several components. We first acquired an extensive collection of generally available mathematical and statistical software. We developed a problem-oriented tree-structured detailed classification scheme to identify the problems the software solved. The organization of the software on the computer facilitates its efficient installation and maintenance. The software documentation takes the form of an inter-library reference. The on-line and the off-line documentation have consistent structure. Finally, a single data base was developed to integrate the maintenance and documentation functions.

The purpose of this paper is to describe in detail the software management problems and the solutions developed at the National Bureau of Standards. Information about how NBS solved these problems may prove useful to other scientific computing centers. Other papers describing earlier versions of this work have focused on statistical software (Howe, 1982, 1983). The focus of this paper is not so much the supported software but rather how that software support is provided.

This paper begins with a brief description of the computing environment at NBS, primarily because our environment has influenced many of our decisions. The features of mathematical and statistical software from a maintainer's point of view are then described, followed by descriptions of the GAMS classification scheme for mathematical and statistical software, and both the printed and the on-line Guide to Available Mathematical Software catalogs. The paper concludes with an overview of our implementations.

2. THE SCIENTIFIC COMPUTING ENVIRONMENT AT NBS

The National Bureau of Standards (NBS) is a multi-disciplinary scientific research laboratory with a staff of 3000. Its mission leads to theoretical and experimental research in the physical and engineering sciences for the purpose of providing the measurement foundation needed by U.S. science and industry.

Computers at NBS include a recently acquired Cyber 180/855 and Cyber 205, and minicomputers, microcomputers, and workstations numbering in the hundreds. Many of these computers and terminals are interconnected via NBSNET, an Ethernet-like local area network.

The Center for Applied Mathematics is responsible for providing software for use on NBS computers, and for informing the user community of both the availability of such software and the information necessary to use the software. Our efforts to date have focused on acquiring, maintaining, and documenting general-purpose mathematical and statistical software which is useful in the scientific disciplines represented at NBS. This focus restricts our attention to approximately ten thousand items, comprising perhaps ten percent of the total available scientific software. Approximately seven staff years have been devoted to bringing the project to its current state in which approximately 2400 user-invokable software items are managed.

3. A SOFTWARE BASE

3.1 Software Acquisition

Widely available general-purpose scientific software commonly takes the form of either Fortran subprograms or Fortran programs, the latter often with tailor-made input languages. An early development in statistical computing was batch-oriented Fortran programs which did not require users to be Fortran programmers. Current versions of these programs have sophisticated languages specifically targeted to statistical data analysis. The more recently developed interactive programs have similarly capable and sophisticated input languages.

The foundation for the development of special-purpose batch and interactive programs has always been Fortran subprograms, because subprograms are the single most important source of implementations of state-of-the-art computational algorithms. The software provided at a research laboratory such as NBS therefore necessarily is an extensive collection of both Fortran subprograms and programs to support the needs of its users.

The software we have acquired is either proprietary, non-proprietary, or a mixture of the two. Leasing of proprietary software libraries, with their extensive capabilities and ease of installation, provides a firm foundation for mathematical and statistical computing.

Non-proprietary software is usually produced in university or government laboratories as a product of research in numerical or statistical methods. This software consists of some narrowly-focused subprogram collections such as LINPACK (Dongarra, et al., 1979) and a plethora of single-purpose programs or subprograms from many authors. There are no restrictions on installing such software, which makes it particularly desirable in a multi-machine environment. Substantial effort may be required, however, to install and test the software on a particular machine and to provide even the most basic support.

Recently, software with non-proprietary and proprietary components has appeared. The non-proprietary component commonly performs the mathematical calculations, while the proprietary software is a graphics software library.

Sources for software documented in GAMS include distribution services, journals, and books. Statistical software is announced or published in periodicals such as The American Statistician, journals such as Communications in Statistics, proceedings from conferences such as the Symposium on the Interface and the Joint Statistical Meetings, and books (e.g., Francis (1981)). Software may also be available from individual authors.

3.2 Software Organization

Over the past several years we have collected software from numerous sources and organized it according to a two-level system. At the higher level are libraries and at the lower level are modules which are collected into libraries. A module is the smallest user-callable problem-solving unit, and may be an individual user-callable subprogram, an individual batch or interactive program, or a command in a large interactive program.

Software from a given vendor usually is a large collection of subprograms or programs, or a large interactive program, and is kept in a library of its own. One special library is the NBS Core Mathematical Library (CMLIB), a library which is partitioned into approximately 50 sublibraries of Fortran subprograms obtained from numerous sources. While CMLIB is partitioned in order to maintain the individual sublibraries, from the user's point of view it is one very large library of portable, non-proprietary software.

The GAMS project currently manages approximately 2400 modules in 20 libraries:

BMDP.......40	statistical programs
CMLIB.....678	mathematical and statistical subprograms
DATAPAC...169	statistical subprograms
ISML......471	mathematical and statistical subprograms
INVAR.......2	interactive regression programs
LINDO.......1	linear programming program
MATHWARE...35	mathematical and statistical subprograms
MATLAB......1	interactive linear algebra program
MINITAB...150	commands in an interactive statistics program
NAG.......487	mathematical and statistical subprograms
PDELIB......3	partial differential equations subprograms
PLOD........1	interactive ordinary differential equations program
PORT......270	mathematical and statistical subprograms

ROSEPACK....1 interactive robust regression program
SIMSCRIPT...1 simulation language
SLDGL......31 ordinary differential equations subprograms
SPECTRLAN...1 interactive spectral analysis program
STARPAC....16 nonlinear regression subprograms
STATLIB....56 statistical subprograms
XMPLIB......2 mathematical programming subprograms

For each user-callable (or executable) module we must maintain the source code for that module and for all non-user-callable modules which that module references, object (either relocatable or executable) code, on-line documentation, and (optionally) test code and results. Where appropriate, the on-line documentation references printed documentation such as manuals. All of these items now number well over 10,000.

4. GAMS: THE GUIDE TO AVAILABLE MATHEMATICAL SOFTWARE

End-users of scientific software are interested in locating software to solve particular problems, and are not interested in how the software is organized for maintenance. The Guide to Available Mathematical Software (GAMS) provides such end-users with a problem-oriented inter-library software reference. This reference takes the form of both a printed catalog and an on-line interactive guide. A detailed problem-oriented classification scheme is fundamental to each form.

4.1 The GAMS Classification Scheme

While each proprietary library documented in GAMS has its own relatively consistent organizational structure, none provides a sufficiently extensive and detailed structure for organizing the whole GAMS software collection. We therefore have developed the GAMS classification scheme (Boisvert, et al., 1983) to synthesize information about the software we support. This classification scheme is a substantial modification of the Bolstad scheme (1975), which in turn evolved from the scheme adopted by the IBM user's group SHARE.

The classes at the highest level of the classification scheme are:

A. Arithmetic, Error Analysis
B. Number Theory
C. Elementary and Special Functions
D. Linear Algebra
E. Interpolation
F. Solution of Nonlinear Equations
G. Optimization
H. Differentiation and Integration
I. Differential and Integral Equations
J. Integral Transforms
K. Approximation
L. Statistics and Probability
M. Simulation and Stochastic Modelling
N. Data Handling
O. Symbolic Computation
P. Computational Geometry
Q. Graphics
R. Service Routines
S. Software Development Tools

These classes generally proceed from fundamental to more advanced topics. Most of these classes are further subdivided, and in these subdivisions core subjects appear before specializations. Consistency has been a goal in developing the scheme, so that, for example, univariate problems appear before multivariate.

The development of the present classification scheme has been strongly influenced by the software at hand. Experience has indicated that projections about scientific software organization in the absence of such software would be highly error-prone. Thus the level of detail varies across the scheme. In having at most about a dozen modules assigned to any class, the scheme also reflects the cómpromise between accuracy and quantity; it would be tedious either to find a few modules in a detailed subtree or to find one useful module among many in a class.

Interrelationships among classes motivated the inclusions of cross-references in the scheme. Thus, for example, class L3, containing software for probability function evaluation, cross-references class C (elementary and special functions). A module which performs several tasks may be assigned to multiple classes; an example is spline approximation. Some user-callable subprograms are almost always used in pairs; for reasons of efficiency, when documentation for one references the other, then only one is classified.

Each module is classified at the lowest appropriate classes. When a module performs tasks in several subclasses of a particular class, however, it is classified at a higher level. This is especially common with statistical software and large interactive programs.

As existing scientific software is added to GAMS, and as new software becomes available, the classification scheme will undergo selective revision. Given its tree structure, however, the scheme itself ought not to undergo radical revision in the near future.

4.2 The Printed GAMS Catalog

From the user's point of view, GAMS manifests itself as a printed catalog and an interactive consultant. The printed catalog is required by those who do not use the computer on which the interactive GAMS resides. These users may well include people not at NBS. The most recent GAMS catalog was released as a 448-page NBS Technical Report (Boisvert, et al., 1984). While we do not distribute the software documented in GAMS, the catalog contains the addresses of the sources which distribute the software. The GAMS catalog is available from the author or NTIS.

In order to satisfy the needs of different users, the catalog is organized in five sections:

A. GAMS Classification Scheme
B. Modules by Class
C. Module Dictionary
D. Library Reference
E. Index

Modules by Class catalogs the software according to the classification scheme. Under each class in the scheme is a list of modules, including a brief description of each module and the library to which it belongs. For higher level classes there may also appear discussions of the types of software found in those classes, along with issues and problems a user should address when selecting software, and references.

The alphabetically organized Module Dictionary contains detailed information about each module in GAMS, including:

* brief description;
* type (e.g., subprogram, batch program);
* proprietary or non-proprietary;
* library (and sublibrary, if appropriate) membership;
* precision (single or double);
* GAMS class(es);
* usage syntax (e.g., call sequence, command syntax);
* location of on-line documentation on an NBS computer;
* location of source on an NBS computer (if not proprietary);
* (optional) location of test programs on an NBS computer;
* (optional) location of sample programs on an NBS computer;
* commands required to access the module on an NBS computer; and
* (optional) names of other modules used with this module.

The contents of each library are summarized in the Library Reference. First, the following general information is given:

* brief description;
* version;
* type (e.g., subprogram library);
* language;
* portability information;
* references; and
* developer and/or distributor.

For each machine on which the library is supported, the following information is given:

* version;
* level of in-house support and a contact person;
* how to obtain on-line documentation; and
* how to access the library.

Similar information is given about each sublibrary in the partitioned library CMLIB.

The index alphabetically organizes keywords and phrases with pointers into the classification scheme.

4.3 The GAMS Interactive Consultant

While the specific details of the on-line version of GAMS are not of interest at sites where it is not available, the general features may be of interest at sites where a similar capability is desirable. The main reason for developing the interactive consultant is timeliness. Whereas the GAMS catalog is printed infrequently, the on-line data are updated regularly, and hence the interactive consultant provides current information.

A user of the interactive consultant may traverse the classification tree. When a node of the tree is visited, a count of the number of modules classified there and a list of the descendent classes are obtained. A user may then obtain information about each modules (similar to that provided by the Module Dictionary) at that node or may move to another node. The consultant is made easy to use by having a simple command syntax and internal help facilities.

Users may constrain their software search by restricting attention to portable software, to software in a particular library, or to software which computes in a particular precision.

Once a user has identified software of interest, on-line detailed instructions on how to use that software is then available.

5. IMPLEMENTATION

The two fundamental components of our software support are the maintenance of the software itself, and the maintenance of information about the software. We have developed software tools to efficiently manage our large software collection.

5.1 Naming Conventions

Developing naming conventions is the first step in automating software management. Such conventions must necessarily conform to the file structure of the computer on which the software is maintained, and we have developed conventions for several computers. For the purpose of this paper, however, our naming conventions are illustrated using UNIX path names.

A module's source code is in the path

Library-name/SOURCE/Sublibrary-name/Module-name.

Relocatable (object) code is in the path

NBS/Library-name/Sublibrary-name/Module-name.

In order to access a module a user need only know the library in which it resides.

A similar format is used for the location of the documentation for an individual module, a sublibrary, and a library, respectively, as follows:

Library-name/DOC/Sublibrary-name/Module-name
Library-name/DOC/Sublibrary-name/SUMMARY
Library-name/DOC/SUMMARY.

Naming conventions for test software, test results, reference materials, and other information are more variable.

5.2 Software Management Procedures

Three categories of frequently-occurring software maintenance activities have warranted the development of software tools. The first is software installation, and tools have been written for Fortran source dispersion, documentation extraction, and Fortran library or sublibrary compilation. The second category is documentation retrieval, for which there exist tools to extract module, sublibrary, and library documentation. Finally, tools have been written to prepare individual Fortran subprograms (and externals), Fortran sublibraries, and CMLIB for redistribution.

5.3 The GAMS Data Base

A single relational data base was chosen to support both software maintenance and documentation functions. This type of data base was chosen for its simplicity and flexibility. It is used to access the data in several ways (e.g., to organize modules either according to the classification scheme or alphabetically for documentation purposes and to organize the modules by library for maintenance purposes) and thus is used to develop the printed catalog, to drive the interactive consultant, and for software maintenance.

A relational database is a collection of tables called relations. The five relations in the GAMS data base are library, sublibrary, module, node (containing the GAMS classification scheme), and tree (containing pointers which describe the tree structure of the classification scheme and identify the modules classified at each node).

Each relation is a matrix in which rows are cases and columns are attributes. The attributes in the library relation are those itemized for the Library Reference (see section 4.2). The attributes in the sublibrary and module relations are similar.

The GAMS data base was constructed using RIM (1982), a relational information management system developed by the Boeing Commercial Airplane Company under contract to NASA. RIM provides both an interactive query system and a Fortran interface. The interactive query system is used to monitor the contents of the data base. The RIM applications program interface is a set of Fortran subprograms which may be used to load data into the database or to retrieve data from it. It has been used to develop the specialized programs which access the GAMS data for the interactive consultant, for the production of the printed GAMS catalog, and for database maintenance.

6. DISCUSSION

The central purpose of the GAMS project has been to provide in integrated system of documented software. The integration has been across many features, not the least of which is the consolidation of mathematical and statistical software under one umbrella. This consolidation facilitates the communication among numerical analysts, statisticians, and other scientists involved in both software development and usage. While the target audience for GAMS products has been NBS scientists, the work has had broader interest. Of particular interest to the statistical computing community, the Committee on Statistical Algorithms of the Statistical Computing Section of the American Statistical Assocation has been involved with the development of that portion of the classification scheme dealing with statistical computations. Current issues under discussion by the committee include providing to the general statistical computing community a more general version of GAMS. This version would on the one hand not contain site-specific information, and on the other would reference software not available on a particular computer (e.g., listings in journals). Of course, it would be desirable to have much of this software provided through a distribution service. Substantial effort may be required to modify the software for portability, prepare on-line documentation, and prepare test software designed to efficiently test whether or not the software has been properly installed.

NBS is currently undertaking the task of converting all of its central computing to the

Cyber 855 and Cyber 205. The GAMS data base will be maintained on a VAX 11/785. As part of these conversions, the GAMS software has become more flexible and more portable. The presence of these large computers has also motivated software consolidation, and as a result, more software will be supported and documented through GAMS. Current efforts involve adding SPSS, Dataplot, and graphics software to the GAMS data base. Future plans include further modifying the data base to fully distinguish among machines in a multi-machine (computers and/or peripherals, especially graphics devices) environment, and providing software specifically vector computers and computers with other interesting architectures.

ACKNOWLEDGEMENTS

I thank Ron Boisvert and David Kahaner for permission to summarize in this paper many of the items discussed in a paper we recently submitted for publication (Boisvert, et al., 1985). I thank the members of the Committee on Statistical Algorithms for their continued support.

DISCLAIMER

Certain software products, both non-proprietary and proprietary, are identified in this paper in order to adequately describe the GAMS work. Identification of such products does not imply recommendation or endorsement by the National Bureau of Standards, nor does it imply that the identified software are necessarily the best available for their purposes. Conversely, the omission of any such software in any edition of GAMS or in this paper does not imply its unsuitability for use.

REFERENCES

_____ (1982). RIM -- Relational Information Management System Version 5.1 User Guide, Boeing Computer Services Co., 7980 Gallows Ct., Vienna, VA.

Boisvert, R. F., S. E. Howe, and D. K. Kahaner (1983). "The GAMS classification scheme for mathematical and statistical software," SIGNUM (Special Interest Group on Numerical Mathematics of the ACM) Newsletter, vol. 18, no. 1, pp. 10-18.

Boisvert, R. F., S. E. Howe, and D. K. Kahaner (1984). The Guide to Available Mathematical Software, PB 84-171305, National Technical Information Service (NTIS), Springfield, VA 22161.

Boisvert, R. F., S. E. Howe, and D. K. Kahaner (1985). "GAMS: A Framework for the Management of Scientific Software," submitted for publication.

Bolstad, J. (1975). "A proposed classification scheme for computer program libraries," SIGNUM Newsletter, vol. 10, nos. 2-3, pp. 32-39.

Dongarra, J. J., C. B. Moler, J. R. Bunch, and G. W. Stewart (1979). LINPACK User's Guide, SIAM, Philadelphia.

Francis, I. (1981). Statistical Software, A Comparative Review, North-Holland, New York, 542 pp.

Howe, S. E. (1982). "The Documentation of Statistical Software in GAMS: The Guide to Available Mathematical Software," Computer Science and Statistics: Proceedings of the 14th Symposium on the Interface, Springer-Verlag, New York, pp. 251-254.

Howe, S. E. (1983). "Cataloging Statistical Software: Current Efforts by NBS and the Committee on Statistical Algorithms," American Statistical Association 1983 Proceedings of the Statistical Computing Section, ASA, Washington, DC, pp. 45-49.

Special Topics in Statistical Computing

Organizer: *Constance L. Wood*

Invited Presentations:

An Iterative Approach to Improving Data Analysis in the Classroom

David P. Kopsco, John D. McKenzie, Jr., and William H. Rybolt

A Statistical Parameterization Model for an Integrated Statistical and Commercial Database

Iheanacho Nwokogba and William H. Rowan, Jr.

Computer Implementation of Matrix Derivatives in Optimization Problems in Statistics

K. G. Jinadasa and D. S. Tracy

Automatic Computation of First and Second Derivatives with Application to Compartmental Models

David E. Gray and David M. Allen

COMPUTER SCIENCE AND STATISTICS:
The Interface, D.M. Allen (ed.)

AN ITERATIVE APPROACH TO IMPROVING DATA ANALYSIS IN THE CLASSROOM

David P. Kopcso, John D. McKenzie, Jr., and William H. Rybolt

Babson College

In the future widely-used statistical software will incorporate many more user-friendly features including some topics usually considered as artificial intelligence. We are currently working on three projects to prepare our students for their future work with such software and to improve our teaching of data analysis. We are monitoring and evaluating user-response patterns. The results of this study will be some suggested enhancements for interactive statistical software and a better understanding of how students analyze data. Other work includes the creation of a preprocessor to assist users in deciding the appropriate commands for their analyses and a report generator designed specifically for introductory applied statistical topics.

Motivation

In contrast to a decade ago, today most colleges and universities use statistical software in their introductory applied statistics courses. In the future the use of such software will be even more prevalent. At our small teaching-oriented college of 1,400 full-time undergraduates and 1,600 full-time and part-time MBA's, we are no exception. In each of our introductory statistics courses, the students are exposed to at least one statistical package. In the past the most commonly taught package was IDA; currently all our students use Minitab. Among the reasons for using such software are the ability to perform calculations more easily and to produce more statistical analyses, including complicated analyses, state-of-the-art analyses, and analyses involving complex data structures. It is often interesting to reflect that just a short time ago there were entire courses devoted to multiple linear regression which contained much the same material now taught in approximately one week.

Too few instructors who use such software recognize that there are weaknesses assocated with its use. For example, at many schools there is only one package available or only one package taught so that the package dictates the material covered in the course. Many students in these single-package courses do not recognize that there are other analyses which are not available on the chosen package. Thus the students use only that package in their subsequent statistical analysis work. New problems such as simultaneous number of introductory applied statistics courses, only a cookbook approach to statistical analysis with the computer is performed. That is, the students are taught that entering the given sequence of commands is the only way to handle a given problem. This lack of teaching the strategy of analysis is a major weakness in the use of statistical software in the classroom today.

The strengths and weaknesses of using statistical software in the classroom will be more pronounced in the future. Statistical packages will become even more powerful and user-friendly in the future. Statistical package developers were slow to realize that users appreciated such features as is shown by the following quote from Francis (1981, p.23) "The notable feature of this table is the group entitled Convenience of User Language. Expressions such as 'ease of use', 'convenience' or 'language' were repeated time and again."

But, today as Kay (1984, p.54) mentions "The user interface was once the last part of a system to be designed. Now it is the first." This "softer software", as Gates calls it, will probably incorporate some of the features now associated with artificial intelligence. (Two excellent summary articles on the interface between artificial intelligence and statistics are Gale and Pregibon (1985) and Hahn et al (1985).)

One characteristic of future statistical software will be a better guidance system of what analyses should be performed. Thus the user will be directed to what command or series of commands should be used in order to analyze a particular set of data. Another future characteristic will be domain-generated output in the form of a readable report. The end-user will not have to translate the package's output. All he or she will have to do is read the package's executive summary of the analysis. The incorporation of both of these features into future statistical software will affect greatly the way applied statistics is taught. In order to prepare ourselves better for our future teaching we are creating both an elementary preprocessor for Minitab to assist the students in selecting the proper commands and report generators for some topics associated with our introductory applied statistics course. Descriptions of our approaches to these two projects may be found at the end of this paper. In order for us to be successful we believe more information is needed about how students analyze data using a statistical package.

Knowledge about how either students or the average user employ a statistical package is quite sparse. Much work in artificial intelligence has centered around how experts feel how average users will employ such packages. Then this expert information is incorporated into the sytem. While we believe it is extremely important that the experts' opinions be placed into future statistical software, we also believe that what the average user does with such software should also be incorporated. Thus our initial work in preparing for the arrival of the statistical software of the future is the monitoring and evaluation of user-response patterns to existing statistical software. From this work we feel that we will gain a better understanding of how our students use a statistical packages and thereby obtain a solid background for beginning our work on the preprocessor and report generators mentioned above. (An immediate byproduct will be an improvement in our teaching of such software.) Moreover we will be able to provide some suggested enhancements to the existing statistical software.

Monitoring and Evaluating User-Response Patterns

Our initial work in monitoring the user-reponse patterns on a data analysis problem was performed on a small group of students taking a second course in applied statistics. Although we designed this work to set the foundation for our later experiments, we did gain some insight into how users employ a statistical package. In addition a discussion of this experiment provides the basis for our later work.

This experiment was given as the take-home portion of the final examination. Here are the instructions given to the students:

1. Examine supplementary exercise 14.47 on pages 669-670 of McClave and Benson (1982).

2. Outline the Minitab commands necessary to complete this exercise.

3. Perform these commands in one (1) run of the statistical package NEWMINI on a hard copy terminal. (To get into NEWMINI, enter NEWMINI at the $ prompt.)

4. Bring your copy of this run, including log-in and log-off information, to the final examination.

5. At the final be prepared to answer questions related to this exercise which may be based upon you NEWMINI run.

Exercise 14.47 is a problem which illustrates that the two independent sample mean problems with unknown variances can be analyzed in three ways. Namely, by a pooled two-sample t-test, a slope test in a simple linear regression model, and a one-way ANOVA test. This exercise from the course's principal text was selected so that the students would use a variety of the NEWMINI commands. The data set for this initial experiment was presented as two columns of eight observations. Thus it could be entered into the machine with no difficulty.

NEWMINI was just a modified version of Minitab provided to us by Minitab, Inc. When someone uses NEWMINI a record of their command entries is placed into a file. This file also included a listing of the Minitab recognized errors encountered along with numerical codes explaining these errors. NEWMINI does not keep track of typographical errors which were corrected before being sent to the main package. Nor does it identify fatal errors. The students were unaware their responses were being monitored.

Note that for this first experiment we asked the students to perform only one run. This unrealistic requirement for modern data analysis in an interactive mode was introduced because our main goal was to test how well NEWMINI worked. That is also the reason we required the students to turn in hard copy runs of their work on NEWMINI. We requested log-in and log-off information for this run in order to guarantee that each student performed the assignment on his or her own account.

At the examination we collected the student's hard copies. Then we compared these runs with the files created on each student's account by NEWMINI. At this point we discovered that we did not have the expected one-to-one correspondence. Somehow a number of NEWMINI files had disappeared. This was probably due to a file saving snafu by our computing center at the end of the semester. Based upon this information we decided to automate further the collection of files generated by NEWMINI for our future experiments.

We wrote a program to process the data in the 18 NEWMINI files we had obtained. This program produced the following 15 pieces of information: student section number, student ID number, session number, type of entered line, command ID number, command string, number of characters in a line, number of entries separated by blanks in the line, whether or not there were subcommands or error messages following the line, the number of Minitab recognized errors along with their error codes, the arguments following HELP and SAVE commands, and the number of data lines following the given line.

For this initial experiment section and session numbers were constant. Entered lines were classified as to whether they were a valid Minitab command, a data line, a Minitab greeting line, or something else (usually an error). For this initial experiment section and session numbers were constant. This information was entered into Minitab for analysis. In this process we created additional variables from the entered information. For example, we created a variable which identified the command classification of each valid command line. This classification scheme was based upon the 20 command categories present in the Minitab documentation. A preliminary summary of this first experiment is presented under the CA0 (Computer Assignment 0) column of Table 1.

From this table we see that 18 students entered 291 lines in performing their 18 runs. Most of these lines were nondata lines. Of these 252 lines, 228 were valid command lines, 7 were invalid

Table 1

Preliminary Summary
of First Two Experiments

Number of:	CA0	CA1
sections	1	4
students	18	150
runs	18	572
command lines	291	12259
nondata lines	252	11750
valid command lines	228	9787
invalid command lines	7	1356
greeting lines	17	607
data lines	39	509
error lines	14	1790
1 common error lines	14	1496
2 common error lines	0	242
3 common error lines	0	11
4 common error lines	0	2
5 common error lines	0	0
individual command error lines	0	39

command lines, and 17 were greeting lines. (The reason there were only 17 greeting lines is that one student emptied the contents of NEWMINI file in order to free up some file space. This also led to our above mentioned decision to automate the data collection process in further analyses.) From Table 1 we see that the 14 errors recognized by Minitab were relatively simple errors in that only one error appeared on each error line. Thirty-five percent of these errors were errors of improper command name designation.

As mentioned above, we had originally intended this experiment as a way to pretest our information gathering process. But to our surprise, we gathered some insights into both potential software improvements and a better understanding of the data analysis process. We noticed that a large number of these students attempted to use Minitab's SET command to enter two variables at a time even though SET is designed to enter one variable at a time. For example, SET C1 is valid, but it is not valid to say SET C1, C2. In addition few student took advantage of the horizontal data entry feature available in SET. We also observed that a lot of these relatively advanced students just restarted after they made a data entry error. This may have been due to the small data set but it pointed out to us the potential wastefulness of repeated data entry. We now believe that we should demonstrate more examples of data correction than we currently do.

Our overall observation was that these students had mastered data analysis using an interactive statistical package fairly well. We saw this by such things as the paucity of errors and the fact the two requested HELP commands were followed by the command on which HELP was requested. This impression can also be seen in Table 2 and Table 3.

Table 2

CA0 Expert Data Analysis Flow Chart

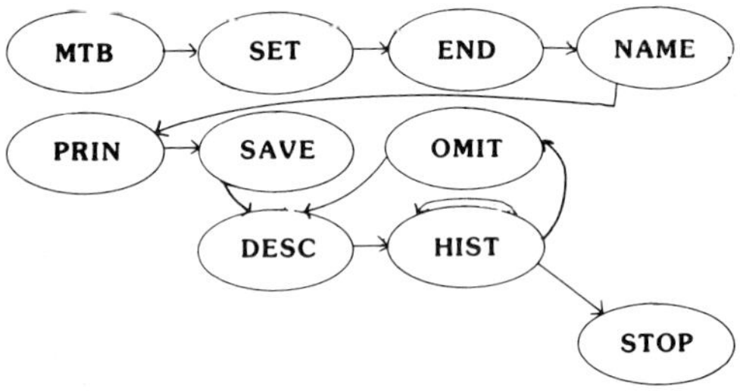

Table 3

CA0 Modal Data Analysis Flow Chart

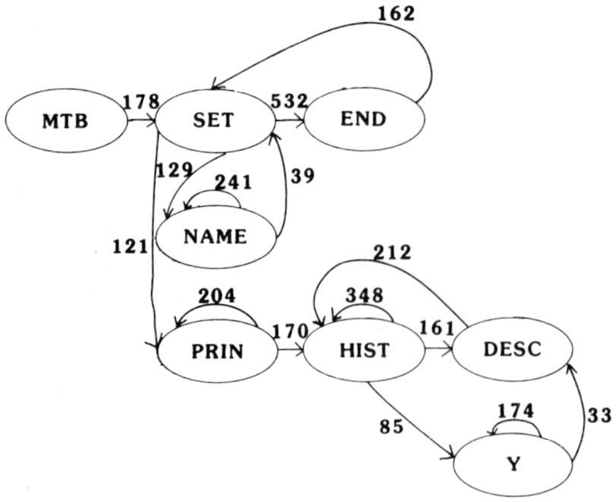

Table 2 is a data analysis flow chart that we constructed. It is our opinion of the expert's sequence of Minitab commands that should have been entered in order to answer exercise 14.47 in one Minitab run. This sequence shows that the student should first enter the two variables, one containing the data and the other indicating the appropriate sample, label these variables, and then print out the labeled variables for verification. Then the student is ready to perform the requested three analyses. In our opinion this is accomplished by performing a two-sample t test with an indicator variable (TWOT), plotting the data, performing a regression after requesting a complete set of output (BRIEF), and performing a one-way ANOVA with an indicator variable (ONEWAY). Finally the student should leave Minitab by STOPping his or her session.

Table 3 is a data analysis flow chart constructed from the students' modal sequence of Minitab commands. In contrast to Table 2 this table also includes the number of students who followed a specific path. For example, eight students started with a READ command while eight students began their run with a SET command. After using these commands 20 students ended their data entry by issuing an END command. The modal next command was PRINT. That is, most of the students printed out their data entries for verification. From the PRINT command the modal command responses were TWOSAMPLE (a two independent sample command involving two columns of data) and REGRESSION. This split probably occurred because a number of students did not perform any regression analysis due to the fact that exercise 14.47 included regression output from SAS. The modal response after TWOSAMPLE was a JOIN command. This command was necessary since the data needed to be restructured in order to perform a REGRESSION. Thus it was followed by a SET command. After REGRESSION the modal response was the AOVONEWAY, an ANOVA command involving two columns of data. This command was then followed by a Minitab STOP command in the modal flow chart. The students' modal responses covered all four parts of this exercise, data entry, two-sample t test, regression, and one-way ANOVA. While their response patterns were not identical to the expert's pattern, their patterns were reasonably close.

The students for our next experiment were members of four sections of an introductory applied statistics course. Most of these students were freshmen who had never dealt with a statistical package before. As for computer expertise all of these students had either taken or were taking an introductory management information course in which they are required to write BASIC programs on Babson's VAX. Thus, while all of these students were Minitab novices, all had some familiarity with the environment in which Minitab resided.

This experiment was given as their first computer assignment. For this project they were asked to describe the data present in Table 4.

Table 4

State and Local Per Capita Tax Burden in Fiscal 1982-1983

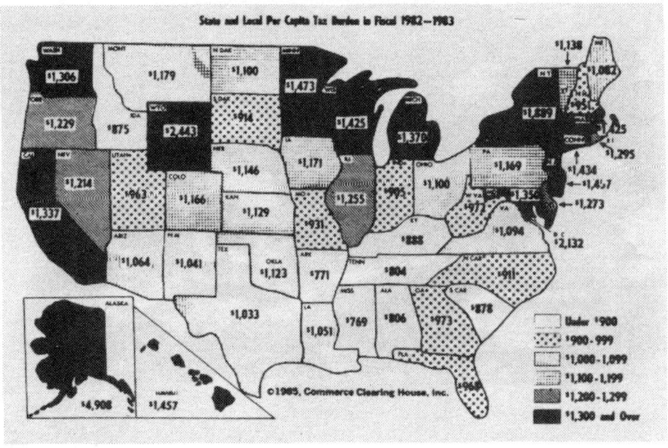

We were interested in monitoring how these students, with less than two weeks of limited Minitab classroom exposure, would enter and manipulate the data in Minitab from the given map. In addition we were interested in learning how these students would describe the data and display the data graphically using the Minitab commands.

To perform this experiment the students were asked to access NEWMINI through a program called CAl on a hardcopy terminal, not a CRT. In that way we obtained hard copies of their runs to check against the monitoring we did using NEWMINI.

The second column in Table 1 presents our initial summary of this experiment. Note that we collected a lot more data in this experiment. This increase was not just due to the larger number of students. It was also due to the nature of the problem (which was much more open-ended than CA0), the nature of the assignment (a one-week assignment, in contrast to a one-run take-home portion of a final examination), and the fact that this experiment involved mainly freshmen. Table 1 also show that the students working on CAl entered many more error lines; something to be expected by this group of naive users.

A number of other interesting observations come forth from our initial examination of the data from this experiment. For instance, the number of runs ranged from 1, by 24 students, to 12, by 2 students. The median number of runs was 3. The median number of continuous data lines entered by the students was 1 due to the large number of students who used the SET command to enter the 51 pieces of data from the map. The two most frequently used commands were PRINT and HISTOGRAM. These commands were entered 9.68% and 9.07% of the time, respectively. Based upon the experiment these responses were no surprise. But it was surprising that 33 different users entered the KRUSKAL-WALLIS command a total of 37 different times for this assignment dealing with descriptive statistics. Another expected response was that the students requested HELP on the HISTOGRAM most frequently. The most common error was the entry of an illegal command which constituted 23.70% of the errors.

From observing items such as the above from the 572 runs of the 150 students who participated in this experiment, we determined six places where the Minitab statistical package might be improved. Most of these improvements deal with increased user-friendliness although our first suggestion might be viewed as making the package less friendly. To our surprise a large number of students were entering lines without any delimiting blanks. For example, DESCCl instead of DESC Cl. While use of this entry method did not cause the user any trouble initially, it led to great difficulty when complex commands appeared. For example, the entering of HISTCl650 250 for HIST Cl 650 250 caused some students frustration. Thus we propose that blanks be required as delimiters in all Minitab commands. We also propose the addition of at least two new options in Minitab: a RANGE command and a LABEL ROW command. Many students tried to use these features. We would like Minitab to recognize synonyms for the STOP command. One leaves Minitab by issuing a STOP command. If one enters an EXIT or QUIT command, they are told to enter STOP. From the frequency of such requests we believe much time can be gained by allowing Minitab users to leave a Minitab session via STOP, EXIT, QUIT, BYE, etc. We also feel that a large number of command errors could be eliminated if Minitab accepted commands in which two characters were transposed. Thus, we would like to see the acceptance of DECS and HIEGHT for DESC and HEIGHT. We believe that Minitab should better publicize the fact that operating system commands can be run from within Minitab by first specifying SYSTEM. Many of these naive students-users tried to obtain a directory listing of their files or to delete an existing file in Minitab. In addition we would like to see Minitab provide a local command warning option so that an individual location could alert their users of Minitab's inability to perform a specific task. Similar to a macro facility, this feature could be used to tell students that they could not run BASIC within Minitab.

We obtained an embryonic understanding of the data analysis process by observing the results from the students' CAl runs. Here are a few of our discoveries. Of the data entries 71.29% were followed by an END command. Thus the majority of the students concluded their data entry in the preferred way. Almost one-third of the students PRINTed our their data after entering it. It appears that even in the first two weeks of the course a surprising large number of students were verifying their entered data. On the negative side most of the students made inefficient use the the Minitab SAVE command. There were 299 SAVE commands and only 305 RETRIEVE commands! In contrast to an expert these naive users only seem to be using their SAVEd data files once. It was also discouraging to see how poor the choice of names of the SAVE files were. We believe that most people would have trouble recalling what was the content of their SAVEd files from the selected names. Finally we noticed that a major error was designating the wrong number of arguments for a command.

We also learned something about the data analysis process by contrasting the data analysis flow chart of an expert (Table 5) with the modal data analysis flow chart of the students (Table 6). According to the expert the sequence of commands to prepare the data for this experiment would be to enter the data by employing the SET and END commands, label that variable by using the NAME command, verify the entered data by issuing a PRINT command, and then place the data into a Minitab file by entering a SAVE command. The expert would DESCRIBE the data, produce a well-constructed HISTOGRAM, OMIT the

Table 5

CAl Expert Data Analysis Flow Chart

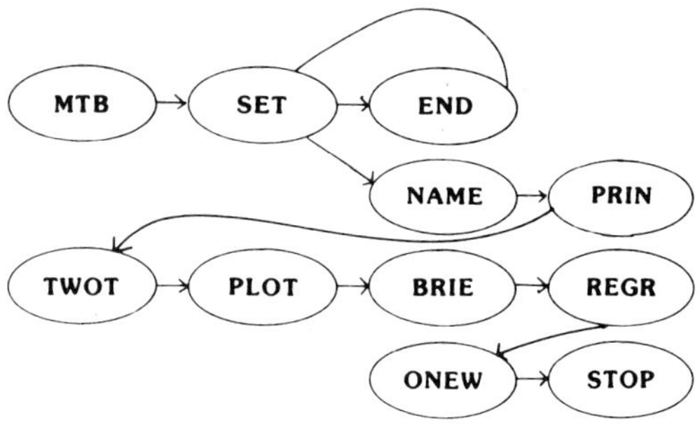

Table 6

CAl Modal Data Analysis Flow Chart

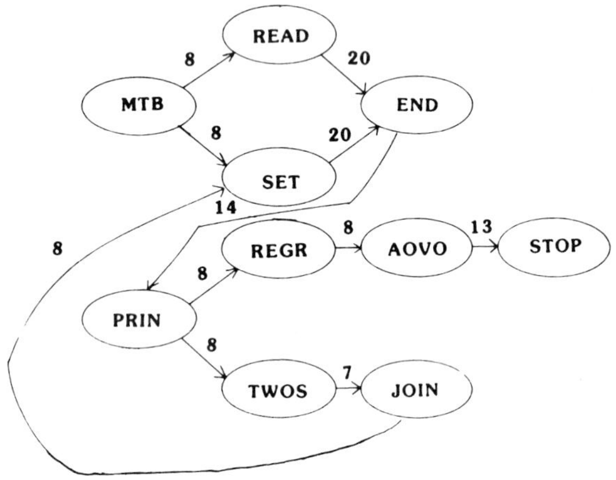

outlying observations, repeat the description and displays, and finally leave Minitab. The data analysis flow for the students was not as straight forward although the students did use basically the same commands. They used the SET and END commands a number of times before issuing the NAME command. (The NAME command is issued repeatedly.) This was probably due to the errors these inexperienced Minitab users introduced. They probably used the PRINT command repeatedly for the same reason. A number of HISTOGRAM commands followed. The next node on the students' modal path is the DESCRIBE command which was usually followed by another HISTOGRAM command. The only other branch of any size from HISTOGRAM was to a Y prompt (probably to guarantee the printing of a long histogram). From the Y prompt most students went back to the DESCRIBE command. Thus this modal data analysis flow chart produces a path which does not reach the STOP command, a big difference from the path taken by the expert.

To determine why the students did not reach the STOP command, we constructed a data analysis flow chart starting at that command. A portion of that chart may be found in Table 7.

Table 7

Path to Stop

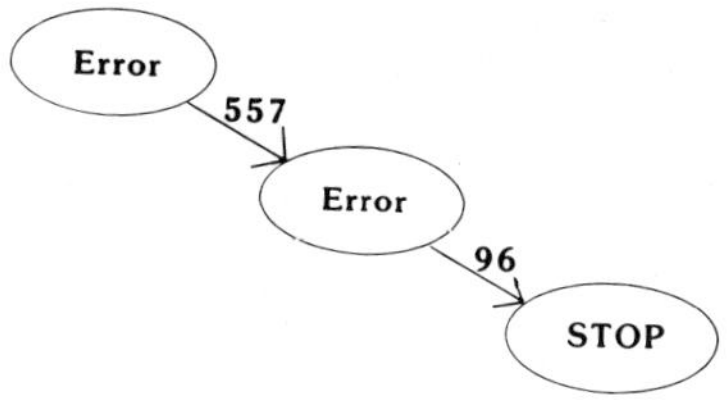

Here we see quite clearly what came before the conclusion of a Minitab session. The modal response was an error. This happens 96 times. And before this error 557 times another error occurred. The way most students concluded their Minitab sessions for this experiment was in frustration.

We are gathering another set of data from the students who are taking their introductory applied statistics course. This experiment deals with a linear regression modeling assignment. In constrast to being the student's initial computer project, it will be their last. We hope to determine what response pattern changes, if any, have occurred in each student over the course of a fifteen-week semester. Here we plan to provide to each student a different, but related, data set. In this way we hope to prevent sharing of commands by a group of students. Initially these data were to be made available to the students in a computer-file to save the students some time, but in order to monitor any difficulties with data entry, we will have the students enter the data into the computer themselves.

Creating an Elementary Preprocessor

As mentioned above we are in the midst of creating a preprocessor to aid our students in the selection of the appropriate Minitab command for a specific analysis. This front-end will be designed similar to the charts prepared by Andrews et al (1981) and the Statpath software outlined by Portier and Lai (1983). Initially we will base our preprocessor on Version 85 of Minitab.

Developing Report Generators for Introductory Applied Statistics Courses

Our report generator plans include backends for the following topics usually found in an introductory applied statistics course: confidence intervals and hypothesis tests for the population mean and for the difference between two population means, simple linear regression, and chi-square tests for independence and for equal proportions. We also plan to produce incorrect reports on these topics so that the student can be taught to criticize the computer-generated output.

Future Directions

We believe the work we have began in monitoring and evaluating user-response patterns will produce a better understanding of the data analysis process; thereby enabling us to better understand the students' techniques and to develop our preprocessor and report generators. In addition we will be able to suggest enhancements to existing and future statistical software. We also believe that there is much more that can be gained by extending our initial work. Four possible extensions are more varied experiments, the introduction of more variables in these experiments, better collection devices, and more complete error analysis.

Additional experiments to understand how users employ statistical software will be based on analyses other than those mentioned. In addition all these experiments could be performed by students at other schools and by non-student users of statistical software. Monitoring and evaluating of user-response patterns could also be performed on other types of statistical software. In contrast to Minitab, an interactive package with command lines, there are batch packages and interactive packages with prompt command lines, cursor menus, and mouse menus. User demographic variables should also be brought into these experiments along with time variables for experiments dealing with interactive systems.

Some ideas for better collection devices include devices which capture all the user entries including errors which the user corrects before sending them to the package. Completely automated devices, possibly part of the software, are another possibility. Finally devices which enable one to examine random samples of the user population are desired extension of our work.

In addition to dealing with the forgiving errors of syntactical or semantic natures complete evaluation of

user-response patterns should analyze typographical errors, fatal errors, and logical errors. Note these errors, especially logical ones, will be difficult to examine.

Acknowledgements

The authors wish to thank Minitab, Inc., the Babson College Computer Center, and Gordon Prichett for their assistance in performing our monitoring experiments.

Selected References

Andrews, Frank M. et al (1981), A Guide for Selecting Statistical Techniques for Analyzing Social Science Data, Second Edition, Ann Arbor: The University of Michigan Survey Research Center.

Brode, John (1978), "Rights and Responsibilities of Statistical Language Users: Language Standards for Statistical Computing", Proceedings of Computer Science and Statistics: 11th Annual Symposium on the Interface, 11, 7-20.

Bubolz, Thomas A. and Gentle, James E. (1979), "The Role of Package Programs in Statistical Methods Courses", Proceedings of the Fourth Annual SUGI Conference, 4, 168-174.

Carr, D. B. et al (1984), "Organizational Tools for Data Analysis Environment", handout at 1984 Joint Statistical Meetings, Philadelphia.

Chambers, John M. (1979), "Designing Statistical Software for the New Computers", Proceeding of the Computer Science and Statistics: 12th Annual Symposium on the Interface, 12, 99-103.

Chambers, John M. (1981), "Some Thoughts on Expert Software", Computer Science and Statistics: Proceedings of the 13th Symposium on the Interface, 13, 36-40.

Chambers, John M. et al (1981), "Expert Software for Data Analysis", Proceedings of the 43rd Session of the International Statistics Institute, 294-308.

Dixon, W. J. and Jennrich, R. I. (1972), "Scope, Inpact, and Status of Packaged Statistical Programs", Annual Review of Biophysics and Bioengineering, 1, 505-528.

Francis, Ivor (1979), "The Statistical Profession and the Quality of Statistical Software", Bulletin of the International Statistical Institute, 46, 212-236.

Francis, Ivor (1977), A Comparative Review of Statistical Software, Voorburg, Netherlands: International Association of Statistical Computing.

Francis, Ivor (1981), Statistical Software: A Comparative Review, New York: North-Holland.

Francis, Ivor and Valliant, Richard (1976), "The Novice with a Statistical Package: Performance Without Competence", Proceedings of Computer Science and Statistics: 8th Annual Interface, 8, 110-114.

Gale, William A. and Pregibon, Daryl (1983), "An Expert System for Regression Analysis", Computer Science and Statistics: Proceedings of the 14th Symposium on the Interface, 14, 110-117.

Gale, William A. and Pregibon, Daryl (1985), "Artificial Intelligence Research in Statistics", The AI Magazine, V, 4, 72-75.

Hahn, Gerald J. et al (1985), "More Intelligent Statistical Software and Statistical Expert Systems", The American Statistician, 39, 1, 1-16.

Hajek, P. and Ivanck, J. (1982), "Artificial Intelligence and Data Analysis", Proceedings in Computational Statistics, 5, 54-60.

Kay, Alan (1984), "Computer Software", Scientific American, 251, 3, 53-59.

Kemeny, John G. (1983), "Finite Mathematics - Then and Now", The Future of College Mathematics: Proceedings of a Conference/Workshop on the First Two

Years of College Mathematics, edited by Anthony Ralston and Gail S. Young, New York: Springer-Verlang.

Kopcso, David P., McKenzie, Jr., John D., and Rybolt, William H. (1984) "Pedagogical Impact of AI on Statistical Software", paper presented at the ASA-IASC-SIAM Conference on the Frontiers in Computational Statistics, Boston.

Lefkowitz, Jerry M. (1985), Introduction to Statistical Computing Packages, Boston: Duxbury.

The Mathematical Association of America (1981), Recommendations for a General Mathematical Sciences Program, Washington: The Mathematical Association of America.

McClave, James T. and Benson, P. George (1982), Statistics for Business and Economics, Second Edition, San Francisco: Dellen Publishing Company.

Miller, John E. and Francis, Ivor (1976), "A Computing Novice with a Statistical Package: Interactive Versus Batch", Proceesings of Computer Science and Statistics: 9th Annual Symposium on the Interface, 9, 268-274.

Muller, Marvin E. (1980), "Aspects of Statistical Computing: What Packages for the 1980's Ought to Do", The American Statistician, 34, 3, 159-168.

National Science Foundation (1982a), The Mathematical Science Curriculum K-12: What is Still Fundamental and What is Not, Washington, U.S. Governament Printing Office.

National Science Foundation (1982b), Today's Problems, Tomorrow's Crises, Washington: U.S. Government Printing Office.

National Science Foundation (1983), A Revised and Intensified Science and Technology Curriculum Grades K-12 Urgently Needed for Our Future, Washington: U.S. Government Printing Office.

Nelder, J. A. (1978), "The Future of Statistical Software", Proceedings in Computational Statistics, 3, 11-19.

Oldford, R. W. and Peters, S. C. (1984), "Building a Statistical Knowledge Based System with Mini-Mycin", MIT Center for Computational Research in Economics and Management Science Technical Report No. 42.

Portier, Kenneth M. and Lai, Pan-Yu (1983), "A Statistical Expert System for Analysis Determination", American Statistical Association 1983 Proceedings of the Statistical Computing Section, 309-311.

Rice, John R. (1976), "Statistical Computing: The Vanguard of the Future of Education", Proceedings of Computer Science and Statistics: 9th Annual Symposium on the Interface, 9, 2-4.

Ryan, Thomas A., Jr., Joiner, Brian L., and Ryan, Barbara F. (1982), Minitab Reference Manual, University Park, Pennsylvania: Minitab Project.

Thisted, Ronald A. (1976), "User Documentation and Control Languages: Evaluation and Comparison of Statistical Packages", American Statistical Association 1976 Proceedings of the Statistical Computing Section, 24-30.

Thisted, Ronald A. et al (1976), "Teaching Statistical Computing Using Computer Pdackages", The American Statistican, 33, 1, 27-35.

Tukey, John (1971), "How Computing and Statistics Affect Each Other", paper prepared for the Babbage Memorial Meeting, London.

Velleman, Paul F. (1980), "Do Statistical Packages Help or Hinder Data Analysis", American Statistical Association 1980 Proceedings of the Statistical Computing Section, 21-26.

Wexelblat, Richard L. (1978), "What is a Language for Statistical Computing", Proceedings of Computer Science and Statistics: 11th Annual Symposium on the Interface, 11, 25-37.

Wilf, Herbert S. (1983), "Symbolic Manipulation and Algorithms in the Curriculum of the First Two Years", The Future of College Mathematics: Proceedings of a Conference/Work on the First Two Years of College Mathematics, edited by Anthony Ralston and Gail S. Young, New York: Springer-Verlang.

Wilke, H. (1982), "Evaluation of Statistical Software Based on Empirical User Research", Proceedings in Computational Statistics, 5, 442-446.

Wyllys, Ronald H. (1978), "Instructional Use of Statistical Program Packages: BMD, IMP, OMNITAB II, and SPSS", Proceedings of Computer Science and Statistics: 10th Annual Symposium on the Interface, 10, 265-270.

COMPUTER SCIENCE AND STATISTICS:
The Interface, D.M. Allen (ed.)

A Statistical Parameterization Model for an Integrated Statistical and Commercial Database

Iheanacho Nwokogba* - William H. Rowan, Jr.**

* Rochester Institute of Technology, Rochester, New York 14623

** C.S. Dept., Vanderbilt University, Nashville, Tennessee 37235

ABSTRACT

Traditional databases accommodate statistical applications - and other applications - at an abstraction level higher than the user level. But statistical analysis possesses exploitable properties that can be used to integrate the realization of statistical functions with the database activity of data acquisition. If statistical functions are parameterized, it is easy to see that they share many common parameters. These parameters are both updatable and additive. Statistical functions and their parameters are explored together with round-off errors resulting from updates.

1. Introduction

Attention has been drawn recently to the inadequacy of current databases in accommodating statistical analysis [2, 12, 24, 31]. The inadequacy arises from the intrinsic structure of statistical analysis and the inability of the underlying models of database systems to capture and correctly model statistical structure efficiently. The proposed model is concerned with exploiting three properties of statistical analysis that result in inefficiencies when realized in traditional databases. They are: (1) the nature of statistical queries (2) the nature of statistical calculations and (3) statistical classifications.

1.1. Statistical Queries

Turner et. al. [35] have distinguished three types of queries - informational, operational and statistical queries. Statistical queries request over 10% of the records in a database while the other types request less. One consequence of this is that response times are high when statistical queries are made. Another important fact is that databases, in both research and practice, have been geared primarily towards informational, and to a lesser extent, operational queries - the concept of primary keys (informational) and secondary keys (operational) and the subsequent theories underlying the various normal forms [3, 10, 11]. Statistical queries have been surbordinated. As a result, man-made attributes such as social security number, employee number, etc., have become the focal point for retrieval in databases. As Turner et. al. have pointed out, statistical queries are based on more natural attributes like 'all females', 'all black employees', 'minority engineers' etc. The consequence of this is that a statistical user is faced with the task of specifying statistical queries in terms of informational and operational queries. Very complex formulations often result [17]. So, we can conclude that the statistical interface in many commercial systems is unfriendly.

1.2. Statistical Calculations

A major source of redundancies in statistical applications is in the function calculations. If statistical functions are parameterized and viewed in terms of these parameters, it will be seen that many of them share common parameters. For example, to calculate the Pearson's correlation coefficient between two variables (attributes) A and B, we need the following parameters: sample size, n; the sum of products of A and B; the sum of squares for A and the sum of squares for B. On the other hand, to calculate the standard deviation for A (or B), the parameters are: n and the sum of squares for A (or B). So, if the parameters for the correlation coefficient are available, no extra retrieval is necessary to compute the standard deviation. The calculations for a multiple regression model begin with a matrix of Pearson's correlation coefficients (whose parameters have been mentioned above). The same matrix is the starting point for factor analysis and discriminant analysis. Thus, thinking of statistical functions in terms of their parameters can save redundant calculations. The parameters of many statistical functions ranging from simple descriptive statistics to multivariate analyses, will be explored.

An interesting property of these statistical parameters is that they are 'updatable' in the sense that if P_n is the parameter for n points, it is possible to compute P_{n+1} given n, P_n and a_{n+1}, where a_{n+1} is the new datum. The updating formulas for some parameters have been known for some time starting with Welford's pioneering paper [37]; and more recently also [6, 19, 22]. The meaning of this is that if all required parameters are known a priori, they can be kept current during data acquisition, i.e., during insertions, deletions and modifications. Updating some parameters introduces additional round-off errors, however. The nature of these errors for some parameters has been investigated [6, 21, 27, 38]. This paper extends these investigations and show how they affect the final function values. It should be mentioned that batch updates are also possible using the additive formulas or transaction type updates [31].

1.3. Statistical Classifications

Classification is an inherent part of statistical analysis. Many classification schemes take the form of 'treatments' or 'categories' by which some metric is grouped. For example: suppose there is a 'category' attribute, RACE, whose domain is {hispanic(h), white(w), black(b), other(o)}, by which different users choose to classify a 'data' attribute, SALARY. Note that a category is a member of the power set of the domain of a category attribute. Let us suppose that we have the following categories: wbh, bh, and o. We shall associate with each category, a set of parameters calculated on the data attribute, that is necessary to realize some statistical function(s) for the category. So if a user (or users) is interested in the mean SALARY, the associated parameters are count and sum. Now, suppose another (or same) user is now interested in the mean SALARY of the category, bho, no additional parameter gathering is necessary since this can be derived by combining the parameters for bh and o. Almost all statistical parameters are 'additive'. The additive operation of sum or count is an arithmetic addition. The additive operation of many other statistical parameters involve many arithmetic operations.

The general rules for deriving new categories are by set union if the categories are disjoint (addition of parameters) and by set difference if the two categories have a super/sub-set relationship (subtraction of parameters). Thus, in the above example, the additional categories that can be derived are wbho, wo and w. Again, it is obvious that redundancies can exist if these deriving rules are not applied.

2. Related Work

Many attempts have been made to accommodate some of the problems that have been stated. The two most common approaches are either to build specialized statistical databases or to integrate statistical analysis tools with commercial databases. Some specialized systems include: the use of inverted file structures, like in TDMS [4]; the use of 'transposed' files such as in RAPID [35]; and the use of special data structures as in SUBJECT [7]. Integrated systems take the form of providing better interfaces between the two systems while providing a rich operational repertoire for each subsystem. Such systems will include REGIS [18], RIGEL [28], etc. There are more examples [31]. The first approach lacks generality. While our approach here is of the second type, it differs from others in that it models the statistical subsystem at all three levels of classical database design, with interfaces between the two subsystems at the two top levels. The preponderance of commercial systems tends to suggest that integration is the preferred means of reaching many statistical users.

As mentioned earlier, the updatability and additivity of many statistical parameters have been known. But these have been mostly limited to means and standard deviation calculations. In SYSTEM R [1], the 'trigger' concept is updatability applied to simple aggregate functions. Triggers lack globality. Koenig and Paige in MADAM [19], allow one to define functions in terms of simpler functions (parameters) - mean defined in terms of sum, for example. No global sharing is apparent.

Sato [30] has given a system of classification and rules of derivation. However, the categories are pairwise disjoint and the database is static. We are considering a more general classification scheme in a dynamic database.

Many investigations [16, 21, 27, 38] about the size of roundoff errors have been carried out based on Wilkinson's work [39]. Most of them have been empirical however. The reason for this is probably due to the tedious nature of the much often desired 'forward' analysis involved in these algebraic processes. But Chan and Lewis [5] have developed theoretical upper bounds for mean and standard deviation calculations.

This paper will first concentrate on the statistical aspect of this model and later, the integrated system.

3. Statistical Parameters

The updatability of some statistical parameters has been investigated [16, 21, 22, 37, 38, 40]. The treatment however, has been to regard these parameters as final function values in their own right rather than as parameters to many statistical functions. The additivity of these parameters also warrants a separate treatment - updatability (a special case of additiviy) is employed during data acquisition to keep parameters current, while additivity is used for the derivation of new categories and/or merging of batch updates into a main database. The updating and additive formulas are derived by simple algebraic manipulations. These formulas are now presented and many statictical functions are parameterized.

In what follows, the beginning letters of the alphabet, A, B, C, ... are single attributes whose values in a table instance of $n-1$ records are $[a_1,a_2,...,a_{n-1}]$, $[b_1,b_2,...,b_{n-1}]$ ... respectively, where [...] is used to denote a multiset. W, X, Y, Z will be used to denote sets of attributes. For instance, $X=A_1A_2 \cdots A_d$ and $d=|X|$, is the dimension for the attribute set X. The records here will be drawn from the cross product of the domains of the attributes in X and corresponding small letters are used to denote the records.

3.1. Updatability

Updatability refers to how to calculate $P(X)_n$ from $P(X)_{n-1}$ and the new datum, x_n. The updating formulas for many parameters - counts; sums; sum squares, sum cubed, ...; product sums (and powers); etc ... - are of the form

$$P(X)_n=P(X)_{n-1}+f(x_n), \quad P(X)_0=0$$

where, $f(x_n)$ is the initial calculation for the term to be added, depending on the parameter. For instance, for count and sum, $f(x_n)$ is the identity function; for sum cubed, $f(x_n)=a_1^3a_2^3 \cdots a_d^3$; etc.

However, the formulas for other parameters require more than one addition operation. For the mean (which can be parameterized), Welford [37], gave the following:

$$P(X)_n=\left(\frac{n-1}{n}\right)P(X)_{n-1}+(1/n)f(x_n), \quad P(X)_0=0.$$

Similarly, the sum of squares is

$$P(A)_n=P(A)_{n-1}+\left(\frac{n-1}{n}\right)(a_n-M(A)_{n-1})^2, \quad P(A)_0=0,$$

and the sum of products is

$$P(X)_n=P(X)_{n-1}+\left(\frac{n-1}{n}\right)\prod_{i=1}^{d}(a_{in}-M(A)_i) \quad P(X)_0=0,$$

where M is the parameter mean. Generally, for the sum of squares and the sum of products, $d=1$ and $d=2$ respectively. To further reduce the roundoff errors from the increased number of operations, Reekan [38] proposed for the mean and sum of squares, the following:

$$P(X)_n=P(X)_{n-1}+(f(x_n)-P(X)_{n-1})/n$$

and

$$P(A)_n=P(A)_{n-1}+(a_n-M(A)_{n-1})^2-(a_n-M(A)_{n-1})^2/n.$$

An additional advantage is that the sum and sum squares can be derived more accurately.

Downdating formulas can also be derived by algebraic manipulation of above formulas - this corresponds to the deletion of a record from a database. West has shown that Reekan's formula gives a better accuracy than other alternative methods. However, in the context of a database, separate parameters can be kept for deleted records using only the updating formulas. Then at the time of actual function computation, the parameters can be 'added' together.

3.2. Additivity

Additivity is explained in the following way. Let $C(X)_i$ be a multiset of size n_i and $P(X)_i$, the associated parameter. Then additivity is: a) given $P(X)_i$, $i=1,...,k$, compute $P(X)^+$, for the multiset $C(X)^+$ of size n, $n=n_1+n_2+\cdots+n_k$. That is, given the parameters of k multisets, find the parameter of the merged set; Or b) given $C(X)_p$ and $C(X)_i$, $i=1,...,k$ such that $C(X)_p \supset \sum_i^k C(X)_i$, compute $P(X)^-$ for the multiset $C(X)^-$ of size m, where $m=n_p-n_1-n_2-...-n_k$ and $C(X)^-=C(X)_p-C(X)_1-C(X)_2-\cdots-C(X)_k$. That is, given the parameters of a set of multisets and the parameter of a multiset containing them, calculate the parameter of the resulting multiset after removing all the smaller ones from the big multiset. The additive formulas for statistical formulas are now given without proofs. The proofs are given elsewhere [24].

For the parameters count; sum; sum square, sum cubed, ...; product sums; etc, the additive formulas are of the form

$$P(X)^+=\sum_{i=1}^k P(X)_i \text{ and } P(X)^-=P(X)_p-\sum_{i=1}^k P(X)_i.$$

For mean, we have that

$$P(X)^+=1/n\sum_{i=1}^k P(X)_i n_i$$

and

$$P(X)^-=1/m\left[P(X)_p n_p-\sum_{i=1}^k P(X)_i n_i\right].$$

The additive formulas for sum of squares are

$$P(X)^+=\sum_{i=1}^k (P(X)_i+n_i M(X)_i^2)-nM(X)^+)^2$$

and

$$P(X)^-=P(X)_p+n_p M(X)_p^2-\sum_{i=1}^k (P(X)_i+n_i M(X)_i^2)$$

$$-m(M(X)^-)^2,$$

where M is the mean.

For the sum of products, we have

$$P(X)^+=\sum_{i=1}^k (P(X)_i+n_i M(A)_i M(B)_i)-nM(A)^+M(B)^+,$$

and

$$P(X)^-=P(X)_p+n_p M(A)_p M(B)_p-\sum_{i=1}^k (P(X)_i$$

$$+n_i M(A)_i M(B)_i)-mM(A)^- M(B)^-.$$

The formula has been given here for $X=AB$, i.e., $d=2$. It can easily be extended for $d>2$.

The condition number, κ, of a multiset is a measure of how well conditioned the numbers are. κ as defined by Chan and Lewis [5] is given by

$$\kappa=||f(C(X))||/((n-1)^{1/2}D),$$

where D is the standard deviation and $||f(C(X)||$ is the Euclidean norm. κ is a parameter used to keep track of how good a computation is given the structure (relative magnitude) of the numbers. κ is updatable. The additive formula is,

$$\kappa^+=((\sum_{i=1}^k (S(A)+n_i M(A)_i^2))/S(A)^+)^{1/2}$$

and

$$\kappa^-=((S(A)_p+n_p M(A)_p^2-\sum_{i=1}^k (S(A)_i+n_i M(A)_i^2))/S(A)^-)^{1/2},$$

where S and M are the sum of squares and mean respectively.

We have now seen both the updating and additive formulas of almost all the statistical parameters that will be needed. Updating is used to keep parameters current while additivity is used for calculating the parameters of derivable categories.

3.3. Statistical functions and their parameters

Some representative statistical functions and their parameters are discussed. These functions can be found in most introductory statistical textbooks and are offered in many statistical packages [23, 29].

Basic screening functions consist of frequency count, sum, mean, variance and standard deviations; and to a lesser extent, skewness and kurtosis. The parameters are counts, sum (or mean) and sum of squares for most of them. For skewness and kurtosis, the parameters are count, sum, sum square, sum cube and sum fourth. A measure of correlation between two attributes frequently used is the Pearson's correlation coefficient and the parameters are counts, sums of squares and sum of products.

Contingency tables are two dimensional tables on the attributes A and B with counts in each cell. Some functions associated with a contingency table are χ^2, ϕ, Cramer's V, the contingency coefficient C, tau B, tau C and Spearman's correlation coefficient. All these functions can be computed from the cell counts of the table - thus, the parameters are counts.

A class of statistical functions are used for parametric and non-parametric hypothesis testing. In the former some information is known about the population. In hypothesizing about a single mean where the standard deviation, σ, is known, the standard normal statistic, Z, is used, where

$$Z=\frac{(M(A)_n-\mu_0)}{\sigma n^{1/2}}$$

and μ_0 is the mean being tested for. If the population is normal, the T statistic is used and the formula is similar to that of Z except for the sample standard deviation substituted for σ. The parameters for both Z and T are counts, sums and sums of squares. The parameters are the same when hypothesizing about two means and one or two variances (F statistic). Sign test is a form of non-parametric hypothesis testing that uses a Z statistic computed from counts. For run tests on the other hand, counts can only be kept incrementally if the input is serialized

Experimental classifications are concerned with the means and variances of k populations - the entire population having been subjected to at least one treatment. One-way classification is concerned with the means and variances of k populations resulting from one treatment. In a database context, a treatment might be the category attribute RACE and the equality (or inequality) about the means of data (or summary) attribute, say SALARY, the point of interest. If the populations are normal and have the same variance, the model equation is given by

$$SST=SSE+SS(Tr),$$

where SST is the sum of squares, SSE is the error sum of squares and $SS(Tr)$ is the treatment sum of squares. Expressing this model within the parametric framework gives,

$$\sum_{i=1}^k\sum_{j=1}^{n_i}(a_{ij}-M(A)^+)^2=\sum_{i=1}^k\left[\sum_{j=1}^{n_i}(a_{ij}-M(A)_i)^2\right]$$

$$+\sum_{i=1}^k n_i(M(A)_i-M(A)^+)^2.$$

Thus, the parameters are the sum of squares, sums and counts. Analagous formulas are derivable for the general n-way classification problem with the same parameters.

Among the class of statistical functions in multivariate analysis, perhaps, the most common are multiple linear regression, factor analysis and discriminant analysis. For this class of problems, the beginning point of computation is the square matrix

$$\left[P_{ij}\right], \quad 1 \leq i, j \leq n,$$

where n is the number of variables (i.e., attributes) and P_{ij} is the Pearson's correlation coefficient for each pair of variables. For linear regression, the variables include both the dependent and independent variables. Discriminant analysis may require more than one such matrix. Thus, the parameters for this class of problems are those for the Pearson's correlation coefficients - sums, sums of squares and counts.

Many statistical functions and their parameters have been discussed. But there are some statistical functions that do not easily lend themselves to this parameterization. They include minimum, maximum (and so, range) and median. The reason why this is so is because each of these functions is an attribute value of a record in the database and so if this particular record is modified (or deleted), the database has to be scanned for a new function value. However, it is possible that for such a function, there may be many records with its value and so counts can be kept together with the function value of such records. Thus, no scanning is necessary until the count becomes zero.

4. Error Analysis

4.1. Overview

Use of the various updating formulas in computation will mean increased floating-point operations. Floating-point operations result in errors due to round-offs and catastrophic cancellations. The size of the error also depends on the computer (word size, guard bits, etc). Thus it is important to understand the nature of these errors while using these updating formulas. A model for studying the size of these floating-point errors has been developed [39]. Most analysis have been empirical however. The reason is probably due to the tedious nature of the much often desired 'forward' analysis in these statistical processes. In forward analysis, we desire to know the pertubation introduced in a function F as a result of the perturbations in a multiset $C(X)$ - i.e., given $\Delta C(X)$, what is ΔF? In particular, the relative error of F $(\Delta F/F)$ is of interest and is generally expressed as $\kappa \Delta C(X)$, where κ is the condition number of the multiset.

Chan and Lewis [5, 6] have a framework from developing theoretical upper bounds for some of these complex statistical formulas. From a set of axioms, upper bounds were developed for means and variances. Some of these bounds are expressed in terms of κ, the condition number (defined above). $\kappa \geq 1$ always and for large n, κ is approximately M/S, the reciprocal of the coefficient of variation. In general, $\kappa=1$ implies the numbers are well conditioned while numbers with $\kappa \gg 1$ are badly conditioned. It is shown that errors for means and variances are proportional to n or $n^{1/2}$.

This approach is used to find the errors for all statistical parameters and functions (excluding those that involve only counts (or sizes) since the computation of the parameters involve only integer arithmetic). It will be assumed that the parameters are accumulated in twice the precision of the final function values to mitigate the errors. Absolute errors are computed for the parameters and relative errors for the function values. In what follows, ϵ and ρ are the smallest representable floating point numbers in single and double precision respectively, on a given computer.

4.2. Errors in Parameters

For sum, T_n,

$$|\Delta T_n| \leq n^{3/2} ||A|| \rho + O(\rho^2).$$

General sum of k-th power parameters of the form

$$|\Delta T_n(k)| = \sum_{i=1}^{n} a_i^k,$$

have errors given by

$$|\Delta T_n(k)| \leq (n + n^{1/2}) ||A||^k \rho + O(\rho^2).$$

This formula is easily generalizable to higher order product sums.

For the mean, many possible updating formulas exist. We give the result for appplying Reekan's formula, which is amongst the least:

$$|\Delta T_n| \leq ((2/3)n^{1/2} + 4) ||A|| \rho + O(\rho^2).$$

Various formulas exist for the computation of the sum of squares. The bound for Reekan's is given. The error is

$$|\Delta T_n| \leq \rho T_n + (3.414 + n + n^{1/2}) ||A||^2 \rho + O(\rho^2)$$

and for the sum of products,

$$|\Delta T_n| \leq (n+3) T_n \rho + O(\rho^2).$$

As pointed out by Chan and Lewis, the choice of the mean computation method does not affect the the error in the sum of products.

4.3. Statistical Functions

With the parameters in double precision ($2t$ digits), these functions will be calculated in double precision and then rounded to t digits. If F_n is a final function value, we are interested in the relative error - $|\Delta F_n|/F_n$. Recall that ϵ is the single precision epsilon.

For sum, the relative error is,

$$\frac{|\Delta F_n|}{F_n} \leq n\rho + \epsilon + O(\rho^2).$$

For the mean (Reekan's), the relative error is,

$$\frac{|\Delta F_n|}{F_n} \leq (.667n + 2)\rho + \epsilon + O(\rho^2).$$

Many statistical functions having the sum of squares as parameter share similar bounds - these include variance and standard deviation. For the variance,

$$\frac{|\Delta F_n|}{F_n} \leq (n+4)\rho + \epsilon + (.94n + 14.5n^{1/2})\kappa\rho + O(\rho^2),$$

using Welford's. This error is proportional to κ. As stated by Chan and Lewis, the error is independent of κ if the standard two-pass method of computing the sum of squares is used (but this method is not updatable) until κ grows larger than $1/\rho^{1/2}$ when the term in $O(\kappa^2\rho^2)$ becomes significant. The error is proportional to κ^2 if some other methods are used [5, 24].

For parametric hypothesis testings, the relative errors for both the Z and T statistics are,

$$\frac{|\Delta Z_n|}{Z_n} \le 3\rho+\epsilon+(2n^3+4)((n-1)^{.5}S$$

$$/(M(A)_n-\mu_0))\kappa\rho+O(\rho^2)$$

and

$$\frac{|\Delta T_n|}{T_n} \le 3\rho+\epsilon+((n+4)\rho+(.94n+4.418n^{1/2})\kappa\rho)^{.5}$$

$$+(.667n^{.5}+4)((n-1)^{.5})S/(M(A)_n-\mu_0))\kappa\rho+O(\rho^2),$$

where M and S are the mean and standard deviation respectively. The errors for all other forms of hypothesis testing - two means, paired sample test and variances - are proportional to κ and are $O(n^{1/2})$ in the worst case [24].

Using Welford's method for the sum of squares in computing Pearson's correlation coefficient, the error is

$$\frac{|\Delta F_n|}{F_n} \le (n+4)\rho+\epsilon+(.667n^{1.25}+4n)((n-1)^{.5}((S(A)_n\kappa(A)$$

$$/(|a_n-M(A)_n|)+(S(B)_n\kappa(B)/(|b_n-M(B)_n|)))\rho$$

$$+O(\rho^2),$$

For some constants a_n and b_n. Again, the error is proportional to κ and n.

For experimental classification, the errors for one-way classification are,

$$\frac{|\Delta SSE|}{SSE} \le k\rho+max|\Delta T_i|+O(\rho^2),$$

$$\frac{|\Delta SS(Tr)|}{SS(Tr)} \le (k+4)\rho+O(\rho^2)$$

and

$$\frac{|\Delta SST|}{SST} \le (n+k+5)\rho+max|\Delta T_i|+O(\rho^2),$$

where T_i is the sum of squares for the i-th treatment group, and k is the number of groups. For two-way classification, only SST and SSE are proportional to κ. For higher order classifications, error bounds are similar. For Latin square and factorial designs the errors are proportional to n^2.

Errors in multivariate analyses are more complicated to analyze. But the errors are related since the computations start with a matrix of Pearson's correlation matrix, P. For linear regresssion, the problem is thus, $P*X=B$. P is symmetric (and positive definite?). It has been shown [34] that,

$$\frac{||\Delta X||}{||X||} \le cond(P)\frac{||\Delta P||}{||P||},$$

if P is perturbed. $||X||$ is the norm (i.e., a measure of magitude) of the matrix X. Since ΔP and P are symmetric, the relative error is $\lambda_u\lambda_m/\lambda_t$, where the λs are respectively, the minimum eigenvalue for P, the maximum and minimum eigenvalues for ΔP. However, it is desirable to express the error bound in terms of n and κ. The errors for factor and discriminant analyses are similarly bound.

4.4. Error Summary

The relative errors are proportional to $n^{1/2}$ in most cases and n^2 in a few. This error is also dependent on κ, the condition number. Some runs similar to the ones performed by Chan and Lewis were performed on a DEC 1099 for validation. The results were generally good since most of the errors occur while performing double precision arithmetic - for instance, for $n=10000$, in the worst case, at least five digits of accuracy was maintained in variance calculations on the average (DECC 1099, can hold 8.429 digits).

With the theoretical upper bounds, it is possible to monitor the type of accuracy (the number of correct leading digits) for each parameter. Intolerable accuracies due to a large n can be rectified by recalculating the particular parameters from the underlying database but this time, breaking up the parameters into say, p, smaller parts. These p parts can then be added (using the additive formulas) before final function values are calculated. Too many insertions and deletions may also warrant occasional recalculation of parameters.

Since errors resulting from down-dating (removal of a datum) are larger, it may be advisable to calculate the parameters for removed data separately. Before function calculation, these parameters are then subtracted to get the actual parameter values - recall, the P^{-}s.

5. The Integrated Model

5.1. Overview

A schematic representation of the model is as shown in Fig. 1.

5.2. User Level

Users are allowed to specify functions of interest. These functions are drawn from a catalog provided by the DBMS. The parameters and their updatable and additive characteristics are known to the system. These statistical functions operate on data attributes, SALARY for example, that are classified by some other category attribute(s) like RACE, SEX, AGE, etc. Thus, users only need to be concerned with function and classification specifications. A distinction is made between a category attribute and a data (or summary) attribute. Turner et. al. have described category attributes as those with small domain sizes and so with great ability to identify. By Steven's typology [33], these correspond to those scales that are at most ordinate (qualitative). We use the term here to include all

COMMERCIAL	STATISTICAL	
user views	user statistical functions and classifications	User
base tables	statistical parameters and base classifications	Conceptual
physical data	stored statistical parameters and classifications	Physical

Fig. 1. The Model

those attributes that are inherently qualitative or have been coded to become so - by grouping, for example. Data attributes on the other hand, have no restrictions although in most cases, they are at least ordinal (quantitative).

The DDL, the data definition language, should include constructs that allow users to make these SPECifications. Rightly, the indirectly SPECified data objects are the categories and their associated parameters. Similarly, the DML, the data manipulation language, provides users the ability to REQuest the values of previously specified functions, as well as the ability to CANcel any previous specifications. For example:

SPEC mean, stddev: SALARY by SEX
SPEC cntgncy: SALARY by SEX and RACE
REQ mean: SALARY by SEX
CAN cntgncy: SALARY by SEX and RACE

The first SPEC states that a user wants the mean and standard deviation of the SALARY for each of the categories, male and female. The second SPEC is for a 2-dimensional contingency table for SALARY. This approach of 'SPECify-before-use' is justified in the sense that statisticians in an enterprise generally know what analyses are of interest, particularly, after an 'explorative' phase has been undertaken [2]. Market research enterprises usually have well defined stable analytic sets that only change periodically. Also, in the case of experimental situations, the factors (category attributes) of interest, the dependent variables (data attributes) as well as the type of analyses are well defined. REQ is an actual REQuest for the calculation of the mean salaries for males and females. CAN is a CANcellation of a previous contigency table SPECification. Both REQ and CAN are part of the DML. It should be noted that these are not exhaustive.

5.3. Conceptual Level

With a knowledge of all specified functions, the DBMS can easily determine the necessary parameters to realize them. Since these parameters are updatable, they can be kept current during data acquisition. It has been suggested [31] that statistical databases are stable in the sense that after the initial data entry and correction, there are few or no updates to the database. While this is true for pure 'statistical data' like census data, it is not necessarily true for general commercial systems. There is no doubt that too many parameters and frequent updates are bound to slow down data acquisition. However, there are applications in which updates are infrequent. Additionally, there are those in which updates come in batches - market surveys, for instance [20]. In this case, straightforward methods of calculating these parameters can be applied to a new batch and these parameters added to those of the original database when the new batch is merged.

Base classifications - a set of categories - are kept at this level to exploit the additivity of the parameters. From the base classifications, the DBMS can determine if newly specified categories are derivable, thus eliminating redundancies arising from user classifications. Redundancies arising from different users can also be eliminated if a base classification is not redundant. Unfortunately, deriving such a base classification is a difficult problem. Also, to determine if a new category is derivable from a given set of categories has been proved to be NP-complete [24]. Thus, it seems that instead of a base classification that is sound and complete with respect to user classifications, a classification that is complete but not necessarily sound (if easy to compute) will be desirable. It should be noted that if user categories are disjoint, then the derivation of the base classification and derivability become polynomially computable.

A classification system CS, roughly, is conceptually a collection of categories and their parameters. A formal model is discussed elsewhere [24, 25]. In an enterprise, many CSs exist at this level, and are chosen to meet the demands of the enterprise. The category and data attribute sets in a CS will be chosen according to the 'target functions' - target functions, being those functions that share common parameters. Hence, for example, one CS may suffice for all functions related to counts such as frequency, contingency tables, etc. Another CS defined on a data attribute set may exist for multivariate analytic functions. The actual number of CSs at this level for an enterprise will depend on the number of target functions and the number of data attribute sets.

Processing at this level will include: the retrieval of parameters; derivability tests [24] for categories; the addition and deletion of categories. Retrieval of parameters will be in response to user REQuests. A user SPECification always implies a derivability test for each of the category in the SPECification. Categories that are not derivable are added to the classification and their initial parameters calculated from the database - may be, with the permission of the DBA. It should be noted that the category attribute set of a user may be a subset of the category attribute set at the conceptual level. In this case, the unspecified attributes are aggregated over [7, 32]. For example, if a user specifies the category 'hispanic', this will become 'hispanic&{female, male}' at the conceptual level if the category attribute set is {RACE, SEX} at this level. Categories are removed in response to user CANcellations when necessary.

5.4. Physical Level

The physical level is concerned with the actual storage of the base classifications and the parameters associated with the categories. Efficient storage and retrieval structures will be needed.

5.5. Interfaces and Mappings

The arrows in Fig. 1 show where the interfaces exist. Besides the two regular interfaces between the three levels, there are two horizontal interfaces. The DBMS has to provide the mappings between the user/conceptual and conceptual/physical interfaces of the statistical subsystem. The former will include mappings from functions to parameters and from user classifications to base classifications (i.e., the way they are derived). The latter interface includes mappings to actual physical storage. The horizontal mapping at the user level exists because classifications are defined on user category attributes and the functions, on user data attributes that are part of a user view. The same explanation holds for the horizontal mapping at the conceptual level except that additionally, the conceptual statistical component may have to make requests directly on the base tables - for instance, when the parameters of an underivable category are to be initially accumulated. No horizontal interface exists at the physical level, implying the independence of the two subsystems at this level. A consequence of this, is that it is possible to answer some statistical queries without the presence of the physical database. This is not to say that the two physical subsystems cannot reside in the same physical device - the independence is logical.

6. Conclusion

A CS captures the essence of statistical queries. By representing different user classifications by a base classification, data sharing is possible. In addition, redundant calculations arising from different function calculations and different users have been reduced. Response time for statistical queries will also be reduced since the intermediate result (parameter) acquisition phase, that involves database scans, has been eliminated. A CS may also be used to model time - time, being a category attribute.

The system can be partially implemented. For instance, a system can be implemented without a sophisticated derivability capability. This may be the case when it is known that all categories are always disjoint - this was the case in an implemented system. The updating capability at a per record basis may be substituted with a higher level merge procedure that uses the direct methods for calculating the parameters, and the additive formulas for the merge. It has also been demonstrated that good algorithms exist - in particular, when category attribute domains are small or ordered.

A CS does not come without a cost. Additional rounding errors are incurred when updating formulas are used in calculating parameter values. However, better accuracy can be achieved if the parameters are kept in twice the precision of the final function values. As pointed out, downdating results in even increased error [35]. Fortunately, many attribute domains are of the same sign and if high accuracy is desired, separate parameters (but same type) can be accumulated for deleted records of the database. Then at function-compute time, these parameters are 'subtracted' from the cumulative ones before function calculation. Additional cost is also incurred in the extra processing required for database activities related to data acquisition - insertion, deletion and modification - if parameters are to be updated. This suggests an environment with infrequent or timed updates as previously mentioned.

Then there are those statistical functions that are expensive to parameterize in the sense that the parameters are not easily updatable. Some clever methods have to be used to update these functions if deletions are a frequent occurence since a deletion will most likely invalidate the parameter, resulting in a scan of the database. Some statisical functions, such as residuals in multivariate analyses, cannot be performed without a database scan.

7. Future Direction

Better algorithms are needed for some of the problems related to derivability. The question of what should constitute a base classification has to be addressed. Minimally, this classification has to be complete. Given such a classification, polynomial time algorithms exist for addition and deletion of categories. The general derivability problem was shown to be NP-complete. More special cases need to be identified - cases that relate to structures in statistical classification schemes.

Security in statistical databases has been a major concern to database designers [8, 15, 36]. Since this model has a collection of 'statistical abstracts', it may be worth it to investigate how this abstract can be used in the various schemes for combating inferential statistical compromise. Or, it may be possible to develop a completely new scheme given these abstracts.

The CS has been discussed without any direct mention of any database model. In the network model, a CS can translate into a DBTG set [9, 26]. The *owner* record type is the record type that contains the category attribute set and the *member* record type is the one that contains the data attribute set. If the same record type contains both the category and data attribute sets, then a *dummy* record type has to be created since the owner and member types are supposed to be distinct. In a relational model, components of a CS correspond directly to one relation and its instance.

In any database model, the classification system can be expanded to include many table schemes - record types or relations. Thus, a category attribute set may span more than one table (so may the data attribute set). It is then necessary to give meaning to this expanded CS in the context of the particular database model - in particular, with regard to the data manipulation operators of the database model. In the case of the relational model, to determine the categories whose parameters are to be updated, given that an insertion has occured in a relation containing part of the category attribute set, is not an easy problem. This problem is directly related to the general problem of updates in a relational system [13, 14]. Updates aside, the initial gathering of parameter values needs proper descriptions of procedures that are not immediately obvious.

Efficient implementation data structures are needed. In a system that has already been implemented, an array linearization scheme was used and since the categories were disjoint derivation was easy to implement. However, a more sophisticated data structure will be required for the more general case of non-disjoint categories.

8. References

[1] Astrahan, et. al., "System R - Relational Approach to Database Management", ACM TODS 1(3), 1976.

[2] Bates, D., et. al., "A framework for research in database management for statistical analysis", ACM SIGMOD, 1982.

[3] Bernstein, P. A., Goodman, N., "What does Boyce-Codd Normal Form Do?", VLDB, 1980.

[4] Bleir, R. E., Vorhaus, A. H., "File organization in the SDC Time-shared Data Management System (TDMS)", TDMS project, Systems Development Corporation, SP-2750, 1967.

[5] Chan, T. F., Lewis, J. G., "Rounding error analysis of algorithms for computing means and standard deviations", Johns Hopkins Univ., Technical Report, #28, 1978

[6] Chan, T. F., Lewis, J. G., "Computing standard deviations: Accuracy", CACM 22(9), 1979.

[7] Chan, P., Shoshani, A., "SUBJECT: A directory driven system for organizing and accessing large statistical databases", VLDB, 1981.

[8] Chin, F. Y., "Security in statistical databases for queries with small counts", ACM TODS 3(1), 1978.

[9] Codasyl Data Base Task Group, April 1971 report. ACM N. Y.

[10] Codd, E. F., "A Relational Model of Data for Large Shared Data Banks", CACM 13(6), 1970.

[11] Codd, E. F., "Further normalization of the database relational model", in Database Systems, Courant Computer

Science Symposia, 6. Prentice Hall: Englewood Cliffs, New Jersey, 1971.

[12] Cohen, Hay, "Why are Commercial Databases Management Systems Rarely Used for Research Data?", Proc. of the 1st LBL workshop on Statistical Database Management, Dec. 1981.

[13] Cosmadakis, S. S., Papadimitriou, C. H., "Updates of Relational Views", Proc. of 2nd ACM Symp. on the Principles of Database systems, 1983.

[14] Dayal, U., Bernstein, P. A., "On the updatability of relational views", VLDB, 1978.

[15] Denning, D. E. et. al., "The tracker: A threat to statistical database security", CACM 4(1), 1979.

[16] Hanson, R. J., " Stably updating mean and standard deviation of data", CACM 18(1), 1977.

[17] Johnson, R. R., "Modelling Summary Data", ACM SIGMOD, 1981.

[18] Joyce, J. D., Oliver, N. N., " REGIS - A relational information system with graphics and statistics", Proc. AFIPS NCC, 1976.

[19] Koenig, S., Paige, R., "A transformational framework for the automatic control of derived data", VLDB, 1981.

[20] Myers, J. H., Tauber, E., *Market Structure Analysis.* American Marketing Association: Chicago, Illinois, 1977.

[21] Neely, M. P., "Comparison of several algorithms for computation of means, standard deviation and correlation coefficients", CACM 9(7), 1966.

[22] Nelson, L. S., "Further remark on stably updating mean and standard deviation estimates", CACM 22(8), 1979.

[23] Nie, N. J., Hull, C. H. et. al. Statistical Packages for the Social Sciences. McGraw-Hill: New York, N. Y., 1975.

[24] Nwokogba, I., "A Statistical Classification Model for a Database System", Ph.D Thesis, Vanderbilt University, 1983.

[25] Nwokogba, I., Rowan, W. H., "A model for an integrated statistical and commercial database", Proc. IEEE COMPSAC, Chicago 1984.

[26] Olle, T. W., *The Codasyl Approach to Database Management.* John Wiley and Sons: New York, N. Y., 1978.

[27] Van Reekan, A. J., "Dealing with Neely's algorithm", CACM 11(3), 1968.

[28] Rowe, L. A., Shoens, K. A., "Data abstraction, Views and updates in RIGEL", ACM SIGMOD, 1979.

[29] SAS Institute Inc. Statistical Analysis User's Guide. SAS: Cary, N. Carolina, 1978.

[30] Sato, H., "Handling summary information in a database: Derivability", ACM SIGMOD, 1981.

[31] Shoshani, A., "Statistical Databases: Characteristics, Problems, and some Solutions", VLDB, 1982.

[32] Smith, J. M., Smith, D. C. P., "Database Abstraction: Aggregation and Generalization", ACM TODS 2(2), 1977.

[33] Stevens, S. S., "On the theory of scales of measurement", Science 103, 1946.

[34] Strang, G. Linear algebra and its applicatins. Academic Press: New York, N. Y., 1976.

[35] Turner, M. H., et. al., "A DBMS for Large Statistical Databases", VLDB, 1979.

[36] Ullman, J. D., *Principles of Database Systems.* Computer Science Press: Potomac, Maryland, 1980.

[37] Welford, B. P., "Notes on a method for calculating corrected sums of squares and products", Technometrics 4(3), 1962.

[38] West, D. H. D., "Updating mean and variance estimates", CACM 22(9), 1979.

[39] Wilkinson, J. H., *Rounding errors in algebraic processes.* National Physical Laboratory: London, England, 1963.

[40] Young, E. A., Cramer, E. M., "Some results relevant to choice of sum and sum of products algorithms", Technometrics 13(3), 1971.

COMPUTER SCIENCE AND STATISTICS:
The Interface, D.M. Allen (ed.)

COMPUTER IMPLEMENTATION OF MATRIX DERIVATIVES IN OPTIMIZATION PROBLEMS IN STATISTICS

K. G. Jinadasa and D. S. Tracy

University of Windsor
Windsor, Ontario, Canada

The usual multivariate estimation problems reduce to optimization problems involving parameter matrices, which may be patterned due to constraints or for reasons of identifiability. Matrix derivatives, where allowance is made for patterns in the argument matrix, are suggested, and permuted identity matrices are extended to cover partitioned matrices. Applications are made to parameter estimation of a factor analysis model. Subroutines are studied for computer implementation.

1. INTRODUCTION

Often the estimation problems in multivariate statistics are optimization problems involving several parameters in the form of matrices, which may be partitioned into blocks. Sometimes the matrices involved are patterned matrices, as in covariance structure models, where conditions are imposed on the parameter matrices for the model to be identifiable. We propose a method to obtain matrix derivatives of a matrix function of matrix arguments, where the argument matrix may be patterned. We also extend the notion of permuted identity matrices to matrices with row blocks or column blocks, in order to take care of partitioned matrices.

We apply some of these notions to the confirmatory inter-battery factor analysis model, where we obtain the generalized least squares estimators of the parameter matrices. Some subroutines are suggested for computer implementation of the method. The method is applicable to other problems where the objective function is a function of matrices, which are possibly patterned due to constraints among the elements.

2. MATRIX DERIVATIVES

Let $Y = f(X)$ be a matrix function of a matrix argument X, where X is $m \times n$ and Y is $p \times q$. Then $\text{vec } X$ denotes the column vector of order mn formed by stacking the columns of X, one above the other, starting with the first column. Matrix Y is similarly "vectorized" to $\text{vec } Y$, and displayed as a row vector $\text{vec}'Y$. If a typical element of $\text{vec}'Y$ is $y_{\alpha\beta}$, and that of $\text{vec } X$ is x_{ij}, then the collection $\left(\frac{\partial y_{\alpha\beta}}{\partial x_{ij}}\right)$ can be represented by a $pq \times mn$ matrix $\frac{\partial \text{vec } Y}{\partial \text{vec } X}$. This becomes a representation of the derivative of the function with respect to the usual bases in the X space and the Y space, and is known as the matrix derivative of Y with respect to matrix X.

When the elements of X have equality or other relationships between them, or some elements are constants, the matrix is said to be patterned. Examples are symmetric or skew-symmetric matrices. This requires a modification of the definition of a matrix derivative. Here we take the k independent and variable elements of X and define a one to one function J on $\mathbb{R}^k$ onto the set of all matrices D with this particular pattern. We take the extension $\tilde{f}(X)$ of the function $f(X)$ to the whole space of all $m \times n$ matrices by ignoring the pattern of X. For any X in D, we have $\tilde{f}(X) = f(X)$, and a corresponding vector x in $\mathbb{R}^k$ such that $J(x) = X$. Now consider the composite function $G(x) = \tilde{f} \circ J(x)$. Since $J(x) = X \in D$, we have $G(x) = f(X)$. Thus we can define the derivative of $G(x)$ by using the chain rule

$$G'(x) = \tilde{f}'(J(x))(J'(x))$$

where $\tilde{f}'(J(x))$ is the derivative of $\tilde{f}$ at the point $J(x)$. By taking the matrix representation of $G'(x)$, we get

$$[G'(x)] = [\tilde{f}'(J(x))][J'(x)].$$

Here $[f'(J(x))]$ is nothing but the matrix derivative obtained by ignoring the pattern of X. Thus the procedure amounts simply to post-multiplication of the matrix derivative, obtained upon ignoring the pattern, by another matrix, which is related to the pattern of X.

For example, let $X = \begin{pmatrix} 1 & x_{21} \\ x_{21} & x_{22} \end{pmatrix}$, symmetric, and $Y = f(X) = |X| = x_{22} - x_{21}^2$. Ignoring the pattern of X, a well known result is $\frac{\partial \text{vec } Y}{\partial \text{vec } X} = |X| \text{ vec}'(X^{-1})$. Letting $x = \begin{pmatrix} x_{21} \\ x_{22} \end{pmatrix}$, define $J(x) = X$, i.c., $J\begin{pmatrix} x_{21} \\ x_{22} \end{pmatrix} = \begin{pmatrix} 1 & x_{21} \\ x_{21} & x_{22} \end{pmatrix}$.

Then $\left[\frac{\partial J(x)}{\partial x}\right] = \begin{pmatrix} 0 & 0 \\ 1 & 0 \\ 1 & 0 \\ 0 & 1 \end{pmatrix}$ and $\frac{\partial \text{vec } Y}{\partial \text{vec } X} = \text{vec}'\begin{pmatrix} x_{22} & -x_{21} \\ -x_{21} & 1 \end{pmatrix} = (x_{22} \;\; -x_{21} \;\; -x_{21} \;\; 1)$. The required derivative is then

$$(x_{22} \;\; -x_{21} \;\; -x_{21} \;\; 1)\begin{pmatrix} 0 & 0 \\ 1 & 0 \\ 1 & 0 \\ 0 & 1 \end{pmatrix} = (-2x_{21} \;\; 1).$$

3. PERMUTED IDENTITY MATRICES

These were introduced in the literature to permute the rows of an identity matrix, and were used to relate vec A and vec A'. We extend these with the purpose of relating the vector of a partitioned matrix and the vectors of the blocks.

Consider the partitions of positive integers m,n as $m = \sum_{i=1}^{r} m_i$ and $n = \sum_{j=1}^{s} n_j$. Then the identity matrix of order mn may be written as

$$I_{mn} = \text{block diag}(\underbrace{I_{m_1},\ldots,I_{m_r}}_{1},\underbrace{I_{m_1},\ldots,I_{m_r}}_{2},\ldots,\underbrace{I_{m_1},\ldots,I_{m_r}}_{n}).$$

We denote by T_r^{mn} the matrix obtained by rearranging the row blocks of the above matrix by taking every r^{th} block starting with the first block, then every r^{th} block starting with the second block, and so on. Interchanging the roles of m and n and taking every s^{th} block, we obtain the matrix T_s^{nm}.

Next we partition I_{mn} as

$$I_{mn} = \text{block diag}(\underbrace{I_{m_1},\ldots,I_{m_1}}_{n},\underbrace{I_{m_2},\ldots,I_{m_2}}_{n},\ldots,\underbrace{I_{m_r},\ldots,I_{m_r}}_{n}).$$

Rearranging the row blocks by taking every n^{th} block starting with the first, we get T_n^{mn}, and interchanging the roles of m and n, we obtain T_m^{nm}. Then we have the relationships

$$T_r^{mn}\,T_n^{mn} = I_{mn}, \quad T_s^{nm}\,T_m^{nm} = I_{mn}.$$

Now consider the partition

$$I_{mn} = \text{block diag}(I_{m_1n_1},\ldots,I_{m_1n_s},I_{m_2n_1},\ldots,I_{m_2n_s},\ldots,I_{m_rn_s}).$$

Rearranging the row blocks by taking every s^{th} block starting with the first, we obtain T_{mn}^{s}. Interchanging the roles of m and n, we obtain T_{nm}^{r}. Then $T_{nm}^{r}T_{mn}^{s} = I_{mn}$.

Let Σ be an $m\times n$ matrix, partitioned into rs blocks, $\Sigma = (\Sigma_{ij})$, where Σ_{ij} is $m_i\times n_j$, $m = \sum_{i=1}^{r} m_i$, $n = \sum_{j=1}^{s} n_j$. We let Rvec denote the formation of vectors of blocks in the row order and Cvec in the column order. Thus

$$\text{Rvec } \Sigma = (\text{vec}'\Sigma_{11},\text{vec}'\Sigma_{12},\ldots,\text{vec}'\Sigma_{1s},\text{vec}'\Sigma_{21},\ldots,\text{vec}'\Sigma_{rs})'$$

$$\text{Cvec } \Sigma = (\text{vec}'\Sigma_{11},\text{vec}'\Sigma_{21},\ldots,\text{vec}'\Sigma_{r1},\text{vec}'\Sigma_{12},\ldots,\text{vec}'\Sigma_{rs})'.$$

We then find relationships like

$$T_r^{mn}\text{vec } \Sigma = \text{Rvec } \Sigma$$

$$\text{vec } \Sigma = T_n^{mn}\text{Rvec } \Sigma$$

$$T_{mn}^{s}T_r^{mn}\text{vec } \Sigma = \text{Cvec } \Sigma$$

$$\text{or} \quad T_{mn,r}^{s}\text{vec } \Sigma = \text{Cvec } \Sigma .$$

Some results on Ⓣ products [5], A Ⓣ $B = [(A_{ij} \otimes B_{k\ell})]$, where $\otimes$ denotes Kronecker product of blocks A_{ij} and $B_{k\ell}$, can be related. If A is $m\times n$ and B is $p\times q$, $p = \sum_{k=1}^{t} p_k$, $q = \sum_{\ell=1}^{u} q_\ell$, we have

$$T_{mp,r}^{t}(B\otimes A) = (B \circledT A)T_{nq,s}^{u}$$

$$T_{np,r}^{t}(B\otimes A)T_q^{nq,s} = B \circledT A$$

If X,Y are random vectors with m,n components respectively and $E(X) = \mu$, $E(Y) = \nu$, $\text{Cov}(Y,X') = \Sigma$, then $E(X \circledT Y) = T_{nm,s}^{r}\text{vec } \Sigma + \mu \circledT \nu$ and for X,Y independent, $\text{Cov}(X,X') = \Sigma^{(1)}$, $\text{Cov}(Y,Y') = \Sigma^{(2)}$,

$$E(XY' \circledT XY') = [T_{mm,r}^{r}\text{vec } \Sigma^{(1)} + \mu \circledT \mu][T_{nn,s}^{s}\text{vec } \Sigma^{(2)} + \nu \circledT \nu]$$

$$\text{Cov}(X \circledT Y,(X \circledT Y)') = (\Sigma^{(1)} + \mu\mu') \circledT (\Sigma^{(2)} + \nu\nu') - \mu\mu' \circledT \nu\nu'.$$

4. A FACTOR ANALYSIS MODEL

We consider the application of the above ideas in the problem of estimation of parameters for the confirmatory interbattery factor analysis model [3]. Here one has two sets of scores x_1 and x_2 of two batteries of tests which have a common factor z. Let

y_1 and y_2 denote the factors specific to batteries 1 and 2, and e_1, e_2 are the error terms. The model is formulated as

$$x_1 = \mu_1 + \Lambda_1 z + \Gamma_1 y_1 + e_1$$

$$x_2 = \mu_2 + \Lambda_2 z + \Gamma_2 y_2 + e_2$$

where μ_1, μ_2 are the two battery means and Λ_1, Λ_2, Γ_1, Γ_2 are the corresponding factor loadings. Further, with $x = \begin{pmatrix} x_1 \\ x_2 \end{pmatrix}$, $y = \begin{pmatrix} y_1 \\ y_2 \end{pmatrix}$, $e = \begin{pmatrix} e_1 \\ e_2 \end{pmatrix}$, it is assumed that $z \sim N(0,\Phi)$, $y_\ell \sim N(0,\Theta_\ell)$, $e_\ell \sim N(0,\gamma_\ell)$, $\ell = 1,2$, and $\mathrm{Cov}(y,z') = 0$, $\mathrm{Cov}(z,e') = 0$, $\mathrm{Cov}(y,e') = 0$, $\mathrm{Cov}(y_1,y_2') = 0$, $\mathrm{Cov}(e_1,e_2') = 0$. Also, γ_1 and γ_2 are assumed to be diagonal matrices and Φ, Θ_1 and Θ_2 are symmetric matrices. The variance covariance matrix $\mathrm{Cov}(x,x')$ of the model becomes

$$\Sigma = \begin{pmatrix} \Lambda_1\phi\Lambda_1' + \Gamma_1\Theta_1\Gamma_1 + \gamma_1 & \Lambda_1\phi\Lambda_2' \\ \Lambda_2\phi\Lambda_1' & \Lambda_2\phi\Lambda_2' + \Gamma_2\Theta_2\Gamma_2' + \gamma_2 \end{pmatrix}$$

$$= \begin{pmatrix} \Sigma_{11} & \Sigma_{12} \\ \Sigma_{21} & \Sigma_{22} \end{pmatrix}$$

In order that the model be identifiable, one has to specify some of the elements in the matrices involved. If Φ is $k \times k$, Θ_1 is $\ell_1 \times \ell_1$, Θ_2 is $\ell_2 \times \ell_2$, then we have to impose k^2 conditions on Λ_1, Λ_2 and Φ, ℓ_1^2 conditions on Γ_1 and Θ_1, and ℓ_2^2 conditions on Γ_2 and Θ_2. In confirmatory factor analysis, one may wish to impose conditions on factor loadings, instead of on the covariance matrices of the factors. Thus, besides the symmetry of Φ, Θ_1 and Θ_2, no other patterns are imposed. However, $\Lambda_1, \Lambda_2, \Gamma_1$ and Γ_2 become patterned matrices. One can use suitable identity matrices as submatrices of the above matrices, so that the required number of specifications is met. For example, we can have I_k as a submatrix of Λ_1 and Λ_2, I_{ℓ_1} as a submatrix of Γ_1 and I_{ℓ_2} as a submatrix of Γ_2.

For the estimation of the unspecified parameters in the model, either the maximum likelihood or the generalized least squares method is generally used. The objective function, which is a function of Σ, to be minimized in the case of the generalized least squares method, is

$$f(\Sigma) = \frac{1}{2}\,\mathrm{tr}[(S-\Sigma)S^{-1}]^2$$

where S is the sample variance covariance matrix. We can write this as

$$f(\Sigma) = \frac{1}{2}\,\mathrm{vec}'(S-\Sigma)(S^{-1} \otimes S^{-1})\mathrm{vec}(S-\Sigma).$$

Since S^{-1} is positive definite, so also is $S^{-1} \otimes S^{-1}$. Denoting its symmetric square root by $U^{\frac{1}{2}}$, we have

$$f(\Sigma) = \frac{1}{2}\,\mathrm{vec}'(S-\Sigma)U^{\frac{1}{2}}U^{\frac{1}{2}}\,\mathrm{vec}(S-\Sigma)$$

$$= \frac{1}{2}\left(h(\Sigma)\right)'h(\Sigma)$$

where $h(\Sigma) = U^{\frac{1}{2}}\,\mathrm{vec}(S-\Sigma)$.

Letting θ denote the vector of all the unspecified distinct parameters involved in the model, the gradient of $f(\Sigma)$ is

$$\frac{\partial f(\Sigma)}{\partial \theta} = [JA(\Sigma)]'h(\Sigma)$$

where

$$JA(\Sigma) = \frac{\partial h(\Sigma)}{\partial \theta} = \frac{\partial}{\partial \theta}U^{\frac{1}{2}}\,\mathrm{vec}(S-\Sigma)$$

$$= U^{\frac{1}{2}}\frac{\partial}{\partial \theta}\mathrm{vec}(S-\Sigma) = -U^{\frac{1}{2}}\frac{\partial \mathrm{vec}\,\Sigma}{\partial \theta}$$

Thus,

$$\frac{\partial f(\Sigma)}{\partial \theta} = -\left(\frac{\partial \mathrm{vec}\,\Sigma}{\partial \theta}\right)'U^{\frac{1}{2}}U^{\frac{1}{2}}\,\mathrm{vec}(S-\Sigma)$$

$$= -\left(\frac{\partial \mathrm{vec}\,\Sigma}{\partial \theta}\right)'S^{-1} \otimes S^{-1}\,\mathrm{vec}(S-\Sigma).$$

Thus, in actual calculation, one does not have to obtain $U^{\frac{1}{2}}$ explicitly. While considering $\frac{\partial \mathrm{vec}\,\Sigma}{\partial \theta}$, we find that because of the pattern of Σ, finding the derivatives of its submatrices is much simpler. Let $\mathrm{ve}\,A$ denote the vector of unspecified parameters involved in a submatrix A of Σ. Clearly $\frac{\partial \mathrm{vec}\,\Sigma_{11}}{\partial \mathrm{ve}\,\Lambda_1}$ is easier to calculate than $\frac{\partial \mathrm{vec}\,\Sigma}{\partial \mathrm{ve}\,\Lambda_1}$, i.e., it is easier to find $\frac{\partial \mathrm{Rvec}\,\Sigma}{\partial \theta}$ or $\frac{\partial \mathrm{Cvec}\,\Sigma}{\partial \theta}$. Using permuted identity matrices, $\frac{\partial \mathrm{vec}\,\Sigma}{\partial \theta} = T\,\frac{\partial \mathrm{Rvec}\,\Sigma}{\partial \theta}$, where T is T_p^{pp}, $p = p_1 + p_2$, p_1, p_2 being the number of tests in the first and the second battery. The gradient of the objective function is

$$\frac{\partial f(\Sigma)}{\partial \theta} = -\left(\frac{\partial \mathrm{Rvec}\,\Sigma}{\partial \theta}\right)'T'\,S^{-1} \otimes S^{-1}\,\mathrm{vec}(S-\Sigma).$$

Clearly, closed form estimates are not

available for this problem, and one has to use a computer approach.

5. COMPUTER IMPLEMENTATION

In most of the optimization routines in practice, the user has to provide the function and the gradient of the function. In the above example, it is easy to program the matrices T and $\frac{\partial \text{Rvec}\,\Sigma}{\partial\theta}$. The latter is a partitioned matrix with 36 blocks, only 16 of which are non-zero blocks. Thus, corresponding to Σ_{11}, the non-zero blocks are

$$(I + I_{p_1,p_1})(\Lambda_1\Phi \otimes I_{p_1})J^{\Lambda_1}, \quad (\Lambda_1 \otimes \Lambda_1)J^{\Phi},$$

$$(I + I_{p_1,p_1})(\Gamma_1\Theta_1 \otimes I_{p_1})J^{\Gamma_1}, \quad (\Gamma_1 \otimes \Gamma_1)J^{\Theta_1}, \text{ and}$$

J^{γ_1}, and so on. The J matrices are the matrices corresponding to the patterns of the parameter matrices indicated.

The implementation of matrix derivatives in a computer program amounts to finding the products of matrices as above. This requires writing subroutines to find the products of matrices, whereas element by element differentiation would have become very cumbersome. Gauss-Newton methods are popular in optimization problems of factor analysis [1],[4].

Usual optimization routines consider general problems with constraints among the variables. In this problem, we have taken care of the constraints in the patterns of the respective matrices. Thus the routine employed should allow the possibility of zero constraints.

We consider the case of 4 tests in the first battery and 5 in the second, with 2 factors common to the two batteries and 2 specific factors for each battery. Thus Σ is 9×9, Λ_1 is 4×2, Λ_2 is 5×2, Γ_1 is 4×2, Γ_2 is 5×2, γ_1 is 4×4, γ_2 is 5×5, and Φ, Θ_1 and Θ_2 are each 2×2. The restrictions require the first two elements of the first column of Λ_1 to be 1 and 0, and the first two elements of the second column of Λ_2 to be 0 and 1 respectively. The upper left 2×2 submatrices of Γ_1 and Γ_2 are I_2. Although there are 89 variables in all, only 42 of them are to be dealt with due to the existence of patterns. There are 3 symmetric and 2 diagonal matrices involved. We have to find 9 J matrices and one T matrix. On the following page, we supply the subroutines to find these.

In order to implement the Gauss-Newton algorithm to this problem, we need to calculate

$$JA(\Sigma) = -U^{\frac{1}{2}}\frac{\partial \text{vec}\,\Sigma}{\partial\theta} = -U^{\frac{1}{2}}\,T_p^{pp}\,\frac{\partial \text{Rvec}\,\Sigma}{\partial\theta}$$

and then

$$\left(JA(\Sigma)\right)'JA(\Sigma) = \frac{\partial \text{Rvec}\,\Sigma}{\partial\theta}'\,T'(S^{-1}\otimes S^{-1})T\,\frac{\partial \text{Rvec}\,\Sigma}{\partial\theta}$$

In our problem $\frac{\partial \text{Rvec}\,\Sigma}{\partial\theta}$ is an 81×42 matrix with 36 blocks, out of which only 16 are non-zero blocks. At each step of iteration,

$$\Delta = -\left[\left(JA(\Sigma)\right)'JA(\Sigma)\right]^{-1}\frac{\partial f(\Sigma)}{\partial\theta}$$

is to be calculated and θ is updated by $\theta+\Delta$ until $\left\|\frac{\partial f(\Sigma)}{\partial\theta}\right\|$ becomes sufficiently small. This method was not possible because $\left[\left(JA(\Sigma)\right)'JA(\Sigma)\right]$ became singular at certain stages.

Thus Marquardt-Levenberg algorithm was used, where the updating procedure at the kth step is to calculate

$$\Delta = -\left[\left(JA(\Sigma)\right)'JA(\Sigma) + u_k I\right]^{-1}\frac{\partial f(\Sigma)}{\partial\theta}$$

where $\{u_k\}$ is a bounded sequence of positive integers. The convergence became very slow. At the last step we used the subroutine ACDPAC [2], providing the subroutines to calculate $f(\Sigma)$ and $\frac{\partial f(\Sigma)}{\partial\theta}$. We had to make slight modifications in the subroutine to handle the case of zero constraints. We find an almost perfect fit.

REFERENCES

[1] Bentler, P.M. and Lee, S.Y., Statistical aspects of a three-mode factor analysis model, Psychometrika 43 (1978) 342-352.

[2] Best, M.J. and Bowler, A., ACDPAC : A FORTRAN subroutine to solve differentiable mathematical programmes. Res. Report CORR 75-26, Univ. of Waterloo, 1976.

[3] Browne, M.W., The maximum-likelihood solution in inter-battery factor analysis, Brit. J. Math. Stat. Psych. 32 (1979) 75-86.

[4] Jöreskog, K.G., Simultaneous factor analysis in several populations, Psychometrika 36 (1971) 409-426.

[5] Singh, R.P., Some Generalizations in Matrix Differentiation With Applications in Multivariate Analysis, Ph.D. Dissertation, Univ. of Windsor (1972).

```
      SUBROUTINE JSYMET(K,M,N,JO)
C
C THIS SUBROUTINE FINDS THE
C J MATRIX CORRESPONDING TO
C ANY SYMMETRIC PATTERN MATRIX.
C
C  K=THE ORDER OF THE MATRIX
C  M=K**2
C  N=K*(K+1)/2
C
      INTEGER JO(M,N)
      DO 30 I=1,M
      DO 30 J=1,N
        JO(I,J)=0
30    CONTINUE
       L=0
      K1=K-1
      DO 100 I=1,K1
      L=L+1
      JO((I-1)*(K+1)+1,L)=1
       KI=K-I
      DO 100 J=1,KI
       L=L+1
      JO((I-1)*(K+1)+J+1,L)=1
      JO((I-1)*(K+1)+K+J+1,L)=1
10    CONTINUE
       JO(M,N)=1
       RETURN
       END
C
C
    SUBROUTINE JTOW(IP,L,LPML,IPL,JTW)
C THIS SUBROUTINE FINDS THE
C J MATRIX CORRESPONDING TO
C THE MATRICES OF THE FORM (I,B)'
C WHERE I IS THE UNIT MATRIX.
C
C  THE GIVEN MATRIX IS OF ORDER (IP,L)
C  THE UNIT MATRIX IS OF ORDER L.
C   LPML=L*(IP-L)
C   IPL=IP*L
C
     INTEGER JTW(IPL,LPML)
     DO 15 I=1,IPL
     DO 15 J=1,LPML
      JTW(I,J)=0
15   CONTINUE
      IPML=IP-L
     DO 25 K=1,L
     DO 25 J=1,IPML
    JTW((K-1)*IP+L+J,(K-1)*IPML+J)=1
25   CONTINUE
     RETURN
     END
C
C
    SUBROUTINE JDIAG(IP,IPP,JDIG)
C THIS SUBROUTINE FINDS THE
C J MATRIX CORRESPONDING TO
C    A DIAGONAL MATRIX.
C
C IP=ORDER OF THE GIVEN MATRIX
C IPP=IP**2
C
      INTEGER JDIG(IPP,IP)
      DO 10 I=1,IPP
      DO 10 J=1,IP
       JDIG(I,J)=0
10    CONTINUE
      DO 20 I=1,IP
    JDIG((I-1)*(IP+1)+1,J)=1
20    CONTINUE
      RETURN
      END
C
C
    SUBROUTINE TMATRX(M,N,MV,IR,MN,ID)
C
C THIS SUBROUTINE FINDS THE
C Tmn,n MATRIX.
C M AND N ARE GIVEN NUMBERS
C PARTITIONED AS
C M=M(1)+M(2)+........+M(IR) AND
C N=N(1)+N(2)+........+N(IS)
C MV IS THE VECTOR OF DIMENSION
C 'IR' CONTAINING THE
C ELEMENTS OF THE PARTITION OF M.
C MN=M*N
C
    INTEGER TMN(MN,MN),MV(IR)
    INTEGER ID(MN,MN)
     DO 100 I=1,MN
     DO 200 J=1,MN
      ID(I,J)=0
20   CONTINUE
      ID(I,I)=1
10   CONTINUE
     DO 300 L=1,MN
     DO 300 I=1,N
      MM=0
     DO 300 K=1,IR
     DO 400 J=1,MV(K)
     TMN(M*(I-1)+MM+J,L)=ID(MM*N+
   &  (I-1)*MV(K)+J,L)
40   CONTINUE
      MM=MM+MV(K)
30   CONTINUE
     RETURN
     END
```

COMPUTER SCIENCE AND STATISTICS:
The Interface, D.M. Allen (ed.)

AUTOMATIC COMPUTATION OF FIRST AND SECOND DERIVATIVES WITH APPLICATION TO COMPARTMENTAL MODELS

David E. Gray and David M. Allen

Department of Statistics
University of Kentucky
Lexington, Kentucky 40506

A method for obtaining computer generated analytic first and second derivatives is presented. These derivatives are used in the fitting of nonlinear models defined by systems of linear differential equations. Second derivative information offers the possibilities of improved convergence and calculation of curvature measures of nonlinearity. The method is illustrated with a C program.

Compartmental models are an important class of mathematical models. They are used in many fields, but a pharmaceutical example will illustrate their characteristics. Suppose D amount of drug is introduced into the blood stream. The drug travels to its site of action, returns to the blood stream and is eliminated. Pictorally this can be shown as follows

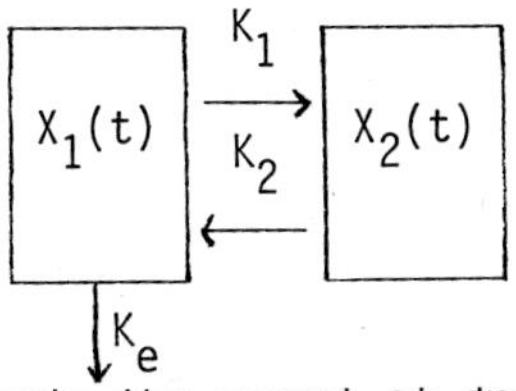

$X_1(t)$ represents the amount of drug in the blood (Compartment 1) at time t and $X_2(t)$ represents the amount of drug at the site of action (Compartment 2) at time t. The K's are called the rate constants. Letting $X_1 = X_1(t)$ and $X_2 = X_2(t)$, we can write

$$\frac{dX_1}{dt} = -(K_1+K_e)X_1 + K_2X_2$$

$$\frac{dX_2}{dt} = K_1X_2 - K_2X_2$$

or more succinctly,

$$\dot{X} = AX \qquad (1)$$

where

$$A = \begin{bmatrix} -(K_1+K_e) & K_2 \\ K_1 & -K_2 \end{bmatrix}$$

$$X = \begin{bmatrix} X_1(t) \\ X_2(t) \end{bmatrix}$$

$$\dot{X} = \frac{d}{dt}X$$

Loosely speaking, this says that in the interval $(t,t+\Delta t)$, Δt small, X_1 decreases (K_1+K_e) $X_1(t)\Delta t$ and increases $K_2X_2(t)\Delta t$. Similarly, in the same interval, X_2 increases $K_1X_1(t)\Delta t$ and decreases $K_2X_2(t)\Delta t$. The concentration of the drug in Compartment 1 is observed by taking blood samples at time t. Statistically we assume the observed concentration has the following structure

$$C = \frac{X_1(t)}{V} + e_t$$

where V is the volume of distribution and e_t is a random variable, usually assumed to be $N(0,\sigma)$. The more general model of Carroll and Ruppert (1984) can also be applied with the obvious changes. Since $X_1(t)$ is a nonlinear function of t and $\underline{K} = (V,K_1,K_2,K_e)$, a nonlinear least squares method is used to estimate $\underline{K}$. To do this, X_1 and its derivatives with respect to $\underline{K}$ are required. X_1 in turn requires solving (1). Both finding X_1 and differentiating X_1 with respect to $\underline{K}$ is tedious and subject to a high probability of error. Various approximate methods for solving X_1 and estimating $dX/d\underline{K}$ exist. We will present a method which solves (1), finds $dX/d\underline{K}$ and, in fact, finds partial derivatives of any order. This method was inspired by Jennrich Bright (1976). This paper presents a somewhat different approach that does not require distinct eigenvalues and gives second derivatives as well as first derivatives. We will also describe some applications of second derivatives and give some details of the method of implementation in the C programming language.

We must solve

$$\dot{X} = AX, \text{ A is cxc}$$

for X. We will assume the characteristic roots of A are all real but they need not be distinct. Herron (1963) shows that this condition will hold for all small practical problems. A numerically stable method for solving (1) is to first find the real Schur decomposition of A,

$$A = QTQ'$$

where T is an upper triangular matrix and Q is

an orthogonal matrix. The diagonal elements of T will be the eigen values of A. Then,

$$Q\dot{X} = QAX = QAQ'QX = Tu$$

where u = QX.

The cth element of $\dot{u}$ satisfies the equation

$$\dot{u}_c = \lambda_c u_c,\ \lambda_i = T_{ii}\ (i=1,\cdots,c).$$

this is easily solved for u and likewise the rest of (1) can be solved by substituting the solutions for u_i that have already been solved and integrating the resulting equation.

It may arise that two or more eigen values will be equal. In that case a power of t will enter in the solution. The general solution can be represented as

$$\sum_{d=0}^{maxdeg} B^d e(\lambda t)$$

where $e(\lambda t)^T = (e(\lambda_1 t), e(\lambda_2 t), \cdots, e(\lambda_c t))$.

For each power of t there is a B^d, d=0,···,maxdeg. The set of B^d can be considered as a three dimensional matrix and will be referred to as B, the solution matrix for $\underline{X}$.

The partial derivatives of $\underline{X}$ with respect to $\underline{K}$ are actually not hard to get. Taking the partial derivative of the system with respect to an element of $\underline{K}$, say K_i,

$$\frac{\partial}{\partial K_i}\frac{dX}{dt} = \frac{\partial}{\partial K_i}(AX) \quad (2)$$

$$= (\frac{\partial}{\partial K_i}A)X + A\frac{\partial X}{\partial K_i}$$

$$= A^{K_i}X + A\frac{\partial X}{\partial K_i}$$

$A^{K_i}_{j\ell}$ is ±1 if $A_{j\ell}$ has $\pm K_i$ in it. If the order of integration of the right hand side of (2) is changed

$$\frac{d}{dt}(\frac{\partial X}{\partial K_i}) = A^{K_i}X + A\frac{\partial X}{\partial K_i}. \quad (3)$$

This is a system of differential equations to be solved for $\partial X/\partial K_i$. Since we already know X, we can rewrite (3) as

$$\frac{d}{dt}\frac{\partial X}{\partial K_i} = A^{K_i}\sum B^d e(\lambda t) + A\frac{\partial X}{\partial K_i}.$$

Premultiplying the above by Q

$$Q\frac{d}{dt}\frac{\partial X}{\partial K_i} = A^{K_i}\sum B^d e(\lambda t) + QA\frac{\partial X}{\partial K_i}$$

$$= QA^{K_i}\sum B^d e(\lambda t) + QAQ'Q\frac{\partial X}{\partial K_i}$$

$$= QA^{K_i}\sum B^d e(\lambda t) + T\frac{\partial u}{\partial K_i}.$$

Allen (1981) showed that the solution of this system for $\partial X/\partial K$ has the form

$$\sum_{d=0}^{maxdeg} C^d e(\lambda t).$$

Second derivatives are found analogously. To find $\partial X/\partial K_i\partial K_j$ differentiate (3) with respect to K_j

$$\frac{d}{dt}(\frac{\partial X}{\partial K_i\partial K_j}) = A^{K_i}\frac{\partial X}{\partial K_j} + A^{K_j}\frac{\partial X}{\partial K_i} + A\frac{\partial X}{\partial K_i\partial K_j}$$

$$= A^{K_i}\sum C_j^d e(\lambda t) + A^{K_j}\sum C_i^d e(\lambda t) + A\frac{\partial X}{\partial K_i\partial K_j}$$

$$= \sum (A^{K_i}C_j^d + A^{K_j}C_i^d)e(\lambda t) + A\frac{\partial X}{\partial K_i\partial K_j}.$$

Multiplying by Q,

$$\frac{d}{dt}\frac{\partial u}{\partial K_i\partial K_j} = Q\sum (A^{K_i}C_j^d + A^{K_j}C_i^d)e(\lambda t) + T\frac{\partial u}{\partial K_i\partial K_j}.$$

Allen's result holds and the solution is of the form

$$\sum D_{ij}^d e(\lambda t).$$

Clearly higher derivatives are possible but the expressions are correspondingly more complicated.

The determination of when the eigenvalues are equal is potentially the hardest part of making the method work. We can tell when two roots are within machine epsilon of one another, but this may not be the criteria we want. For statistical purposes, we may deem two eigenvalues equal long before a numerical analyst would. The analogy in the linear regression case is that the X matrix may be nonsigular with respect to machine precision but the confidence intervals for the parameters are so large as to be useless.

We haven't determined a satisfactory criteria yet. In the C subroutine we present, we sidestep this by introducing a function cmplam. This function returns 1 if the eigenvalues are determined to be equal whatever criteria we are currently using and Ø if not equal.

It should be noted that the eigenvectors of A are not found. In the presence of equal or near equal eigenvalues , this can be an unstable calculation. Also, the Schur decomposition of A is only done once per iteration to calculate all orders of partial derivatives.

APPLICATIONS

Second derivative information is useful for many problems. We will briefly describe two of them.

The classical Gauss method of function minimization requires the second derivative matrix of the objective function. The objective function in our case is the residual sum of squares. It's second derivative matrix is

$$\dot{r}'\dot{r} + r'[\ddot{r}]$$

where

$$r = y - X$$

$$\dot{r} = \frac{\partial r}{\partial K}, \ n \times p$$

$$\ddot{r} = \frac{\partial^2 r}{\partial K \, \partial K} \text{ is } n \times p \times p.$$

The bracket notation is adopted from Bates and Watts [1980] where it means to sum over the sample space index.

We may not want to perform the Gauss step at every iteration but perhaps every p^{th} step as a restart for the Gauss Newton. It could also be invoked based on the basis of convergence criteria or line search failure.

Another use of second derivative information are the nonlinearity measures of Bates and Watts. The coordinates of the second derivative matrix are calculated relative to an orthogonal basis of the tangent plane. These coordinates are used to construct the nonlinearity measures and are useful in themselves in understanding the nature of the nonlinearity in the problem. To calculate these coordinates, the second derivative only has to be formed after the Gauss Newton method has converged.

IMPLEMENTATION

The preceeding method is only a part of a larger program for estimating $\underline{K}$. We need some way of keeping track of the solution matrices. For a problem with c compartments and p rate constants, 1 solution matrix is required for the system, p for the partial derivatives and p(p+1)/2 for the second derivatives. One way to keep track of these matrices is to keep track of their addresses. In the programming language C we can define an array of pointers. For example, for second order partials we can define

double * D[i][j].

This declares D to be a two dimensional array of pointers to double. A double is a single precision real in C. We could have defined D to be an array of pointers to char or int or anything since pointers almost always have the same length even if the things they point to don't. The address of the (i,j) solution matrix is assigned to D[i][j]. Referencing the (i,j) element of D is the same as referencing the ij solution.

We could predefine all the matrices we might at compile time and then assign those addresses to the pointer arrays. Or we can allocate matrices as we need them. There are C library functions for demand allocation of storage, but they only return a block of memory, not any specific type such as an array. One must then make this memory area look like a known type to the compiler. For arrays this requires setting up dope vectors containing the right information to simulate a compiler generated array.

For a problem with 3 compartments and 5 parameters, the total storage for all the solution would be around 20K bytes assuming 8 byte reals. A few years ago this might be one half to one third of all the storage available to a typical microcomputer. For today's and future microcomputers, this is really an insignificant amount of storage so we don't try to manage it as efficiently as we could.

C is a fast and flexible language. The lack of type checking has good and bad effects. The ability to manipulate pointers allows great freedom but can also produce obscure bugs. However, assembly language programmers have lived without benefits of any protections since the beginnings of computers.

The programs we have written are also portable. By being compatible with UNIX*, C compilers are also compatible with each other. We have transferred C code written for a Motorola 68000 processor to a VAX running 4.2BSD UNIX and have compiled and executed without change.

To improve the efficiency of a computation intensive program like this one, significant portions should be written in assembly language. Languages with claimed small overhead over assembly language achieve that small overhead only when the assembly language programmer must play by the same rules as the compiler. Given complete freedom, the assembly written program will usually be much faster than that written in a higher level language. Unfortunately, the code is not as portable.

The following subroutine solves the system of differential equations. If nullb is 1 then system (1) will be solved; else system (3) will be solved.

```
solvesys(a,b,c,x0,lam,nc,deg,nullb)

REAL a[MAXNC][MAXNC],
b[MAXDEG][MAXNC][MAXNC],
c[MAXDEG][MAXNC][MAXNC];
REAL x0[MAXNC], lam[MAXNC];
int nullb, nc, *deg;
{
 int i,p,d,r,k,g,tmd;
 REAL temp, fact, powk, lamdif;

 for (p=nc-1;p>=0;p--)
 {
  if ((p==nc-1)&&(nullb))
  {
   c[0][p][p]=x0[p];
   continue;
  }
  tmd= *deg;
  for(k=0;k<nc;k++)
  {
   for(d=0;d<=tmd;d++)
   {
```

*UNIX is a trademark of AT&T.

```
      temp=0.0;
      for(g=p+1;g<nc;g++)
       temp += a[p][g]*c[d][g][k];
      if (!nullb) temp+=b[d][p][k];
      if (temp==0.0) continue;
      lamdif=lam[k]-lam[p];
      if (cmplam(lam,k,p))
      {
       c[d+1][p][k] += temp/(d+1);
      }
      else
      {
       fact=1;
       powk=lamdif;
       for(r=0;r<=d;r++)
       {
        if (r==0) c[d-r][p][k] += temp/powk;
        else
        {
         powk *= lamdif;
         fact *= (d-r+1);
         c[d-r][p][k]+=(fact*temp)/powk;
        }
       }
      }
     }
    }
   if (nullb) c[0][p][p] = x0[p];
   for(i=0;i<nc;i++)
    if (i!=p) c[0][p][p]-=c[0][p][i];
  }
 }
```

REFERENCES

[1] Allen, D.M., Fitting pharmacokinetic models to data, Technical Report 187, Dept. of Statistics, University of Kentucky (January 1982). (To appear in Biometrics.)

[2] Bates, D.M. and Watts, D.G., Relative curvature measures of nonlinearity, Journ of the Royal Statistical Society, Series B, 42, (1980) 1-25.

[3] Carroll, R.J. and Ruppert, D., Power transformations when fitting theoretical models to data, Journal of the American Statistical Association, 79 (1984) 321-328.

[4] Herron, H.J., Theorems on linear systems, Annals of the New York Academy of Science, (1963) 108.

[5] Jennrich, R.I. and Bright, P.B., Fitting systems of linear differential equations using computer generated exact derivatives, Technometrics, Vol. 18, No. 4 (1976)